ATLAS
ICBM
MISSILE WEAPON SYSTEM
TECHNICAL MANUAL

DECLASSIFIED

USAF

THE TECHNICAL ORDERS REPLACED BY THIS PUBLICATION ARE LISTED ON PAGE vi.

COMMANDERS ARE RESPONSIBLE FOR BRINGING THIS PUBLICATION TO THE ATTENTION OF ALL AIR FORCE PERSONNEL CLEARED FOR THE OPERATION OF SUBJECT MISSILE WEAPON SYSTEM.

T. O. 21M-HGM16F-1

TECHNICAL MANUAL

OPERATION

USAF MODEL
HGM-16F
MISSILE WEAPON SYSTEM

OPERATION MANUAL

SQUADRON COMPLEXES
OSTF-2	576-D, E
556	577
550	578
551	579

**LATEST CHANGED PAGES SUPERSEDE
THE SAME PAGES OF PREVIOUS DATE**
Insert changed pages into basic
publication. Destroy superseded pages.

COMMANDERS ARE RESPONSIBLE FOR BRINGING THIS PUBLICATION
TO THE ATTENTION OF ALL AIR FORCE PERSONNEL CLEARED
FOR OPERATION OF SUBJECT MISSILE WEAPON SYSTEM.

PUBLISHED UNDER AUTHORITY OF THE SECRETARY OF THE AIR FORCE

THIS PUBLICATION REPLACES T. O. 21M-HGM16F-1SS-2 DATED
15 JULY 1964, T. O. 21M-HGM16F-1SS-4 DATED 29 JULY 1964 AND
T. O. 21M-HGM16F-1S-1 DATED 25 AUGUST 1964 AND IS INCOM-
PLETE WITHOUT SUPPLEMENT 21M-HGM16F-1AA DATED
15 NOVEMBER 1963.

1 APRIL 1964
CHANGED 21 OCTOBER 1964

AF Carpenter Litho Co.,Inc. Springfield,O.   2-65   2,500

Reproduction for nonmilitary use of the information or illustrations contained in this publication is not permitted without specific approval of the issuing service. The policy for use of Classified Publications is established for the Air Force in AFR 205-1.

LIST OF EFFECTIVE PAGES

INSERT LATEST CHANGED PAGES. DESTROY SUPERSEDED PAGES.

NOTE: The portion of the text affected by the changes is indicated by a vertical line in the outer margin of the page.

TOTAL NUMBER OF PAGES IN THIS PUBLICATION IS 542 CONSISTING OF THE FOLLOWING:

Page No.	Issue	Page No.	Issue
* Title	21 Oct 64	1-19 thru 1-20A	15 Apr 64
* A thru D	21 Oct 64	1-20B Blank	15 Apr 64
* E Blank	21 Oct 64	* 1-21	21 Oct 64
i thru iii	Original	1-22	15 Apr 64
iv Blank	Original	* 1-23	21 Oct 64
* 1-1	21 Oct 64	1-24 thru 1-25	Original
1-2	15 Apr 64	1-26 thru 1-26A	15 Apr 64
1-3 thru 1-4	Original	1-26B Blank	15 Apr 64
* 1-5	21 Oct 64	* 1-27	21 Oct 64
1-6 thru 1-8	15 Apr 64	1-28 thru 1-36	Original
1-9	Original	* 1-37	21 Oct 64
1-10 thru 1-10A	15 Apr 64	1-38	Original
1-10B Blank	15 Apr 64	* 1-39 thru 1-40	21 Oct 64
1-11	15 Apr 64	1-41 thru 1-43	15 Apr 64
* 1-12 thru 1-12A	21 Oct 64	1-44	Original
* 1-12B Blank	21 Oct 64	1-45	15 Apr 64
1-13	15 Apr 64	1-46 thru 1-48	Original
* 1-14	21 Oct 64	* 1-49	21 Oct 64
1-15 thru 1-18	Original	1-50 thru 1-51	Original
1-18A	15 Apr 64	1-52 thru 1-53	15 Apr 64
1-18B Blank	15 Apr 64	1-54 thru 1-64	Original
		1-65	15 Apr 64

CURRENT MISSILE CREW CHECKLISTS

T. O. No.	Basic Date	Change Date
21M-HGM16F-1CL-1	1 April 1964	21 Oct 1964
21M-HGM16F-1CL-2	1 April 1964	21 Oct 1964
21M-HGM16F-1CL-3	1 April 1964	21 Oct 1964
21M-HGM16F-1CL-4	1 April 1964	21 Oct 1964
21M-HGM16F-1CL-5	1 April 1964	21 Oct 1964
21M-HGM16F-1CL-6	1 April 1964	21 Oct 1964

*The asterisk indicates pages changed, added, or deleted by the current change.

ADDITIONAL COPIES OF THIS PUBLICATION MAY BE OBTAINED AS FOLLOWS: B-5

USAF ACTIVITIES—In accordance with T.O. 00-5-2.   USAF

A

LIST OF EFFECTIVE PAGES

Page No.	Issue	Page No.	Issue
1-66 Blank	Original	1-143	15 Apr 64
1-67 thru 1-77	Original	1-144 thru 1-148	Original
1-78 Blank	Original	*1-149	21 Oct 64
1-79	Original	1-150 thru 1-153	Original
1-80 Blank	Original	1-154 Blank	Original
1-81	Original	1-155 thru 1-171	Original
1-82 Blank	Original	*1-172	21 Oct 64
1-83 thru 1-91	Original	*1-173 thru 1-178 Deleted	21 Oct 64
1-92 Blank	Original	1-179 thru 1-183	Original
1-93	Original	1-184 thru 1-185	15 Apr 64
1-94 Blank	Original	*1-186 Added	21 Oct 64
1-95 thru 1-98	Original	2-1	15 Apr 64
1-99	15 Apr 64	2-2	Original
1-100 Blank	Original	2-3	15 Apr 64
1-101	Original	2-4 Blank	15 Apr 64
1-102 Blank	Original	2-5	15 Apr 64
1-103	Original	2-6 Blank	15 Apr 64
1-104 Blank	Original	2-7	15 Apr 64
1-105	Original	2-8 Blank	15 Apr 64
1-106 Blank	Original	3-1	15 Apr 64
1-107	Original	3-2	Original
1-108 Blank	Original	*3-3	21 Oct 64
1-109	Original	3-4	Original
1-110 Blank	Original	*3-5 thru 3-6	21 Oct 64
1-111 thru 1-117	Original	3-7	Original
1-118 Blank	Original	*3-8 thru 3-10	21 Oct 64
1-119	Original	*3-10A Added	21 Oct 64
1-120 Blank	Original	*3-10B Blank Added	21 Oct 64
1-121	Original	*3-11 thru 3-14	21 Oct 64
1-122 Blank	Original	*3-15	21 Oct 64
*1-123	21 Oct 64	3-16 Blank	Original
1-124	15 Apr 64	3-17	Original
1-125	Original	*3-18 thru 3-21	21 Oct 64
1-126 Blank	Original	3-22	Original
1-127	Original	*3-23 thru 3-24	21 Oct 64
1-128 Blank	Original	3-25 thru 3-27	Original
1-129	Original	3-28	15 Apr 64
1-130 Blank	Original	3-29 thru 3-30	Original
1-131	Original	*3-31	21 Oct 64
1-132 Blank	Original	3-32	Original
1-133	Original	*3-33 thru 3-35	21 Oct 64
1-134 Blank	Original	3-36 thru 3-37	Original
1-135 thru 1-137	Original	*3-38	21 Oct 64
1-138	15 Apr 64	3-39 thru 3-43	Original
1-139 thru 1-141	Original	3-44	15 Apr 64
1-142 Blank	Original	*3-45	21 Oct 64

* The asterisk indicates pages changed, added, or deleted by the current change.

LIST OF EFFECTIVE PAGES

Page No.	Issue	Page No.	Issue
3-46	Original	4-14	15 Apr 64
* 3-46A	21 Oct 64	* 4-14A	21 Oct 64
* 3-46B Blank	21 Oct 64	4-14B	15 Apr 64
3-47	15 Apr 64	* 4-15	21 Oct 64
3-48 thru 3-49	Original	4-16 thru 4-18	Original
3-50	15 Apr 64	4-19	15 Apr 64
3-51 thru 3-60	Original	4-20	Original
* 3-61	21 Oct 64	* 4-21	21 Oct 64
3-62 thru 3-63	Original	4-22 thru 4-22A	15 Apr 64
* 3-64	21 Oct 64	4-22B Blank	15 Apr 64
3-65 thru 3-66	15 Apr 64	* 4-23 thru 4-26	21 Oct 64
* 3-67 thru 3-69	21 Oct 64	* 4-26A Added	21 Oct 64
3-70 thru 3-98	Original	* 4-26B Blank Added	21 Oct 64
* 3-99 thru 3-101	21 Oct 64	4-27	Original
3-102 thru 3-105	Original	4-28	15 Apr 64
3-106	15 Apr 64	* 4-29 thru 4-30	21 Oct 64
* 3-107	21 Oct 64	4-31	15 Apr 64
3-108 thru 3-109	Original	4-32	Original
* 3-110	21 Oct 64	* 4-33	21 Oct 64
3-111 thru 3-113	Original	4-34 thru 4-35	Original
3-114 thru 3-115	15 Apr 64	4-36 thru 4-36A	15 Apr 64
3-116 thru 3-117	Original	4-36B Blank	15 Apr 64
3-118	21 Aug 64	* 4-37 thru 4-39	21 Oct 64
3-119 thru 3-120	Original	4-40	Original
* 3-121 thru 3-122	21 Oct 64	* 4-41 thru 4-42	21 Oct 64
3-123 thru 3-125	15 Apr 64	* 4-42A Added	21 Oct 64
* 3-126	21 Oct 64	* 4-42B Blank Added	21 Oct 64
3-127 thru 3-130	15 Apr 64	4-43	15 Apr 64
3-131	Original	4-44	Original
3-132	15 Apr 64	* 4-45	21 Oct 64
3-133 thru 3-134	Original	4-46 thru 4-48	15 Apr 64
* 3-135 thru 3-136	21 Oct 64	4-49	Original
3-137 thru 3-138	15 Apr 64	* 4-50	21 Oct 64
3-139	Original	4-51	Original
* 3-140 thru 3-142	21 Oct 64	4-52	15 Apr 64
3-143 thru 3-145	Original	* 4-53 thru 4-56	21 Oct 64
* 3-146	21 Oct 64	* 4-56A Added	21 Oct 64
3-147	15 Apr 64	* 4-56B Blank Added	21 Oct 64
* 3-148	21 Oct 64	4-57	15 Apr 64
3-149	Original	* 4-58 thru 4-60	21 Oct 64
3-150 Blank	Original	4-61 thru 4-64A	15 Apr 64
4-1	15 Apr 64	4-64B Blank	15 Apr 64
* 4-2 thru 4-7	21 Oct 64	4-65 thru 4-66	Original
4-8 thru 4-9	Original	* 4-67	21 Oct 64
4-10 thru 4-11	15 Apr 64	4-68	Original
* 4-12 thru 4-13	21 Oct 64	4-69	15 Apr 64

* The asterisk indicates pages changed, added, or deleted by the current change.

━ LIST OF EFFECTIVE PAGES ━

Page No.	Issue	Page No.	Issue
* 4-70 thru 4-70A	21 Oct 64	4	Original
* 4-70B Blank	21 Oct 64	5	15 Apr 64
* 4-71 thru 4-72	21 Oct 64	6	Original
* 4-72A Added	21 Oct 64	7	15 Apr 64
* 4-72B Blank Added	21 Oct 64	8 thru 9	Original
4-73	15 Apr 64	10 Blank	Original
4-74 thru 4-77	Original		
4-78	15 Apr 64		
* 4-79	21 Oct 64		
4-80	15 Apr 64		
4-81 thru 4-88 Deleted	15 Apr 64		
4-89	15 Apr 64		
* 4-90 thru 4-91	21 Oct 64		
4-92	15 Apr 64		
4-93 thru 4-94	Original		
4-95	15 Apr 64		
* 4-96	21 Oct 64		
4-97 thru 4-105	Original		
4-106	15 Apr 64		
* 4-107	21 Oct 64		
4-108 thru 4-110	15 Apr 64		
* 4-111 thru 4-112	21 Oct 64		
4-113 thru 4-116	Original		
5-1 thru 5-3	Original		
* 5-4 thru 5-5	21 Oct 64		
5-6	15 Apr 64		
* 5-7	21 Oct 64		
5-8	15 Apr 64		
* 5-9 thru 5-10	21 Oct 64		
* 5-10A Added	21 Oct 64		
* 5-10B Blank Added	21 Oct 64		
* 5-11	21 Oct 64		
5-12 thru 5-27	Original		
* 5-28	21 Oct 64		
6-1	Original		
* 6-2	21 Oct 64		
6-3	15 Apr 64		
6-4	Original		
6-5 thru 6-6	15 Apr 64		
6-7	Original		
6-8	Original		
7-1 thru 7-2	Original		
G-1 thru G-2	15 Apr 64		
1	15 Apr 64		
2	Original		
3	15 Apr 64		

* The asterisk indicates pages changed, added, or deleted by the current change.

ALPHABETICAL INDEX

ALPHABETICAL INDEX (Continued)

ALPHABETICAL INDEX (Continued)

ALPHABETICAL INDEX (Continued)

ALPHABETICAL INDEX (Continued)

ALPHABETICAL INDEX (Continued)

ALPHABETICAL INDEX (Continued)

ALPHABETICAL INDEX (Continued)

ALPHABETICAL INDEX (Continued)

T.O. 21M-HGM16F-1

TABLE OF CONTENTS

A PERSONAL MESSAGE TO YOU — THE CREW MEMBER

SCOPE. This manual contains the necessary information for safe and efficient operation of the HGM16F weapon system. These instructions provide you with a general knowledge of the missile, aerospace ground equipment, real property installed equipment, and specific normal and emergency operating procedures. Your training and background are recognized, and therefore basic principles have been avoided. Although the manual contains procedures for peacetime operations, it is tactically oriented toward "on alert" status and tactical launch procedures.

SOUND JUDGEMENT. The instructions contained in this manual are designed to provide for the needs of a combat crew experienced in the operation of the weapon system. The manual provides the best possible operating instructions under most circumstances, but cannot be used without sound judgment. Multiple emergencies, as an example, require close scrutiny by the crew commander and individual crew members prior to initiating a given emergency procedure. Each crew member is a specialist in his own areas and crew commanders should make full use of the entire crew at all times. Crew commanders must be fully knowledgable of the command policies regarding the weapon system. Command policy may require that crew commanders secure outside assistance in emergency situations in peacetime once a safe condition is reached, and prior to initiating certain procedures in this manual. Command disaster control procedures may be required in conjunction with technical procedures contained herein. Sound judgment must be exercised at all times.

PERMISSABLE OPERATIONS. The operation manual takes a positive approach and normally states only what you can do. Unusual operations or configurations are prohibited unless specifically covered herein. Clearance must be obtained from the headquarters of the using command before attempting any operation that is not specifically permitted in this manual or other USAF technical orders.

FORMAT AND CONTENT. The manual is divided into seven sections to simplify its reading and use as a reference manual. Section I, III, IV, and VI must be thoroughly read and fully understood before operation of the weapon system is attempted. The remaining sections provide information for safe and efficient mission accomplishment.

READ . . .
THE PERSONAL MESSAGE ON THESE PAGES TO GAIN THE GREATEST BENEFIT FROM THIS MANUAL!

CHECKLISTS. The operation manual contains only amplified procedures. Abbreviated checklists have been issued as separate technical orders. See the back of the operation manual title page for T.O. number and current date of your checklist. Line items in the operation manual and the checklists are identical with respect to arrangement and item number. The crew member is expected to use the abbreviated checklists in performing alert and launch duties and thus must have complete understanding of the amplified procedures contained herein. The amplified procedures are designed to tell who, what, when, where, why, and how. The abbreviated checklists are intended to tell only who, what, when, and where. Each crew member is responsible for ensuring that his checklists are available and properly located for immediate use when required.

A PERSONAL MESSAGE TO YOU — THE CREW MEMBER (cont)

MEMORY ITEMS. Certain information and procedures in this manual and the attendant checklists are of such critical nature that they must be committed to memory. In some cases, emergency procedures must be performed from memory because time does not permit reference to the checklists. In other cases, the information must be committed to memory because it is of a prohibitive nature and must not be accomplished or loss of life, and/or damage to equipment would result. Those steps in the amplified procedures and the abbreviated checklists that are considered critical and must be committed to memory, are presented in distinctive print (BOLD TYPE). The responsible crew member must be capable of performing these steps in the correct order, from memory, and without reference to the checklists. Upon completion of the memory items, he shall refer to the checklist and accomplish it in its entirety, including a check to ensure that all memory items have been completed. Combat crew members must be knowledgable of all warnings and cautions. The using command will identify the warnings and cautions throughout the manual and other information in narrative form in Section IV that is considered critical and to be committed to memory.

WARNINGS, CAUTIONS, NOTES. The following definitions apply to warnings, cautions, and notes found throughout the operation manual and abbreviated checklists:

WARNING

Operating procedures that must be strictly complied with or personnel injury or loss of life may result.

CAUTION

Operating procedures that must be strictly complied with or equipment damage may result.

NOTE

Operating procedures of such importance that they must be emphasized.

SAFETY SUPPLEMENTS. Information involving safety shall be promptly forwarded to you in the form of safety supplements. Information of a critical nature that must be brought to the attention of crew members will be issued by electrical transmission and followed up by a formal safety supplement. See T.O. 00-5-1 for the details on safety supplements.

CREW COMPOSITION. The HGM16F combat crew is composed of the following five crew members. (See Section VII for crew duties.)

Missile Combat Crew Commander — MCCC

Deputy Missile Combat Crew Commander — DMCCC

Ballistic Missile Analyst-Technician Specialist — BMAT

Missile Facilities Technician/Specialist — MFT

Electrical Power Production Technician/Specialist — EPPT

YOUR RESPONSIBILITY. Each missile combat crew member will be issued his personal copy of the operation manual, normal and emergency procedure checklists and safety supplements. It is your personal responsibility ot properly post and maintain your copies in accordance with T.O. 00-5-1. Every effort is made to keep the oepration manual and checklists up to date. However we cannot correct an error unless we know of its existence. In this regard it is essential that you do your part. If you discover an error or deficiency, promptly submit an AFTD Form 22 (T.O. System Publication Deficiency Report) in accordance with TO 00-5-1 and your command directives. One last word — This manual and the attendant checklists are intened to tell you how to safely and efficiently operate the weapon system — how to do your job — so, to be professional you must read and fully understand these documents.

READ THE PAPER

The following note was extracted from a technical manual dated 20 March 1918:

For the good of the service you will

thoroughly familiarize yourself with

the contents of this BOOK

???OUTDATED???

SECTION I

WEAPON SYSTEM DESCRIPTION

TABLE OF CONTENTS

Changed 21 October 1964

TABLE OF CONTENTS (Continued)

LIST OF ILLUSTRATIONS

LIST OF ILLUSTRATIONS (continued)

LIST OF ILLUSTRATIONS (continued)

LIST OF TABLES

1-1. SCOPE.

1-2.   This section presents a physical and functional description of the HGM16F Missile Weapons System. Included are descriptions of the HGM16F Missile, launch complex, missile and re-entry vehicle handling and transport, supporting squadron maintenance area, missile checkout operations, maintenance plan, systems, and a listing of personnel. The descriptions are presented to familiarize personnel with the major characteristics of the weapons system. No detailed functional theory or detailed physical descriptions are presented. For detailed descriptions, refer to applicable system organizational maintenance manuals.

1-3.   GENERAL. The HGM16F Missile Weapons System is composed of an integrated system of equipment, personnel, and facilities to support, maintain, test check out and launch the HGM16F Missile. Supply and maintenance support are provided by Systems Support Manager Prime Air Materiel Areas, and by the contractors as required. Additional logistical support is provided by the parent airbase of the facility. An operational strategic missile squadron consists of 12 launch complexes located varying distances from the parent airbase and one squadron maintenance area located at the parent airbase. Additional launch complexes (one each) are Operational System Test Facility — 2 (OSTF-2), 576-D, and 576-E located at Vandenberg Air Force Base (VAFB), California. Tables 1-1 and 1-2 present the nomenclature of major equipment and leading particulars of the Weapons System.

1-4.   LAUNCH COMPLEX.

1-5.   A typical launch complex (figure 1-1) consists of two underground structures (launch control center and silo.) In addition, the OSTF-2 Launch Complex also contains an instrumentation building and utility building. (See figure 1-2.) The structures within the launch complex contain the equipment required to perform tests, checkout, and launch functions on the missile. The launch complex is capable of performing propellant loading exercises (wet countdowns), simulated countdowns, and tactical countdowns. Training launches may be performed, in addition, from the Vandenberg complexes only. In propellant loading exercises, the missile and launch countdown equipment are cycled through the countdown sequence except for starting the rocket engines. The exercise is aborted in the final phase of countdown. In simulated countdowns, no missile systems are tested as countdown responses normally received from the missile are simulated by control monitor group 3 and 4 of 4.

1-6.   At Launch Complexes 576-D and 576-E, and all operational launch complexes, the launch control center is connected to the silo by a tunnel running from the lower level to level 2 of the silo. (See figure 1-3.) Entrance to the launch control center is by means of a stairway within an entry tunnel leading from the ground level down to a vestibule.

Table 1-1.   Federal Supply Classification Nomenclature to Common Nomenclature

PART NUMBER	FSC NOMENCLATURE	COMMON NOMENCLATURE	LOCATION
106205-111 (Hallamore Division of Seigler CORP)	Launch Control Console	Missile launch control console or launch control console	Launch control center
0165R0268 (General Electric CO)	Power Remote Control Panel	Power remote control panel	Launch control center

Table 1-1. Federal Supply Classification Nomenclature to Common Nomenclature (Continued)

PART NUMBER	FSC NOMENCLATURE	COMMON NOMENCLATURE	LOCATION
F-14001-3 (Aerojet General CORP)	Missile Destruct System Checkout and Control Console OA-3195/GSW	Missile flight safety system checkout console	Launch control center (VAFB only)
27-68746-213 27-68746-215 (OSTF-2)	Control-Monitor Group 1 of 4 2 of 4 3 of 4 4 of 4	Relay logic unit NO. 1 (2 of 2) Relay logic unit NO. 1 (1 of 2) Relay logic unit NO. 2 Signal responder unit NO. 1 Signal responder unit NO. 2	LCC Silo Silo Silo Silo
27-49910-851 27-49910-853 27-49910-855 27-49910-857	Guided Missile Silo Launcher Platform	Launcher platform or launch platform	Silo
DM-21-E1-13 (Minneapolis Honeywell)	Pneumatic System Manifold Regulator	Pneumatic distribution unit	Silo
GM-4-S1-17 (Minneapolis Honeywell)	Helium Control Charging Unit	Helium charge unit	Silo
GS-46-S1-11 (Minneapolis Honeywell)	Pressure System Control	Pressurization control unit	Silo
27-87160-897 27-87160-899 (OSTF-2)	Hydraulic Pumping Unit	Hydraulic Pumping Unit	Silo
27-08657-5 (Sprague Engineering)	Hydraulic Pumping Unit A/E27A-10	Hydraulic Pumping Unit	Silo
MD-2 (Kurz-Root CORP)	Motor Generator, Type MD-2	400-CPS motor generator	Silo
5301021-9 (Kurz-Root CORP)	Power Supply-Distribution Set PP-2537/GSW	28-VDC power supply	Silo
27SE3005 (General Dynamics/Fort Worth)	Distribution Box	AC distribution box	Silo

Table 1-1. Federal Supply Classification Nomenclature to Common Nomenclature (Continued)

PART NUMBER	FSC NOMENCLATURE	COMMON NOMENCLATURE	LOCATION
27SE3004-1 (General Dynamics/Fort Worth)	Remote Switching Control C-3183/GSW	Power control unit	Silo
(OSTF-2) 18200002 18200001 (Interstate Engineering CORP)	Interconnecting Box	Umbilical junction box	Silo
226E662G1 (General Electric CO)	Re-entry Vehicle Prelaunch Monitoring Set A/E24T-23XC-1	Re-entry vehicle prelaunch monitor	Silo
(American Bosch Arma CORP)	Missile Guidance Set AN/DJN-3	Missile guidance set or system	Missile
2-00031-539 (American Bosch Arma CORP)	Missile Guidance Computer CP-488/DJN-2	Computer	Missile
2-00029-005 (American Bosch Arma CORP)	Inertial Guidance Sensing Platform OA-2183/DJN-2	Platform	Missile
2-00044-085 (American Bosch Arma CORP)	Missile Guidance Control C-3483/DJN-3	Control unit	Missile
27-99094-863	Missile Lifting Launch Platform Drive Assembly HLU-50/E	Launch platform drive	Silo
27-55063-847 thru 861	Semitrailer Mounted Missile Electrical Systems Checkout Test Station	Checkout test station	Mobile (MAMS or Launch Area)

Changed 15 April 1964

Table 1-2. Leading Particulars, HGM16F Missile Weapon System

EQUIPMENT	PARTICULARS
HGM16F Missile:	
Length	Approximately 80 feet
Diameter	Tank section: 10 feet, tapering to 70.5 inches (48 inches with re-entry vehicle adapter)
	Booster section: 10 feet, flaring to 16 feet
Propellant capacities:	
Fuel tank	Approximately 12,000 gallons
Liquid oxygen tank	Approximately 19,000 gallons
Propellants	RP-1 fuel and liquid oxygen

Propulsion system	Five rocket engines:
	    Two booster engines: 330,000 pounds thrust (sea level)
	    One sustainer engine: 57,000 pounds thrust (sea level)
	    Two vernier engines: 2000 pounds thrust (sea level)

Guidance system	Inertial guidance
Range	Greater than 5500 miles

Launch complex:	Underground hardsite:
	    One silo
	    One launch control center

Silo:	
NO. of crib levels	8
Depth	173.5 feet
Diameter	52 feet
Purpose	Provides facilities for test, checkout, countdown, and launch of the HGM16F Missile
Launch control center:	
NO. of floor levels	2
Height	33 feet
Diameter	44 feet
Purpose	Provides personnel facilities, communications equipment, and equipment for checkout, countdown, and launch control

1-7. LAUNCH CONTROL CENTER.

1-8. The launch control center (figures 1-4 and 1-5) consists of two floor levels (crib) that are suspended from the ceiling of a concrete structure and air-cushioned to absorb ground shocks. The suspension system is composed of four air cylinder spring supports attached from the ceiling of the structure to the first floor level and four level-detecting devices mounted between the second floor level and the concrete base. Should the floor level lower or tilt, the level detecting devices sense the change. Solenoid-operated valves are then actuated to allow compressed air to enter or to bleed air from the respective air cylinders. (See figure 1-6.) The first level (upper floor) contains a medical supply room; rest room; heating, ventilation, and air conditioning equipment room; and a training-briefing room. The second level (lower floor) containing the launch control center is divided into four main rooms; a battery room, office, communications equipment room, and a launch control room. Entrance to the launch control center is gained through a blast door and stairway. An escape hatch is also provided for emergency exit. The launch control room contains the equipment to monitor and control countdown and launch of the missile and equipment to monitor power, hazardous conditions, and facility status. Controls and monitoring equipment consist of panels, consoles, and television. The television monitors missiles condition within the silo or may also be connected to external (above ground) cameras.

1-9. LAUNCH CONTROL CONSOLE.

1-10. The launch control console is located on the second level of the launch control center in the launch control room. A panel on the console (figure 1-7) contains the controls and indicators necessary for the missile combat crew commander (MCCC) to initiate a countdown and launch the missile. Arranged in various functional platches, the indicators display the summary status of the aerospace ground equipment (AGE) and missile systems at standby and during a countdown. The information displayed enables the MCCC to monitor the progress of a countdown, maintain a safe missile condition, and make the required decisions in the event of a subsystem malfunction. A communications subpanel provides the various telephone line connections required by the MCCC.

1-11. During countdown, all relay logic subsystems are remotely controlled from the launch control console. Signals from the console energize circuits in the countdown panels (figure 1-8) of the countdown control system. The countdown control system, in turn, energizes and controls circuits in the other relay logic systems. Signals from the control-monitor group 1 and 2 of 4 then actuate and control the airborne and AGE systems. The responses are interlocked in the relay logic unit as required for comparison and further sequencing. Certain critical status responses are displayed on the front panels of the control-monitor group 1 and 2 of 4 to provide information for fault isolation and local control operations. Control-monitor group 1 and 2 of 4 send summary status signals to the launch control console for display.

1-11A. PNEUMATIC LOCAL CONTROL PANEL.

1-11B. The pneumatic local control panel (PLCP), located on the left side of the launch control console, contains the controls and indicators to sequence the pneumatics end-to-end (PETE) test. (See figure 1-100.) The PETE test is conducted periodically and verifies the functional integrity of missile pneumatic and pressurization systems, both ground and airborne. Indicators on the panel display the operation and sequencing of missile system valves, pressure switches, and regulators while the PETE test is being performed.

1-12. GROUND COMMUNICATIONS.

1-13. The ground communications systems available at the launch complex include the following: the direct line telephone, the research and development system (OSTF-2), the administrative dial telephone, the missile flight safety system (Vandenberg AFB), the public address (PA), and the launch maintenance conference network. (See figure 1-9.) The direct line communications system is the primary mode of communication used during missile countdown and launch. It provides direct communication between consoles and from consoles to other specific stations, with no switchboard intervening. Depressing a line selected pushbutton on a console connects the attendant's headset to the direct line station selected. The command post and the alternate command post console operators can, by depressing a preset conference pushbutton, originate a conference between their stations and launch control consoles. The command post or alternate command post attendant can, by depressing an arbitrary conference pushbutton, originate a conference between his station and any combination of the 12 launch control consoles. Also the command post attendant can include the central security control console in his arbitrary conference. Alarm circuits integrated with this equipment provide both audible and visible alarms in the event of abnormal or faulty equipment operation. The administrative dial telephone serves as a backup communication system for the direct line telephone. Dial lines on the console enable the console operator to call areas not connected by the direct line telephone. It provides an alternate communication patch in the event of a temporary direct line circuit failure.

1-14. There are 34 nonexplosionproof communication panels (figure 1-10) within the weapon complex. Twelve of these communication panels are within the missile enclosure area, 17 are outside the missile enclosure area in the silo and launch control center and five are on the silo cap. Each communications panel provides accommodations for two headsets and the capability of selecting any one of four communications nets for each headset. Each panel has two signalling buttons (call signal at launch control console) and two net release pushbuttons. To establish a conference with the launch control console, one of the four net pushbuttons is depressed, then depress the signal pushbutton. This causes the corresponding net selected pushbutton on the launch control console (figure 1-7) to flash, and an audible signal to sound. If the LCC operator does not desire to remain on the net he is using, at the time he is signalled, he can depress the Hold pushbutton or he may release it and then depress the signalling net pushbutton, thus placing his console in communication with the calling station. Depressing or releasing the HOLD bar will not affect the continued operation of net communication between the stations on that net. To make contact with another station, the launch control console operator can use the public address system to request the specific station to come up on a specified net (one of four). When both stations have depressed the desired net pushbuttons, a conference between them is possible. After terminating a conference, the release pushbutton should always be depressed and the microphone key released.

1-15. The launch control console (figure 1-7) communication system patch has several capabilities which include DIAL 1 pushbutton (for access to standard dial telephone system), DIAL 2 pushbutton (for access to auxiliary standard dial telephone system), and direct lines to the following points:

   a. Command post (OPS).

   b. Alternate command post (AOPS).

   c. Job control (maintenance) (JC).

   d. Central security control (CSC).

   e. Gate house (GH).

   f. Semitrailer mounted missile electrical systems checkout test station when located on silo cap.

   g. Relay logic units, guidance countdown group, and re-entry vehicle prelaunch monitor (LU).

   h. ALCO COMM/CONTROL panel (launch control center level 1).

1-16. The launch control console (figure 1-7) communication system patch has the capability of monitoring any of the communication nets (one at a time). Depressing the HOLD bar on the communication system patch allows placing of any one of the lines or nets on hold for an indefinite period.

1-17. There are two explosion proof communications panels (figure 1-11). One is located on the silo cap and one on silo level 7. These panels have the capability of providing communication on four conference nets. The panels require a special plug to maintain their explosionproof qualities.

1-18. There are three explosionproof jacks (figure 1-12) located in the silo. These jacks require a special headset plug to maintain their explosionproof qualities. The explosionproof jacks have but one capability, that of signaling and conferring with the launch control center on the maintenance safety net. These jacks are normally used when hazardous conditions (high concentration of RP-1 fuel vapor, gaseous oxygen, or diesel fuel vapor) exist in the silo.

1-19. The research and development (R & D) communication system (OSTF-2) provides the facilities to support all R & D tasks such as engineering safety, training, testing, installation, and maintenance. The system consists of 145 stations and is capable of handling up to 25 lines in a single conference with 10 conferences occurring simultaneously. The R&D communication panels are located throughout the launch complex and consist of two network lines terminating in a pushbutton. Additional pushbuttons are provided for dial and direct lines. The R&D communication control panel (13, figure 1-5) is located in the communication room of the launch control center.

1-20. Normal communications between the launch control center and the silo, except the two explosionproof communications panels and the three explosionproof jacks, are automatically disabled from countdown start to abort complete. The disabling feature reduces possible explosion hazards. When communications are disabled, a C O M M U N I-C A T I O N D I S A B L E indicator on the communications disconnect panel (23, figure 1-5) is illuminated white. If emergency communications are required during countdown or abort, a key-operated switch on the communications disconnect panel may be turned to the COMM OVERRIDE position. In the COMM OVERRIDE position, all communications are enabled and a COMMUNICATIONS OVERRIDE indicator flashes red. After abort is complete, the key-operated switch may be returned to the RESTORE COMM position and the COMMUNICATION DISABLE indicator will be extinguished.

1-21.   The missile flight safety system (MFSS) communications (VAFB only) provide the control circuits to integrate MFSS instrumentation sites and launch complex operator positions. The MFSS supports launch operations by providing missile flight safety system control for protection of populated mainland and off-shore island areas. The control point for the MFSS communication system is located at the missile destruct system checkout and control console (32, figure 1-5.) The test and training monitor console (24) and the launch complex safety officer's console (20) provide interface functions with the Pacific Missile Range Operations Center at Point Arguello, California.

1-22.   Installed at each site is a public address system that consists of a network of loudspeakers located on each level of the silo (figure 1-13), on the silo cap, and in the launch control center. This system permits the missile combat crew commander to summon individuals to the launch maintenance conference networks throughout the site. Microphone connections into the public address (PA) system are available at the communication system patch on the launch control console (figure 1-7) and at remote PA microphone panels located at the silo cap and silo levels 2 and 7. The remote microphones, located inside panels marked LCOSS MICROPHONE ACCESS, are of the push-to-talk type. Strike tone signals are sounded over the PA system. The tone signifies that missile combat crew members will report to their countdown positions as soon as practicable. Pushbuttons used to sound the strike tone are located at the communication system patch on the launch control console (figure 1-7), the silo cap, and silo levels 2 and 7. The remote pushbuttons, marked ALERT, are located on boxes adjacent to the remote PA microphone panels.

1-23.   SITE SECURITY TELEVISION.

1-24.   A television monitor and camera (figures 1-14 and 1-15) are provided to monitor the entrance to the launch control center. The television camera monitors the entrance at an intruder entrapment enclosure located at the bottom of the stairway leading from the above-ground entrance. The image is sent to the television monitor (figure 1-14) located directly in front of the missile launch control console. The entrapment area doors are remotely controlled from an entrapment area door control panel (figure 1-14) that contains lock and unlock pushbuttons and associated indicators to display the locked or unlocked status of the doors.

1-25.   CLOSED CIRCUIT TELEVISION MONITORING SYSTEM

1-26.   The closed circuit television monitoring system consists of portable and permanently installed television sets to observe operation both above ground and within the silo enclosure. One set located at silo level 3 is permanently installed. The remaining sets are portable units dispatched from the MAMS as required.

1-27.   The camera for the permanent TV set is located within the missile enclosure at level 3, quadrant III. It transmits a picture of the missile boiloff valve, electrical umbilical disconnects frost level, and other objects within its field of view. The unit may be used to check whether liquid oxygen is being dumped overboard during the commit sequence and whether the boiloff valve is open or closed during an abort operation. This latter condition is pictured indirectly by the presence or absence of fumes escaping from the boiloff valve.

1-28.   The three portable camera locations are as follows:

a.   Surface camera with pan and tilt assembly mounted on a tripod-dolly. The tripod-dolly is positioned on a concrete pad near the end of the spare conduit west of the silo and approximately 40 feet southwest of the launch control center. It views personnel in the area, top-side operations, and the missile during and after raising.

Changed 21 October 1964

b. Level 7 camera with pan tilt assembly mounted permanently or on a tripod-dolly. The camera is positioned in a manner to provide optimum surveillance of liquid oxygen system transfer components.

c. Level 8, quadrant I, within the missile enclosure between the missile and the personnel elevator. This camera may be used to view the pneumatic, fuel and cryogenic fluid disconnects, the LO_2 and fuel tank pressure gages on top of the pneumatic system manifold regulator, fuel leveling prefab, pneumatic system manifold regulator, pneumatic system controller, umbilical loop, LN_2 evaporator and the bottom of the launcher platform.

d. Level 6A (if applicable) within the missile enclosure wall between the missile and counterweights, near the corner in quadrant IV. This camera is used to view the LO_2 disconnect fitting. Failure of this quick disconnect fitting to close permits liquid oxygen to be dumped overboard during the commit sequence and topping operation.

1-29. The permanent closed circuit television monitoring set consists of a camera with pan and tilt assembly and an equipment rack comprised of a monitor, camera control unit, accessory control panel, power panel (for lighting control and communications power), synchronizing generator, and associated interconnecting cables. Each portable set consists of a camera with pan and tilt assembly and one equipment rack with monitor and camera control unit. In addition, the equipment rack of one of the portable sets has a synchronizing generator with changeover switch to serve as a spare synchronizer. The cameras are equipped with zoom lens, variable focus and iris. Camera operations are remotely controlled from the equipment racks located in the launch control center near the launch control console.

1-30. Two floodlights are attached to each camera installed in the silo (no lights are furnished with the ground level camera.) The floodlights are attached to the pan and tilt mechanism so that the light beams follow the camera viewing area. Light intensity is controlled by a switch on the power panel in the equipment rack. The lighting control transformer which provides power to the floodlights is located on the underside of the silo level 3 floor, quadrant II.

1-31. Communications equipment associated with the television monitoring system consists of six telephones and jacks with interconnecting cables running to a 28-volt power supply in equipment rack NO. 1 in the launch control center. Telephone jacks are provided in each camera junction box, in the lighting control transformer case, and in the 28-volt power supply panel.

1-32. LAUNCH ENABLE SYSTEM.

1-33. The launch enable system serves to prevent the untimely launch of a missile by imposing a disabling signal on the launch control equipment. It is controlled from the command post and alternate command post. The disabling signal prevents the final countdown phase (commit) from taking place until it is removed by the command post and alternate command post.

1-33A. ALCO COMM/CONTROL PANEL.

1-33B. A second safeguard against inadvertent missile launch is provided by the ALCO COMM/CONTROL panel (figure 1-101), located on level 1 of the launch control center. Table 1-2A lists the switch and indicators involved and their functions. Shortly before missile and ground systems are ready for the final countdown commit sequence, the DMCCC leaves the launch control console, proceeds upstairs, and takes his place before the ALCO COMM/CONTROL panel. A green M I S S I L E READY indication on this panel shows that all systems are ready for commit. When the LCO COMMIT indicator is illuminated, signifying that the MCCC has actuated the COMMIT START key switch on the launch control console, the DMCCC turns the COMMIT SWITCH clockwise within 3 seconds and the commit start command is enabled. Illumination of the COMMIT IN PROGRESS indicator shows that the commit sequence has been initiated. A telephone handset is mounted on the ALCO COMM/CONTROL panel for communication between the DMCCC and MCCC.

1-34. LAUNCH CONTROL CENTER
ENVIRONMENTAL CONTROL.

1-35. The heating, ventilating, and air conditioning system controls temperature and humidity, and also provides air circulation within the launch control center (LCC.) Air temperature is maintained between 70°F and 80°F with 70 percent maximum relative humidity. It is composed of an air conditioning system, chilled water system and a hot water system. (See figure 1-16 for simplified flow diagram of the LCC air conditioning system.)

1-36. Located on the top floor of the launch control center, the air conditioning unit consists of supply and exhaust fans, heating and cooling coils, motor operated dampers, dust and CBR filters, as well as the necessary ducting and controls. (See table 1-3 for LCC associated indicators on FRCP.) Chilled water is supplied by the silo water chillers, which supply water to the cooling coils. (See figure 1-17.)

1-37. The facility terminal cabinet NO. 2 (figures 1-18 and 1-19) located on silo level two provides additional controls and indicators necessary for the operation of this equipment. The hot water portion of this system is a closed loop in which water is pumped from the heat recovery silencers to the thrust section heating coil, launch platform heating coil, and the launch control center.

1-38. LAUNCH CONTROL CENTER
ELECTRICAL SYSTEM.

1-39. Electrical power for the launch control center is obtained from the diesel generators (40 and 46, figure 1-20) located in the silo. A 460-volt AC,

Table 1-2A. ALCO COMM/CONTROL Panel Indicators and Control

INDICATOR OR CONTROL	FUNCTION
MISSILE READY indicator	Illuminated green when missile and ground systems are ready for commit.
LCO COMMIT indicator	Illuminated green when MCCC actuates the COMMIT START key switch on the launch control console.
COMMIT SWITCH	Energizes commit start ready when turned fully clockwise within 3 seconds of actuation of COMMIT START key switch on launch control console.
COMMIT IN PROGRESS indicator	Illuminated green when commit sequence starts and POWER INTERNAL indicator illuminates green on launch control console.

3-phase, 60-cycle line from the generators, running through the utility tunnel, provides the power input to a 480-volt control center (29 figure 1-4.) From a 600-ampere bus in this control center, 460-volt power is supplied to the intake and exhaust fans in the ready room, sewage pumps in the vestibule, water supply well pumps and utilities, facility remote control panel (19) and to the lighting distribution transformer (31.) Commercial power is also furnished to the OSTF-2 launch complex as a backup supply. (Reference figure 1-102 for simplified flow diagram of complex power distribution.

Table 1-3. Facility Remote Control Panel Indicators and Controls

INDICATOR OR CONTROL	FUNCTION
Blast closures and doors:	
LCC AIR INTAKE OPEN and LCC AIR INTAKE CLOSED indicators	OPEN indicator extinguishes and CLOSED indicator illuminates when LCC air intake is closed. OPEN indicator illuminates and CLOSED indicator extinguishes when LCC air intake is open.
LCC VESTIBULE EXTERIOR BLAST DOOR OPEN and LCC VESTIBULE EXTERIOR BLAST DOOR CLOSED indicators (not included at OSTF-2)	OPEN indicator illuminates and CLOSED indicator extinguishes when LCC vestibule exterior blast door is open. OPEN indicator extinguishes and CLOSED indicator illuminates when LCC vestibule exterior blast door is closed.
LCC VESTIBULE INTERIOR BLAST DOOR OPEN and LCC VESTIBULE INTERIOR BLAST DOOR CLOSED indicators (not included at OSTF-2)	OPEN indicator illuminates and CLOSED indicator extinguishes when LCC vestibule interior door is open. OPEN indicator extinguishes and CLOSED indicator illuminates when LCC vestibule interior blast door is closed.
LCC AIR EXHAUST OPEN and LCC AIR EXHAUST CLOSED indicators	OPEN indicator extinguishes and CLOSED indicator illuminates when LCC air exhaust is closed. OPEN indicator illuminates and CLOSED indicator extinguishes when LCC air exhaust is open.
LCC STAIRWELL AIR EXHAUST OPEN and LCC STAIRWELL AIR EXHAUST CLOSED indicators	OPEN indicator extinguishes and CLOSED indicator illuminates when LCC stairwell air exhaust is closed. OPEN indicator illuminates and CLOSED indicator extinguishes when LCC stairwell air exhaust is open.
LCC SEWER VENT OPEN and LCC SEWER VENT CLOSED indicators	OPEN indicator extinguishes and CLOSED indicator illuminates when LCC sewer vent is closed. OPEN indicator illuminates and CLOSED indicator extinguishes when LCC sewer vent is open.
LCC GRADE ENTRY DOOR OPEN and LCC GRADE ENTRY DOOR CLOSED indicators	OPEN indicator illuminates and CLOSED indicator extinguishes when grade entry door is open. OPEN indicator extinguishes and CLOSED indicator illuminates when grade entry door is closed.

Table 1-3. Facility Remote Control Panel Indicators and Controls (Continued)

INDICATOR OR CONTROL	FUNCTION
Blast closures and doors (CONT)	
LCC ESCAPE HATCH DOOR OPEN and LCC ESCAPE HATCH DOOR CLOSED indicators	OPEN indicator illuminates and CLOSED indicator extinguishes when LCC escape hatch door is open. OPEN indicator extinguishes and CLOSED indicator illuminates when LCC escape hatch door is closed.
SILO AIR INTAKE CLOSURES OPEN and SILO AIR INTAKE CLOSURES CLOSED indicators	OPEN indicator extinguishes and CLOSED indicator illuminates when silo air intake closures are closed. OPEN indicator illuminates and CLOSED indicator extinguishes when silo air intake closures are open.
SILO AIR EXHAUST CLOSURES OPEN and SILO AIR EXHAUST CLOSURES CLOSED indicators	OPEN indicator extinguishes and CLOSED indicator illuminates when silo air exhaust closures are closed. OPEN indicator illuminates and CLOSED indicator extinguishes when silo air exhaust closures are open.
SILO BLAST DOOR OPEN and SILO BLAST DOOR CLOSED indicators	OPEN indicator illuminates and CLOSED indicator extinguishes when silo blast doors are open. OPEN indicator extinguishes and CLOSED indicator illuminates when silo blast doors are closed.
LCC BLAST CLOSURES MANUAL OPERATION OPEN and CLOSE pushbuttons	Opens or closes LCC blast closures when depressed.
BLAST CLOSURE EMERGENCY OPEN pushbutton	If depressed opens any closed blast closures after LP is up and locked.
BLAST CLOSURE test close pushbutton	Closes SILO and LCC blast closures when depressed.
SILO GRADE ENTRY DOOR CLOSED and SILO GRADE ENTRY DOOR OPEN indicators (O S T F - 2 only)	Open indicator extinguishes and CLOSED indicator illuminates when silo grade entry door is open. OPEN indicator illuminates and CLOSED indicator extinguishes when silo grade entry door is closed.
Launch platform air conditioning system:	
LAUNCH PLATFORM EXHAUST FAN ON and LAUNCH PLATFORM EXHAUST FAN OFF indicators	ON indicator extinguishes and OFF indicator illuminates when launch platform exhaust fan is off. ON indicator illuminates and OFF indicator extinguishes when launch platform exhaust fan is on.
LAUNCH PLATFORM FAN COIL UNIT ON and LAUNCH PLATFORM FAN COIL UNIT OFF indicator	ON indicator extinguishes and OFF indicator illuminates when launch platform fan coil unit is off. ON indicator illuminates and OFF indicator extinguishes when launch platform fan coil unit is on.

Table 1-3. Facility Remote Control Panel Indicators and Controls (Continued)

INDICATOR OR CONTROL	FUNCTION
Launch platform air conditioning system: (CONT)	
LAUNCH PLATFORM DAMPERS OPEN and LAUNCH PLATFORM DAMPERS CLOSED indicators	OPEN indicator extinguishes and CLOSED indicator illuminates when launch platform dampers are closed. OPEN indicator illuminates and CLOSED indicator extinguishes when launch platform dampers are open.
RP-1 & FIRE FOG SYSTEM DAMPERS OPEN and RP-1 & FIRE FOG SYSTEM DAMPERS CLOSED indicators	OPEN indicator extinguishes and CLOSED indicator illuminates when RP-1 and fire fog system dampers are closed. OPEN indicator illuminates and CLOSED indicator extinguishes when RP-1 and fire fog system dampers are open.
Air compressor control system:	
STARTING AIR RECEIVER NORMAL PRESS indicator	Extinguished when starting air receiver pressure is low. Illuminates green when starting air receiver pressure is normal.
INSTRUMENT AIR RECEIVER LOW PRESSURE indicator	Extinguished when instrument air receiver pressure is normal. Illuminates red when instrument air receiver pressure is low (less than 1100 PSI)
Primary electrical power and battery system:	
DIESEL GENERATOR D-60 ON and DIESEL GENERATOR D-61 ON indicators (D-61 ON indicator not included at OSTF-2)	Display operating status of diesel generators. Indicates diesel generators are on when illuminated green. Indicates diesel generators are off when extinguished.
DIESEL GENERATOR D-60 OVERSPEED LO LUBE OIL PRESS, HI-TEMP and DIESEL GENERATOR D-61 OVERSPEED, LO LUBE OIL PRESS, HI-TEMP (D-61 indicator not included at OSTF-2)	Indicates (red) malfunction of diesel generator due to overspeed (over 850 RPM), or low oil pressure (less than 15 PSI) or high water temperature (greater than 200 F.) **NOTE** Diesel generator will shut down when DIESEL GEN D-60 (or D-61) OVERSPEED, LOW LUBE OIL PRESS, HI-TEMP indicator illuminates red due to overspeed or low lube oil pressure (less than 10 PSI.)
BATTERY CHARGER FAILURE	Extinguished when battery charger is normal. Illuminates red when battery charger has failed.
OFF SITE POWER ON indicator (OSTF-2 only)	Illuminates green when commercial power is available. Extinguishes when commercial power is not available.

Table 1-3. Facility Remote Control Panel Indicators and Controls (Continued)

INDICATOR OR CONTROL	FUNCTION
Missile enclosure fog:	
MISSILE ENCLOSURE FOG ON and OFF pushbuttons	When depressed, the ON pushbutton opens, or the OFF pushbutton closes the missile enclosure fog system.
MISSILE ENCLOSURE FOG ON and OFF indicator	Indicates missile enclosure fog system is on or off. OFF indicator illuminates green and ON indicator extinguishes when system is off. ON indicator illuminates red and OFF indicator extinguishes when system is on.
Warning light and horn:	
PERSONNEL WARNING RED ON AND LIGHT switch (not included at VAFB)	Illuminates red personnel warning light on warning light pole when placed in ON position.
PERSONNEL WARNING switch (VAFB only)	Illuminates red, amber, and green personnel warning lights on warning light pole when placed in 1, 2, and 3 positions respectively.
HORN pushbutton	Sounds warning horn on warning light pole when depressed.
PERSONNEL EVACUATION WARNING ON pushbutton (VAFB)	Sounds personnel warning evacuation signal when depressed.
Gaseous oxygen detector system:	
STORAGE AREA OXYGEN 19% indicator	Extinguished when oxygen content of storage area (silo levels 7 and 8) is normal. Illuminates red when oxygen content is low.
STORAGE AREA OXYGEN 25% indicator	Extinguished when oxygen content of storage area (silo levels 7 and 8) is normal. Illuminates red when oxygen content is high.
Diesel fuel vapor detection system:	
DIESEL VAPOR HIGH LEVEL indicator	Extinguished when diesel fuel vapor is not high. Illuminates red when diesel fuel vapor is high.
Water system (OSTF-2 only)	
SPRAY PUMP NORMAL indicator	Extinguished when spray pump is not normal. Illuminates green when spray pump is normal.
SETTLING TANK NORMAL	Extinguished when settling tank is not normal. Illuminates green when settling tank is normal.

Table 1-3. Facility Remote Control Panel Indicators and Controls (Continued)

INDICATOR OR CONTROL	FUNCTION
Utility water system: UTILITY WATER PRESSURE indicator	Extinguished when utility water pressure is normal. Illuminates red when utility water pressure is low.
Silo sump and LCC sewage pump system: SILO SUMP HIGH LEVEL indicator	Extinguished when silo sump level is not high. Illuminates red when silo sump level is high (high level float switch in silo pump closed.)
LCC SEWAGE PUMP HIGH LEVEL indicator	Extinguished when LCC sewage pump level is not high. Illuminates red when LCC sewage pump level is high (high level alarm switch at sewage pump closed.)
Control cabinet system: SILO CONTROL CABINET HI TEMP indicator	Extinguished when control cabinet temperature is normal. Illuminates red when control cabinet temperature is high (greater than 100°F.)
Heat exhaust system: MAIN EXHAUST FAN NOT OPER-ATING indicator	Extinguished when main exhaust fan EF-30 is operating. Illuminates red when main exhaust fan EF-30 is not operating.
Well control system: WATER STORAGE TANK LOW LEVEL indicator	Extinguished when water storage tank level is not low. Illuminates red when water storage tank level is too low (low level switch at water storage tank closed.)
WATER STORAGE TANK HIGH LEVEL indicator	Extinguished when water storage tank level is not high. Illuminates red when water storage tank level is too high (high level switch at water storage tank closed.)
Air conditioning system: AIR WASHER DUST COLLECTING UNITS NOT OPERATING indicator	Extinguished when air washer dust collecting units are operating. Illuminates red when air washer duct collecting units fail to operate.
SILO WATER CHILLER UNITS MAL-FUNCTION	Extinguished when silo water chiller units are operating. Illuminates red when silo water chiller units fail to operate.

Table 1-3.   Facility Remote Control Panel Indicators and Controls (Continued)

INDICATOR OR CONTROL	FUNCTION
Missile pod air conditioning system: MISSILE POD AIR CONDITIONER MALFUNCTION indicator	Extinguished when the temperature, pressure, and humidity of missile pod air are within limits. Illuminated red when missile pod air supply is out of tolerance. Nature of malfunction is shown by the following three indicators.
MISSILE POD AIR HI TEMPER-ATURE indicator	Extinguished when missile pod air temperature is normal. Illuminated red when air temperature in the missile pod cooling duct is high.
MISSILE POD AIR LO PRESSURE indicator	Extinguished when missile pod air pressure is normal. Illuminated red when pressure in the missile pod cooling duct is low.
MISSILE POD HI HUMIDITY indicator	Extinguished when humidity of missile pod air is normal. Illuminated red when humidity in the missile pod cooling duct is high.
Condenser water system: EMERGENCY WATER PUMP P-32 ON indicator (NA OSTF-2)	Extinguished when pump P-32 is in standby status. Illuminated red when pump is supplying water to the condenser water system.

Table 1-3.   Facility Remote Control Panel Indicators and Controls (Continued)

INDICATOR OR CONTROL	FUNCTION
Gaseous oxygen vent system: GASEOUS OXYGEN VENT CLOSED and GASEOUS OXYGEN VENT OPEN indicators	OPEN indicator extinguishes and CLOSED indicator illuminates when gaseous oxygen vent is closed. OPEN indicator illuminates and CLOSED indicator extinguishes when gaseous oxygen vent is open.
GASEOUS OXYGEN VENT FAN ON indicator	Extinguished when gaseous oxygen vent fan is off. Illuminates when gaseous oxygen vent fan is on.

1-40.   The lighting distribution transformer, rated at 45 KVA, 3-phase, 480-120/208 volts, supplies power to distribution panel D. From the 225-ampere bus of distribution panel D, power is distributed to lighting panel A, communication equipment distribution panel C, communication equipment, all located in the launch control room; and to the blast detection system equipment rack in the battery room.

1-41.   BLAST DETECTION SYSTEM.

1-42.   A blast detection system (figure 1-21) is provided (except at OSTF-2) to help protect personnel and equipment from the effects of a nuclear explosion. (See table 1-3 for LCC and silo blast closure indicators on FRCP.) The system actuates the closing of air inlets and air outlets in the launch control center and silo. Closure is initiated either by high intensity light or by electromagnetic waves.

1-43.   Two optical sensor units located above the launch control center send a signal to an amplifier when subjected to high intensity light. The amplified signal passes through a series of relays which complete circuits required to actuate closing of the air inlets and outlets, illuminate indicators, and place the missile guidance system on memory if guidance is in aligned status.

1-44.   The same function can also be performed by three loop antennae located ten feet below the surface of the earth in the area of the launch control center. When subjected to electromagnetic waves,

the antennae transmit a signal to the amplifier. The signal then accomplishes the same functions described in paragraph 1-43.

1-45.   SILO.

1-46.   The silo contains eight floor levels (suspended crib) and a guided missile silo launcher platform (figure 1-22) which supports, raises, and lowers the missile. The silo is constructed of reinforced concrete and is covered with steel and concrete doors which automatically open during a countdown. Equipment contained in the silo consists of facilities equipment; electrical power production and distribution equipment; gas, propellant and fuel supply tanks; and checkout, monitoring, and launch equipment. Entrance to the silo is gained through a tunnel from the LCC, at all sites except OSTF-2. At OSTF-2 entrance to the silo is gained through an above ground door and stairway connected to the second level. (See table 1-3 for LCC door monitoring indicators on FRCP.) A spiral staircase and elevator are provided to gain access to each level of the silo.

1-47.   CRIB.

1-48.   The crib is an octagonal steel structure containing the eight silo floor levels. A certain amount of space is left between the silo walls and the crib to allow motion of the crib. The crib is suspended from the silo walls by a suspension system (figure 1-6) that uses shock struts and horizontal dampers. Crib suspension is used to minimize equipment damage in the event of ground shocks due to explosive blasts. The horizontal dampers are provided to act as brakes to prevent crib oscillation. The missile is stored on the guided missile silo launcher platform (figure 1-22) in a 21-foot square missile enclosure that extends the full length of the crib. The launch platform raises and lowers the missile for installation, removal, checkout, and launch operations. The launch platform contains four levels upon which are mounted the missile and equipment that must be operated up to the point of missile release or during checkout with external equipment. An interconnecting box (11) (checkout cable junction box) is provided on the launcher platform for missile checkout operations. A flame deflector (4) is mounted immediately below the launcher pedestal (1) to deflect the engine flame from the equipment mounted in the lower levels. Electrical connections from the launcher platform to the missile (umbilical cable connections) are made from an interconnecting box (13) mounted behind the flame deflector.

1-49. SILO WORK PLATFORMS.

1-50. To enable missile maintenance functions to be performed on the lowered missile, hydraulically actuated work platforms are provided. These work platforms fold down from the sides of the missile enclosure and provide access to essential maintenance areas on the missile. The work platforms are electrically interlocked, preventing the launcher platform from being raised while the work platforms are extended.

1-51. The facility elevator safety platform, (figure 1-23) located on level one of the silo, is used to facilitate the transfer of large heavy equipment from outside the silo into the facility elevator. The safety platform is constructed with steel frame and grating and will withstand a 6000-pound load distributed over a 7-½ square foot area. The facility elevator safety platform is powered by a pneumatic-hydraulic system. A control panel on silo level one is provided to regulate the raising and lowering of the safety platform by extending or retracting the cylinder rods. The facility elevator safety platform is not mechanically or electrically interlocked to prevent the raising of the launcher platform. Prior to starting the launcher platform raise cycle, the facility elevator safety platform must be in the stowed position.

1-52. The stretch mechanism is stored on work platform 1B and 1D, and must be stowed before the work platform can be retracted. In the event the missile pressure fails or is shut off, or if maintenance is to be performed on the missile, the stretch mechanism is manually engaged. The mechanism consists of a folding stretch mechanism pedestal, two pins that engage opposite sides of the missile structure, two hydraulic cylinders, a hand pump, and a hydraulic pressure gauge.

1-53. Work platform 1 (15, figure 1-24), located on silo level 2, consists of three sections, each of which is operated by two hydraulic cylinders. The six cylinders receive hydraulic fluid from one solenoid-operated hydraulic valve. Limit switches indicate work platform position to the logic circuitry. The work platform side leaf must be stowed manually to prevent damage to equipment before the work platform is retracted.

1-54. Work platform 2 (11, figure 1-25), located on silo level 5, is similar to work platform 1 with the exception that work platform 2 has one section, one hydraulic cylinder, and linkage.

1-54A. Work platform 2B (VAFB only) (5, figure 1-26), located on silo level 5, provides access to the instrumentation and range safety system antenna and to the missile destruct package. It is mechanically connected to work platform 3B and is automatically extended or retracted by actuation of platform 3B. A side leaf on work platform 2B must be stowed manually before the platform is retracted to prevent damage to equipment. A ladder is provided to connect platforms 2B and 3B.

1-55. Work platforms 3 and 4 (figure 1-26), located on silo level 5A, are mechanically connected by two cascade mechanisms and are extended by either of two key switches located on silo level 5A. The platforms are extended by energizing one electrically controlled hydraulic valve which sends high-pressure fluid to the six cylinders. When the five sections of work platform 3 are extending, a cascade mechanism also extends the two sections comprising work platform 4. Work platform section 3D includes a manually extended side leaf which must be manually retracted and stowed prior to retraction of work platforms. Failure to retract and stow this leaf may cause damage to equipment when the work platforms are hydraulically retracted. An additional work platform is located on level 6. (See 9, figure 1-27 or 1-28.)

1-56.   The silo mouth platform provides walkways for personnel access to the missile for attachment of missile umbilical cables and attaching or detaching erection links. The platform also provides access to the launcher pedestals for engaging or retracting the missile holddown latches. A ladder provides access between the silo mouth platform and the engine access platform.

1-57.   ENVIRONMENTAL CONTROL.

1-58.   Due to the underground configuration of the launch complexes, an extensive environmental control system is provided consisting of water chiller, pumps, fins, heaters, hazards detectors, dust collector, and ducting. (See table 1-3 for silo associated air conditioning indicators on FRCP.) The intake air of the silo is washed and cooled and partly ducted to the missile enclosure to maintain the enclosure temperature at 70°F. (See figure 1-29 for simplified flow diagram of the silo air intake and exhaust system.) A constant temperature of the enclosure is required

due to the RP-1 fuel stored in the missile fuel tank because an increase in temperature causes an increase in volume due to expansion of the fuel. The remainder of the chilled air is ducted to the diesel compartments in levels five and six where it is warmed by heat generated by the 500KW diesel generators (40 and 46, figure 1-20.) The warmed air is diffused throughout the silo and maintains a temperature of 50 to 100°F depending on ambient temperatures. An exhaust fan and ducting expel the warm air from the launching silo.

1-59.   Cooling air must be supplied to components in the missile equipment pods when missile systems are energized. Air Conditioner A/E32C-5 (pod cooling unit) (figures 1-30 and 1-31) mounted on the launcher platform receives chilled water from a water chiller and supplies and recirculates 40°F dehumidified air to the missile equipment pod electronic equipment to maintain equipment within operating temperatures. Table 1-3 lists the associated missile pod air conditioner malfunction indicators on the facility remote control panel (FRCP) (figure 1-35 or 1-36). These indicators monitor missile pod air temperature, pressure, and humidity. These

data provide the MCCC with the information required to decide whether power must be removed from the missile guidance system and abort initiated in the event of a pod air conditioner malfunction, or whether countdown should be continued.

1-60. During missile countdown, engine valves and components in the thrust section of the missile are heated to prevent adverse temperature effects due to liquid oxygen and liquid nitrogen temperatures. These components are heated by a thrust section heater (16, figure 1-32 or 1-33) located on silo level 8. (See figure 1-29 for simplified flow diagram of the silo air conditioning system.) A thrust section heater malfunction during countdown causes an alarm to sound and the THRUST SECTION Heaters indicator on the launch control console (figure 1-7) to be illuminated amber. Countdown may be continued at the discretion of the MCCC.

1-61. To reduce explosion hazards due to excessive oxygen content in the missile enclosure, a gaseous oxygen vent system is provided to expel at high velocity the exhaust from the liquid oxygen boiloff valve on the missile. The gaseous oxygen vent system is retracted from the missile during launcher platform rise and extends when the launcher platform is lowered. Hazard detection systems are also provided to detect excessive oxygen content and fuel (diesel fuel) vapor in the silo. Gaseous oxygen detectors (4 and 6, figure 1-34) sample the oxygen content from sampling stations located on the seventh and eighth levels of the silo. When the oxygen content of the air rises above 27 percent by volume or falls below 17 percent by volume the detectors initiate both audible and visible alarms at the detector units, level seven, level eight, and facility remote control panel (FRCP) (figures 1-35 and 1-36.) At OSTF-2, the gaseous oxygen concentration read by the detection system is also shown on a meter located in launch control center, level two. The diesel fuel vapor detector (12, figure 1-34) samples the diesel fuel vapor content of the air at the diesel room exhaust and also initiates an audible and visible alarm at the detection unit and on the FRCP when predetermined limits are reached. The FRCP contains indicators and a warning buzzer to monitor the hazardous conditions in the silo and also monitors power equipment status, door positions, and other environmental control functions (fans,

vents, dampers.) Controls are provided on the panel for personnel warning, to control the missile enclosure fog system, and to manually open or close the blast closures. (See table 1-3 for FRCP indicators and controls.)

1-62. MISSILE ENCLOSURE.

1-63. The missile fuel tank is maintained at a constant temperature of approximately 70°F with 65 percent relative humidity. To reduce the volume of air which must be controlled for this fuel storage requirement, the missile and its launcher platform are enclosed within the silo shaftway by an insulated metal envelope. This enclosure is cooled by air conditioning or heated by hot water coils, depending upon ambient temperatures. (See figure 1-29 for simplified flow diagram of silo air conditioning system.) The enclosure also serves to lessen the area exposed to hazardous fumes, and gases escaping from the missile.

1-64. The enclosure is made up of metal panels attached to the crib at each level and is large enough to accommodate the various retractable missile work platforms. The panels are removable for missile maintenance and have gas-tight doors for access to work platforms.

1-65. WATER DISTRIBUTION SYSTEM.

1-66. The water distribution system consists of utility water, demineralized water, dust collector water, condenser water, hot water, and chilled water systems.

1-67. Water is supplied to the water distribution system from water wells located at each launch complex or commercial sources. A building containing a water treatment system may be located near the wells, depending on the purity of the well water supply. (See figure 1-37 for simplified flow diagram of the water treatment system.) Water from the wells is pumped to a valve box sunk into the ground adjacent to four underground storage tanks. A line connects the valve box to one of the water storage tanks which are interconnected and provides a total capacity of approximately 100,000 gallons. Indicators on the FRCP (figure 1-35 or 1-36) monitor the storage tank water levels. Water supply pumps are automatically started when the water level in the tanks drops to 90,000 gallons. A water line con-

Changed 21 October 1964

nects the storage tanks to the water distribution equipment located on level four of the silo.

1-68. The utility water system supplies water from the storage tanks to the drinking water dispensers, fog system nozzles, fire hose stations, emergency eyewash and shower installations, and the LCC domestic water system. (See figure 1-38 for simplified flow diagram of the utility water system.) The utility water system consists of turbine-type water pumps, hydropneumatic tank, and necessary valves and controls. An automatic control system maintains the required air pressure in the hydropneumatic tank. An indicator on the FRCP (figure 1-35 or 1-36) monitors the utility water pressure.

1-69. The demineralized water supply system supplies makeup water to the chilled water system and to the hot water system. See figure 1-40 for simplified flow diagram of the demineralized water system. It also supplies makeup water to the diesel engine water jackets. The system consists of water storage tank, water pump, and necessary valves and controls. The water storage tank is periodically replenished at a silo cap stubup with demineralized water to maintain the required water level. Water level control on the hot water and cold water expansion tanks will operate the demineralized water pump to maintain the required water level in the tanks.

1-70. The condenser water system is a closed loop system which supplies cooled water to the condenser units of the water chillers and to the diesel engine heat exchangers. (See figure 1-39 for simplified flow diagram of the condenser water system.) Cooling water is also supplied to the instrument air prefab. The system consists of a water cooling tower, water circulating pumps, and necessary valves and controls. Cooling water is circulated from the water cooling tower to the condenser units and diesel engine heat exchangers by the condenser water pumps and then recirculated back through the water cooling tower. Makeup water is obtained from the utility water system through a chemical pot feeder. If the primary and standby pumps which supply condenser water from the cooling tower fail, or if water flow from the cooling tower fails, an emergency water pump is automatically activated. Water is drawn from the storage tanks by the emergency pumps, circulated through the condenser units and diesel engine heat exchangers and then pumped directly to the silo sump. The cooling tower will be bypassed and the

system will no longer be in a recirculating operation. The status of the emergency water pump is monitored by means of an indicator on the FRCP (figure 1-35), at all sites except OSTF-2. Table 1-3 lists the indicator and its function.

1-71. The hot water system is a closed loop system which supplies hot water to the heating coils of the thrust section heater and the launch platform fan coil unit. See figure 1-39 for simplified flow diagram of the hot water system. Hot water is also supplied to the launch control center. The system consists of a hot water expansion tank, circulating pumps, heat recovery silencers, electric boiler, and necessary valves and controls. The water is pumped to the heat recovery silencers where it is heated, through the electric boiler for further heating if required, and then circulated to the heating coils and launch control center. Makeup water for the hot water system is obtained from the demineralized water supply system.

1-72. The chilled water system is a closed loop system which supplies chilled water to the cooling coils of the launch platform fan coil unit, control cabinet fan coil unit, and the pod air conditioning unit. (See figure 1-40 for simplified flow diagram of the chilled water system.) Chilled water is also supplied to the launch control center. The system consists of a cold water expansion tank, two circulating pumps (7 and 8, figure 1-17), two chiller units (refrigeration units complete with compressors, chillers, and condensers), and necessary valves and controls. Either chiller unit is selected as the primary unit from a local control. The secondary unit is cycled on and off automatically by a thermostatic control in the chilled water return line between the pumps and the water chiller units to provide additional cooling capacity during periods of peak load. Water in the system is pumped to the chiller units where it is cooled and then circulated to the cooling coils of the launch control center. Makeup water is obtained from the demineralized water supply system.

1-73. The dust collector water system is a closed loop system which supplies water to the silo intake air dust collectors. (See figure 1-29 for simplified flow diagram of the dust collector water system.) The system consists of a water makeup tank, circulating pumps, settling tank, and necessary valves and controls. The water is circulated through dust collectors and a settling tank. Makeup water for the dust col-

Changed 15 April 1964

lector water system is obtained from the utility water system. Following a nuclear blast certain valves in the dust collector water system will assume a new position causing supply water to come directly from the storage tanks to be circulated through the dust collectors and then directly pumped out of the silo. This operation will continue for 30 minutes following a nuclear blast, and then the dust collector water system will return to a recirculation operation.

1-74. POWER PRODUCTION AND DISTRIBUTION.

1-75. Power production equipment (primary) for the launch complex consists of two 500-KW diesel generator (40 and 46, figure 1-20) located on the fifth and sixth levels of the silo at all operational sites. Either generator may be selected locally at the 480-volt switchgear or remotely from the power remote control panel (PRCP) (21, figure 1-4) A single 500-KW diesel generator (5, figure 1-25) is located on level five at OSTF-2. Fuel and lubricating oil for the diesel engines are supplied from a diesel day tank (10) and clean lube oil tank (7). (Reference figure 1-102 for simplified flow of complex power distribution system.) A dirty lube oil tank (4) stores dirty oil which is discharged at the top of the silo by a transfer pump. The output of each diesel generator (460-volt, 3-phase, 60-cycle). is supplied to the 480-volt switchgear (1) for distribution to launch complex equipment. See table 1-3 for diesel generator associated indicators on FRCP. Feeder disconnect switches, located on the switchgear, control power distribution to 480-volt control centers (1, 16, and 17, figure 1-20) in the silo and launch control center. The control centers (figures 1-41 and 1-42) distribute power to equipment requiring 460-volt, 3-phase 60-cycle AC power. Power for lighting, communications, and other equipment requiring 120/208-volt, 60-cycle, AC power is supplied by step down transformers. Control power, to operate facility system motor starters located in the essential and nonessential motor control centers, is

supplied by the 120-VAC control voltage system. The system receives 460-VAC power from the motor control center 600-ampere essential bus and converts it into 120-VAC, single-phase power, with stepdown transformers. An automatic transfer switch (10, figure 1-24) in the system transfers to a standby transformer in the event of an operating transformer malfunction. The load side of the transfer switch is connected to an 18-circuit, single-phase circuit breaker panel that feeds the control voltage circuits. A 28-VDC battery (12, figure 1-43) supplies emergency power to the launch control system for shutting down launch operations in the event of a power failure. A 48-volt battery rack (2, figure 1-27 or 3, figure 1-28) is provided as backup for the 48-volt battery charger to provide power for the safety circuit, tripping circuit breakers, indicator lights on the 480-volt switch gear, PRCP, and for the diesel generator governor motor controls. In addition to the 500-KW diesel generator, provisions are made at OSTF-2 for connecting a power company line as an alternate power source. The power company line feeds the main power transformer (7, figure 1-28) which feeds 460-VAC to the 480-volt switchgear. Either power source (diesel generator or power company) may be selected locally at the 480-volt switchgear or remotely from the power remote control panel (30, figure 1-5). The power remote control panel (figure 1-44 or figure 1-45) contains voltmeters, ammeters, and controls to monitor power and start and stop the diesel generators or connect the power company line. (See table 1-4.) Twenty-eight VDC power and 400-cycle AC power are supplied by Power Supply-Distribution Set PP-2537/GSW (9, figure 1-43) and 400-cycle Motor Generator type MD-2 (10). The 400-cycle motor generator control panel (figure 1-46) provides the controls and indicators necessary for the operation of the motor generator. The voltage output of the power supply-distribution set and the motor generator may be monitored at MISSILE GROUND POWER (PANEL 2) (figure 1-47).

Table.1-4.   Power Remote Control Panel Controls and Indicators

CONTROL OR INDICATOR	FUNCTION
RUNNING VOLTAGE and INCOMING VOLTAGE meters or GENERATOR VOLTMETER A. C. (OSTF-2 only)	Indicate voltage output of diesel generators

Table 1-4. Power Remote Control Panel Controls and Indicators (Continued)

CONTROL OR INDICATOR	FUNCTION
GENERATOR NO. 1 AMMETER AC and GENERATOR NO. 2 AMMETER AC meters or GENERATOR AMMETER A. C. (OSTF-2 only)	Indicates load current when generators are connected to the line
RUNNING FREQUENCY and INCOMING FREQUENCY meters or GENERATOR FREQUENCY A.C. meter (OSTF-2 only)	Indicate frequency of generator output voltage
GENERATOR NUMBER 1 AMMETER SWITCH and GENERATOR NUMBER 2 AMMETER SWITCH or GENERATOR AMMETER SWITCH (OSTF-2 only)	Selects either phase A, B, or C current for indication on PRCP ammeters
SYNCHROSCOPE and SYNCHRONIZING indicators (not included at OSTF-2)	Indicates phase relation between running and incoming generators
GENERATOR NO. 1 MAIN BREAKER CONTROL SWITCH and GENERATOR NO. 2 MAIN BREAKER CONTROL SWITCH or GENERATOR MAIN BREAKER CONTROL SWITCH (OSTF-2 only)	Connects generator D-60 and D-61 to line **NOTE** Generator D-61 is not included at OSTF-2
Main breaker control switch indicators	Indicate position of GENERATOR NO. 1 MAIN BREAKER CONTROL SWITCH, GENERATOR NO. 2 MAIN BREAKER CONTROL SWITCH, and GENERATOR MAIN BREAKER CONTROL SWITCH (OSTF-2 only.) Red indicator illuminates when main breaker control switch is closed. Green indicator illuminates when main breaker control switch is opened.
GENERATOR NUMBER 1 S Y N C H R O N I Z I N G SWITCH and GENERATOR NUMBER 2 SYNCHRONIZING SWITCH (not included at OSTF-2)	Control switches for synchronizing both generators for parallel diesel operation
GENERATOR NUMBER 1 GOVERNOR MOTOR CONTROL SWITCH and GENERATOR NO. 2 GOVERNOR MOTOR CONTROL SWITCH or GENERATOR GOVERNOR MOTOR CONTROL SWITCH (OSTF-2 only)	Control switches for locally controlling diesel generator governors
ENGINE NUMBER 1 START STOP SWITCH and ENGINE NUMBER 2 START STOP SWITCH or ENGINE START STOP SWITCH (OSTF-2 only)	Control switches for locally starting and stopping diesel engine
	Indicators display engine started or stopped. Diesel is started when red indicator is illuminated. Diesel engine is stopped, and is capable of being started from the PRCP when green indicator is illuminated

Table 1-4. Power Remote Control Panel Controls and Indicators (Continued)

CONTROL OR INDICATOR	FUNCTION
FEEDER NO. 3 NON-ESSENTIAL BUS CONTROL SWITCH	Control switch for locally applying or removing nonessential power
FEEDER NO. 3 NON-ESSENTIAL BUS CONTROL SWITCH indicators	Indicate non-essential power is on when red indicator is illuminated. Indicates that non-essential power is off when green indicator is illuminated
EMERGENCY SHUTDOWN pushbuttons	Pushbuttons for locally shutting down diesel engines and simultaneously tripping GENERATOR MAIN BREAKER CONTROL SWITCH
NOTE The following indicators and controls are contained on the PRCP at OSTF-2 only.	
POWER CO. LINE AMMETER A. C.	Indicates commercial power load current when commercial power is being used
POWER CO. LINE VOLTMETER A. C.	Indicates commercial power voltage
POWER CO. LINE FREQUENCY A.C.	Indicates commercial power frequency
POWER CO. LINE AMMETER SWITCH	Selects either phase A, B, or C current for indication on POWER CO. LINE AMMETER A. C.
POWER CO. LINE VOLTMETER SWITCH	Selects either phase A, B, or C voltage for indication on POWER CO. LINE VOLTMETER A. C.
POWER CO. LINE MAIN BREAKER CONTROL SWITCH	Control switch for locally connecting commercial power
POWER CO. LINE MAIN BREAKER CONTROL SWITCH indicator	Indicates Position of POWER CO. LINE MAIN BREAKER CONTROL SWITCH. Red indicator illuminates when commercial power is connected. Green indicator illuminates when commercial power is not connected.

1-76. PROPELLANT AND GASES.

1-77. Gases (nitrogen and helium), liquid nitrogen, and liquid oxygen are stored in tanks located on level eight of the silo. (See figure 1-32, 1-33 and 1-48.) The storage tanks are initially filled and replenished as required from supply trailers connected to ground level charging connections. Nitrogen gas is used during standby periods to pressurize the missile fuel and liquid oxygen tanks, to perform fuel leveling and transfer operations, to charge hydraulic accumulators, to operate the liquid oxygen topping control unit (20, figure 1-48 and figure 1-49), to back up instrument air, and to supply the nitrogen control unit (1, figure 1-48) for test, checkout, and purging operations. Helium is used during countdown to charge the missile helium bottles and to pressurize the missile fuel and liquid oxygen tanks. Liquid nitrogen is used during countdown for chilling the shrouded missile helium bottles and loading lines to enable a greater amount of helium to be stored in the airborne bottles. The helium is cooled by passage through a liquid nitrogen/helium heat exchanger tank (14) prior to loading. Loading of liquid nitrogen is accomplished during countdown by pressurizing the liquid nitrogen storage tank (13) with gaseous nitrogen from the pneumatic system manifold regulator (5, figure 1-48 and figure 1-50). The pneumatic system manifold regulator receives helium and nitrogen and controls distribution to the missile, helium control charging unit (2, figure 1-48 and figure 1-51) and the pressure system control (4, figure 1-48.) Compressed air for pneumatic valve operations is also received by the pneumatic system figure 1-51), and the pressure system control (4, manifold regulator, from the instrument air prefab (18) and distributed to the liquid nitrogen prefab (15, figure 1-48 and figure 1-52), pressure system control, pressurization prefab (19, figure 1-48 and figures 1-53 and 1-54), liquid oxygen control prefab (16, figure 1-48 and figure 1-55) and the liquid oxygen fill prefab (17, figure 1-48.)

1-78. The instrument air prefab consists of an air compressor, air receiver, drier, and coolers. (See figure 1-56 for simplified flow diagram of the instrument air system.) In addition to supplying compressed air to the pneumatic system manifold regulator, the instrument air prefab supplies instrument and control air pressure to equipment in the launch control center, to the utility water system, diesel air start system, and blast closure (air intake and exhaust) mechanism. A facility compressed air system,

consisting of a facility air compressor and regulator, is provided for squadron complexes at VAFB, 550, 556 and 579. The compressed air system provides compressed air for controlling components of the silo heating, ventilating, and air conditioning systems. At squadron complexes 551, 577, and 578, these components of the silo heating, ventilating, and air conditioning system are electrically operated and a facility compressed air system is not provided. (See figure 1-96 for a simplified flow diagram for complexes at VAFB and 550; and figure 1-97 for a simplified flow diagram for squadron complexes 556 and 579.) In the event loss of facility compressed air occurs, the instrument air system supplies emergency instrument air to the facility air compressor and regulator station.

1-79. Missile tank pressurization is maintained by the pressure system control during standby periods and during countdown. Nitrogen gas is used during the standby periods while helium is used after countdown start. During the commit sequence, the pressurization of the missile tanks is switched to the internal pressurization system where helium is supplied by the airborne helium bottles. Loading of helium into the airborne bottles is controlled by the pneumatic system manifold regulator (figure 1-50.) The helium control charging unit (figure 1-51) ensures that the airborne bottles are fully charged up to the point of missile launch by maintaining the bottles at 3000 PSI during launcher platform rise.

1-79A. The helium control charging unit also serves as an emergency pressurization source for the missile tanks. Helium for this purpose is supplied to the helium control charging unit from the inflight helium storage tanks, through a valve in the pressure system control, and through an umbilical loop hose. After the launcher platfrom leaves the down and locked position, the pressure system control no longer maintains missile tank pressures. If the launcher platform cannot be lowered, manual pressurization of the missile oxidizer tank will be required if all the liquid oxygen or liquid nitrogen is allowed to boiloff. Manual valves on the helium control charging unit enable manual pressurization to be accomplished. Fuel tank pressurization during this period is automatically maintained by the helium control charging unit. However, manual valves for controlling the fuel tank pressure are also provided on the helium control charging unit. The helium control charging unit also serves as a backup to the pressure system control while the launcher platform is down and locked. Pressure gages on the helium control charging unit enable the operator to monitor missile fuel and oxidizer tank pressures.

1-80. PROPELLANT LOADING SYSTEM.

1-81. Propellants, RP-1 fuel and liquid oxygen, are fully loaded in their respective missile tanks before the missile is elevated to the ground level firing position. Fuel is loaded from a tank truck or the silo catchment tank, through a fuel purification unit, when the missile is first installed in the silo. If the temperature of the RP-1 fuel varies from that in the silo missile enclosure, considerable volumetric change takes place within the fuel tank as the fuel temperature stabilizes to the missile enclosure tem-perature (65 to 75°F.) Fuel volume in the missile tank is adjusted to the full level during this period of temperature stabilization by draining fuel into a leveling tank or by pumping fuel from the leveling tank to the missile fuel tank using controls and indicators on FUEL TANKING (PANEL 2) (figure 1-57.)

1-81A. At VAFB only, a fuel fill-and-drain line shutoff plate (figure 1-99) is installed in the fuel fill-and-drain line below the riseoff disconnect on level 1 of the launcher platform. One end of the shutoff plate has a hole to allow passage of fuel; the other end is blank. The blank end of the shutoff plate is installed in the fuel fill-and-drain line before a training launch to seal the line and protect it form the blast of the missile engines.

1-82. During countdown, launch control logic circuits sequentially operate selected valves in the pressurization and liquid oxygen subsystems. Stored liquid oxygen is forced by pneumatic pressure into the missile liquid oxygen tank. Liquid level sensing transducers in the tank respond to the presence of liquid oxygen as it attains successive levels within the missile tank, generating electrical signals which enable the launch control logic to control the sequence of propellant loading.

1-83. The propellant loading system consists of three subsystems: fuel, liquid oxygen, and pressurization; and that part of the instrument air system needed to operate the liquid oxygen subsystem. Propellant loading equipment is located on several levels within the silo. On the seventh level are the liquid oxygen control prefab, (16, figure 1-48) the liquid oxygen fill prefab (17), the liquid oxygen topping control unit (20), the pressurization prefab (19), and the instrument air prefab (18.) The liquid oxygen storage tank (11) and the liquid oxygen topping tank (12) are suspended between the seventh and eighth levels. Seven gaseous nitrogen storage tanks (7) are suspended between the seventh and eighth levels. The fuel loading prefab (3), the pressure system control (4), and the pneumatic system manifold regulator (5) are located on the eighth level within the missile enclosure area. Connections to the liquid oxygen fill-and-drain line, the liquid oxygen topping line, and the missile fuel fill-and-drain line are located near the first level of the launcher platform. The fuel catchment tank is located ten feet underground adjacent to the silo cap.

1-84. The trailer mounted fuel purification unit is used to remove water and impurities from the RP-1 fuel during initial loading and during a missile recycle. Since fuel is permanently stored in the missile fuel tank, all free water must be removed. The main components of the fuel purification unit are a pump, a filter-separator, and a dehydrator. The RP-1 fuel is pumped through the filter-separator, then through the dehydrator where silica gel absorbs water molecules. The fuel is constantly monitored at the outlet line of the dehydrator by a moisture monitor. The dehydrator is able to process approximately 45,000 gallons of fuel before requiring replacement.

1-85. The liquid oxygen subsystem transfers liquid oxygen from the ground tanks to the missile liquid oxygen tank during countdown. It consists of a liquid oxygen storage tank, topping tank fill prefab, topping control unit, and interconnecting piping and valves. Main transfer is accomplished through a 10-inch fill and drain line connected from the liquid oxygen storage tank, through the liquid oxygen control prefab, to the missile tank. Secondary transfer is provided by a line connected from the liquid oxygen topping tank, through the topping control unit, to the missile booster pumps. Gaseous nitrogen from the pressurization subsystem applied to the storage and topping tanks provides the pressure to make the transfer. The timing of liquid oxygen loading is determined by time delay relays in LO$_2$ TANKING (PANEL 1 and PANEL 2) (figure 1-60) working in conjunction with liquid oxygen level sensing transducers in the missile tank.

1-86. During standby, the status of the propellant loading system is monitored and displayed on the launch control console (figure 1-7) by a summary signal from the various subsystem panels that illuminates the LO$_2$ & FUEL indicator. Liquid oxygen loading does not start during countdown until the pneumatic system enters the phase II cycle. When this happens, the fuel tank pressure is raised to approximately 63.2 PSI and the liquid oxygen boiloff valve is placed on automatic to regulate the liquid oxygen tank pressure to approximately 3.6 PSI. With these conditions established, a 40-second line chilldown sequence is initiated to pre-cool the transfer lines and valves. The LO$_2$ LINE FILLED indicator in the countdown patch of the launch control console illuminates green following the conclusion of the line chilldown sequence, and rapid loading of liquid oxygen starts. Rapid loading continues until a propellant level sensing transducer senses that the tank is 95 percent full. When this occurs, the RAPID LO$_2$ LOAD indicator lights green and the fine loading sequence is started.

1-87. The fine loading sequence continues until the oxidizer tank is 99 percent full. A line draining sequence is then initiated to drain the main 10-inch liquid oxygen fill and drain line. The LO$_2$ READY indicator on the countdown patch of the launch control console illuminates green at this point denoting that the liquid oxygen system is ready for the commit sequence. An on-off topping sequence continues to replace the loss of liquid oxygen due to boiloff and maintains the tank at 99.25 percent or above.

1-88. During the commit sequence, after missile power has been transferred to internal, the liquid oxygen boiloff valve is closed and rapid topping occurs to bring the tank up to the 100 percent level. The LO₂ COMMIT indicator on the launch control console commit patch illuminates green indicating the LO₂ system is ready for launch. Simultaneously, the LO₂ commit signal initiates a pulsating reset signal to the Computer Assembly CA-108B (propellant utilization computer assembly), which continues until either an abort is initiated or missile lift-off has been accomplished.

1-89. PROPELLANT LOADING SYSTEM STATUS INDICATORS. The propellant loading system status indicators, located on the launch control console, are monitored during standby and countdown by the missile combat crew commander or by his designated alternate. Table 1-5 lists the indicators involved and their functions.

1-90. SILO HYDRAULICS.

1-91. Silo hydraulics (figure 1-61) are used in the launcher platform drive (clutch and brake) system, crib locking, silo door opening, and work platform actuation. The major components of the system are two hydraulic pumps (13, figure 1-61) and reservoir (12), hydraulic accumulators (14), and nitrogen pressure tanks (15.) A 1-HP hydraulic pump supplies the hydraulic pressure required during standby periods and a 40-HP hydraulic pump supplies the hydraulic pressure required during abort and checkout operations. During the commit sequence, hydraulic pressure is supplied by the hydraulic accumulators that are charged with nitrogen from the nitrogen pressure tanks.

1-92. The overhead doors (1) are opened by two door closure cylinders (3) and four breakaway cylinders (2.) The four breakaway cylinders assist the

Table 1-5. Propellant Loading System Indicators on Launch Control Console

INDICATOR	FUNCTION
LO₂ & FUEL indicator	Illuminates green when all of the following conditions exist: a. SYSTEM IN STANDBY indicator on PROPELLANT LEVEL (PANEL 2) (figure 1-58) illuminated green b. SYSTEM IN STANDBY indicator on FUEL TANKING (PANEL 1) (figure 1-59) illuminated green c. SYSTEM IN STANDBY indicator on LO₂ TANKING (PANEL 1) (figure 1-60) illuminated green Illuminates red when any of the above conditions are not as specified
FUEL LEVEL indicator	Illuminates amber when either of the following conditions exists: a. During standby when FUEL switch on PROPELLANT LEVEL (PANEL 2) is in LOCAL position and fuel tank level is too high b. Within 5 seconds after countdown starts and fuel tank level is too high
LO₂ LEVEL SENSING indicator	Forty seconds after countdown is started, illuminates amber to indicate single sensor failure or red to indicate a double sensor failure. One hundred and twenty seconds after start of rapid LO₂ load the ability to detect a LO₂ level sensor failure is discontinued

door cylinders by exerting an initial upward thrust to overcome the weight of the overhead doors and to break any ice formations. To prevent movement of the crib during launcher platform raising and lowering, the crib structure is locked to the silo structure by the horizontal and vertical crib locks (4 and 8.) If the crib has tilted in relation to the silo walls, the vertical crib locks align the crib so that it becomes parallel to the silo walls and positioned in the center. Centering of the crib is required to ensure proper alignment of the launcher platform to silo cap locks (9) in order that the launcher platform will lock to the silo cap when fully raised.

1-93.   The work platforms in the missile enclosure are operated by the work platform actuators (11) which are controlled from key switch control panels (figure 1-62.) The launcher platform drive is mechanically engaged and hydraulically disengaged by a launcher platform drive clutch and brake (5, figure 1-61) and enables the launcher platform to be stopped at intermediate points between the fully raised and fully lowered positions.

1-94.   The missile support hydraulic system is used to fill and bleed the missile hydraulic system, provide hydraulic power for missile hydraulic system testing, and furnish ground hydraulic power to the missile hydraulic system during countdown until the airborne pumps take over. It consists of Hydraulic Pumping Unit A/E27A-10 (9, figure 1-22 and figure 1-63) and the lines necessary to deliver hydraulic fluid to the missile. Composed of two independent pumping units, the first stage of the hydraulic pumping unit serves the missile booster hydraulic subsystem; the second stage serves the sustainer-vernier hydraulic subsystem.

1-95.   The hydraulic pumping unit is housed in a cabinet on the third level of the launch platform. Steel tubing is used to connect the pressure and return fittings on the hydraulic pumping unit to the missile riseoff couplings located on the top level of the launcher platform. The hydraulic pumping unit is air cooled and requires a gaseous nitrogen input of 1000 to 1500 PSIG, and electrical inputs of 440 volts, 60 CPS, 3 phase AC (60 amperes approximately) and 28 volts DC (30 amperes approximately.) It may be operated either from the control panel of the unit or remotely from control monitor group 2 of 4 (5, figure 1-43) in the launch silo.

1-96.   MISSILE LIFTING SYSTEM.

1-97.   The missile lifting system is composed of pneumatic, hydraulic, and electrical subsystems which control all missile lift, door closure, crib locking, launch platform locking, and work platform actuation operations. The hydraulic system provides the hydraulic pressures for the above operations on command of the missile lift system logic units or from local control sources. A local control hydraulic panel (7, figure 1-24 and figure 1-64) located on level 2 of the silo monitors system hydraulic and pneumatic (nitrogen) pressures and enables local (manual) control for checkout of the hydraulic missile lift system. The electrical missile lifting control system (6, figure 1-65), receives and sends signals to the control monitor group and distributes and controls all internal drive voltages. The missile lift system motor control center (7) converts 440-volt AC power to voltage levels required by the system. The missile lift logic racks (part of electrical missile lifting control system) control actuation of the missile lift system in a predetermined sequence and contain indicators and controls to perform checkout, test, and self-test operation. A fault register chassis is also provided in the logic racks, and displays, on punched tape, malfunctions of specific loops within the missile lift system.

1-98.   The launcher platform is raised or lowered by missile lifting launch platform drive assembly (12, figure 1-65) consisting of two electric motors, reduction gears, and drive sheaves. During countdown fast launcher platform lifting is accomplished by a high speed motor since the weight of the fully loaded missile and launcher platform is counterbalanced by the launcher platform counterweight (15, figure 1-24.) When a fully loaded missile is not installed on the launcher pedestal (1, figure 1-22), the drive system must overcome the resulting unbalance. Thus empty launcher platform lifting and lowering are accomplished at a greatly reduced speed. The low speed launcher platform drive motor is used during slow speed operations through an auxiliary gear reducer.

1-99.   The missile lifting system operates in one of three modes: Sequence I, Sequence II, and a manual mode controlled from the control station manual operating level (CSMOL) figure 1-66.) Sequence I and II are fully automatic sequences, initiated by launch control countdown commands. Sequence I is the sequence of events required to position a loaded missile at ground level for launch. Sequence II is the sequence of events required to lower the empty launcher platform to the bottom of the silo after launch, or to lower the loaded missile to the bottom of the silo in the event of an abort. When Sequence II is completed, the weapon system is restored to a safe condition. Before Sequence I is initiated, the following conditions must exist:

a.   Silo doors fully closed.

b.   Vertical crib locks unlocked.

c.   Horizontal crib locks unlocked.

d.   Launcher platform down and locked.

e.   Facility elevator safety platform stowed (no logic interlock.)

f.   Work platforms 1, 2 and 3 retracted.

g.   28 VDC circuit breakers on.

h.   480 VAC circuit breakers on.

i.   120 VAC circuit breakers on.

j.   Launcher platform drive coupling disengaged.

k.   All inertial guidance (AIG) pod handling fixture retracted.

l.   Power pack drain valve closed.

m.   Hydraulic return line pressure greater than 17 PSIG.

n.   Hydraulic subsystem N_2 supply recharged.

o.   Hydraulic subsystem accumulator vent valves open.

p.   Hydraulic distribution not at operating pressure.

q.   Logic system test plugs stored.

r.   Sequence I receiver reset energized.

s.   Pressure switch manual valves open.

1-100.   When the propellants and gases are loaded aboard the missile and the summaries for missile commit have been completed, the commit sequence is initiated. When all guidance and missile summaries required for missile launch are completed, a missile lift signal is sent to the missile lift system. This signal is stored by the missile lift logic auto programmer. The auto programmer simultaneously energizes the drive

memory relays for launch platform locks in forward motion, open silo doors, and lock vertical locks. When the launch platform wedge locks are retracted, the launcher platform main locks retract. Upon extension of the vertical crib locks, the drive memory for horizontal crib locks is energized and the horizontal locks extend. The launcher platform will not start to rise, however, until the silo doors are fully open.

1-101.   With the work platforms retracted, and the conditions in paragraph 1-99 met, a signal is sent to the up run drive mechanism and the launcher platform starts to rise. An interlock will prevent the launcher platform from lifting in the event that work platforms on levels 2, 5, or 6 are not retracted. No interlock is provided to stop platform lift if the facility elevator safety platform is extended. Therefore, before the launcher platform is raised, ensure that the facility elevator safety platform is stowed. The up drive of the launcher platform is controlled by a reference voltage speed control which causes the high speed motor to run at 1620 RPM which is equivalent to a lifting rate of 33 inches per second. As the platform continues to rise, an overspeed sensing device monitors launcher platform travel speed for correct velocity. Upon sensing an overspeed condition the unit sets the brake and stops the drive motor.

1-102.   As the launcher platform approaches the top of the silo, decelerate limit switches are activated and slow the platform to a rate of one inch per second rise. When the launcher platform actuates any two diagonally opposed fully up limit switches, the main locks start to extend. The high speed motor continues to drive the launcher platform until the main locks are fully extended and the wedge locks start to extend. Completion of Sequence I provides a signal to the control monitor group resulting in a MISSILE LIFT UP AND LOCKED green indication on the launch control console. At any time during the operation of Sequence I, Sequence II may be initiated by depressing the ABORT pushbutton.

1-103.   When the launcher platform is up and locked or when Sequence I is stopped, Sequence II is initiated by depressing the ABORT pushbutton on the launch control console, located in the launch control center. At the start of Sequence II, a signal is sent to engage the low speed drive motor to the main speed reducer. When the down run command is present, a launcher platform locks reverse motion signal is sent and the launcher platform wedge locks are retracted which in turn retracts the main locks. When all the locks are retracted and the launcher platform drive interlocks summary is completed, the launcher platform drive system is enabled. The drive command controls the slow speed motor causing it to run at 1620 RPM, driving the launcher platform down at a rate of 3 inches per second. The down motion will continue at 3 inches per second until the down decelerate limit switch is actuated. This switch when actuated, will command the launcher platform drive to slow down to approximately 1/16 inch per second to prevent damage to the crib and down lock striker. When the launcher platform actuates one north and one south fully down limit switch, the main locks are extended, which in turn extends the wedge locks. The launcher platform drive motor will continue to drive the launcher platform down until two diagonally opposite launcher platform fully down limit switches are actuated. Upon completion of launcher platform down and locked,

a signal is sent to close the silo doors. After completion of silo door closing, a signal is sent to unlock horizontal crib locks. When the horizontal crib locks are retracted, the signal is sent to unlock the vertical crib locks. Completion of this operation provides a signal to the launch control console indicating a SITE HARD green indication. Hydraulic recharge will then occur and nitrogen will be vented on level 2.

1-104   MISSILE LIFTING SYSTEM STATUS INDICATORS. The status indicators on the CSMOL panel (figure 1-66) are monitored during sequences I and II during manual modes of operation by the missile facilities technician. Table 1-6 lists the indicators and controls on the CSMOL panel and their functions. Status indicators located on the launch control console (figure 1-7) are monitored during standby, countdown, and abort by the missile combat crew commander, or by his designated alternate. Table 1-6 also lists the missile lifting system status indicators on the launch control console and their functions.

Table 1-6.   Missile Lift System Indicators and Controls on Launch Control
Console and Control Station Manual Operating Level

INDICATOR OR CONTROL	FUNCTION
LAUNCH CONTROL CONSOLE	
FACILITIES & MISSILE LIFT indicator	Illuminated green when the following two conditions exist: a. SYSTEM IN STANDBY indicator on MISSILE LIFTING PLATFORM (PANEL 1) (figure 1-67), illuminated green b. SYSTEM POWER indicator on FACILITY (PANEL 1) (figure 1-68), illuminated green Illuminated red when either of above conditions are not as specified
MISSILE LIFT FAIL indicator	Illuminated red when missile lifting system malfunction occurs
CONTROL STATION MANUAL OPERATING LEVEL (CSMOL)	**NOTE** No position indicators will be illuminated unless one of the following conditions exists: a. The 40-HP hydraulic pump is on and missile lifting system hydraulic system has reached operating pressure. b. The missile lifting hydraulic system has reached

Table 1-6.   Missile Lift System Indicators and Controls on Launch Control
Console and Control Station Manual Operating Level (Continued)

INDICATOR OR CONTROL	FUNCTION
	operating pressure during an automatic sequence. c. The missile lifting system logic unit is in checkout mode (test plugs installed.) **NOTE** No pushbuttons, except HYDRAULIC 40 HP PUMP ON, are operative unless both of the following conditions exist: a. The 40-HP hydraulic pump is on and the missile lifting system has reached operating pressure. b. The missile lifting system logic unit is not in automatic sequence.
HYDRAULIC 40 HP PUMP ON pushbutton	Starts missile lifting system 40-HP hydraulic pump when depressed. Pushbutton is illuminated after being depressed and while pump drive memory relay and pump motor contactor are energized.
HYDRAULIC 40 HP PUMP OFF pushbutton	Stops missile lifting system 40-HP hydraulic pump when depressed and extinguishes HYDRAULIC 40-HP PUMP ON pushbutton
HYDRAULIC 40 HP PUMP PRESSURE indicator	Illuminated green when missile lifting hydraulic system is at operating pressure
R E S E T PROGRAMMER key switch	Enables local control from control station manual operating level (CSMOL) panel when positioned to ON during an automatic sequence. Performs the following functions when positioned to ON: a. Blocks commit and abort sequence commands from launch control system. b. Resets magnetic latching automatic sequence command memory and in-cycle memory relays, thereby terminating the automatic sequence. Enables automatic missile lifting system sequence to be initiated from launch control system when positioned to OFF.
M/L STOP indicator	Illuminated red when pressurization system is in emergency mode or when missile oxidizer tank pressure is less than 8.0 PSI after pressurization system has been in emergency mode. The indicator is disabled (cannot become illuminated) from 5 seconds after start of countdown until pneumatics internal and from launcher platform down and locked until abort complete.

Table 1-6.   Missile Lift System Indicators and Controls on Launch Control
Console and Control Station Manual Operating Level (Continued)

INDICATOR OR CONTROL	FUNCTION
CRIB VERTICAL LOCK and CRIB HORIZONTAL LOCK indicators	Illuminated green when crib vertical and horizontal locks are fully extended and locked and with the following conditions present: a. Final position limit switches actuated. b. Final status relay energized. c. Command memory relay for opposite direction (unlocked) not pulsed.
CRIB VERTICAL LOCK and CRIB HORIZONTAL L O C K pushbuttons	Extend and lock crib vertical and horizontal locks when depressed. CRIB VERTICAL LOCK pushbutton is enabled when HYDRAULIC 40 HP PUMP PRESSURE indicator is illuminated green. CRIB HORIZONTAL LOCK pushbutton is enabled when CRIB VERTICAL LOCK indicator is illuminated green.
CRIB VERTICAL UNLOCK and CRIB HORIZONTAL UNLOCK indicators	Illuminated green when CRIB VERTICAL and HORIZONTAL LOCKS are retracted and with the following conditions present: a. Final position limit switches actuated. b. Final status relay energized. c. Command memory relay for opposite direction (locked) not pulsed.
CRIB VERTICAL UNLOCK and CRIB HORIZONTAL UNLOCK pushbuttons	Retract crib vertical and horizontal locks when depressed. CRIB HORIZONTAL UNLOCK pushbutton is enabled when LAUNCHER PLATFORM DOWN COMPLETED RUN AND LOCKED indicator is illuminated green. CRIB VERTICAL UNLOCK pushbutton is enabled when CRIB HORIZONTAL UNLOCK and SILO DOORS CLOSE indicators are illuminated green.
SILO DOORS CLOSE indicator	Illuminated green when both silo overhead doors are fully closed and with the following conditions present: a. Fully closed position limit switches actuated. b. Final status relay energized. c. Command memory relay for opposite direction (silo doors open) not pulsed.
SILO DOORS OPEN indicator	Illuminated green when both silo overhead doors are fully open with the following conditions present: a. Fully open position (95 degrees) limit switches actuated. b. Final status relay energized. c. Command memory relay for opposite direction (silo doors close) not pulsed.

Table 1-6.   Missile Lift System Indicators and Controls on Launch Control
Console and Control Station Manual Operating Level (Continued)

INDICATOR OR CONTROL	FUNCTION
SILO DOORS CLOSE and SILO DOORS OPEN pushbutton	Close and open silo overhead doors and actuate personnel warning klaxon horn when depressed and held. SILO DOORS OPEN pushbutton is enabled when CRIB VERTICAL LOCK and CRIB HORIZONTAL LOCK indicators are illuminated green. SILO DOORS CLOSE pushbutton is enabled when LAUNCHER PLATFORM DOWN COMPLETED RUN AND LOCKED indicator is illuminated green.
LAUNCHER PLATFORM UP COMPLETED RUN AND LOCKED indicator	Illuminated green when launcher platform is fully up and locked with the following conditions present: a. Fully up position limit switches actuated. b. Main locks extended limit switches actuated. c. Wedge locks pressurized pressure switches actuated. d. Down run command memory relay not pulsed.
LAUNCHER PLATFORM DOWN COMPLETED RUN AND LOCKED indicator	Illuminated green when launcher platform is fully down and locked with the following conditions present: a. Fully down position limit switches actuated. b. Main locks extended limit switches actuated. c. Wedge locks pressurized pressure switches actuated. d. Up run command memory relay not pulsed.
LAUNCHER PLATFORM DOWN RUN pushbutton	Drives launcher platform down when depressed. Pushbutton is enabled when launcher platform is not down and locked.
LAUNCHER PLATFORM UP RUN pushbutton	Drives launcher platform up when depressed. Pushbutton is enabled when SILO DOORS OPEN, CRIB HORIZONTAL LOCK, and CRIB VERTICAL LOCK indicators are illuminated green.
LAUNCHER PLATFORM STOP pushbutton	Stops launcher platform motion (up or down) when depressed during local operations from CSMOL panel. Does not stop launcher platform during automatic sequences.
LAUNCHER PLATFORM UP CREEP and LAUNCHER PLATFORM DOWN CREEP pushbuttons	Enable up creep and down creep of launcher platform when depressed, to position launcher platform at critical positions. LAUNCHER PLATFORM UP CREEP pushbutton is enabled any time drive interlocks summary is completed with the launcher platform below the low speed raising deceleration limit switch. LAUNCHER PLATFORM DOWN CREEP pushbutton is enabled any time the drive interlocks summary is completed.

Table 1-6. Missile Lift System Indicators and Controls on Launch Control
Console and Control Station Manual Operating Level (Continued)

INDICATOR OR CONTROL	FUNCTION
	NOTE L A U N C H E R PLATFORM U P CREEP and LAUNCHER PLATFORM DOWN C R E E P push-buttons are disabled 30 seconds after being depressed and remain disabled for 5 minutes.
CREEP DISABLE indicator	Illuminated red when launcher platform drive interlocks relay is not energized and while 5-minute creep disable timer is running after creep operation has been initiated.
EQUIPMENT IN CHECKOUT OPERATIONS DEACTIVATED indicator	Illuminated amber when missile lifting system is in checkout configuration (test plugs installed in panel A4A7 of missile lifting system logic unit.)

1-105. MISSILE GUIDANCE SYSTEM CHECK-
OUT AND ALIGNMENT EQUIPMENT.

1-106. To provide an accurate means of calibrating and checking the guidance alignment equipment, bench marks with azimuth reference prisms and a polaris sighting tube are provided. The alignment group (collimator) is periodically checked against the star Polaris as required. The collimator, along with the countdown group, maintains the missile guidance system in continuous alignment. (See figure 1-69.) A sight tube extends from the collimator to the missile guidance pod sighting port to provide an unobstructed path for the light beam from the alignment group and the reflected light beam from an alignment prism mounted in the platform assembly of the missile guidance system. The reflected light beam is indicative of the azimuth orientation (alignment) of the platform. If the reflected light beam indicates that the platform is not aligned, a signal is generated and sent to the countdown group. The countdown group then sends a signal to correct the azimuth alignment of the platform. When the launcher platform is raised, the alignment group sight tube is rotated upward, clear of the launcher platform.

1-107. The countdown group is also used to test, calibrate, and monitor the missile guidance system,

and programs the guidance system countdown. An amplifier assembly is provided in the umbilical junction box to prevent loss of critical low-level signals from the missile guidance system to the countdown group. The alignment-countdown group contains six modes of operation (warm up, missile guidance system checkout, system integration test, countdown, commit and self-check.) The warmup mode heats the components of the missile guidance system until the warmup period is completed. The missile guidance system checkout mode automatically tests, calibrates, and aligns the missile guidance system. After completion of the checkout mode, the missile guidance system is in a ready status. The last step of the checkout mode levels the stable platform and orients it to the selected target. The system integration test is performed by the semitrailer mounted electrical systems checkout test station and determines whether the guidance system is properly integrated with the remaining systems. The countdown and commit modes of operation are normally sequenced during an actual countdown but may be performed for test purposes from the countdown group. These modes of operation ensure that the system is capable of performing its intended functions during an actual countdown. The self-check mode of operation repeats the first part of the missile guidance system checkout mode. This mode is used only during maintenance operations to check the alignment-countdown group circuits. In general, the alignment-countdown group ensures a successful operation of the missile guidance set by performing the following functions:

a. Checkout of all missile guidance system and guidance aerospace ground equipment to ensure that they are within allowable limits.

b. Countdown, alignment, test, and monitoring of the missile guidance system.

c. Support maintenance by calibration, alignment, and automatic fault isolation of major replaceable units on a system basis.

1-108. LAUNCH CONTROL, MISSILE LIFT, AND RE-ENTRY VEHICLE MONITORING AND CHECKOUT EQUIPMENT.

1-109. Launch control and monitoring equipment, consisting of control monitor group 1 of 4 and 2 of 4 (5, figure 1-43) and launch control console (figure 1-7), are located on the third level of the silo and in the launch control center. The control monitor group 1 and 2 of 4 sequence the countdown when a countdown start signal is received from the launch control console. The control monitor group performs logic functions that initiate sequences, coordinate sequencing of related launch equipment, and check that all sequencing events occur in the proper order and at proper time intervals. The control monitor group panels contain indicators and controls to monitor system status and to perform specific checks on the monitored systems. The launch control console provides the indicators and controls that monitor conditions during standby periods, countdown operations, and abort operations. Controls are provided to start countdown, start missile commit sequence, start abort sequence, and to set in and select target data. Controls are also provided to manually pressurize the missile tanks, to perform lamp tests, and to enable communications. The test and training monitor console (24, figure 1-5) contains communication facilities and also a LAUNCH ENABLE pushbutton and an ABORT pushbutton to support countdown operations.

1-110. Control monitor group 3 of 4 and 4 of 4 (8, figure 1-43), located on level three of the silo, checks control monitor group 1 of 4 and 2 of 4 and the launch control console by simulating responses

normally received during an actual missile countdown. This type of countdown (simulated countdown) exercises control monitor group 1 of 2 and 2 of 2 (5) to determine their operational capability in performing an actual missile countdown. Simulated countdowns are performed by disconnecting control-monitor group 1 of 4 and 2 of 4 from the missile and aerospace ground equipment and connecting them to control-monitor group 3 of 4 and 4 of 4. A self-test capability is also incorporated into control-monitor group 3 of 4 and 4 of 4.

1-111. Re-entry vehicle monitoring and checks are accomplished by the re-entry vehicle prelaunch monitor and control unit (figure 1-95.) This unit contains indicators to monitor re-entry vehicle status and controls to perform check on the re-entry vehicle system. The monitor unit is checked by using a re-entry vehicle simulator which isolates malfunctions to the re-entry vehicle, missile, or to the monitor.

1-112. FIRE DETECTION AND WARNING SYSTEM.

1-113. Fire detectors (fixed temperature and rate of temperature rise) are located on each level of the silo, in the missile enclosure, and on each level of the launch control center. A fire alarm panel (16, figure 1-4 or 21, figure 1-5) is located on the second level of the launch control center and contains indicators, alarm buzzer, ammeter, and pushbuttons to monitor and test the fire alarm system. The fire alarm panel (annunciator) indicates the location of a fire through illuminating an indicator lamp for one or more of the nine fire zones within the launch site. A manual fire alarm system is also provided as well as the automatic system. Manual fire alarm boxes (figure 1-71) are located on levels 2, 4, 6, and 8 of the silo. These alarms may be operated manually any time a fire is observed in the area. When either the manual or the automatic fire alarm system is actuated, the alarm buzzer on the fire alarm panel, bell and red warning lamps at fire alarm stations indicate the detection of a fire. The bells are silenced by depressing the ACKNOWLEDGE pushbutton on the fire alarm panel. A power set, in conjunction with the fire alarm panel (annunciator) supplies 24 VDC power to the system. For standby power failure operation, the power set is equipped to sustain the fire alarm system in an alert status for 30 hours by means of a wet-cell storage battery.

1-114. FIRE FOG SYSTEM.

1-115. The fire fog system (see figure 1-72) lessens the hazard of RP-1 fuel vapor concentrations, and fire in the missile enclosure area through the use of concentrated water spray from fire fog nozzles located at silo levels 5, 6, 7, and 8 in the missile enclosure area. See figure 1-38 for a simplified flow diagram of the utility water system. The fire fog system can be manually activated at the facility remote control panel (19 Figure 1-4) when explosive vapors or a fire is detected in the missile enclosure area.

1-116. FIREX SYSTEM (VAFB ONLY).

1-117. A firex system is provided for missile washdown, which prevents damage to the missile and launch platform due to propellant leakage and fires; to flush the silo cap in the event of propellant spillage; and to flush leakage or spillage of propellants on level eight of the silo. The system is controlled from firex pushbutton panels (28, figure 1-5) located in the launch control room, at the silo cap, and at level 7 of the silo. The firex control panel, in the launch control room, is capable of controlling the firex system for missile washdown, silo cap flushing, and eighth level flushing. The firex control panel, at the silo cap, is capable of controlling the silo cap flushing and missile washdown only, while the control panel on the seventh level of the silo controls only eighth level flushing. The firex control panel, at the pumping station, contains system valve test and pump start-stop pushbuttons. Each of the other control stations also contains a pump start-stop pushbutton.

1-118. LAUNCH COMPLEX EMERGENCY EQUIPMENT.

1-119. Launch complex equipment, provided for personnel safety and fire fighting, consists of the emergency shower system, the eyewash system, various items of protective clothing, breathing apparatus, portable fire extinguishers, the fire hose station system, and the fire fog system.

1-120. EMERGENCY SHOWER SYSTEM.

1-121. The emergency shower system provides a quick water wash for personnel who have received harmful contaminants on their clothing or person (See figure 1-72.) Two emergency showers are available, one being located on silo level 7 and the other one on silo level 8. Each emergency shower is equipped with a quick opening push type valve that remains open until manually closed.

1-122. EMERGENCY EYEWASH SYSTEM.

1-123. The eyewash system provides a quick water wash for personnel who have received harmful contaminants in or around the eyes. (See figure 1-72.) Two eyewash fountains are available, one being located on silo level 7 and the other one on silo level 8. Each eyewash fountain is located next to the emergency shower at the station and is equipped with a quick opening push type valve that remains open until manually closed.

1-124. PORTABLE FIRE EXTINGUISHERS.

1-125. Portable fire extinguishers are provided for immediate use by personnel in the event of fire. The extinguishers are located at standard locations (figure 1-70) throughout the complex and consist of three types: water extinguisher for class A fires, 15-pound capacity CO_2, and 50-pound capacity CO_2 for class B and C fires. Minimum requirements are: one water-type extinguisher located on level one of the launch control center, one 50-pound CO_2 extinguisher located on the silo cap, and one located on each of silo levels 5 and 6. One 15-pound CO_2 extinguisher located on level one of the launch control center, one in the tunnel near the silo blast door, one in the sump area, and one on each of levels 2, 3, and 4 of the launcher platform. Two 15-pound CO_2 extinguishers are located on level two of the launch control center, facility elevator, and each level of the silo.

1-126. FIRE HOSE STATION SYSTEM.

1-127. The fire hose station system furnishes water to silo levels 7 and 8 in the event of fire. (See figure 1-72.) The system consists of two fire hose rack units, each located on a wall adjacent to the facility elevator entrance on silo levels 7 and 8. A typical fire rack unit consists of a 6-pound valve, rack nipple, pin-type rack, drain and vent, 75 feet of linen hose, couplings, and a 1-1/2-inch combination nozzle.

1-128. PERSONNEL SAFETY AND EMER-
GENCY RESCUE EQUIPMENT.

1-129. Designated storage areas for personnel
safety and rescue equipment are shown in figure
1-73. The safety equipment located in the safety
equipment storage closet on launch control center
level 1, is listed in table 1-7. This equipment is used
on a daily work requirement basis. The door to the
storage closet is normally locked and access to this
safety equipment is controlled by the MCCC. Equipment shall be inventoried daily. After use, it must
be inspected to insure usable condition prior to
placing the equipment into the closet for future use.
An inventory sheet and location log will be attached
to the closet door. The location log sheet will reflect
the items and location of all equipment removed
for use.

1-130. Emergency rescue and safety equipment,
stored in a steel cabinet (figure 1-73) on launch
control center level 2, is listed in table 1-8. This
equipment is to be used for emergencies only. The
doors to the cabinet are secured with lead seals and
are not locked. After equipment from this cabinet is
used, it must be inspected and recharged (if necessary), prior to placing the equipment back into the
cabinet for future use.

1-131. A minimum of 19 breathing apparatus (5
to 7 minute Scot-Pak or equivalent) are located
within the silo for emergency use. (See figure 1-73.)
An additional five (5 to 7 minute Scot-Pak or
equivalent) breathing apparatus will be located in
the safety equipment closet, used to replace those
removed from the silo for repair. Five self-contained
breathing apparatus, with 30-minute supply air
tanks, are located on launch control center level 2
for use when hazardous breathing conditions exists.
Both types may be recharged from air supply cylinders, with refiller hose assembly attached, located
on the stairwell landing above launch control center
level 1. (See figure 1-73.) Breathing apparatus shall
be inspected daily.

1-132. A portable combustible gas indicator and
oxygen deficiency indicator are located on top of
the blast detection cabinet on level 2 of the launch
control center. These indicators are used for con-

ducting silo safety inspections after a PLX or launch
and anytime a suspected abnormal atmospheric
condition exists in the silo. This equipment will be
inspected daily.

1-133. 6 VDC EMERGENCY LIGHTING.

1-134. The emergency lights are self-contained
6-volt emergency units utilizing sealed beam lamps
(figure 1-74.) The emergency lighting is actuated
by normally open relays which close in the event
of a power failure, actuating the individual battery
supply of each emergency unit.

1-135. The emergency units are located at critical
points, such as entrances and exits, and operate
in the event of power failure in either the silo or
the launch control center. There are no emergency
procedures for operation of the lighting system. In
case of scheduled or unscheduled power outage at
the 480-volt switchgear, the 6 VDC emergency lighting will be energized automatically.

1-136. The emergency lighting system can be
checked with the regular area lighting either on or
off. To energize all emergency lights in the silo,
open circuit breaker NO. 1 on panel LD. To energize
all emergency lights in the launch control center,
open circuit breaker NO. 10 on lighting panel A.

1-137. MISSILE.

1-138. The HGM16F missile (figure 1-75) is
capable of delivering a thermo-nuclear warhead or
other re-entry vehicle payloads to a target area
farther than 5500 miles away. Boosting and guiding
of the re-entry vehicle into the desired ballistic trajectory are accomplished by the basic missile with
five rocket engines, a flight control system, an inertial
guidance subsystem, and associated subsystems.

1-139. The missile consists of three major sections: a Re-entry Vehicle (1), a booster section (6),
and a tank section (2). The nose section consists
of the re-entry vehicle and adapter and houses the
payload of the missile. It also contains circuits and
components required to accomplish re-entry vehicle
separation, arming, and fuzing.

1-140. The tank section consists of a thin walled stainless steel monocoque structure which maintains its rigidity by pressurization and is separated into two tanks by an intermediate bulkhead. A sensor stillwell assembly of the propellant utilization system is located within each tank. The forward end of the tank section contains provisions for mounting the re-entry vehicle. The aft end contains provisions for mounting components of the hydraulic and propulsion subsystems. Equipment pods (3 and 10) are located externally on the tank section to house missile electrical equipment (battery and inverter), flight control system equipment, retarding rockets, guidance system equipment, propellant utilization computer assembly, and instrumentation equipment. Two additional pods (11) contain penetration devices. These devices, which include decoys, are jettisoned at preset intervals during flight to cause false indications in detection equipment attempting to plot the course of the missile. The forward end of the tank section contains a bulkhead and access plate to allow entry into the oxidizer tank. Also mounted on the bulkhead are liquid oxygen boiloff valve and the re-entry vehicle adapter.

1-141. The booster section contains assemblies of the propulsion, pressurization, and hydraulic subsystems used during the booster stage of missile flight. It also contains rise-off disconnect fittings to connect hydraulic, pneumatic, and propellant lines to the missile from ground supply sources. The booster section is coupled to the tank section by four special staging disconnect couplings. Thrust is trans-

Table 1-7. Safety Equipment Storage Closet Requirements

ITEM NOMENCLATURE	QUANTITY
Block, Double Sheave	2
Lifeline, ½-inch, 50-foot	2
Lifeline, ¾-inch, 200-foot	2
Hook, Hoist	3
Goggles, Plastic	6 Pairs
Respirator, Air Filtering	2
Spectacles, Safety	6 Pairs
Belt, Safety, 33 to 44-inch, Adjustable	3
Ear Muff, Sound Barrier	4
Breathing Apparatus (5 to 7 minutes)	5
Faceshield, Frame	3
Faceshield, Visor	3
Harness, Suspension, Bos'n	1
Tail line, 6 foot	3
Flashlight, Explosion-Proof	10
Gloves, Rubber, Solvent Resistant	6 Pairs
Lantern, Electrical, 6-Volt	2

mitted from the booster section to the tank section by a thrust ring which is welded to the aft end of the tank section. The booster section is jettisoned during the staging portion of missile flight. Staging is accomplished approximately 120 seconds after missile launch.

1-142. PROPULSION SYSTEM.

1-143. The missile propulsion system is used to propel the missile from the launcher platform into the desired ballistic trajectory. The system consists of two booster engines (7 and 9, figure 1-75). two vernier engines (5), one sustainer engine (9), a vernier solo tank system, and associated subsystems. All engines are of the single-start type and use RP-1 fuel and liquid oxygen. The combined thrust of the five engines at sea level is approximately 389,000 pounds.

1-144. Starting of the engines is accomplished during countdown by the launch control system. For starting and progression into mainstage operation, the following sequence of events occurs. The booster

Table 1-8. Emergency Rescue and Safety Kit Storage Cabinet Requirements

ITEM NOMENCLATURE	QUANTITY
Sling, Nylon Web	4
Life line, ½-inch, 50-foot	1
Hook, Hoist	3
Blanket, Fire, Asbestors	1
Belt, Safety, 33 to 44-inch, adjustable	2
Tail Line, Steel 6-foot	4
Cutter, Cable	1
Bar, Wrecking	1
Blanket, Electrical, Rubber	2
Lantern, Electrical, 6-Volt	2
Litter, Stokes, Portable (Outside Cabinet)	1
Gloves, Electrician, Rubber	2 pairs
Gloves, Electrician, Leather	2 pairs
Hood, Fireman	2
Coat, Aluminized, Fire	2
Trousers, Aluminized, Fire	2
Boots, Fireman, Rubber	2 pairs
Gloves, Cryogenic	2 pairs
Voice Gun, Portable	1
Hood, Rocket Fuel	2
Coverall, Safety LO_2	2

and sustainer engines each utilize a solid propellant gas generator and are controlled by a fuel pressure ladder sequence. During start, an engine start signal fires the solid propellant gas generator initiators, forcing hot gases produced by the burning of the solid propellant to the turbopump turbines. (See figure 1-76.) This provides the initial drive to the turbopumps. The flow of hot gases from the solid propellant gas generators causes the pumps to force liquid propellants from the missile tanks directly to the thrust chambers and liquid propellant gas generators. Electrical signals and fuel pressure control the various propellant valves. Combustion is initiated by hypergolic combination in all thrust chambers. Pyrotechnic igniters, electrically fired, ignite the propellants as they flow into the liquid propellant gas generators. The hot gases produced by the liquid gas generators sustain mainstage operation after starting and until the engines are shut down.

1-145. The vernier engines (5, figure 1-75) start shortly after the booster (7 and 9) and sustainer engines (8) reach mainstage (approximately 3.5 seconds) and operate on propellants supplied from the sustainer engine. Pressurized spherical vernier solo tanks contain and supply the small amounts of fuel and oxidizer necessary for vernier solo operation.

1-146. During standby, the solid propellant gas generators are monitored and their status displayed on the launch control console (figure 1-7) by a summary signal from the engine subsystem panels that illuminates the ENGINES AND GROUND POWER indicator. The only indicators directly related with the propulsion system during countdown but prior to commit are HEATERS ON and ENGINES & MISSILE POWER READY indicators in the countdown patch. When these indicators have illuminated green and all of the other system conditions have been satisfied, the READY FOR COMMIT indicator will illuminate green. Initiating commit starts the final sequence if the LAUNCH ENABLED indicator on the commit patch illuminates green. When the rocket engines enter the ignition stage, the ENGINE START indicator on the commit patch will illuminate green. From this point, all sequencing and operation of the booster and sustainer engine components are governed by a fuel pressure ladder sequence and, therefore, are independent of the launch control center.

The ability to shut down the propulsion system with direct circuitry is maintained until umbilical separation at launch. However, after ENGINE START, only a malfunction will shut down the propulsion system; no manual control is possible. When the propulsion system develops sufficient thrust to lift the missile from the launcher platform, a signal is sent to eject umbilicals and illuminate the MISSILE AWAY indicator on the commit patch of the launch control console.

1-147. After certain programed flight conditions are met, signals from the missile guidance system (figure 1-77) cut off the booster engines and separate the booster section (6 figure 1-75) from the missile. The sustainer engine is similarly cut off at a later time, but remains attached to the missile. Vernier engines continue to operate on propellants from the vernier solo tank which have been filled and pressurized before sustainer cutoff. Upon completion of the vernier correction phase, the vernier engines are cut off and the missile is separated from the re-entry vehicle by a signal from the missile guidance system.

1-148. Each booster engine (figure 1-76) is a fixed-thrust engine consisting of a solid propellant gas generator and liquid propellant gas generator driven turbopump, and a lightweight, tubular-wall, gimbal-mounted thrust chamber. The turbopump is fastened to a support bracket that incorporates mounting points for attachment to the missile and thrust chamber gimbal. The various control valves, gas generator, and interconnecting ducting and electrical cables are fixed in position to the turbopump and thrust chamber.

1-149. The turbopump delivers liquid oxygen and fuel to the thrust chamber, and consists of two centrifugal pumps driven by a high speed turbine through a reduction geartrain. The initial power to turn the turbopump and bring it up to sufficient speed to supply propellants to the thrust chamber and the liquid propellant gas generator is supplied by a solid propellant gas generator. The high-velocity gas that provides the sustained power for the turbine is supplied by the liquid propellant gas generator. The gas generator consists of a combuster, where liquid propellants are burned, and the valves that control propellant flow. The propellants are supplied under pressure and are originally ignited in the liquid propellant gas generator by pyrotechnic igniters. The resultant combustion gas is directed to the turbo-pump turbine impellers. The turbopump gears and bearings are lubricated with oil supplied under pressure from an oil pump. The oil supply tank mounted on the thrust chamber is pressurized with helium to ensure adequate oil flow to the oil pump.

1-150. The booster thrust chamber is bell shaped. Tubes run lengthwise from the top of the thrust chamber to the bottom of the skirt, forming a cylindrical combustion chamber converging into a narrow throat and then diverging to a wide-mouthed expansion chamber. During countdown, inert fluid is injected into the booster thrust chamber by the inert fluid injection system. The inert fluid reduces the pressure surge that occurs when the booster engines are started. Before fuel is fed into the combustion chamber, it passes through the tubes and cools the walls of the chamber. A circular injection plate at the top of the thrust chamber introduces both fuel and liquid oxygen into the combustion chamber. Fuel for ignition passes through an ignition fuel valve and a hypergolic fluid container to six orifices in the injector plate. The propellants are ignited in the combustion chamber by contact of hypergolic fluid and liquid oxidizer. A gimbal mount transmits thrust loads, developed in the thrust chamber, to the missile. The gimbal mount permits both yaw and pitch axis thrust chamber movement.

1-151. The sustainer engine (figure 1-76) is an integral, gimbal-mounted, fixed-thrust unit installed on the thrust cone of the missile fuel tank. The engine consists of a liquid propellant gas generator, turbopump, thrust chamber, and control and lubricating components. The sustainer solid propellant gas generator, liquid propellant gas generator, and turbopump are similar to the corresponding booster components. The turbopump supplies propellants to the vernier engines and to the sustainer engine during operation. The sustainer thrust chamber is similar to the booster

thrust chamber, except that the sustainer combustion chamber is smaller.

1-152. The two vernier engines (figure 1-76) are bi-thrust (two rated thrust level) units installed on the missile as separate and complete engines. The vernier thrust chambers are double walled, with the inner and outer walls being separated by a solid spiral coil that forms a passageway for fuel flow. The passageway allows fuel to flow from a fuel inlet manifold at the exit nozzle of the chamber to the injector housing at the head, cooling the inner wall of the chamber. A pneumatically actuated propellant valve controls propellants to the thrust chambers. The thrust chambers are mounted on gimbal shafts that turn in response to hydraulic actuators.

1-153. The vernier solo tank subsystem includes the vernier solo oxidizer and fuel tanks and the necessary vernier control components. The subsystem provides propellants for vernier engine solo operation and control of the vernier propellant valves after sustainer cut-off. After the sustainer engine reaches mainstage, the vernier engines are started. The vernier engines then use propellants supplied from the main missile tanks by the sustainer turbopump. During mainstage, the vernier solo tanks are filled with propellants from the sustainer turbopump. The filled solo tanks provide the propellants used by the vernier engines during solo operation.

1-154. The missile helium supply furnishes the pneumatic power for actuating the vernier engine control valve and for pressurizing the vernier solo tanks and engine lubricating oil tanks during operation. Missile hydraulic power is used to operate the sustainer propellant flow regulating components.

1-155. PROPULSION SYSTEM STATUS INDICATOR. The propulsion system status indicator, located on the launch control console, is monitored during standby and countdown by the missile combat

Table 1-9. Propulsion System Indicator on Launch Control Console

INDICATOR	FUNCTION
ENGINES AND GROUND POWER indicator	Illuminates green when both of the following conditions exist: a. SYSTEM IN STANDBY indicator on ENGINE (PANEL 2) (figure 1-78) illuminated green b. SYSTEM IN STANDBY indicator on MISSILE GROUND POWER (PANEL 1) (figure 1-79) illuminated green Illuminates red when either of the above conditions is not as specified

crew commander or by his designated alternate. Table 1-9 lists the indicator involved and its function.

1-156. PROPELLANT UTILIZATION SYSTEM.

1-157.   When a missile is launched, the propellant tanks contain fuel and liquid oxygen at a specified weight ratio. During missile acceleration, the heavier (more dense) liquid oxygen has a tendency to be supplied to the engines at a faster rate than the less dense RP-1 fuel. This tendency is equalized by the propellant utilization system which regulates propellant flow in such a manner that both tanks empty evenly.

1-158.   The propellant utilization system monitors the propellant level ratio and controls the propellant flow ratio to the sustainer engine during flight. (No such control is needed for booster and vernier engines.) To provide indications of propellant consumption, ultrasonic sensors are installed on stillwells at discrete propellant levels (stations) in both the fuel and oxidizer tanks. (See figure 1-98.) These sensors relay changes in impedance as the propellant levels pass each sensor station. The stations are located so that similarly numbered sensors in both tanks are uncovered at the same time when the flow ratio is within limits.

1-159.   Time differences in the signals received from a matching pair of sensors as they are uncovered, indicate a discrepancy in the volume of flow. The propellant utilization computer assembly receives the impedance changes from the sensors and generates correcting signals, which are compared to servocontrol feedback signals from the propellant utilization valve transducer. A resulting control command signal is then sent to the propellant utilization servocontrol valve, which adjusts the fuel flow rate to produce the proper level in the tank at the next sensor station. To maintain a constant propellant flow, any change in the fuel flow rate causes an inverse change in the liquid oxygen flow rate through activation of the head suppression valve.

1-160. FLIGHT CONTROL SYSTEM.

1-161.   The flight control system (figure 1-77) consists of the autopilot subsystem and the inertial guidance subsystem. It has the function of steering and stabilizing the missile during the powered por-

tion of flight so that when the re-entry vehicle separates from the tank section and enters free fall, it will continue in a ballistic trajectory to intersect the designated target area. The autopilot controls missile attitude during flight by issuing steering commands to the servovalves of the engine thrust chamber hydraulic actuators. The hydraulic actuators, in turn, regulate the angular displacement of the thrust chambers, thereby controlling the direction of forces acting upon the missile.

1-162.   The combined actions of the flight control system control the four flight phases of the missile. The powered portion of flight consists of the booster, staging, sustainer, and vernier phases. During the first portion of this period, the inertial guidance subsystem supplies the roll voltage to the autopilot subsystem. The missile rotates about its roll axis to establish the yaw axis of the missile in a vertical plane through the desired cutoff trajectory. At the end of the roll maneuver, vernier engine NO. 2 is oriented on the side facing the target. A missile pitchover, which lasts until staging, is then programed to direct missile toward the target.

1-163.   The inertial guidance subsystem continuously corrects the missile flight attitude in order to select, incrementally, the optimum trajectory for successful accomplishment of the mission. Targets are preselected while the missile is on the ground by inserting specific target constants boards into the missile guidance computer. (See figure 1-77.)

1-164.   The autopilot subsystem (figure 1-77) consists of a programmer package, a gyroscope package, and a filter-servoamplifier package. The programmer package contains electronic timing and switching circuits which generate signals to control the programed functions of missile flight. Assisted in part by discrete commands from the inertial guidande subsystem, the preset switching circuits of the autopilot programmer control the sequencing of the following major inflight operations: missile roll maneuver and pitch program; booster engine cutoff and booster section jettison; sustainer and vernier engine cutoff; re-entry vehicle separation; firing retarding rockets; and destruction of the missile tank section.

1-165.   There are a total of six displacement and rate gyroscopes in the gyroscope package. Displacement gyroscopes provide signals proportional to the amount of missile angular displacement from the yaw, roll, and pitch reference planes; rate gyroscopes provide signals proportional to the rate (angular velocity) of displacement from the yaw, pitch, and roll reference planes. Pitch, yaw, and roll displacement gyroscopes, and the roll rate gyroscope, are contained in a canister in the B-2 equipment pod (10, figure 1-75). The pitch and yaw rate gyroscopes are housed in a canister mounted in the cable fairing, forward of the B-2 equipment pod, to compensate for bending modes in the pitch and yaw planes.

1-166. Missile attitude deviations from the programed flight path are sensed by one or more of the rate and displacement gyroscopes. Each displacement gyroscope senses deviations in its particular reference plane (pitch, yaw, or roll). Displacement gyroscope reference planes are controlled by torque signal amplifiers, which are changed for steering purposes. Rate gyroscopes are in fixed reference planes; they sense the rate of missile attitude changes. Gyroscope output, in the form of error voltages, is fed to the filter-servoamplifier package.

1-167. The filter-servoamplifier package contains the electronic circuitry to convert gyroscope error signals into input data for the servo-amplifiers. These servoamplifiers operate the servovalves of the engine actuators, enabling the hydraulic system to gimbal the engines in the proper direction to counteract the deviations sensed by the gyroscopes. Feedback transducers on the actuators send back signals to the filter-amplifier package which are proportional to engine displacement. Such a feedback signal opposes the gyroscope output signal, causing the engine to return to the null position during an attitude correction sequence.

1-168. The airborne inertial guidance subsystem (figure 1-80) consists of a missile guidance control, an inertial guidance sensing platform, and a missile guidance computer. Aerospace ground equipment components, which include the Inertial Guidance Alignment-Countdown Set AN/GJQ-12, maintain the airborne equipment in a state of readiness so that its objectives can be achieved.

1-169. The missile guidance control is located in the lower portion of the missile guidance pod. It functions as an interconnecting and switching point between the airborne elements of the subsystem and also provides amplification and power for the operation of the inertial guidance sensing platform.

1-170. Located centrally in the guidance pod, the sensing platform comprises a platform upon which are mounted two gyroscopes, three accelerometers, and four pendulums. The axis of the three accelerometers are mutually perpendicular and define the X, Y, and Z axis of a reference coordinate system. (See figure 1-80.) The coordinate system is established and maintained prior to missile launch by leveling the platform with the pendulums and aligning the X axis accelerometer to the target azimuth. The Y and Z axis accelerometers are thus oriented to sense lateral and vertical accelerations respectively, while the X axis accelerometer senses down range accelerations.

1-171. Stabilization of the sensing platform assembly during standby is necessary to prevent the accelerometers from sensing inflight accelerations that are not parallel to the axes of the reference coordinates. Two 2-degree-of-freedom gyroscopes maintain this orientation after missile launch. The gyroscopes are at their null position when the accelerometer axes correspond with the axes of the reference coordinates. Any tendency of the platform to deviate from its original launch point orientation is sensed by the gyroscopes, which dispatch signals to servoamplifiers, which power servomotors to maintain the stable position. Electrical resolvers mounted on the sensing platform, sense any relative motion between the missile air frame and the platform, and provide reference signals for missile azimuth, roll, and pitch control. While the missile is in flight, the accelerometers continuously measure missile acceleration along the three axes and provide signals to the computer that are a function of that acceleration.

1-172. The missile guidance computer is located in the upper portion of the missile guidance pod. The X, Y, and Z output signals from the accelerometers and the elapsed time furnish the basic data from which the computer calculates the re-entry vehicle release point. To determine missile position and velocity, the computer processes these output signals, together with precomputed target constants and time. Computer output signals (discrete and steering) are sent to the autopilot subsystem to accomplish yaw steering, booster staging, sustainer engine cutoff, and vernier engine cutoff. A prearm signal is dispatched to the re-entry vehicle when certain prerequisite conditions are met.

1-173. During powered flight, from the moment of launch until vernier engine cutoff, the computer continuously calculates missile position and velocity. Position is considered in the three dimensions (X, Y, and Z). From missile position and velocity calculations, the computer determines where the re-entry vehicle would strike the earth's surface if all engines were shut off and the re-entry vehicle were allowed to enter free fall. The computed impact point is compared to the desired impact point. Continuous calculations of range error functions (REF) and cross-range error functions (CEF) are made during flight. Cross-range error function (CEF) is the error in azimuth which would result if the missile maintained its course. To bring the missile back on course when CEF varies from the programmed flight path, yaw steering signals are generated.

1-174. While the engines are operating, the velocity of the missile is constantly increasing, changing the calculated impact point. The calculated impact point thus comes closer to target point and REF decreases as the missile traverses its trajectory.

1-175. Engine cutoff occurs in three stages. When the missile guidance computer determines that missile range velocity (V_x) has attained a specified value, it generates a signal to accomplish booster engine cutoff and staging. The sustainer and vernier engines continue to propel the missile. As the free fall point (re-entry vehicle separation) comes closer, and at a specified value of range error function, the computer generates a signal for sustainer engine cutoff. Missile velocity increases very slowly after sustainer engine cutoff since only the vernier engines are still operating. When the calculated range error functon approaches a predetermined point, the computer generates a vernier engine cutoff signal. The re-entry vehicle then separates and enters a free fall trajectory to the target.

1-176. The airborne components of the inertial guidance subsystem are calibrated, checked, and maintained in a launch ready state by the inertial guidance alignment countdown set (See figure 1-69.) The sensing platform alignment group (C) establishes a known azimuth against which the X axis accelerometer can be accurately aligned to the target azimuth. Countdown group (A) maintain the proper orientation of the inertial guidance sensing platform in conjunction with the alignment group and monitors, checks, and calibrates the airborne inertial guidance subsystem. If a ground shock should occur, the subsystem is automatically placed in a memory mode, holding alignment in its last remembered position.

1-177. During standby, the flight control and interrelated re-entry vehicle system status is monitored and displayed on the launch control console (figure 1-7) by a summary signal from various subsystem panels that illuminates the FLIGHT CONTROL & R/V indicator. A green indication shows that these systems are ready for countdown. If either system is not functioning properly, the indicator illuminates red and an audible alarm sounds. The source of trouble may be isolated by observing the GUIDANCE FAIL, R/V SAFE, and AUTOPILOT FAIL indicators in the malfunction patch.

1-178. The indicators directly related with the flight control system during countdown but prior to commit are AUTOPILOT ON, AUTOPILOT TEST, GUIDANCE READY, and FLIGHT CONTROL & R/V READY indicators in the countdown patch. These indicators show that the inertial guidance subsystem countdown is complete and ready for the go on memory command, the autopilot subsystem gyroscopes are running and are stabilized at a null position, the engine positioning actuators are in a nulled position, and the flight programmer is reset and safe. When these and all other system indicators have illuminated green, the READY FOR COMMIT indicator will illuminate green. Turning the COMMIT START key switch and then the COMMIT SWITCH on the ALCO COMM/CONTROL panel initiates the final sequence providing that the LAUNCH ENABLED indicator on the commit patch is illuminated green.

1-179. All remaining steps during countdown are performed automatically under launch control system command. When missile guidance goes on memory, the gyroscope stabilized sensing platform is maintained in position by the platform gimbal torque motors, and when the autopilot flight programmer has been armed, the PROGRAMMER ARMED and GUIDANCE COMMIT indicators illuminate green, completing the flight control system portion of the countdown.

1-180. FLIGHT CONTROL SYSTEM STATUS INDICATORS. The status indicators located on the launch control console, related to the flight control system, are monitored during standby and countdown by the missile combat crew commander, or by his designated alternate. Table 1-10 lists the indicators involved and their functions.

Table 1-10. Flight Control System Indicators on Launch Control Console

INDICATOR	FUNCTION
FLIGHT CONTROL & R/V indicator	Illuminates green when all of the following conditions exist: a. SYSTEM IN STANDBY indicator on RE-ENTRY VEHICLE (PANEL 1) (figure 1-81), illuminated green b. SYSTEM IN STANDBY indicator on GUIDANCE (PANEL 1) (figure 1-82), illuminated green c. SYSTEM IN STANDBY indicator on AUTOPILOT (PANEL 1) (figure 1-83), illuminated green Illuminates red when any of the above conditions are not as specified
R/V SAFE indicator	Illuminates red when re-entry vehicle is not safe
AUTOPILOT FAIL indicator	Illuminates amber when all of the following conditions exist: a. Gyroscope fine heaters have failed b. Autopilot has not failed c. Internal commit sequence has not started Illuminates red when either of the following conditions exist: a. Autopilot has failed b. Abort sequence has not started and command has not been given to arm flight programmer

1-181. RE-ENTRY VEHICLE.

1-182. The Mark IV re-entry vehicle houses the missile payload, protects it during re-entry into the atmosphere, fuzes and arms the warhead, and transports the payload to the target area during the final portion of flight. It is of the ablative type, structurally consisting of a nose section, center section, and aft section. Functional systems include a separation system and an arming-fuzing system. Prearming of the warhead occurs during flight upon receipt of a signal from the flight control system when the missile has reached a predetermined position. When the ideal release point is reached, the re-entry vehicle separates from the missile tank section for its final plunge to the target.

1-183. The re-entry vehicle is held in place on the missile airframe by tension applied through the separation mechanism to a tension cone. This cone firmly grasps the flare of the re-entry vehicle and provides for electrical interface with the missile airframe. When the separation command is received, the separation mechanism releases the tension on the cone, causing it to spring apart and free the re-entry vehicle. The electrical interface is broken at this time to complete the break.

1-184. After separation, the re-entry vehicle is rotated on its longitudinal axis at a slow spin rate for stability. Retarding rockets in the missile tank section are fired to decelerate the airframe to prevent the possibility of interference with the planned trajectory of the re-entry vehicle. The warhead (if installed) is detonated above the target (air burst) or on the target (ground burst), as determined by preselection. The ground burst capability also serves as a backup in the event of an airburst selection failure.

1-185.   Before the start of countdown, controls in the target selection patch of the launch control console (figure 1-7) permit the choice of either target A or target B. The HIGH and LOW indicators show the type of re-entry vehicle installed. During standby, the status of the re-entry vehicle and interrelated flight control system is monitored and displayed on the launch control console by a summary signal from various subsystem panels that illuminates the FLIGHT CONTROL & R/V indicator. The lack of a red R/V SAFE indicator condition in the malfunction patch denotes that the re-entry vehicle arming and fuzing circuits and the warhead safety circuits are in a safe configuration.

1-186.   At countdown start, the re-entry vehicle sequencer initiates a verification check of the re-entry vehicle prelaunch monitor set and re-entry vehicle circuits and starts a battery warmup cycle which ensures that the re-entry vehicle battery is at the proper operating temperature prior to launch. An R/V BATTERY TEMPERATURE green indication denotes that all circuits are satisfactory. The FLIGHT CONTROL & R/V indicator illuminates green when these systems are ready for the final countdown commit sequence. No changes occur in the re-entry vehicle status during commit.

1-187.   RE-ENTRY VEHICLE STATUS INDICATORS. The status indicators located on the missile launch control console, related to the re-entry vehicle, are monitored during standby and count-

down by the missile combat crew commander, or by his designated alternate. Table 1-11 lists the indicators involved and their function.

1-188.   MISSILE HYDRAULIC SYSTEM.

1-189.   The missile hydraulic system furnishes power to the engine thrust chamber actuators to provide pitch, roll, and yaw control of the missile during flight. Hydraulic pressure from the system is also used to operate the sustainer propellant valves and gas generator valves during sustainer engine operation. There are three separate hydraulic subsystems operational during the various flight phases: the booster subsystem, the sustainer-vernier subsystem, and the vernier solo subsystem.

1-190.   The booster hydraulic subsystem (figure 1-77) provides hydraulic pressure to gimbal the two booster engine thrust chambers for pitch, yaw, and roll control from missile launch until staging. During this period the sustainer-vernier hydraulic subsystem provides pressure to gimbal the vernier engines for roll control only. Following booster engine cutoff the sustainer engine is gimbaled briefly for pitch and yaw control, then it is locked on center to permit jettisoning of the booster section. The sustainer-vernier hydraulic subsystem is again enabled after staging and until sustainer engine cutoff to gimbal the sustainer engine for pitch and yaw control and the verniers for roll control. After sustainer engine

Table 1-11.   Re-entry Vehicle Indicators on Launch Control Console

INDICATOR	FUNCTION
FLIGHT CONTROL & R/V indicator	Illuminates green when all of the following conditions exist: a. SYSTEM IN STANDBY indicator on RE-ENTRY VEHICLE (PANEL 1) (figure 1-81), illuminated green b. SYSTEM IN STANDBY indicator on GUIDANCE (PANEL 1) (figure 1-82), illuminated green c. SYSTEM IN STANDBY indicator on AUTOPILOT (PANEL 1) (figure 1-83), illuminated green Illuminates red when any of the above conditions are not as specified
R/V SAFE indicator	Illuminates red when re-entry vehicle is not safe

T.O. 21M-HGM16F-1

cutoff the vernier solo subsystem supplies pressure to gimbal the vernier engines for the remaining seconds of powered flight.

1-191. Figure 1-77 shows the booster hydraulic subsystem only. Operation of the sustainer-vernier and vernier solo subsystems is similar. The booster and sustainer-vernier hydraulic pumps are operated by the booster and sustainer engine turbopumps respectively. Vernier solo hydraulic actuators are operated, after sustainer engine cutoff, by a hydraulic-pneumatic accumulator which has been pressurized with gaseous nitrogen. The hydraulic accumulator and hydraulic pumps supply hydraulic fluid to the servovalves and hydraulic pistons which gimbal the engine thrust chambers. Feedback transducers, physically connected to the hydraulic actuators, sense the amount of thrust chamber movement from the null position and feed back a cancellation signal to the flight control system. The hydraulic tank serves as a fluid reservoir and is pressurized with gaseous nitrogen prior to flight, and helium during flight, to supply a positive head, preventing system cavitation. Accumulators, pressurized with gaseous nitrogen, are provided to prevent momentary line pressure drops in the event system demand exceeds pump capacity and also to absorb pressure drops.

1-192. There are two hydraulic actuators mounted on each engine thrust chamber for gimbaling purposes. Each hydraulic actuator consists of a hydraulic controller and a linear-motor feedback transducer. The booster and vernier engine controllers contain an electro-hydraulic servovalve, hydraulic actuator, and a hydraulic flow limiter valve.

Sustainer engine hydraulic controllers consists of an electro-hydraulic servovalve and a hydraulic actuator. The hydraulic actuators utilize hydraulic energy to produce mechanical motion. One end of an actuator is secured to the missile structure and the other end is attached to a thrust chamber outrigger. The piston rods of booster and sustainer actuator assemblies are connected to lever-arm linkages which transfer piston movement to thrust chamber movement. Vernier actuators consist of two single-ended pistons connected by a rack. The rack turns a pinion gear on the shaft, which rotates the vernier engine thrust chambers.

1-193. The hydraulic actuator servovalves are two-stage, electro-hydraulic, four-way, transfer valves, which control a flow of hydraulic fluid proportional to the amount of net direct current received from the filter-servoamplifier canister of the flight control system. Direction of valve motion is controlled by the direction of armature deflection; direction of armature movement is controlled by the direction of net current flow in the valve coil. Flow limiter valves on the vernier and booster electro-hydraulic servovalves regulate the flow of hydraulic fluid to prevent the actuators from moving too rapidly.

1-194. During standby, the status of the hydraulic, pneumatic, liquid nitrogen, and helium systems is monitored and displayed on the launch control console panel (figure 1-7) by a summary signal from the various subsystem panels that illuminates the HYD-PNEU & LN$_2$-HE indicator. The indicators directly related with the missile hydraulic system during countdown but prior to commit are the HYDRAULIC PRESSURE and HYD-PNEU &

Table 1-12.   Hydraulic System Indicator on Launch Control Console

INDICATOR	FUNCTION
HYD-PNEU & LN$_2$-HE indicator	Illuminates green when all of the following conditions exist: a. SYSTEM IN STANDBY indicator on HYDRAULIC (PANEL 1) (figure 1-84), illuminated green b. SYSTEM IN STANDBY indicator on PNEUMATICS (PANEL 1) (figure 1-85), illuminated green c. SYSTEM IN STANDBY indicator on LN$_2$-HELIUM (PANEL 1) (figure 1-86), illuminated green Illuminates red when any of the above conditions are not as specified

LN$_2$-HE READY indicators in the countdown patch. These indicators illuminate green when hydraulic pressure is within limits and the pneumatics, liquid nitrogen, and helium systems are also ready.

1-195.   The commit internal signal initiates evacuation of the airborne hydraulic fluid tanks.   The reservoir shutoff valves close and the evacuation chamber shutoff valves are opened. The application of 2000-PSI hydraulic pressure forces the evacuation chamber piston to the nitrogen end of the chamber. This action draws approximately 65 cubic inches of fluid from each missile hydraulic fluid tank, providing space for fluid expansion or pressure surges. In normal countdown sequence, the turbopumps that drive the missile hydraulic fluid pumps start a few seconds after evacuation is completed.

1-196.   HYDRAULIC SYSTEM STATUS INDI-CATOR. The status indicator located on the missile launch control console, related to the missile hydraulic system, is monitored during standby and countdown by the missile combat crew commander, or by his designated alternate.

1-197.   The HYD-PNEU & LN$_2$-HE indicator summarizes the status of the hydraulic, pneumatic, liquid and nitrogen and helium system. Table 1-12 lists the indicator and its function.

1-198.   MISSILE PNEUMATIC SYSTEM.

1-199.   The missile pneumatic system (figure 1-87) provides inflight pressurization for the fuel and liquid oxygen tanks from the time pressurization is transferred to internal during the commit sequence of countdown through missile flight. Pressures in the propellant tanks are regulated and maintained at flight pressure by pressure regulators and the missile helium supply. The missile pneumatic system is protected from over-pressurization by relief valves.

1-200.   The missile pneumatic system comprises two basis subsystems, the primary supply subsystem and the secondary supply subsystem. The primary supply subsystem consists of six shrouded helium bottles, control valves, and interconnecting tubing. It controls and distributes helium at a regulated pressure to the propellant tanks during the booster stage only, and is jettisoned at staging with the booster

section. The secondary supply subsystem, which consists of a single airborne ambient helium bottle, valves, and interconnecting tubing, supplies helium to activate the vernier control valves.

1-201.   At launch, when the missile separates from ground connections, missile lines are sealed by spring-loaded valves in all but two of the missile halves of the connectors. The two liquid nitrogen connectors have no valves; consequently, these lines remain open allowing liquid nitrogen in the shrouds of the helium bottles to drain. After rise-off, the airborne helium supply furnishes inflight pressurization for the propellant tanks, airborne hydraulic fluid tanks, and the engine lubrication tanks. The helium bottles are cooled by the liquid nitrogen filled shrouds prior to launch, enabling a greater volume of the gas to be contained in the bottles. An airborne heat exchanger, located in the NO. 2 booster engine turbopump exhaust, heats and expands the chilled helium as it passes through the heat exchanger coil. The helium is then routed through regulators to the propellant tanks for pressurization.

1-202.   The oxidizer tank pressure sensing line actuates a pressure regulator to furnish helium pressure, as needed, to maintain the oxidizer tank at approximately 26 PSI. In similar fashion, fuel tank pressurization is maintained at approximately 62 PSI. Helium pressure, also regulated to approximately 60 PSI, is supplied to pressurize the hydraulic fluid tanks and engine lubrication tanks.

1-203.   The shrouded helium bottles, heat exchanger, oxidizer tank pressure regulator, and the fuel tank pressure relief valve and pressure regulator are located in the booster section. When the booster section stages, these components are jettisoned. Jettisoning is accomplished by the release of separation latches, which are actuated by pressure from the ambient helium bottle. The four connectors in the propellant tank lines and the one connector in the ambient helium bottle line contain spring-loaded valves which close at staging, sealing the pressure in the sustainer section systems. Liquid oxygen tank pressure is maintained by boiloff vapor. The fuel tank has sufficient pressure for the remainder of flight. After staging, the sustainer engine lubrication tank and hydraulic tanks continue to receive pressure from the fuel tank pressure line. The vernier pneumatic manifold is pressurized throughout flight by the amient helium bottle in the sustainer section.

1-204. The status of the airborne missile pneumatic system is monitored and displayed on the launch control console (figure 1-7) during standby by a summary signal from the various subsystem panels that illuminates the HYD-PNEU & LN$_2$-HE indicator. Indicators directly related with the pneumatic system during countdown but prior to commit are the PNEUMATICS IN PHASE II, LN$_2$ LOAD, HELIUM LOAD, and HYD-PNEU & LN$_2$-HE READY indicators in the countdown patch. When these indicators have illuminated green and all other missile system conditions are ready for the commit sequence, the READY FOR COMMIT indicator will illuminate green.

1-205. The transfer of airborne pneumatics to internal control starts when POWER INTERNAL indicator on the commit patch is illuminated green. When the ground helium countdown pressurization subsystem and the helium control charging unit complete their functions and missile pressurization components take over, the PNEUMATICS INTERnal indicator on the commit patch is illuminated green and the pneumatic system is ready for launch.

1-206. PNEUMATIC SYSTEM STATUS INDICATORS. The status indicators located on the missile launch control console, related to the missile pneumatic system, are monitored during standby and countdown by the missile combat crew commander, or by his designated alternate. Table 1-13 lists the indicators involved and their functions.

1-207. MISSILE ELECTRICAL SYSTEM.

1-208. The missile electrical system (figure 1-88) consists of the missile battery (5), 400-cycle missile inverter (1), power changeover switch (7), and interconnecting cabling and harnesses. It distributes AC and DC power to the various missile systems during standby, countdown, and flight. During standby and the first part of countdown, the power required by the missile is supplied from ground sources (except power for the re-entry vehicle which is supplied by a separate battery located in the nose section). The missile battery is activated and the 400-cycle inverter is started during countdown. After the power changeover switch transfers power from ground to internal (airborne) during the commit

Table 1-13. Pneumatic System Indicators on Launch Control Console

INDICATOR	FUNCTION
HYD-PNEU & LN$_2$ - HE indicator	Illuminates green when all of the following conditions exist: a. SYSTEM IN STANDBY indicator on HYDRAULIC (PANEL 1) (figure 1-84), illuminated green b. SYSTEM IN STANDBY indicator on PNEUMATIC (PANEL 1) (figure 1-85), illuminated green c. SYSTEM IN STANDBY indicator on LN$_2$-HELIUM (PANEL 1) (figure 1-86), illuminated green Illuminates red when any of the above conditions are not as specified
INFLIGHT HE SUPPLY indicator	Illuminates amber when all of the following conditions exist: a. Countdown has started b. Helium supply NO. 1 is more than 1450 PSI but less than 3600 PSI c. Helium supply NO. 2 is more than 1450 PSI but less than 4000 PSI Illuminates red when both of the following conditions exist: a. Helium supply NO. 1 and NO. 2 pressure is less than 1450 PSI b. Internal commit sequence has not started

sequence, the missile battery provides DC power and the battery inverter combination provides AC power to missile systems. Both the battery voltage and the 400-cycle missile inverter output are checked prior to switching the missile to internal power.

1-209.   The main missile battery consists of 20 silver-zinc-oxide cells, a battery activation mechanism, and a thermostatically controlled heater. Enclosed in a rectangular canister, the battery is located in the B-2 equipment pod, along with the 400-cycle inverter and the power changeover switch. The battery is stored in a dry-charge state. It contains a squib activation circuit which permits remote energizing of the battery cells during countdown. Activation occurs when a signal from the launch control system is applied across the pins of the activation squib, resulting in the production of pressurized gas which ruptures two diaphragms, allowing the electrolyte to flow to the battery cells. The battery is capable of supplying 28 volts at 170 amperes for a maximum period of 10 minutes.

1-210.   The 400-cycle missile inverter is contained in a sealed canister and comprises a motor-driven inverter, a magnetic amplifier, a voltage regulator, and frequency regulator. It converts 28 VDC from the ground source or the main missile battery to 115-volt, 3-phase, 400-cycle, AC power.

1-211.   The missile power changeover switch assembly is housed in a sealed canister and consists of a motor-driven switch and associated circuitry, seven receptacles, and grounding plate. It distributes power to the airborne systems whether the source is from ground or missile generation equipment. The 28-VDC motor in the unit drives the switching mechanism to close make-before-break DC contacts and break-before-make AC contacts on the switching mechanism, providing smooth transition from ground to missile electrical system power.

1-212.   Ground power reaches the missile power changeover switch assembly through an umbilical connector. The umbilical plug contains a solenoid-triggered, spring-loaded ejector mechanism that ejects the plug from the missile receptacle at rise-off. Should this electrical signal fail, a lanyard triggers the mechanism to eject the plug from the receptacle as the missile rises. (See figure 1-89 or 1-90 for umbilical connections.)

1-213.   The frequency of the alternating current that operates missile loads must be held to a close tolerance. The two sensors that determine the frequency of the power applied to these loads is a part of the launch control equipment. One is connected to the missile ground power AC source. After the airborne inverter has been started on ground power, the output of the frequency sensor is automatically switched to the launch control indicator. When the inverter output is sensed to be within tolerance, the launch control equipment sends the changeover signal.

1-214.   The motor-driven power changeover switch connects the missile battery to all DC loads on the missile, with the exception of the re-entry vehicle system, and then disconnects the missile ground power DC source from the loads. It next disconnects the missile ground power AC source, and a few milliseconds later connects the loads to the inverter. Then it sends a signal to the ground, indicating that these sequences have been completed. Main missile loads are powered by the missile battery and by the inverter during the period between changeover and the firing of retarding rockets, following separation of the re-entry vehicle.

1-215.   The status of the missile power control subsystem, associated AC and DC ground power sources, and also the missile engines, is monitored and displayed on the launch control console (figure 1-7) during standby by a summary signal from the various subsystem panels that illuminates the ENGINES AND GROUND POWER indicator. During countdown but prior to commit, the indicators in the countdown patch concerned with the electrical systems are the MISSILE POWER, MISSILE BAT. ACTIVATED, ENG & MISSILE POWER READY, and the R/V BATTERY TEMPERATURE indicators. The READY for COMMIT indicator will light green when these indicators have illuminated green and all other system conditions have been satisfied. Two seconds after start of the commit sequence, a timer runs out resulting in a signal being sent to the missile power changeover switch to transfer power to internal. The POWER INTERNAL indicator in the commit patch illuminates green when the missile inverter voltage and frequency are within tolerance and the internal power changeover signal is received from the missile. The missile electrical system is now ready for launch.

1-216. ELECTRICAL SYSTEM STATUS INDI-CATORS. The status indicators located on the missile launch control console, related to the missile electrical system, are monitored during standby and countdown by the missile combat crew commander, or by his designated alternate. Table 1-14 lists the indicators involved and their functions.

1-217. MISSILE EXPLOSIVES.

1-218. The missile is equipped with explosive and hypergolic devices to initiate various functions in a predetermined sequence during countdown, launch and flight. These functions include activating the main missile battery, starting rocket engines, initiating booster separation, retarding the missile tank section after re-entry vehicle separation, and destroying the missile tank section following re-entry vehicle separation, or (VAFB only) destroying the missile if range safety is threatend during a training launch. Figure 1-91 shows the location of missile explosive assemblies.

1-219. The main missile battery is energized at the start of countdown by a squib. Activation of the squib results in the production of gas, which upon expansion ruptures the plastic diaphragms contain-ing the battery electrolyte, forcing it under gaseous pressure into the cells. After the electrolyte is forced into the cells, the gas escapes through a small orifice to a vent. The battery cannot be recharged. If the launch is aborted, the battery must be removed from the missile within ten hours following activation.

1-220. Four types of devices are used to start missile engines: solid propellant gas generators, solid propellant gas generator initiators, gas generator igniters, and hypergolic igniters. The solid propellant gas generators provide power to start the turbopumps supplying propellants to the booster and sustainer engines. Ignition of the gas generators takes place when an engine start command is sent from the launch control system, firing the solid propellant gas generator initiators, which produce hot particles and gases to fire the solid propellant gas generators. The solid propellant gas generators, in turn, produce hot gas which drives the booster and sustainer turbopumps up to the speed and pressure required to feed liquid propellants to the engine thrust chambers and also to the liquid propellant gas generators. The liquid propellant gas generators power the engine turbopumps after the solid propellant gas generators have burned out.

Table 1-14.   Electrical System Indicators on Launch Control Console

INDICATOR	FUNCTION
ENGINES AND GROUND POWER indicator	Illuminates green when both of the following conditions exist: a. SYSTEM IN STANDBY indicator on ENGINE (PANEL 2) (figure 1-78), illuminated green b. SYSTEM IN STANDBY indicator on MISSILE GROUND POWER (PANEL 1) (figure 1-79), illuminated green Illuminates red when either of the above conditions is not as specified
400 CYCLE POWER indicator	Illuminates red when missile ground power supply 400-cycle, 3-phase, AC voltage or frequency is out of tolerance
MISSILE INVERTER indicator	Illuminates red when all the following conditions exist: a. Missile inverter is started and is on for 60 seconds. b. Internal AC voltage or frequency is out of tolerance

1-221. Gas generator igniters are pyrotechnic devices that start the burning of propellant mixtures in the combustion chambers of the liquid propellant gas generators. Two igniters are installed in each liquid propellant gas generator. They are fired electrically by signals from the launch control system. Sustainer gas generator igniters are fired simultaneously with the sustainer engine lock-in signal, and the booster gas generator igniters are fired simultaneously with the solid propellant generators that start the booster turbopumps. The pyrotechnic material from the igniter cartridges is already burning when the gas generator propellant valves admit the flow of liquid propellants, causing immediate ignition.

1-222. Hypergolic igniters start combustion in the booster, sustainer, and vernier engine thrust chambers. Each igniter consists of a casing containing a slug of hypergolic fuel held in place by burst diaphragms. The hypergolic fuel burns spontaneously upon contact with oxygen. When the turbopumps start, a portion of the fuel being pumped to the engines is sent through ignition fuel valves to the hypergolic igniters. There, fuel pressure ruptures the burst diaphragms, forcing a slug of hypergol into each thrust chamber. The hypergol combines with the liquid propellants which instantly ignite, starting the engines.

1-223. Booster section separation at staging is activated by the separation explosive valve assembly, consisting of two explosive valves mounted and wired in parallel. The firing of either valve activates the booster separation mechanism, which consists of four hook-type latches on the skin structure of the booster section that mate with four stirrups on the tank section. At staging, an electrical impulse from the flight control autopilot subsystem programmer fires the separation explosive valves, driving a plunger through a diaphragm, which releases pressurized helium gas for distribution through a manifold to the four latches. All latches open simultaneously and the booster section separates from the missile body. The booster engines are shut off just before actuation of the separation explosive valves.

1-224. A single retarding rocket is located in a fairing forward of each vernier engine, and two retarding rockets are installed in the forward end of the B-2 equipment pod. The thrust chambers of these rockets point toward the front of the missile. The retarding thrust generated by the rocket slows the forward velocity of the missile body, enabling the re-entry vehicle to move away from it without interference. Two seconds after re-entry vehicle separation, the retarding rockets fire upon receipt of a signal from the flight programmer. The missile body is fragmented by a tank fragmentation explosive assembly; the re-entry vehicle continues on its planned trajectory to the target.

1-225. The tank fragmentation explosive assembly consists of a metal case, two squibs, two booster charges, the main explosive charge, and associated wiring. A destruct command signal received from the automatic pilot subsystem fires the explosive assembly at a preselected time after re-entry vehicle separation. Detonation of the unit ruptures the intermediate bulkhead between the fuel and oxidizer tank, fragmenting the missile.

1-226. MISSILE FLIGHT SAFETY SYSTEM.

(VAFB ONLY)

1-227. A missile flight safety system is provided in training launch missile for intentionally destroying the missile should the missile have an erratic flight which may endanger populated areas. Transducers aboard the missile monitor critical pressures, temperatures, velocities, and attitudes. The output signals of the transducers are adapted for transmission to ground control stations for evaluation. Beacon tracking signals along with electronic (radar) surveillance data, is used by the range safety officer in reaching a missile destruct decision. Should the range safety officer decide to destroy the missile, a destruct pushbutton located on the range safety officer console is depressed which causes the command destruct transmitter to send a coded signal to the missile, igniting primers of an explosive charge mounted near the fuel and oxidizer tank bulkhead. Detonation of this explosive charge destroys the missile.

1-228. MISSILE AND RE-ENTRY VEHICLE HANDLING AND TRANSPORT.

1-229. MISSILE HANDLING AND TRANSPORT.

1-230. The missile is received mounted on a missile semitrailer. The semitrailer is constructed of welded trusses and beams and has an overall length of approximately 70.5 feet and width of approximately 11 feet. The trailer, with missile, is towed overland by truck-tractor or may be carried aboard a C-133 aircraft. The missile semitrailer contains provisions for missile fuel and liquid oxygen tank pressurization and control; missile longitudinal stretch; inter-communications; gyro heating; booster section separation from tank section; trailer stabilization, tiedown, and support; and hydraulic, pneumatic, and electrical power.

1-231. The missile may be transferred from one semitrailer to another by use of lifting fixtures and two 15-ton capacity cranes. The missile is transferred from the missile semitrailer to the launcher platform (figure 1-22) by a missile erector vehicle. The erector vehicle is towed by a truck-tractor and consists principally of an erector boom, erector screw, and erector motor and gear box, stabilizing outriggers (jacks), and tie-downs. Transfer of the missile to the launch platform is accomplished by aligning the missile semitrailer and erector vehicle with the launcher pedestal (1, figure 1-22), connecting the erector boom to the missile, raising the missile free of the semitrailer, and rotating the missile to the vertical position with the erector boom.

1-232. RE-ENTRY VEHICLE HANDLING AND TRANSPORT.

1-233. Re-entry vehicle handling, transport, mating, and demating are accomplished with a re-entry vehicle trailer, cradle and carriage assembly, hoisting yoke, and mobile crane. The trailer is used to transport the re-entry vehicle to and from the munitions area. The cradle supports the re-entry vehicle and provides the means of handling the re-entry vehicle during assembly, transport, mating, and demating operations. The carriage is roller mounted and supports the cradle while mounted on the re-entry vehicle trailer or on a work stand. The re-entry vehicle is mated to the missile after the missile is lowered into the silo. The hoisting yoke attaches to the cradle and when lifted by the mobile crane, provides the means of rotating the re-entry vehicle from the horizontal to the vertical position for mating.

1-234. SQUADRON MAINTENANCE AREA.

1-235. The squadron maintenance area supporting the launch complex consists of the missile assembly and maintenance shops (MAMS) and a munitions section. Mobile checkout vehicles supporting the launch complex are dispatched from the missile assembly and maintenance shops as required. The major portion of organizational and field level maintenance activities are conducted at the squadron maintenance area. Additional maintenance support for the launch complex and the squadron maintenance area is provided by the parent air base and Air Force depots. Maintenance and modifications requiring special skills or facilities not available within the using command may be performed, upon request, by contractor personnel and facilities.

1-236. MISSILE ASSEMBLY AND MAINTENANCE SHOPS.

1-237. In the missile assembly and maintenance shops (MAMS), located remotely from the launch complex, the major portion of inspection, checkout and repair operations is accomplished. The MAMS is divided into various areas and rooms to perform aerospace ground equipment maintenance, missile maintenance, engine maintenance, guidance system maintenance, and component checkout, calibration, and repair. Areas are also provided for administration and supply functions. The M A M S and associated personnel are capable of receiving the missile and performing all inspection, assembly, and checkout operations to determine if the missile is ready for transport to the launch complex and installation on the launch platform. The MAMS is also capable of repair and checkout of malfunctioning components or end items.

1-238. MUNITIONS SECTION.

1-239. The munitions section for squadron complexes is located remotely from the MAMS and consists of warehouses, storage magazines, and a surveillance and inspection building. The surveillance and inspection building is divided into various areas for administration functions; assembly, test, and checkout operations; equipment and spares. The building also provides the facility for complete processing of the warhead, re-entry vehicle, and associated equipment. The upper floor contains the administrative offices, briefing room, and conference room. Storage areas are provided by the four warehouses and storage magazines. The storage magazines house the re-entry vehicle warheads and are constructed of reinforced concrete. Roof ventilators and louvered blowout panels provide ventilation for each storage compartment of the magazines.

1-240. PERSONNEL.

1-241. Operating, maintaining and supporting the launch complex require a wide range of air force specialty codes and skill levels. Personnel required include surveillance and launch personnel, maintenance and checkout personnel, and support personnel. Surveillance and launch operations are performed by a missile combat crew consisting of the following members.

   a. Missile Officer (2), Air Force Specialty Code (AFSC) 3124B.

   (1) Missile combat crew commander (MCCC).

   (2) Deputy missile combat crew commander (DMCCC).

   b. Ballistic Missile Analyst Technician/Specialist, AFSC 312X4D.

   c. Electrical Power Production Technician/Specialist, AFSC 543X0.

   d. Missile Facilities Technician/Specialist, AFSC 541X0D.

1-242. The missile combat crew commander (MCCC) is assigned the responsibility of ensuring that the launch complex is maintained in emergency war order (EWO) readiness condition and implementing alert messages when received. The MCCC also ensures that missile combat crew proficiency is maintained in regards to safety, launch operations, and standby surveillance of equipment. It is the responsibility of the Sector Commander or his designated representative to ensure that all inspection, servicing, and checkout of equipment are performed as scheduled. The deputy missile combat crew commander (DMCCC) assists the MCCC in the assigned responsibilities.

1-243. The BMAT, EPPT, and MFT assist the MCCC and DMCC maintaining EWO readiness by monitoring, inspecting, checking, servicing, and repairing launch complex equipment. During missile launches, the BMAT, EPPT, and MFT assist launch operations by monitoring designated equipment and advising the MCCC of abnormal or hazardous conditions, and by performing emergency procedures when required and as directed by the MCCC.

1-244. Additional personnel required to support training launch operations at Vandenberg AFB include missile flight safety, Pacific Missile Range, command post, instrumentation personnel, launch complex safety officer, and missile accident and emergency team.

1-245. Mobile checkout and maintenance (MOCAM) crews support launch complex maintenance, checkout, and launch operations. Personnel specialties associated with the MOCAM crews include the following:

   a. Missile Maintenance Technician, AFSC 44370A.

   b. Ballistic Missile Analyst Technician, AFSC 31274D.

   c. Ballistic Missile Analyst Specialist, AFSC 31254D.

   d. Ballistic Missile Inertial Guidance Maintenance Specialist, AFSC 31252A.

   e. Missile Pneudraulics Specialist, AFSC 44250A.

   f. Missile Facilities Technician, AFSC 54170D.

   g. Missile Electrical Repairman, AFSC 44150A.

   h. Missile Engine Technician, AFSC 44371A.

   i. Missile Engine Mechanic, AFSC 44351A.

   j. Electrical Technician, AFSC 54270D.

   k. Liquid Fuel Specialist Missile Systems, AFSC 54650D.

1-246. MISSILE CHECKOUT OPERATIONS.

1-247. Missile checkout operations are performed at both the launch complex and at the squadron maintenance area. The re-entry vehicle is received by the munitions section where assembly-checkout operations are performed. The basic missile is received and inspected at the MAMS and separately packaged items are installed. Missile inspections and a complete checkout of all missile systems are then performed. The major portion of checkout operations are performed using the alignment countdown set inertial guidance and mobile checkout vehicles (semitrailer-mounted pneumatic test set and semitrailer-mounted missile electrical systems checkout test station). The pneumatic test set contains fuel and liquid oxygen ullage simulation tanks, pneumatic disconnect panel (figure 1-92), and control panels (figures 1-93 and 1-94) to perform calibration and checkout of the missile pneumatic system. The checkout test station contains a major portion of missile systems checkout equipment and is capable of checking the pneumatic test set and performing dynamic tests of the missile when used with the pneumatic test set. Data for the test is punched into business machine cards which are used by a programmer to control the entire test sequence. Punched card decks are provided to perform self-test, missile systems survey, and checkout of individual missile systems. The self-test deck verifies that the checkout equipment is operationally ready to perform missile systems test. The individual system card decks verify each system and are capable of detailed malfunction analysis. Results of all tests are printed on a paper tape and are displayed (GO - NOGO) on an indicator panel in the checkout test station.

1-248. After completion of all missile systems checks, the missile and checkout equipment are secured and the trailer and missile are prepared for transport to the launch complex. The trailer is towed to the launch complex where the missile is erected on the launch platform using the erector vehicle. The installed missile and support systems in the launch area are then validated and checked, utilizing in part the mobile checkout vehicles.

1-249. COUNTDOWN EXERCISES.

1-250. Six different and separate launch exercises may be performed with the weapon system: (a) Tactical launch countdown, (b) Training launch countdown, (c) Propellant loading exercise countdowns, (d) maintenance countdown, (e) Simulated EWO countdown, and (f) Launch signal responder countdown. Procedures used by the missile combat crew during the countdown, commit and abort sequence during all launch exercises are contained in a common countdown and abort checklist. Basic differences in the exercises are in the complex preparation and desired mission to be accomplished.

1-251. TACTICAL LAUNCH COUNTDOWN.

1-252. A TACTICAL LAUNCH countdown shall be initiated at an alert status complex upon receipt of a valid and authenticated launch order. The complex is prepared for alert status using appropriate maintenance flow diagrams and maintained on alert through daily inspection, servicing, and repair as required.

1-253. TRAINING LAUNCH COUNTDOWN.

1-254. A training launch countdown may be initiated only after appropriate preparation. These exercises are performed to check functional integrity of the launch complex, missile, and combat crew personnel. Preparation for countdown and mission are different than that of a tactical launch, however, the countdown sequence is identical.

1-255. PROPELLANT LOADING EXERCISE COUNTDOWN (PLX)

1-256. A propellant loading exercise countdown may be initiated only after appropriate preparation. These exercises are performed to check the (a) Functional integrity of a launch complex and missile on EWO alert status, (b) Evaluation of combat ready crews in performance of countdown procedures and, (c) Training of combat crew personnel. Countdown sequence to the point of launch is identical to that of a tactical countdown. The MISSILE AWAY indicator in the launch control console will not illuminate.

1-257. MAINTENANCE COUNTDOWN.

1-258. A maintenance countdown may be initiated only after appropriate preparation. These exercises are performed in support of maintenance functions to check functional integrity of a complex and missile. Training and evaluation of missile combat crew personnel may also be accomplished. Countdown sequence is identical to that of a PLX countdown.

1-259. LAUNCH SIGNAL RESPONDER COUNTDOWN (LSR).

1-260. A launch signal responder countdown may be initiated only after appropriate preparations. These exercises are performed in support of maintenance and evaluation/training of missile combat crew personnel. Countdown sequence as they appear in the launch control center are identical to a tactical or PLX countdown, although only the countdown logic is being exercised. The RESPONDER MODE indicator on the launch control console will be illuminated AMBER.

1-261. SIMULATED COUNTDOWN (TANK THROUGH).

1-262. A Simulated Countdown may be initiated at any time with no special preparation, as no systems are activated. These exercises are performed to train and maintain countdown proficiency of the missile combat crew members. All actions and procedures are simulated. Announcement of actions and steps will be made as appropriate.

1-263. POSTLAUNCH AND POSTPROPELLANT LOADING EXERCISE OPERATIONS.

1-264. POSTLAUNCH OPERATIONS.

1-265. Postlaunch operations are performed after missile launch. The silo equipment is inspected for damage and repaired or replaced as required. Checkout operations are then performed to ensure that the complex is ready to receive a missile. The missile is transported from the missile assembly and maintenance shops and installed on the launch platform. Checkout, test, and set up operations are then performed to establish ready state A.

1-266. POSTPROPELLANT LOADING EXERCISE OPERATIONS.

1-267. Postpropellant loading exercise operations are performed after the abort sequence is completed. Missile rocket engines are serviced, the missile is checked for leakages and other abnormal conditions, and the silo fluid and gas supplies are replenished. Ready state A (tactical launch or training launch) is established by installing and checking the missile ordnance items.

1-268. MAINTENANCE PLAN.

1-269. MAINTENANCE INSPECTIONS.

1-270. Maintenance inspections are divided into two categories; preventive maintenance, and control. Preventive maintenance consists of visual and servicing type inspections and generally requires no sequencing which will place equipment out of commission. Control inspections generally require that the status of missile or aerospace ground equipment be changed from in-commission to out-of-commis-sion. The control inspections involve operating equipment and thus must be sequenced to prevent damage to equipment or injury to personnel.

1-271. MAINTENANCE.

1-272. Requirements for two types of maintenance (scheduled and unscheduled) will be generated. Scheduled maintenance is that maintenance which must be performed at specific time intervals to insure that weapon system equipment is maintained in operational condition. Unscheduled maintenance is that maintenance which cannot be specifically predicted, such as equipment malfunction or damage.

1-273. LEVELS OF MAINTENANCE.

1-274. Maintenance levels are divided into three categories: organizational, field, and depot. Maintenance functions beyond the capability of organizational or field level maintenance are normally performed at depot level maintenance facilities (Air Force Depots, contractor facilities, if required, or depot maintenance teams).

1-275. Organizational or field level maintenance is performed at the launch complex missile assembly and maintenance shops, and the munitions section. Repair of end items, including the missile, at the launch complex will be determined by the test equipment and facilities required, dangers of contamination and damage, and time required. In general, organizational or field level maintenance at the various areas will include the following.

    a. Launch complex:

    (1) Missile installation, checkout, and prelaunch inspections.

    (2) Daily monitoring and inspection of missile, aerospace ground equipment, and real property installed equipment.

    (3) Periodic and special inspections of missile, aerospace ground equipment, and real property installed equipment.

    (4) Periodic functional checkout of missile and aerospace ground equipment.

    (5) Malfunction correction.

    (6) Servicing and calibration.

    (7) Re-entry vehicle mating and demating.

(8) Missile removal.

b.   Missile assembly and maintenance shops:

(1) Receiving inspection and assembly of missile as received from factory or launch complex.

(2) Missile processing (checkout).

(3) Periodic and special inspection of aerospace ground equipment and real property installed equipment.

(4) Functional checkout of aerospace ground equipment and real property installed equipment.

(5) Repair, checkout, and calibration of components and end items.

(6) Modification to missile, aerospace ground equipment, and real property installed equipment as required.

(7) Missile storage and monitoring.

c.   Munitions section:

(1) Re-entry vehicle inspection, handling, and storage.

(2) Re-entry vehicle maintenance and checkout.

(3) Explosives inspection, handling, and storage.

1-276.   TECHNICAL ORDERS.

1-277.   Technical orders (manuals) are provided to the using command to accomplish inspections, checkout, maintenance, and launch operations. Organizational maintenance manuals are provided to accomplish maintenance and checkout operations on individual systems such as hydraulic, pneumatic, guidance, and launch control. One organizational maintenance manual is provided for each system and contains the following information:

a.   Description (physical and functional) of equipment within the system.

b.   Checkout and maintenance procedures.

c.   Trouble analysis.

d.   Component removal and installation procedures.

1-278.   Field maintenance and depot overhaul manuals are also provided for repair, reshipment, and checkout of individual equipment within a system, such as nitrogen control unit, pressure system control, helium control charging unit, and air conditioner. These equipment oriented manuals contain the following information:

a.   Physical and functional description of components within the equipment.

b.   Maintenance and reshipment procedures.

c.   Theory of operations and system tie-in.

d.   Field maintenance procedures (repair, replace, and checkout).

e.   Depot level maintenance procedures.

f.   Illustrated parts breakdown for equipment covered by manual.

1-279.   Checklist manuals, inspection requirement manuals, and inspection work cards are also provided to implement the operational requirements at the missile assembly and maintenance shops and at the launch complex. These manuals present all tasks required in processing a missile from receipt by the squadron through launch of the missile and post-launch. The inspection work cards include all tasks involved in daily, periodic, and special inspections.

1-280.   Checkout manuals are also provided for operational use of the checkout test station. One checkout manual is provided for each card deck (missile systems checkout, checkout test station self-test, and hydraulic fill and bleed). Separate checkout manuals are provided for the launch complex and the missile assembly and maintenance shops. These checkout manuals contain the following information:

a.   Equipment requirement during checkout operations.

b.   Preparation for checkout.

c.   Card by card instructions (function of card, readout, manual actions required, and backout instructions).

d.   Equipment shutdown procedures.

e.   Troubleshooting procedures or procedure reference for malfunctions occuring during checkout.

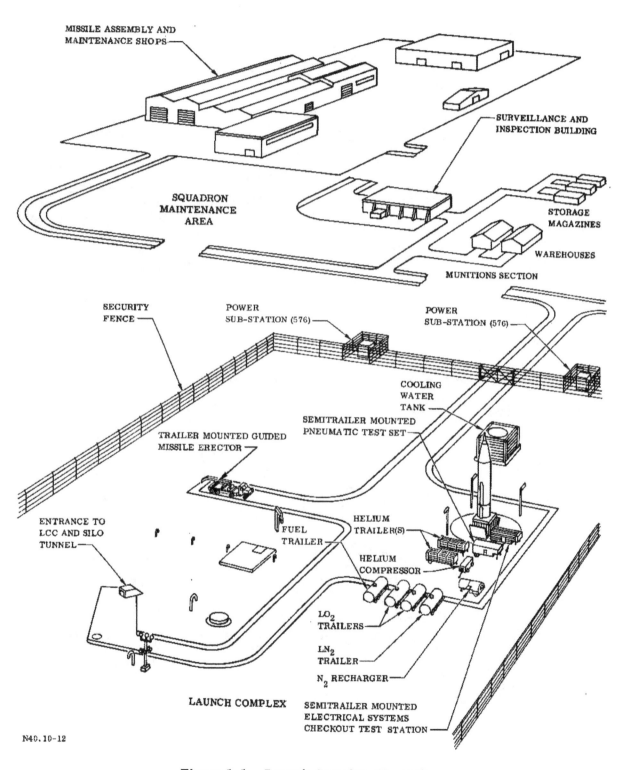

MISSILE ASSEMBLY AND
MAINTENANCE SHOPS

SURVEILLANCE AND
INSPECTION BUILDING

SQUADRON
MAINTENANCE
AREA

STORAGE
MAGAZINES

WAREHOUSES

MUNITIONS SECTION

SECURITY
FENCE

POWER
SUB-STATION (576)

POWER
SUB-STATION (576)

COOLING
WATER
TANK

SEMITRAILER MOUNTED
PNEUMATIC TEST SET

TRAILER MOUNTED GUIDED
MISSILE ERECTOR

ENTRANCE TO
LCC AND SILO
TUNNEL

FUEL
TRAILER

HELIUM
TRAILER(S)

HELIUM
COMPRESSOR

LO_2
TRAILERS

LN_2
TRAILER

N_2 RECHARGER

LAUNCH COMPLEX

SEMITRAILER MOUNTED
ELECTRICAL SYSTEMS
CHECKOUT TEST STATION

N40.10-12

Figure 1-1. Launch Complex (Typical)

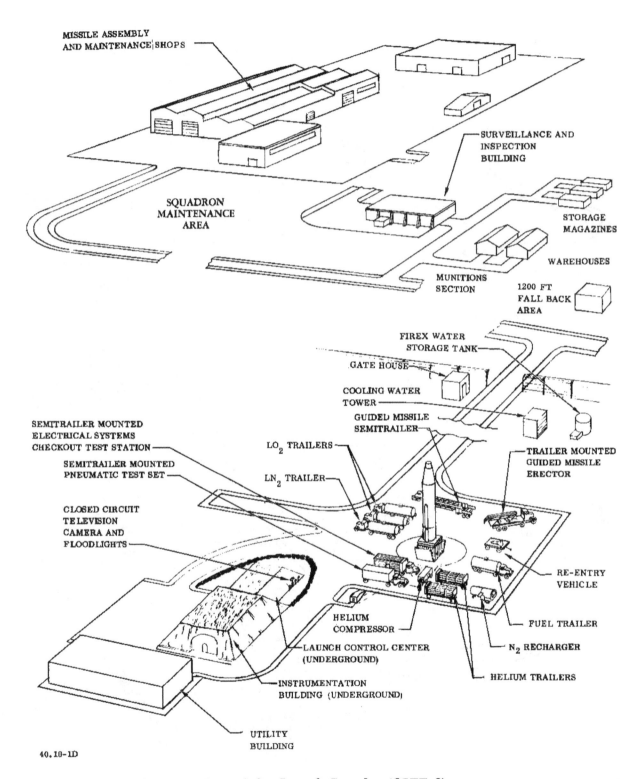

40.10-1D

Figure 1-2. Launch Complex (OSTF-2)

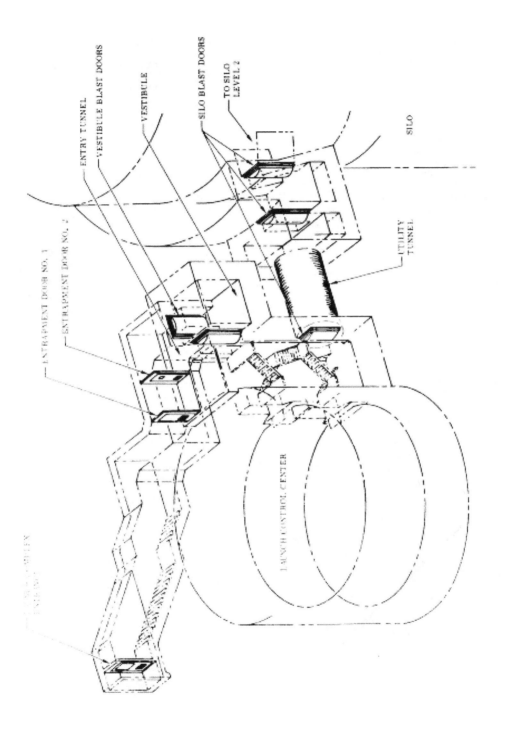

ENTRY TUNNEL
VESTIBULE BLAST DOORS
VESTIBULE
SILO BLAST DOORS
TO SILO LEVEL 2
SILO
UTILITY TUNNEL
ENTRAPMENT DOOR NO. 1
ENTRAPMENT DOOR NO. 2
LAUNCH CONTROL CENTER

Figure 1-3. Launch Complex Entrance

1	TELEPHONE TERMINAL CABINET	18	COMM ANNUNCIATOR
2	SIGNALLING SYSTEM CABINET	19	FRCP
3	MAIN DISTRIBUTION FRAME	20	COMMUNICATIONS DISCONNECT PANEL
4	INTERIM PA BAY & LES CONTROL BOX	21	PRCP
5	L/M BAY	22	LO_2 TANKING PANELS 1 & 2
6	D/L BAY	23	LC CONSOLE
7	COMM RACKS	24	CSMOL
8	COMMUNICATIONS PANEL C	25	TV MONITOR
9	LES "J" BOX	26	ENTRAPMENT TV MONITOR
10	ANNUNCIATOR PANEL	27	GATE AND DOOR CONTROL PANEL
11	UHF AND VHF SYSTEMS	28	DISTRIBUTION PANEL "A"
12	JUNCTION BOX	29	480-VOLT CONTROL CENTER
13	BLAST DETECTION CABINET	30	DISTRIBUTION PANEL "D"
14	COMMUNICATIONS CONSOLE	31	440-VOLT TRANSFORMER
15	FIRE ALARM BATTERY BOX	32	BATTERY BANK
16	FIRE ALARM PANEL	33	CHARGER BAY
17	NOTIFIER PANEL		

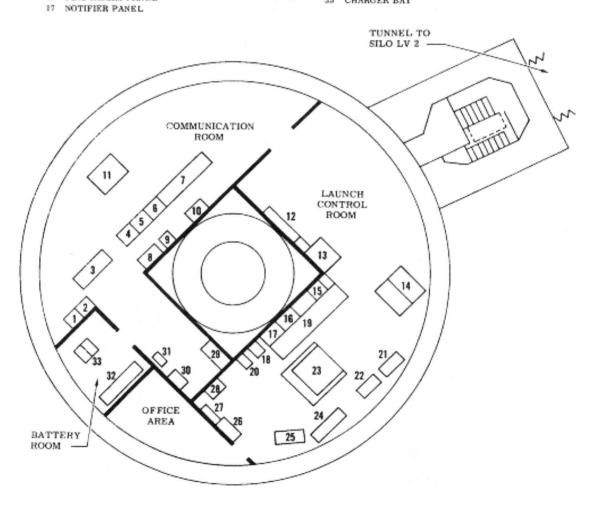

Figure 1-4. Launch Control Center (Typical)

40.10-137

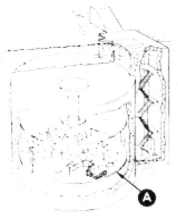

1 TELEPHONE TERMINAL CABINET
2 PUBLIC ADDRESS SYSTEM CABINET
3 COMMUNICATIONS EQUIPMENT
   POWER DISTRIBUTION PANEL C
4 480 VOLT CONTROL CENTER
5 LIGHTING DISTRIBUTION
   TRANSFORMER
6 MAIN DISTRIBUTION FRAME
7 COMMUNICATIONS PANEL C
8 MISCELLANEOUS TRUNK BAY
9 POWER BOARD
10 MISCELLANEOUS RELAY RACK
11 ADMINISTRATION DIAL SYSTEM
12 LAUNCH COMPLEX SAFETY
   OFFICER EQUIPMENT BAY
13 R&D EQUIPMENT BAY AND R&D
   COMMUNICATION CONTROL PANEL
14 VOICE RECORDER BAY
15 FILE CABINET
16 R&D EQUIPMENT STATION 50
17 R&D EQUIPMENT STATION 49
18 FIRE ALARM TROUBLE INDICATOR
   PANEL

19 R&D EQUIPMENT STATION 114
20 LAUNCH COMPLEX SAFETY OFFICERS
   CONSOLE
21 FIRE ALARM PANEL
22 FACILITY REMOTE CONTROL PANEL
23 COMMUNICATIONS DISCONNECT PANEL
24 TEST AND TRAINING MONITOR CONSOLE
25 CCTV CAMERA
26 MONITOR CONSOLE (BMD/ST1)
27 LIGHTING PANEL A
28 FIREX PUSHBUTTON PANEL
29 LAUNCH CONTROL CONSOLE
30 POWER REMOTE CONTROL PANEL
31 TV MONITOR
32 MISSILE DESTRUCT SYSTEM CHECKOUT
   AND CONTROL CONSOLE
33 DESK
34 BATTERIES AND RACK
35 CHARGER BAY
36 CONTROL STATION MANUAL OPERATING
   LEVEL

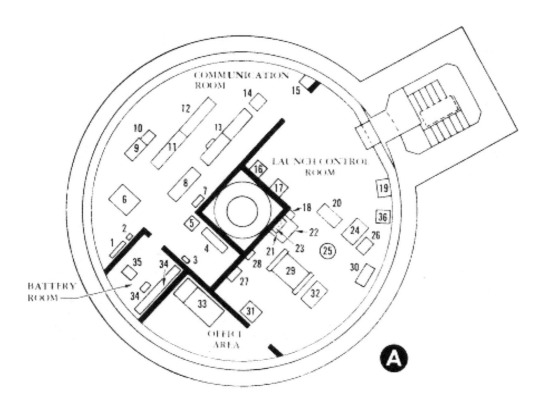

Figure 1-5. Launch Control Center (OSTF-2)

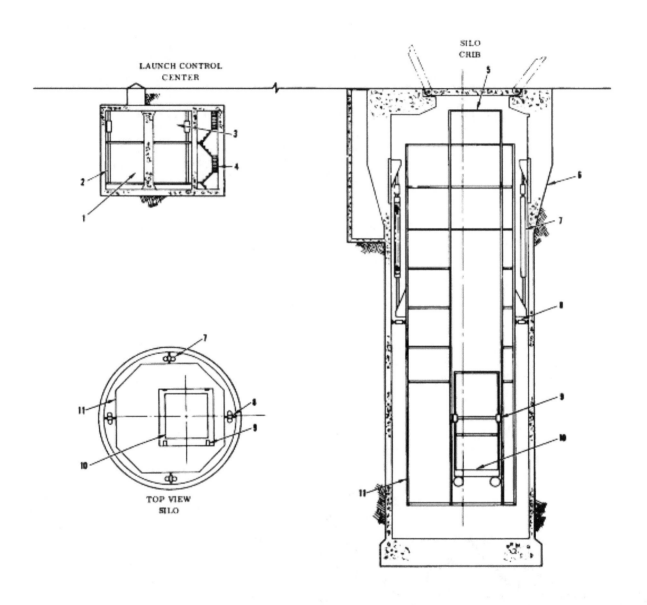

1 SECOND LEVEL	6 SILO WALL
2 SUSPENSION STRUT	7 SHOCK STRUT (2 AT 4 PLACES)
3 FIRST LEVEL	8 HORIZONTAL DAMPER (1 AT 4 PLACES)
4 STAIRWAY	9 LAUNCH PLATFORM DOWN LOCK
5 MISSILE ENCLOSURE	10 LAUNCH PLATFORM
	11 CRIB

40.10-8B

Figure 1-6. Launch Control Center and Silo Crib Suspension System

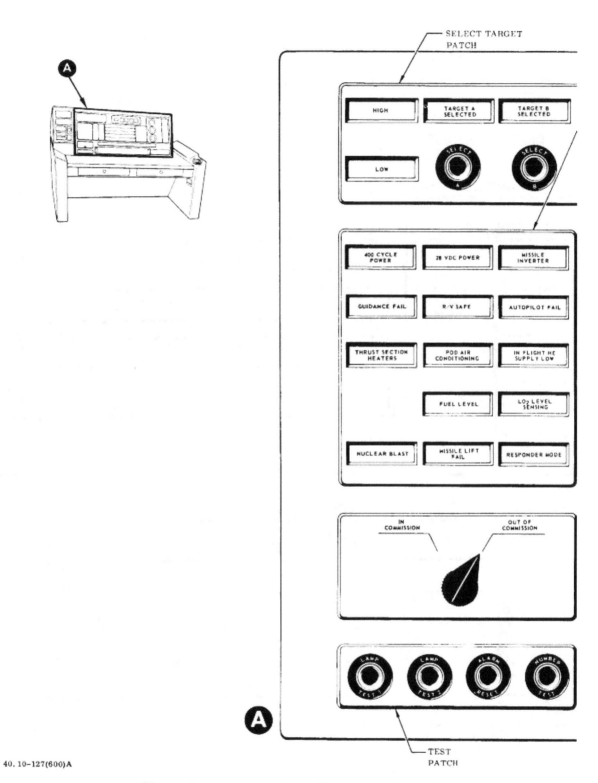

Figure 1-7.   Launch Control Console (Typical)

Changed 15 April 1964

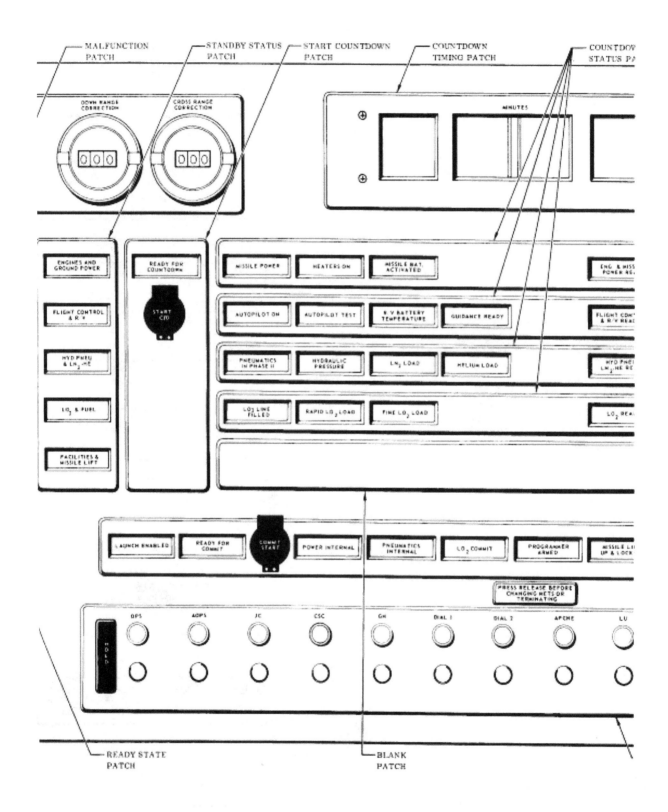

MALFUNCTION PATCH

STANDBY STATUS PATCH

START COUNTDOWN PATCH

COUNTDOWN TIMING PATCH

COUNTDOWN STATUS PATCH

DOWN RANGE CORRECTION

CROSS RANGE CORRECTION

MINUTES

ENGINES AND GROUND POWER

READY FOR COUNTDOWN

START CTD

MISSILE POWER

HEATERS ON

MISSILE BAT. ACTIVATED

ENG. & MISS POWER RE

FLIGHT CONTROL & R.V.

AUTOPILOT ON

AUTOPILOT TEST

R.V. BATTERY TEMPERATURE

GUIDANCE READY

FLIGHT CONT & R/V READ

HYD PNEU & LN₂ HE

PNEUMATICS IN PHASE II

HYDRAULIC PRESSURE

LN₂ LOAD

HELIUM LOAD

HYD PNEU LN₂ HE RE

LO₂ & FUEL

LO₂ LINE FILLED

RAPID LO₂ LOAD

FINE LO₂ LOAD

LO₂ REA

FACILITIES & MISSILE LIFT

LAUNCH ENABLED

READY FOR COMMIT

COMMIT START

POWER INTERNAL

PNEUMATICS INTERNAL

LO₂ COMMIT

PROGRAMMER ARMED

MISSILE LI UP & LOCK

PRESS RELEASE BEFORE CHANGING NETS OR TERMINATING

HOLD

OPS   AOPS   JC   CSC   GH   DIAL 1   DIAL 2   APCHE   LU

READY STATE PATCH

BLANK PATCH

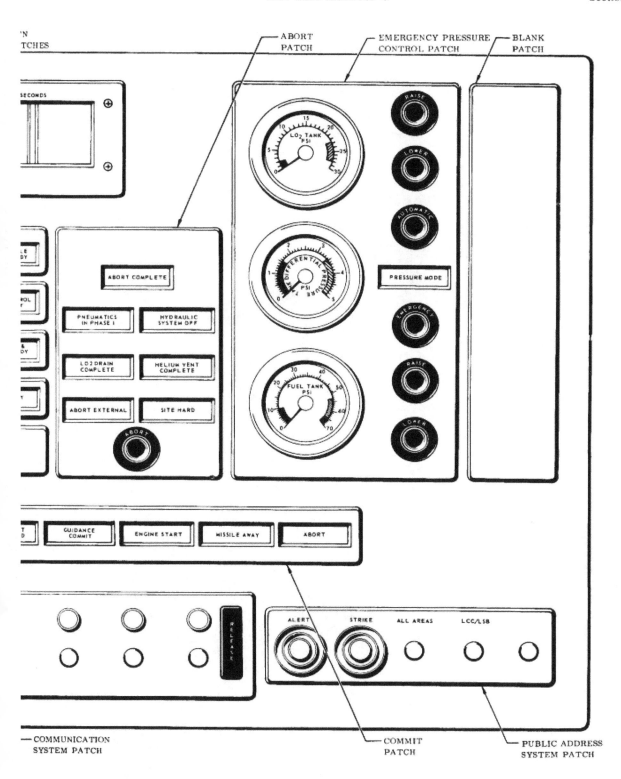

ABORT PATCH

EMERGENCY PRESSURE CONTROL PATCH

BLANK PATCH

COMMUNICATION SYSTEM PATCH

COMMIT PATCH

PUBLIC ADDRESS SYSTEM PATCH

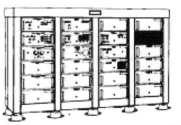

CONTROL-MONITOR GROUP 2 OF 4

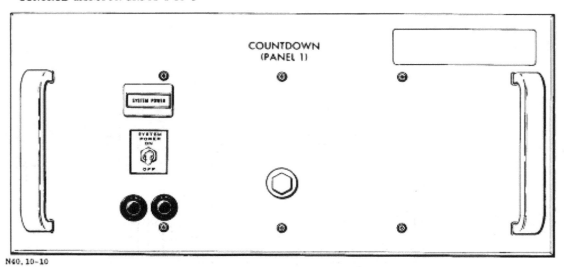

N40, 10-10

Figure 1-8.   Countdown (Panel 1)

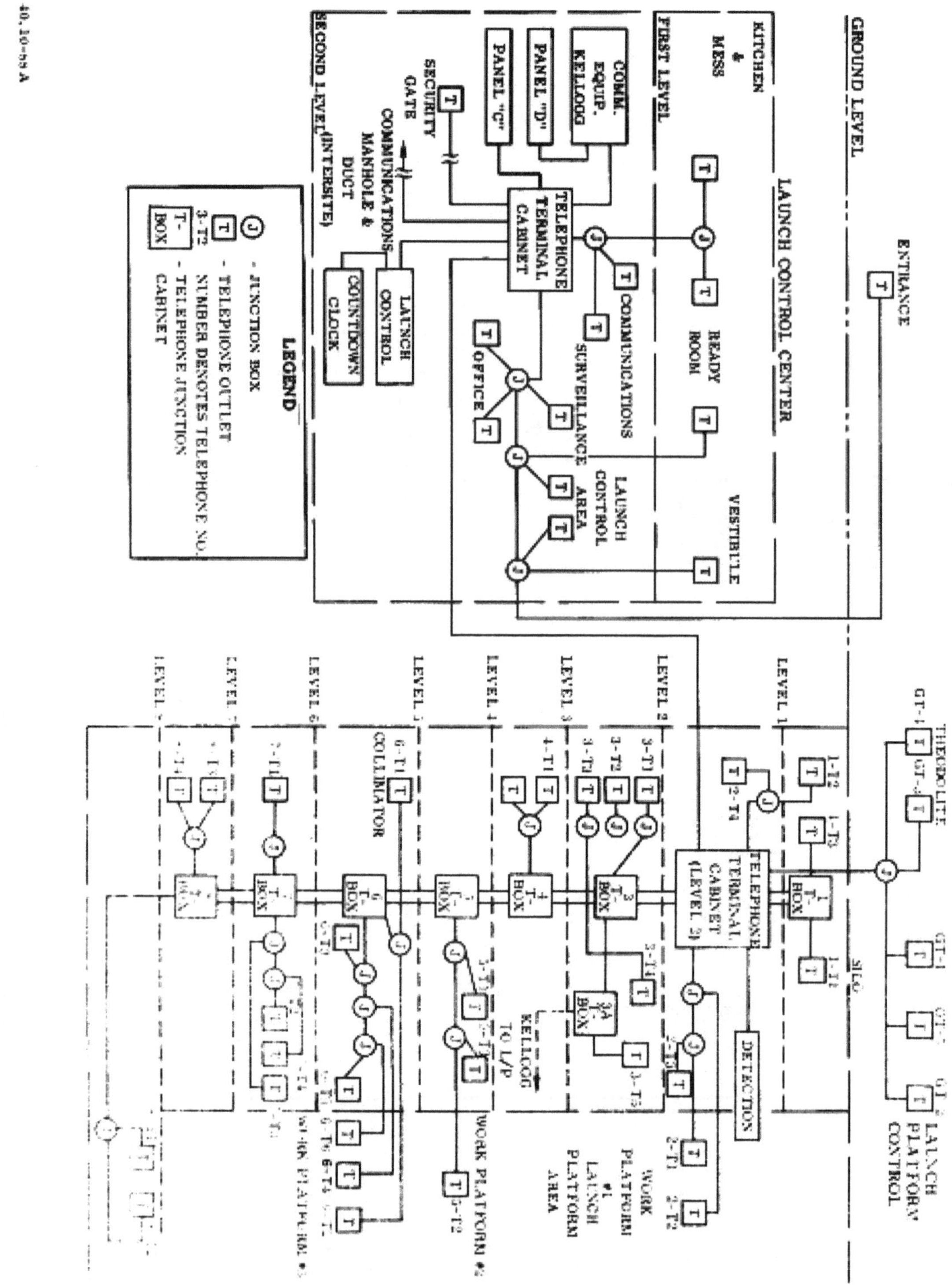

Figure 1-9. Launch Complex Communications Location Diagram

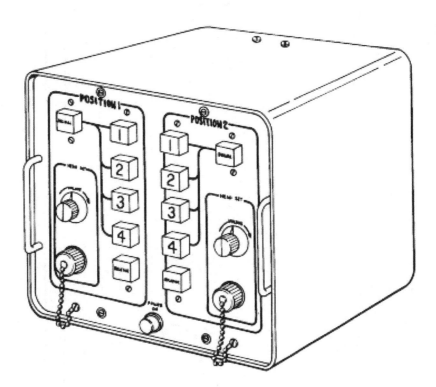

40. 10-93

Figure 1-10. Nonexplosionproof Communications Panel (Typical)

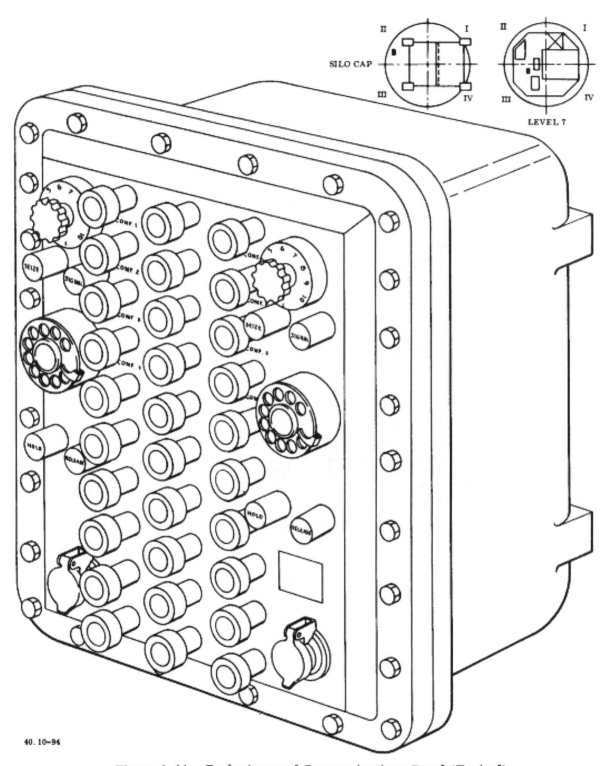

40.10—94

Figure 1-11. Explosionproof Communications Panel (Typical)

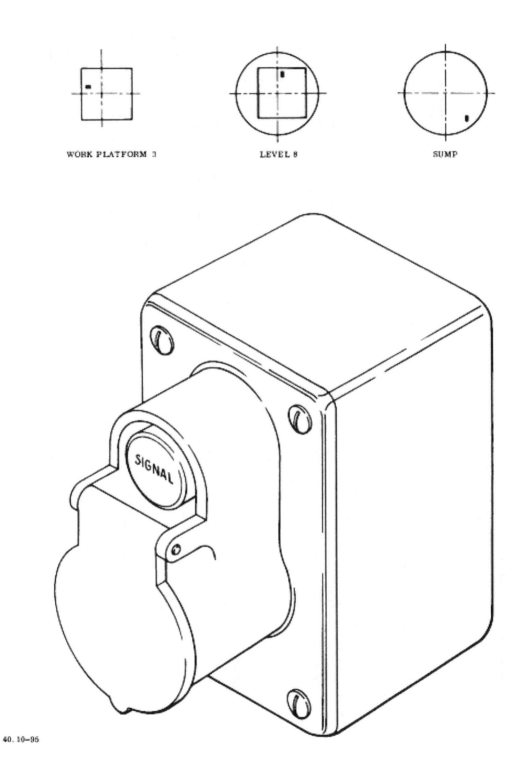

40.10—95

Figure 1-12.   Explosionproof Jacks (Typical)

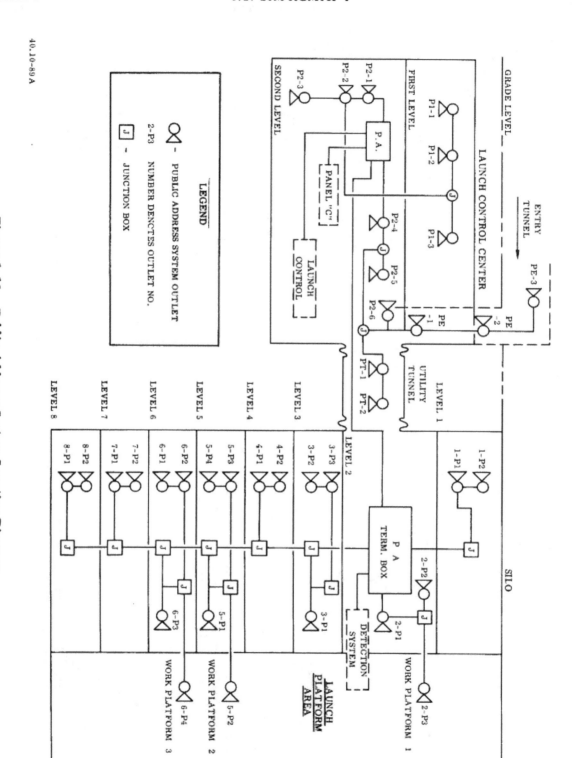

Figure 1-13. Public Address System Location Diagram

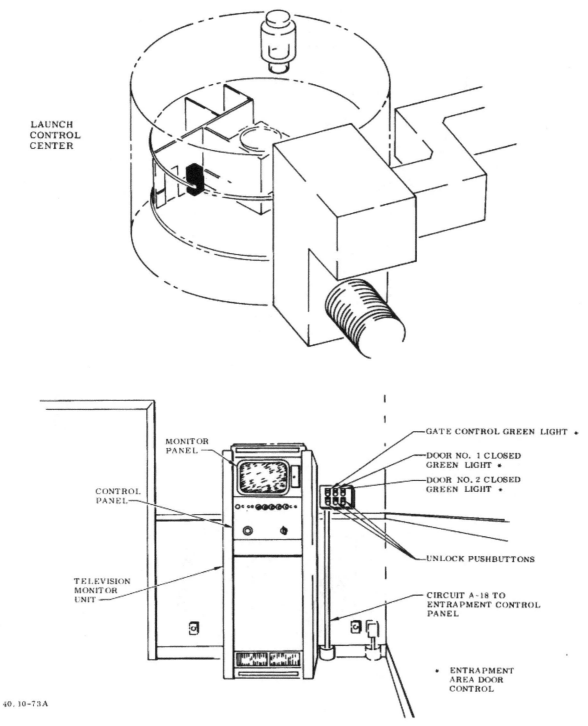

40.10-73A

Figure 1-14.   Television Monitor

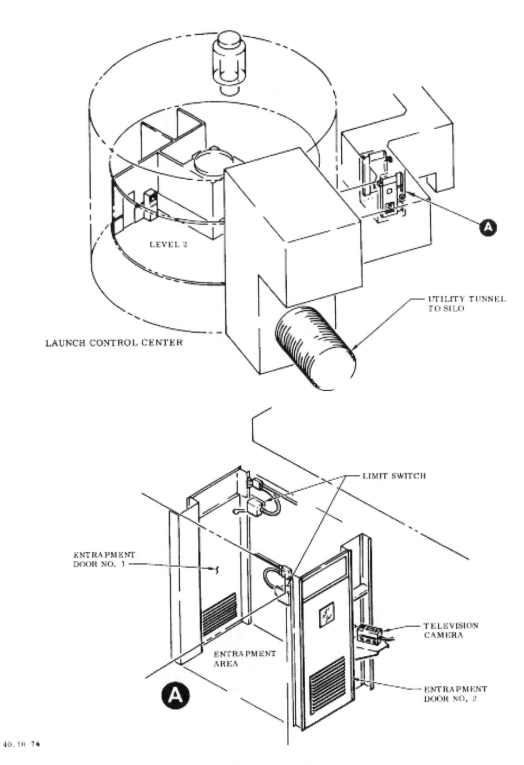

LEVEL 2

UTILITY TUNNEL
TO SILO

LAUNCH CONTROL CENTER

LIMIT SWITCH

ENTRAPMENT
DOOR NO. 1

TELEVISION
CAMERA

ENTRAPMENT
AREA

ENTRAPMENT
DOOR NO. 2

40. 10-74

Figure 1-15. Launch Complex Entrapment Area

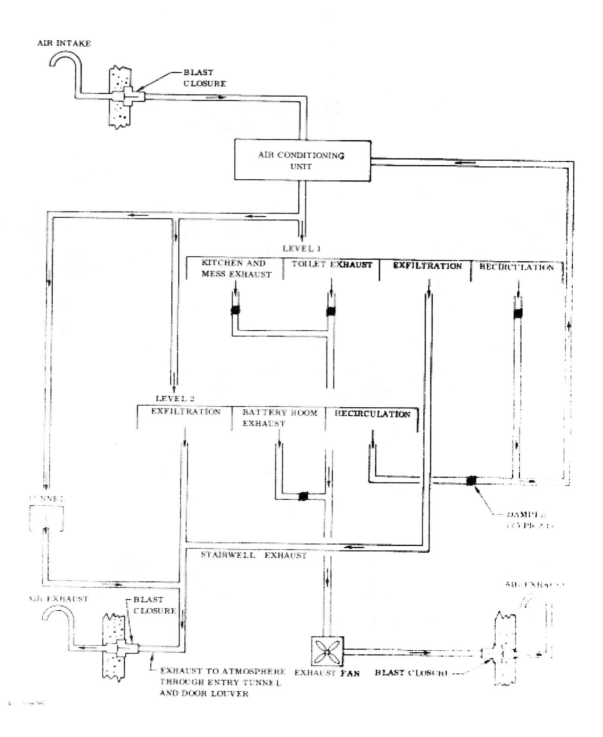

Figure 1-16. Launch Control Center Air Conditioning System Flow Diagram

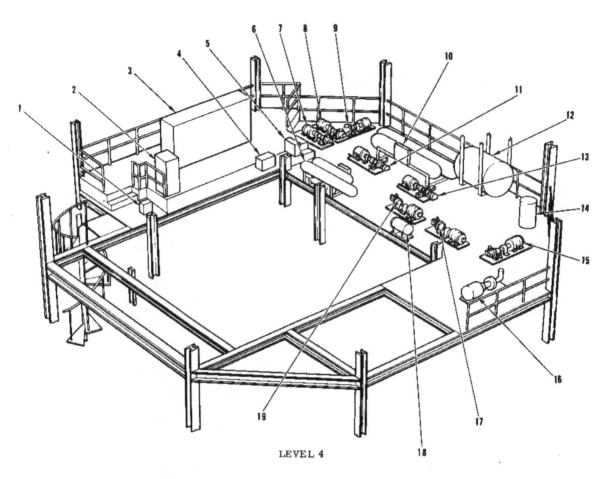

LEVEL 4

1 JUNCTION BOX ASSY IR56 SILO CHECKS (576 AND OSTF-2)
2 INSTRUMENTATION CABINET (OSTF-2)
3 INSTRUMENTATION CABINET (OSTF-2)
4 LIGHTING PANEL LB
5 LIGHTING PANEL LB
6 CONDENSER, WATER CHILLER, AND REFRIGERATIONN
   COMPRESSOR
7 CHILLED WATER PUMP P-51 (P-50 FOR OSTF-2)
8 CHILLED WATER PUMP P-50 (P-51 FOR OSTF-2)
9 EMERGENCY WATER PUMP P-32
10 CONDENSER, WATER CHILLER, AND REFRIGERATION
   COMPRESSOR

11 CONDENSER WATER PUMP P-31
12 HYDROPNEUMATIC UTILITY WATER TANK 80
13 CONDENSER WATER PUMP P-30
14 AIR TANK (OSTF-2)
15 WATER PUMP P-80
16 UTILITY WATER PUMP P-81
17 HOT WATER PUMP P-61
18 HOT WATER EXPANSION TANK 63
19 HOT WATER PUMP P-60

40.10-117

Figure 1-17. Silo Level 4 Equipment Location

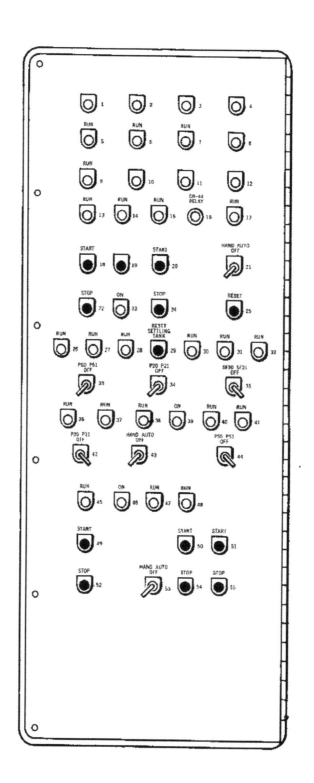

Fig

LEGEND	
DESCRIPTION	
 ⊙	RED INDICATOR
 ⦿	PUSHBUTTON
 ⦸	SELECTOR SWITCH
 ◎	GREEN INDICATOR

1 SPARE
2 SPARE
3 SPARE
4 SPARE OR SAND SETTLING TANK
5 LAUNCH PLATFORM EXHAUST FAN EF-40
6 LAUNCH PLATFORM PURGE EXHAUST FAN EF-41
7 LOWER SILO SUPPLY FAN SF-22
8 GASEOUS OXYGEN FAN ON OR SPARE
9 LAUNCH PLATFORM PURGE SUPPLY FAN SF-41
10 SPARE
11 SPARE
12 SPARE
13 THRUST SECTION PRESSURE FAN PF-70
14 OXYGEN PURGE
15 LAUNCH PLATFORM FAN COIL UNIT FC-40
16 CR-44 RELAY
17 MAIN EXHAUST FAN EF-30
18 THRUST SECTION PRESSURE FAN PF-70
19 OXYGEN PURGE RESET
20 LAUNCH PLATFORM FAN COIL FC-40
21 MAIN EXHAUST FAN EF-30
22 THRUST SECTION PRESSURE FAN PF-70
23 THRUST SECTION HEATING COIL EC-71

24 LAUNCH PLATFORM FAN COIL UNIT FC-40
25 CR-44 RELAY RESET OR OXYGEN PURGE RESET
26 HOT WATER PUMP P-60
27 STANDBY HOT WATER PUMP P-61
28 SPRAY PUMP P-20
29 RESET SETTLING TANK
30 STANDBY SPRAY PUMP P-21
31 SUPPLY FAN SF-20
32 SUPPLY FAN SF-21
33 HOT WATER PUMP P-61
   HOT WATER PUMP P-60
34 SPRAY PUMP P-21
   SPRAY PUMP P-20
35 SUPPLY FAN SF-21
   SUPPLY FAN SF-20
36 CONDENSER WATER PUMP P-30
37 CONDENSER WATER PUMP P-31
38 COOLING TOWER EXHAUST FAN EF-31
39 COOLING TOWER ELECTRIC HEATER EH-30
40 CHILLED WATER PUMP P-50
41 CHILLED WATER PUMP P-51
42 CONDENSER WATER PUMP P-30
   CONDENSER WATER PUMP P-31
43 COOLING TOWER EXHAUST FAN EF-31
44 CHILLED WATER PUMP P-50
   CHILLED WATER PUMP P-51
45 CONTROL CABINET FAN COIL UNIT FC-10
46 CONTROL CABINET ELECTRICAL HEATER COIL (EC-10-1)
   CONTROL CABINET ELECTRICAL HEATER COIL (EC-10-2)
   CONTROL CABINET ELECTRICAL HEATER COIL (EC-10-3)
47 DEMINERALIZED WATER PUMP P-90
48 EMERGENCY WATER PUMP P-32
49 CONTROL CABINET FAN COIL UNIT FC-10
50 EMERGENCY WATER PUMP P-32
51 CHILLED WATER PUMP P-50
52 CONTROL CABINET FAN COIL UNIT FC-10
53 DEMINERALIZED WATER PUMP P-90
54 EMERGENCY WATER PUMP P-32
55 CHILLED WATER PUMP P-50

Figure 1-18. Facilities Terminal Cabinet No. 2 (Typical)

LEVEL 2

WCU-51 OIL FAIL WCU-50 OIL FAIL LOWER SILO SUPPLY FAN SF-22

OFF AUTO

HAND

MAIN EXHAUST FAN EF-30

WCU-51 FREEZE UP WCU-50 FREEZE UP L/P EXHAUST FAN EF-40

START

STOP

LAUNCH PLATFORM FAN COIL UNIT

WCU-51 HI-10 WCU-50 HI-10 L/P PURGE EXHAUST FAN EF-43

OXYGEN PUMPS

THRUST SECTION HEATING COIL HC-71 L/P PURGE SUPPLY FAN SF-41

START STOP

THRUST SECT PRESS FAN

40.10-123(600)B

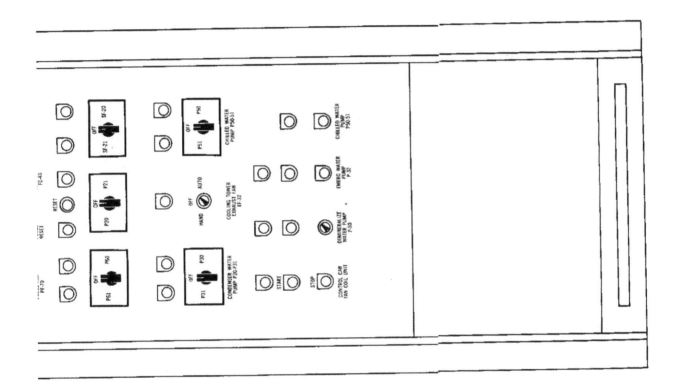

Figure 1-19. Facilities Terminal Cabinet NO. 2 (OSTF-2)

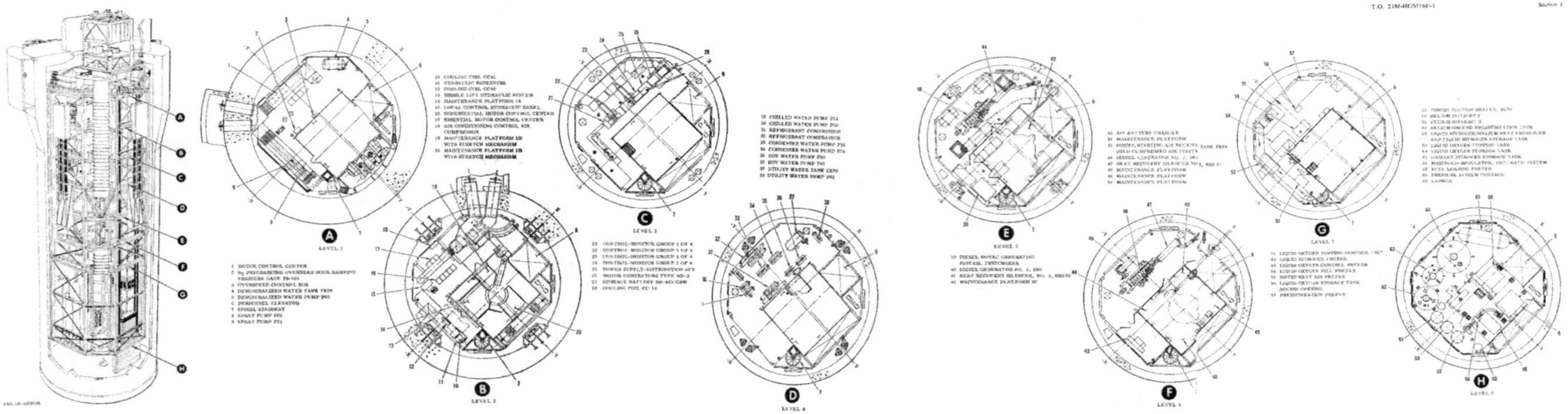

Figure 1-29. Launch Site Equipment Location

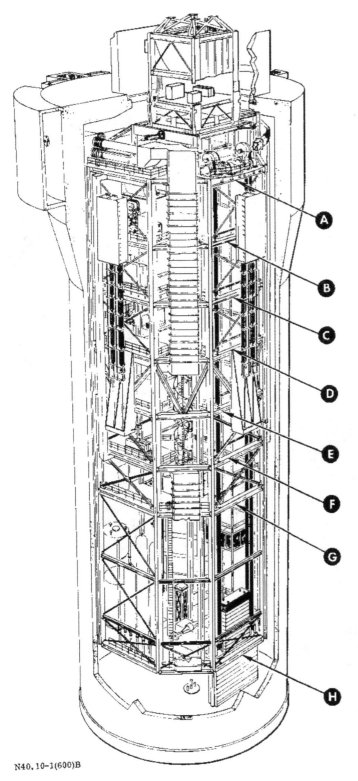

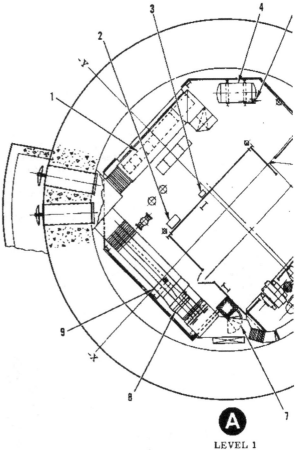

A

LEVEL 1

1 MOTOR CONTROL CENTER
2 N_2 PRECHARGING OVERHEAD DOOR DAMPING
   PRESSURE GAGE PI-993
3 OVERSPEED CONTROL BOX
4 DEMINERALIZED WATER TANK TK90
5 DEMINERALIZED WATER PUMP P90
6 PERSONNEL ELEVATOR
7 SPIRAL STAIRWAY
8 SPRAY PUMP P20
9 SPRAY PUMP P21

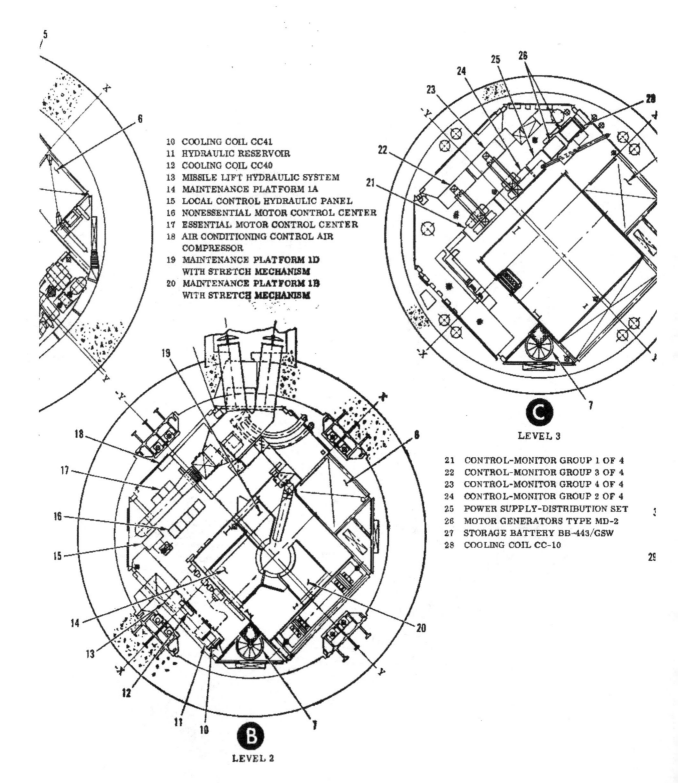

10 COOLING COIL CC41
11 HYDRAULIC RESERVOIR
12 COOLING COIL CC40
13 MISSILE LIFT HYDRAULIC SYSTEM
14 MAINTENANCE PLATFORM 1A
15 LOCAL CONTROL HYDRAULIC PANEL
16 NONESSENTIAL MOTOR CONTROL CENTER
17 ESSENTIAL MOTOR CONTROL CENTER
18 AIR CONDITIONING CONTROL AIR
   COMPRESSOR
19 MAINTENANCE PLATFORM 1D
   WITH STRETCH MECHANISM
20 MAINTENANCE PLATFORM 1B
   WITH STRETCH MECHANISM

C

LEVEL 3

B

LEVEL 2

21 CONTROL-MONITOR GROUP 1 OF 4
22 CONTROL-MONITOR GROUP 3 OF 4
23 CONTROL-MONITOR GROUP 4 OF 4
24 CONTROL-MONITOR GROUP 2 OF 4
25 POWER SUPPLY-DISTRIBUTION SET
26 MOTOR GENERATORS TYPE MD-2
27 STORAGE BATTERY BB-443/GSW
28 COOLING COIL CC-10

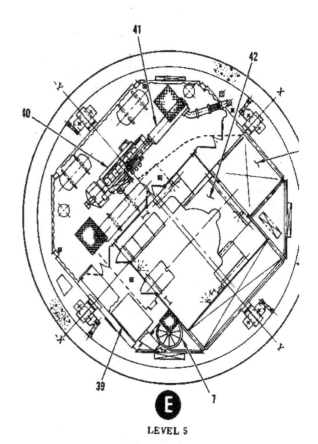

29 CHILLED WATER PUMP P51
30 CHILLED WATER PUMP P50
31 REFRIGERANT COMPRESSOR
32 REFRIGERANT COMPRESSOR
33 CONDENSER WATER PUMP P31
34 CONDENSER WATER PUMP P30
35 HOT WATER PUMP P60
36 HOT WATER PUMP P61
37 UTILITY WATER TANK TK80
38 UTILITY WATER PUMP P81

E

LEVEL 5

39 DIESEL 480VAC GENERATING
   SYSTEM, SWITCHGEAR
40 DIESEL GENERATOR NO. 1, D60
41 HEAT RECOVERY SILENCER, NO. 1, HRS 60
42 MAINTENANCE PLATFORM 3C

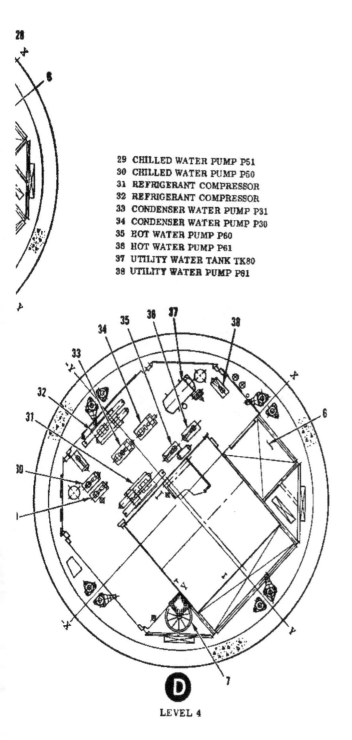

D

LEVEL 4

4

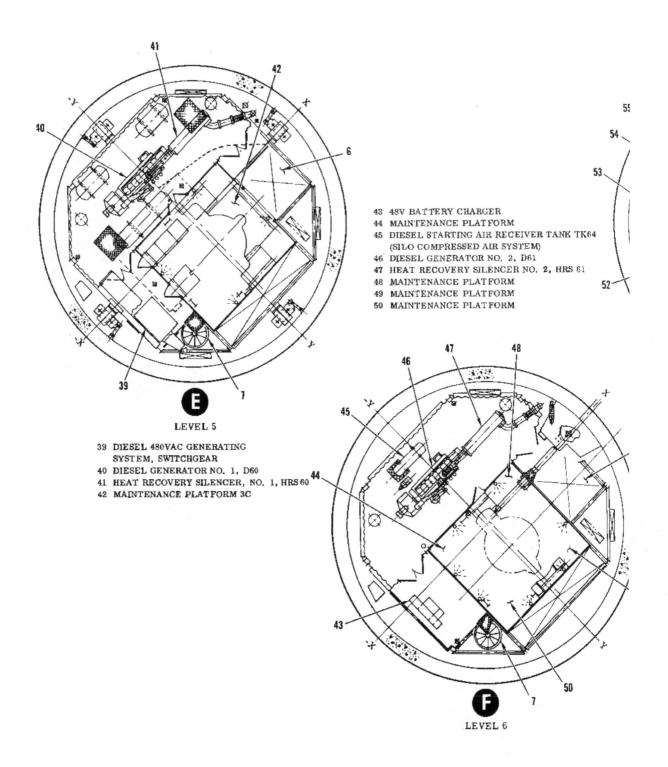

43 48V BATTERY CHARGER
44 MAINTENANCE PLATFORM
45 DIESEL STARTING AIR RECEIVER TANK TK64
    (SILO COMPRESSED AIR SYSTEM)
46 DIESEL GENERATOR NO. 2, D61
47 HEAT RECOVERY SILENCER NO. 2, HRS 61
48 MAINTENANCE PLATFORM
49 MAINTENANCE PLATFORM
50 MAINTENANCE PLATFORM

E

LEVEL 5

39 DIESEL 480VAC GENERATING
    SYSTEM, SWITCHGEAR
40 DIESEL GENERATOR NO. 1, D60
41 HEAT RECOVERY SILENCER, NO. 1, HRS 60
42 MAINTENANCE PLATFORM 3C

F

LEVEL 6

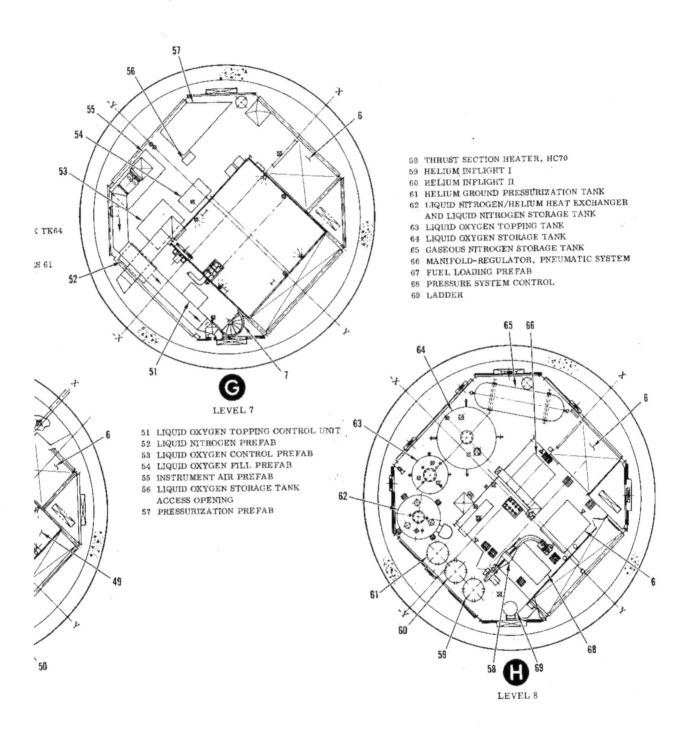

58 THRUST SECTION HEATER, HC70
59 HELIUM INFLIGHT I
60 HELIUM INFLIGHT II
61 HELIUM GROUND PRESSURIZATION TANK
62 LIQUID NITROGEN/HELIUM HEAT EXCHANGER
    AND LIQUID NITROGEN STORAGE TANK
63 LIQUID OXYGEN TOPPING TANK
64 LIQUID OXYGEN STORAGE TANK
65 GASEOUS NITROGEN STORAGE TANK
66 MANIFOLD-REGULATOR, PNEUMATIC SYSTEM
67 FUEL LOADING PREFAB
68 PRESSURE SYSTEM CONTROL
69 LADDER

G

LEVEL 7

51 LIQUID OXYGEN TOPPING CONTROL UNIT
52 LIQUID NITROGEN PREFAB
53 LIQUID OXYGEN CONTROL PREFAB
54 LIQUID OXYGEN FILL PREFAB
55 INSTRUMENT AIR PREFAB
56 LIQUID OXYGEN STORAGE TANK
    ACCESS OPENING
57 PRESSURIZATION PREFAB

H

LEVEL 8

Figure 1-20. Launch Silo Equipment Location

1

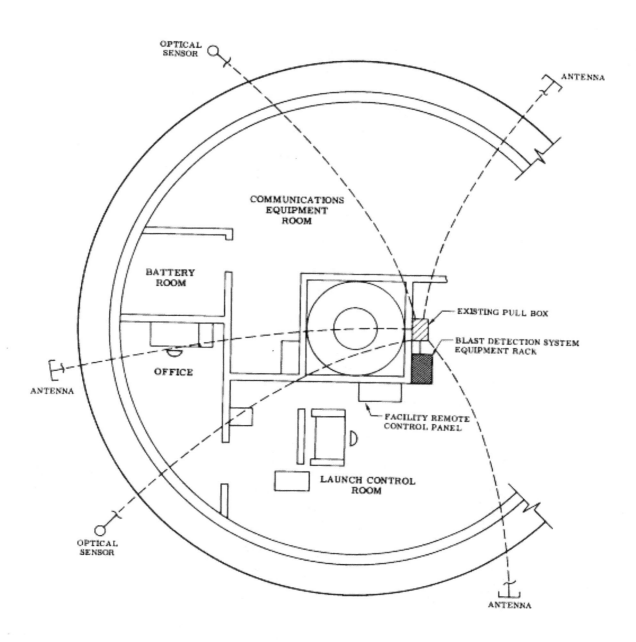

40.34-2

Figure 1-21. Blast Detection System

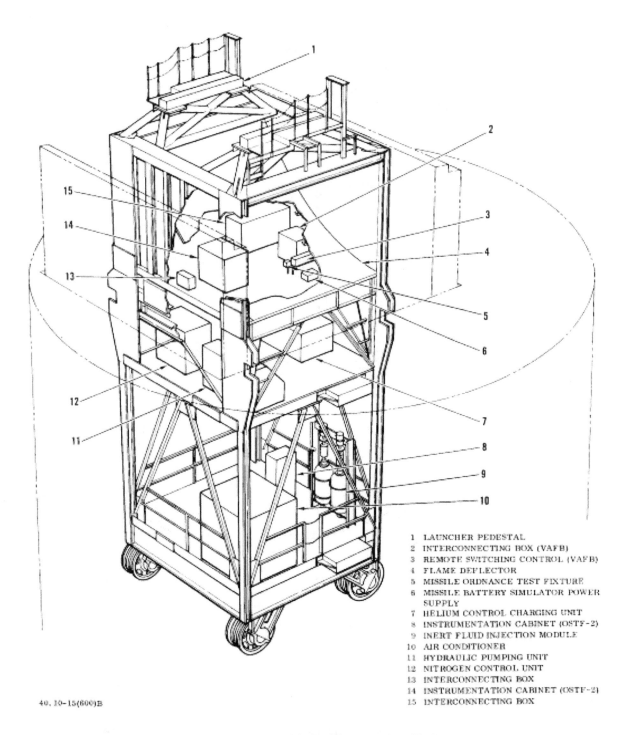

1   LAUNCHER PEDESTAL
2   INTERCONNECTING BOX (VAFB)
3   REMOTE SWITCHING CONTROL (VAFB)
4   FLAME DEFLECTOR
5   MISSILE ORDNANCE TEST FIXTURE
6   MISSILE BATTERY SIMULATOR POWER
    SUPPLY
7   HELIUM CONTROL CHARGING UNIT
8   INSTRUMENTATION CABINET (OSTF-2)
9   INERT FLUID INJECTION MODULE
10 AIR CONDITIONER
11 HYDRAULIC PUMPING UNIT
12 NITROGEN CONTROL UNIT
13 INTERCONNECTING BOX
14 INSTRUMENTATION CABINET (OSTF-2)
15 INTERCONNECTING BOX

40.10-15(600)B

Figure 1-22.   Guided Missile Silo Launcher Platform

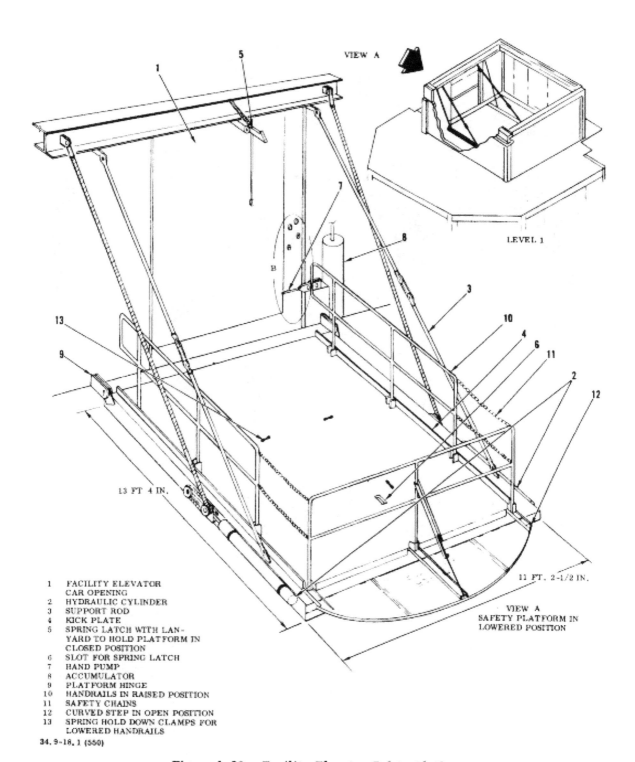

1   FACILITY ELEVATOR
      CAR OPENING
2   HYDRAULIC CYLINDER
3   SUPPORT ROD
4   KICK PLATE
5   SPRING LATCH WITH LAN-
      YARD TO HOLD PLATFORM IN
      CLOSED POSITION
6   SLOT FOR SPRING LATCH
7   HAND PUMP
8   ACCUMULATOR
9   PLATFORM HINGE
10   HANDRAILS IN RAISED POSITION
11   SAFETY CHAINS
12   CURVED STEP IN OPEN POSITION
13   SPRING HOLD DOWN CLAMPS FOR
      LOWERED HANDRAILS

34.9-18.1 (550)

Figure 1-23.   Facility Elevator Safety Platform

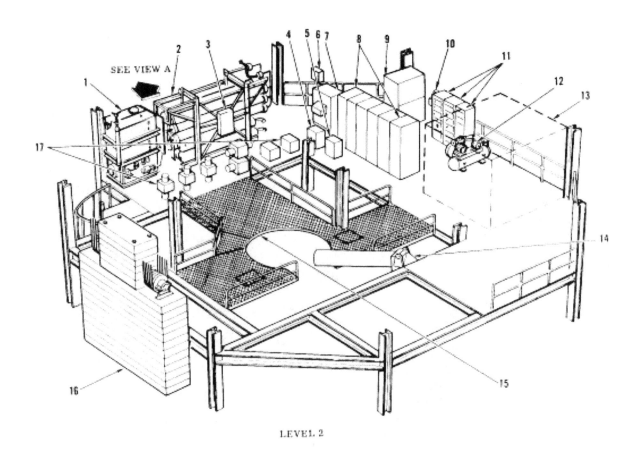

LEVEL 2

1 HYDRAULIC PUMP AND RESERVOIR
2 HYDRAULIC ACCUMULATOR AND
   GASEOUS NITROGEN PRESSURE TANKS
3 INTERCONNECTING JUNCTION BOX
4 LIGHTING PANEL LA
5 LIGHTING PANEL LD
6 30-KVA TRANSFORMER
7 LOCAL CONTROL HYDRAULIC PANEL
8 NONESSENTIAL MOTOR CONTROL CENTER
9 COMPRESSED AIR SYSTEM REGULATOR PANEL
10 120-VAC CONTROL VOLTAGE AUTOMATIC
   TRANSFER SWITCH
11 ESSENTIAL MOTOR CONTROL CENTER
12 FACILITY AIR COMPRESSOR
13 SILO EXHAUST FAN AND PLENUM
14 GASEOUS OXYGEN VENT
15 WORK PLATFORM 1
16 LAUNCHER PLATFORM COUNTERWEIGHT
17 MANIFOLD ASSEMBLY WORK PLATFORMS CRIB
   LOCKS SILO OVERHEAD DOORS
18 FILTERS

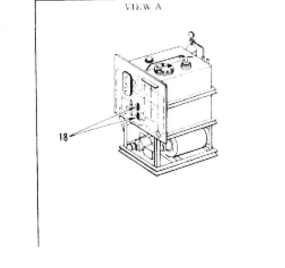

VIEW A

10, 10-116(600)JI

Figure 1-24.   Silo Level 2 Equipment Location

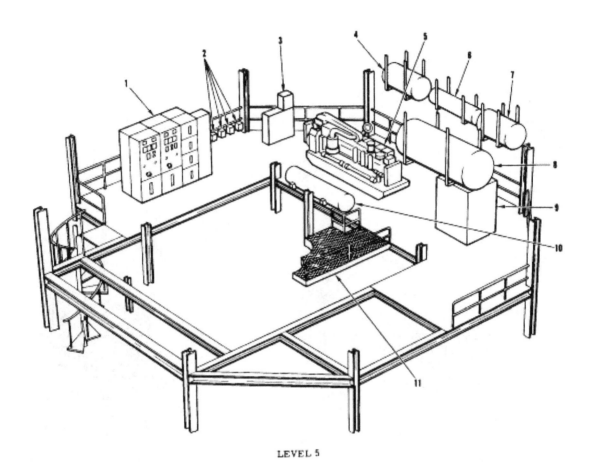

LEVEL 5

1 460-VOLT SWITCHGEAR	7 CLEAN LUBE OIL TANK
2 INSTRUMENTATION BOXES (OSTF-2)	8 HEAT RECOVERY SILENCER
3 SURGE PROTECTION PANEL (EXCEPT OSTF-2)	9 WATER HEATER (OSTF-2)
4 DIRTY LUBE OIL TANK	10 DIESEL DAY TANK
5 500 KW DIESEL GENERATOR	11 WORK PLATFORM 2
6 AIR RECEIVER (OSTF-2)	

40.10-118A

Figure 1-25. Silo Level 5 Equipment Location

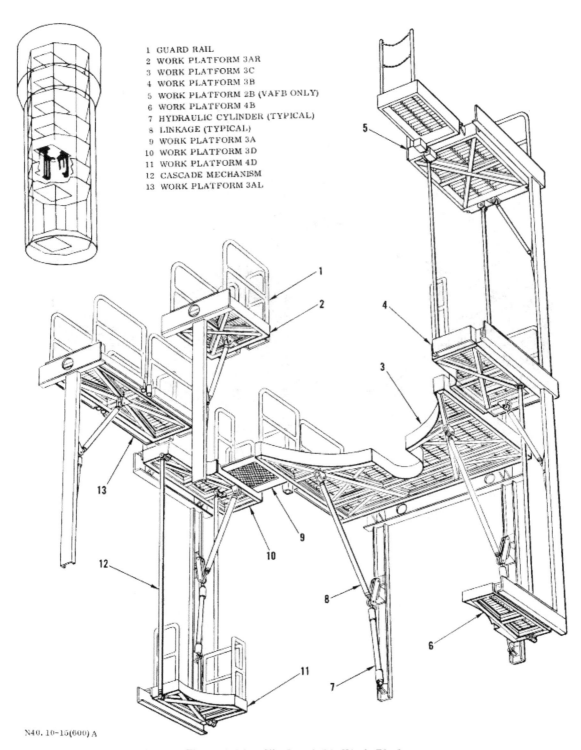

1 GUARD RAIL
2 WORK PLATFORM 3AR
3 WORK PLATFORM 3C
4 WORK PLATFORM 3B
5 WORK PLATFORM 2B (VAFB ONLY)
6 WORK PLATFORM 4B
7 HYDRAULIC CYLINDER (TYPICAL)
8 LINKAGE (TYPICAL)
9 WORK PLATFORM 3A
10 WORK PLATFORM 3D
11 WORK PLATFORM 4D
12 CASCADE MECHANISM
13 WORK PLATFORM 3AL

N40. 10-15(600) A

Figure 1-26.   Silo Level 5A Work Platform

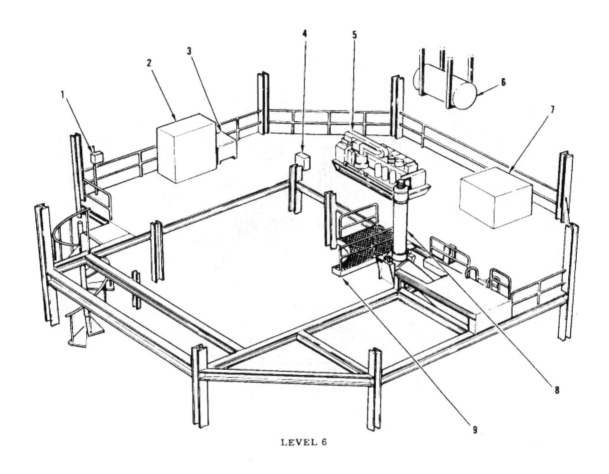

LEVEL 6

1 48-VOLT DC DISTRIBUTION PANEL	6 AIR RECEIVER
2 48-VOLT BATTERY RACK	7 WATER HEATER
3 48-VOLT BATTERY CHARGER	8 ALIGNMENT GROUP SIGHT TUBE
4 INTERCONNECTING JUNCTION BOX (VAFB)	9 WORK PLATFORM
5 500 KW DIESEL GENERATOR	

40.10-122

Figure 1-27. Silo Level 6 Equipment Location (Typical)

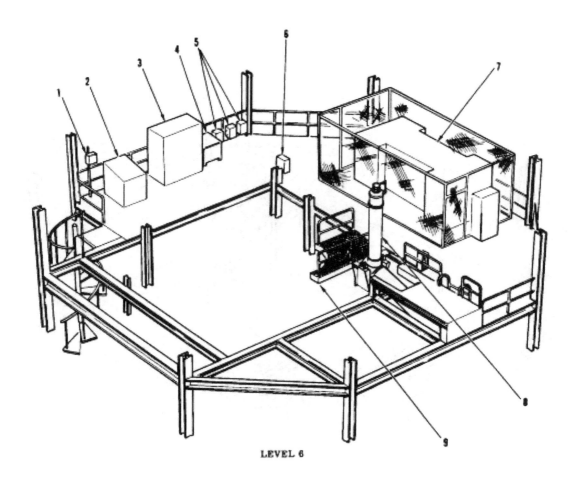

LEVEL 6

1	48-VOLT DC DISTRIBUTION PANEL	6	INTERCONNECTING JUNCTION BOX
2	ELECTRICAL EQUIPMENT CABINET	7	MAIN POWER TRANSFORMER
3	48-VOLT BATTERY RACK	8	ALIGNMENT GROUP SIGHT TUBE
4	48-VOLT BATTERY CHARGER	9	WORK PLATFORM
5	INSTRUMENTATION BOXES		

40, 10-124

Figure 1-28. Silo Level 6 Equipment Location (OSTF-2)

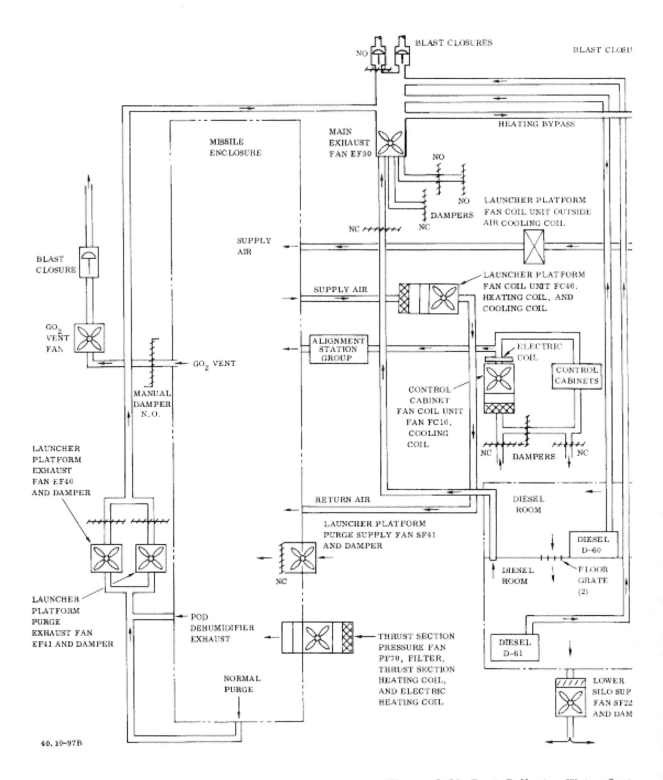

Figure 1-29. Dust Collector Water Syst

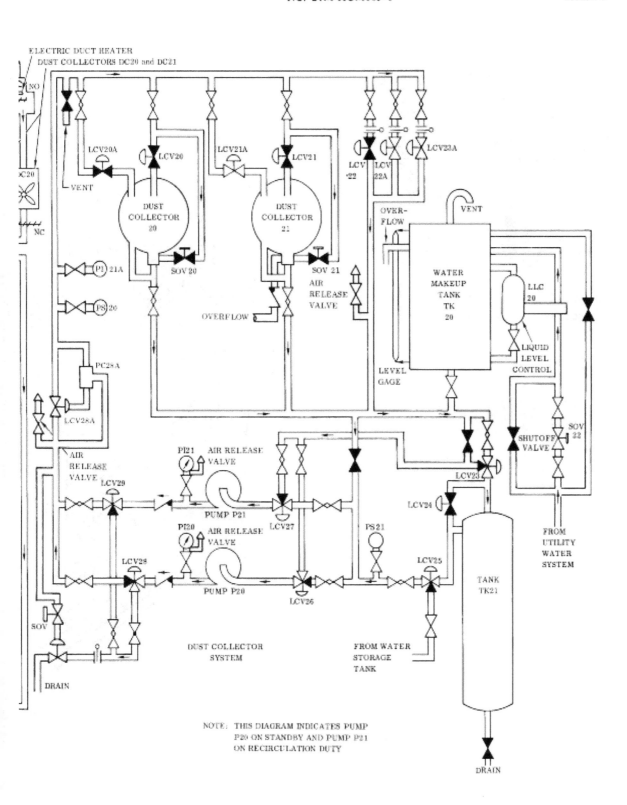

NOTE: THIS DIAGRAM INDICATES PUMP
      P20 ON STANDBY AND PUMP P21
      ON RECIRCULATION DUTY

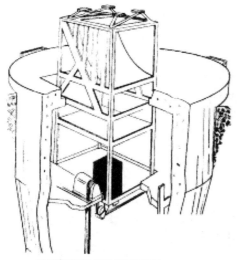

ROTATED 180 DEGREES

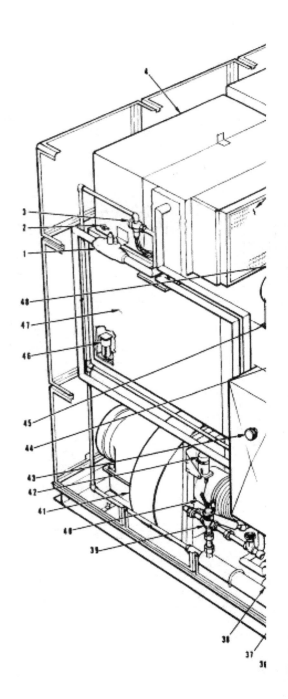

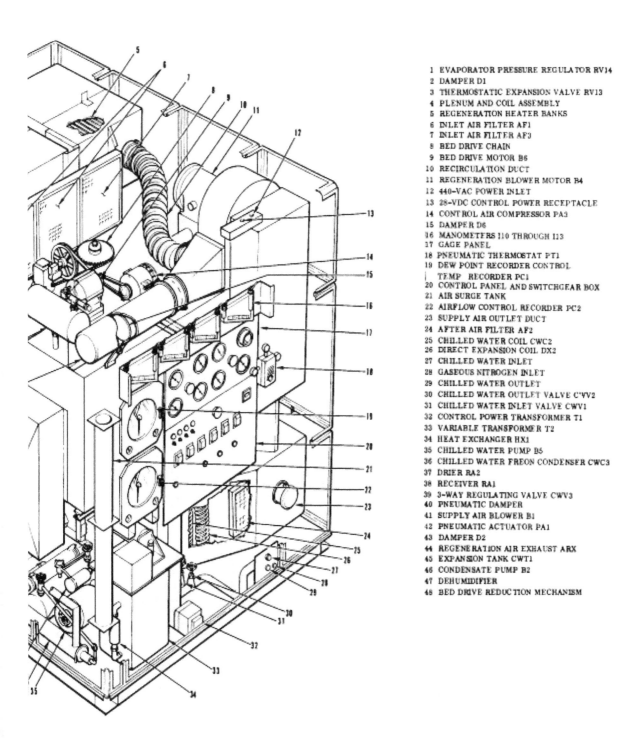

1 EVAPORATOR PRESSURE REGULATOR RV14
2 DAMPER D1
3 THERMOSTATIC EXPANSION VALVE RV13
4 PLENUM AND COIL ASSEMBLY
5 REGENERATION HEATER BANKS
6 INLET AIR FILTER AF1
7 INLET AIR FILTER AF3
8 BED DRIVE CHAIN
9 BED DRIVE MOTOR B6
10 RECIRCULATION DUCT
11 REGENERATION BLOWER MOTOR B4
12 440-VAC POWER INLET
13 28-VDC CONTROL POWER RECEPTACLE
14 CONTROL AIR COMPRESSOR PA3
15 DAMPER D6
16 MANOMETERS I10 THROUGH I13
17 GAGE PANEL
18 PNEUMATIC THERMOSTAT PT1
19 DEW POINT RECORDER CONTROL
   TEMP RECORDER PC1
20 CONTROL PANEL AND SWITCHGEAR BOX
21 AIR SURGE TANK
22 AIRFLOW CONTROL RECORDER PC2
23 SUPPLY AIR OUTLET DUCT
24 AFTER AIR FILTER AF2
25 CHILLED WATER COIL CWC2
26 DIRECT EXPANSION COIL DX2
27 CHILLED WATER INLET
28 GASEOUS NITROGEN INLET
29 CHILLED WATER OUTLET
30 CHILLED WATER OUTLET VALVE CWV2
31 CHILLED WATER INLET VALVE CWV1
32 CONTROL POWER TRANSFORMER T1
33 VARIABLE TRANSFORMER T2
34 HEAT EXCHANGER HX1
35 CHILLED WATER PUMP B5
36 CHILLED WATER FREON CONDENSER CWC3
37 DRIER RA2
38 RECEIVER RA1
39 3-WAY REGULATING VALVE CWV3
40 PNEUMATIC DAMPER
41 SUPPLY AIR BLOWER B1
42 PNEUMATIC ACTUATOR PA1
43 DAMPER D2
44 REGENERATION AIR EXHAUST ARX
45 EXPANSION TANK CWT1
46 CONDENSATE PUMP B2
47 DEHUMIDIFIER
48 BED DRIVE REDUCTION MECHANISM

Figure 1-30. Air Conditioner

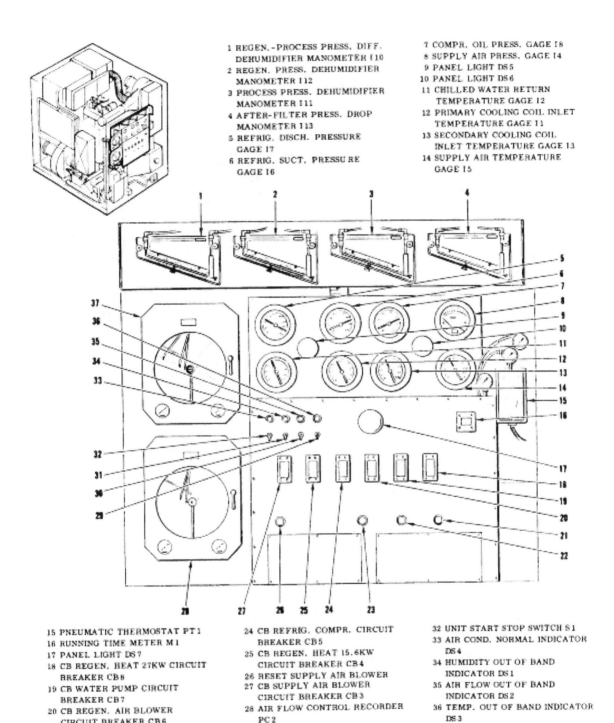

1 REGEN.-PROCESS PRESS. DIFF.
  DEHUMIDIFIER MANOMETER I 10
2 REGEN. PRESS. DEHUMIDIFIER
  MANOMETER I 12
3 PROCESS PRESS. DEHUMIDIFIER
  MANOMETER I 11
4 AFTER-FILTER PRESS. DROP
  MANOMETER I 13
5 REFRIG. DISCH. PRESSURE
  GAGE 17
6 REFRIG. SUCT. PRESSURE
  GAGE 16

7 COMPR. OIL PRESS. GAGE I 8
8 SUPPLY AIR PRESS. GAGE I 4
9 PANEL LIGHT DS 5
10 PANEL LIGHT DS 6
11 CHILLED WATER RETURN
   TEMPERATURE GAGE I 2
12 PRIMARY COOLING COIL INLET
   TEMPERATURE GAGE I 1
13 SECONDARY COOLING COIL
   INLET TEMPERATURE GAGE I 3
14 SUPPLY AIR TEMPERATURE
   GAGE I 5

15 PNEUMATIC THERMOSTAT PT 1
16 RUNNING TIME METER M 1
17 PANEL LIGHT DS 7
18 CB REGEN. HEAT 27KW CIRCUIT
   BREAKER CB 8
19 CB WATER PUMP CIRCUIT
   BREAKER CB 7
20 CB REGEN. AIR BLOWER
   CIRCUIT BREAKER CB 6
21 RESET WATER PUMP
22 RESET REGEN. AIR BLOWER
23 RESET REFRIG. COMPR.

24 CB REFRIG. COMPR. CIRCUIT
   BREAKER CB 5
25 CB REGEN. HEAT 15.6KW
   CIRCUIT BREAKER CB 4
26 RESET SUPPLY AIR BLOWER
27 CB SUPPLY AIR BLOWER
   CIRCUIT BREAKER CB 3
28 AIR FLOW CONTROL RECORDER
   PC 2
29 AUX. HEAT 7.8KW SWITCH S 4
30 REMOTE CONTROL SWITCH S 3
31 PANEL LIGHTS SWITCH S 2

32 UNIT START STOP SWITCH S 1
33 AIR COND. NORMAL INDICATOR
   DS 4
34 HUMIDITY OUT OF BAND
   INDICATOR DS 1
35 AIR FLOW OUT OF BAND
   INDICATOR DS 2
36 TEMP. OUT OF BAND INDICATOR
   DS 3
37 DEW POINT RECORDER CONTROL
   TEMP. RECORDER PC 1

40.30-106(OSTF-2)

Figure 1-31. Air Conditioner Panel Controls and Indicators

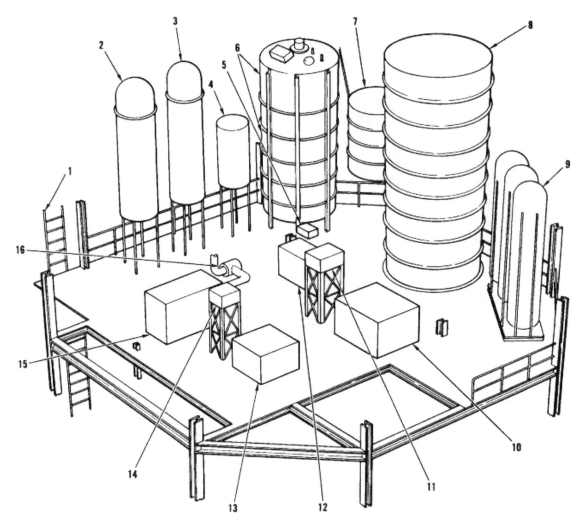

LEVEL 8

1 LADDER TO LEVEL 7	9 GASEOUS NITROGEN TANKS
2 INFLIGHT HELIUM SUPPLY TANK NO. 1	10 PNEUMATIC SYSTEM MANIFOLD REGULATOR
3 INFLIGHT HELIUM SUPPLY TANK NO. 2	11 COLD DISCONNECT PANEL
4 GROUND PRESSURIZATION SUPPLY TANK	12 LN_2 EVAPORATOR
5 VACUUM PUMP	13 FUEL PREFAB
6 LN_2 STORAGE TANK AND HEAT EXCHANGER	14 HOT DISCONNECT PANEL
7 LO_2 TOPPING TANK	15 PRESSURE SYSTEM CONTROL
8 LO_2 STORAGE TANK	16 THRUST SECTION HEATER

40.10-68(576D/E)B

Figure 1-32. Silo Level 8 Equipment Location

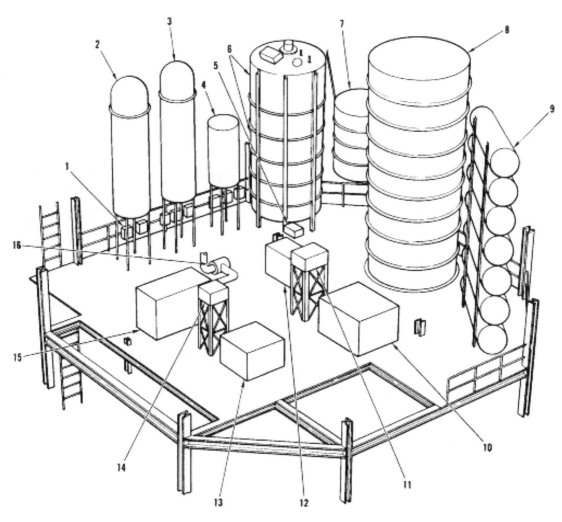

LEVEL 8

1 INSTRUMENTATION BOXES	9 GASEOUS NITROGEN TANKS
2 INFLIGHT HELIUM SUPPLY TANK NO. 1	10 PNEUMATIC SYSTEM MANIFOLD REGULATOR
3 INFLIGHT HELIUM SUPPLY TANK NO. 2	11 COLD DISCONNECT PANEL
4 GROUND PRESSURIZATION SUPPLY TANK	12 LN_2 EVAPORATOR
5 VACUUM PUMP	13 FUEL PREFAB
6 LN_2 STORAGE TANK AND HEAT EXCHANGER	14 HOT DISCONNECT PANEL
7 LO_2 TOPPING TANK	15 PRESSURE SYSTEM CONTROL
8 LO_2 STORAGE TANK	16 THRUST SECTION HEATER

40.10-68A

Figure 1-33. Silo Level 8 Equipment Location (OSTF-2)

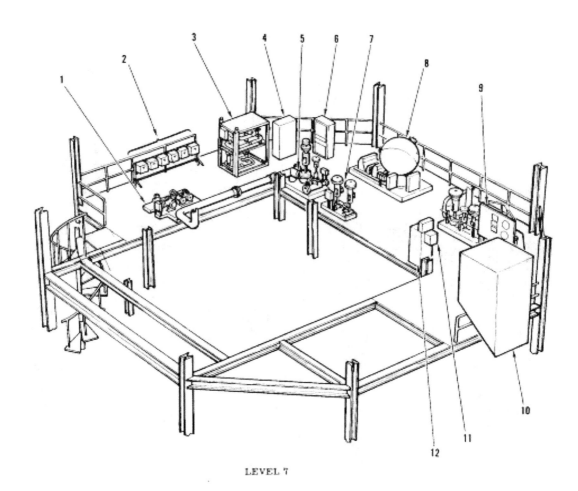

LEVEL 7

1 LO₂ TOPPING CONTROL UNIT	8 INSTRUMENT AIR PREFAB
2 INSTRUMENTATION BOXES (OSTF-2)	9 PRESSURIZATION PREFAB
3 LN₂ PREFAB	10 ALIGNMENT GROUP ENCLOSURE
4 GASEOUS OXYGEN DETECTOR	ALIGNMENT GROUP
5 LO₂ CONTROL PREFAB	BENCH MARKS
6 GASEOUS OXYGEN DETECTOR	11 FIREX CONTROL PANEL (VAFB)
7 LO₂ FILL PREFAB	12 DIESEL FUEL VAPOR DETECTOR

40.10-67(576 D/E)B

Figure 1-34. Silo Level 7 Equipment Location

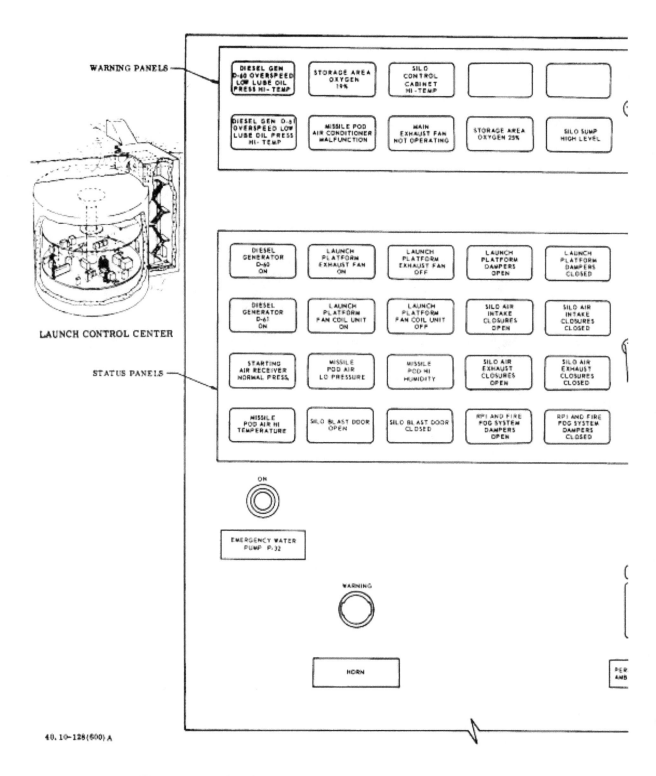

WARNING PANELS

DIESEL GEN D-60 OVERSPEED LOW LUBE OIL PRESS HI-TEMP	STORAGE AREA OXYGEN 19%	SILO CONTROL CABINET HI-TEMP		
DIESEL GEN D-61 OVERSPEED LOW LUBE OIL PRESS HI-TEMP	MISSILE POD AIR CONDITIONER MALFUNCTION	MAIN EXHAUST FAN NOT OPERATING	STORAGE AREA OXYGEN 25%	SILO SUMP HIGH LEVEL

DIESEL GENERATOR D-60 ON	LAUNCH PLATFORM EXHAUST FAN ON	LAUNCH PLATFORM EXHAUST FAN OFF	LAUNCH PLATFORM DAMPERS OPEN	LAUNCH PLATFORM DAMPERS CLOSED
DIESEL GENERATOR D-61 ON	LAUNCH PLATFORM FAN COIL UNIT ON	LAUNCH PLATFORM FAN COIL UNIT OFF	SILO AIR INTAKE CLOSURES OPEN	SILO AIR INTAKE CLOSURES CLOSED
STARTING AIR RECEIVER NORMAL PRESS.	MISSILE POD AIR LO PRESSURE	MISSILE POD HI HUMIDITY	SILO AIR EXHAUST CLOSURES OPEN	SILO AIR EXHAUST CLOSURES CLOSED
MISSILE POD AIR HI TEMPERATURE	SILO BLAST DOOR OPEN	SILO BLAST DOOR CLOSED	RP1 AND FIRE FOG SYSTEM DAMPERS OPEN	RP1 AND FIRE FOG SYSTEM DAMPERS CLOSED

LAUNCH CONTROL CENTER

STATUS PANELS

ON

EMERGENCY WATER PUMP P-32

WARNING

HORN

40.10-128(600)A

Figure 1-35. Facility Remote Control Panel (Typical)

Changed 15 April 1964

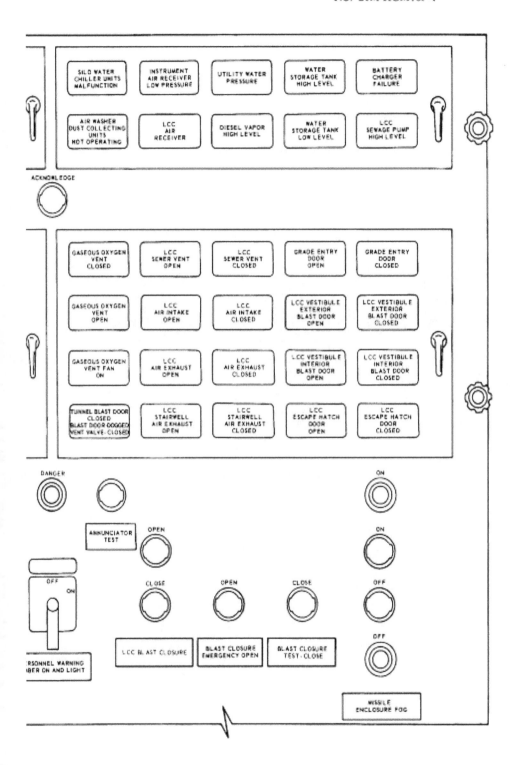

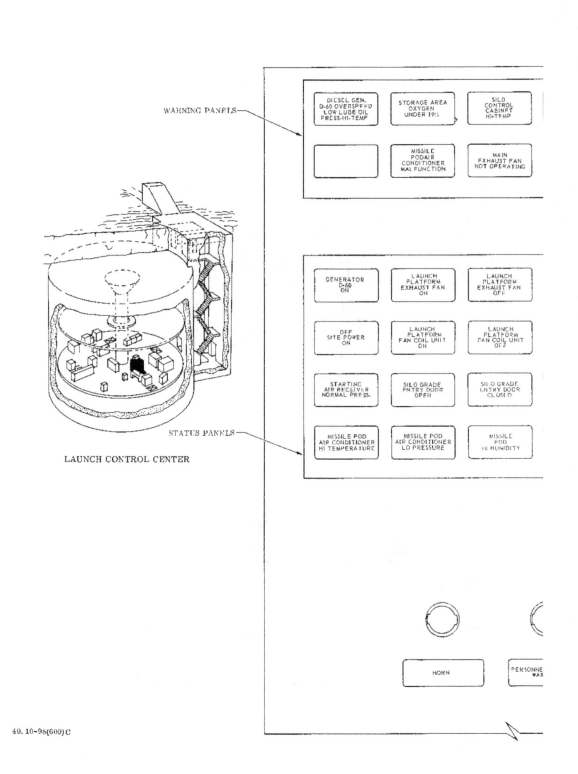

WARNING PANELS

DIESEL GEN. D-60 OVERSPEED LOW LUBE OIL PRESS-HI-TEMP	STORAGE AREA OXYGEN UNDER 19%	SILO CONTROL CABINET HI-TEMP
	MISSILE POD AIR CONDITIONER MALFUNCTION	MAIN EXHAUST FAN NOT OPERATING

GENERATOR D-60 ON	LAUNCH PLATFORM EXHAUST FAN ON	LAUNCH PLATFORM EXHAUST FAN OFF
OFF SITE POWER ON	LAUNCH PLATFORM FAN COIL UNIT ON	LAUNCH PLATFORM FAN COIL UNIT OFF
STARTING AIR RECEIVER NORMAL PRESS.	SILO GRADE ENTRY DOOR OPEN	SILO GRADE ENTRY DOOR CLOSED
MISSILE POD AIR CONDITIONER HI TEMPERATURE	MISSILE POD AIR CONDITIONER LO PRESSURE	MISSILE POD HI HUMIDITY

STATUS PANELS

LAUNCH CONTROL CENTER

HORN

PERSONNEL WAR

40. 10-98(600) C

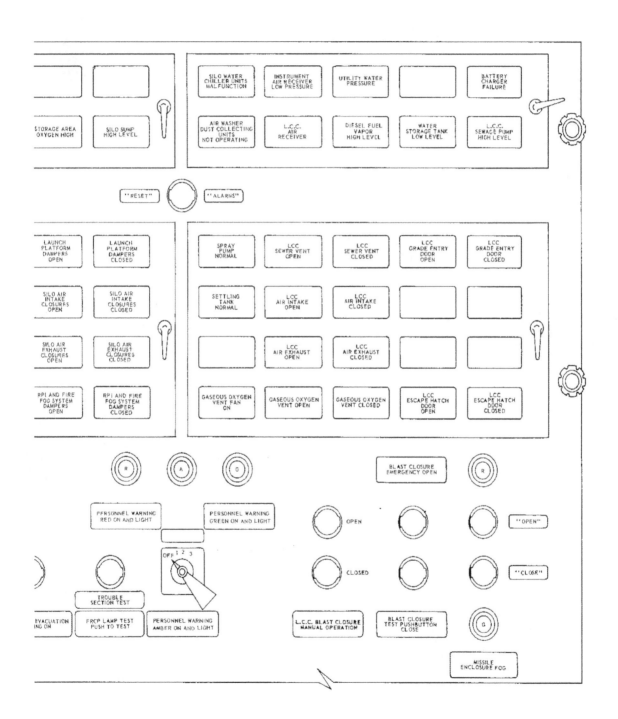

Figure 1-36. Facility Remote Control Panel (OSTF-2)

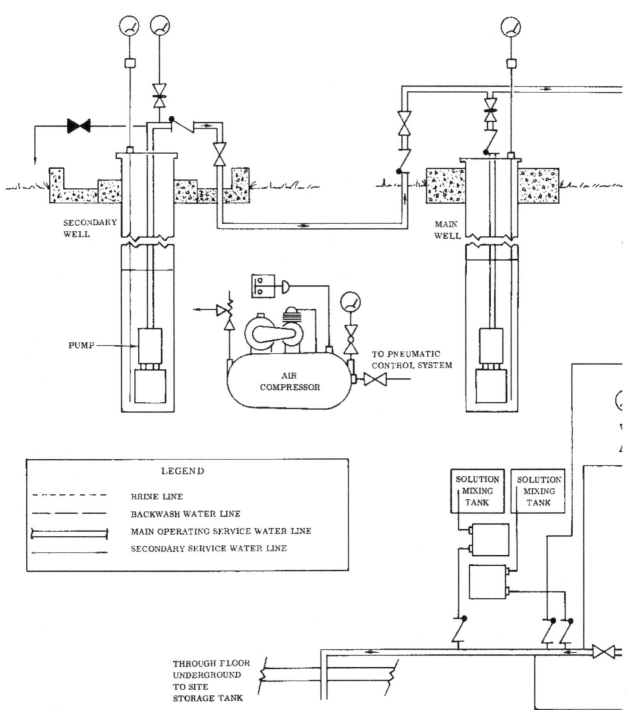

SECONDARY
WELL

MAIN
WELL

PUMP

AIR
COMPRESSOR

TO PNEUMATIC
CONTROL SYSTEM

LEGEND

- - - - - - - BRINE LINE

- - - - BACKWASH WATER LINE

MAIN OPERATING SERVICE WATER LINE

SECONDARY SERVICE WATER LINE

SOLUTION
MIXING
TANK

SOLUTION
MIXING
TANK

THROUGH FLOOR
UNDERGROUND
TO SITE
STORAGE TANK

40.10-99A

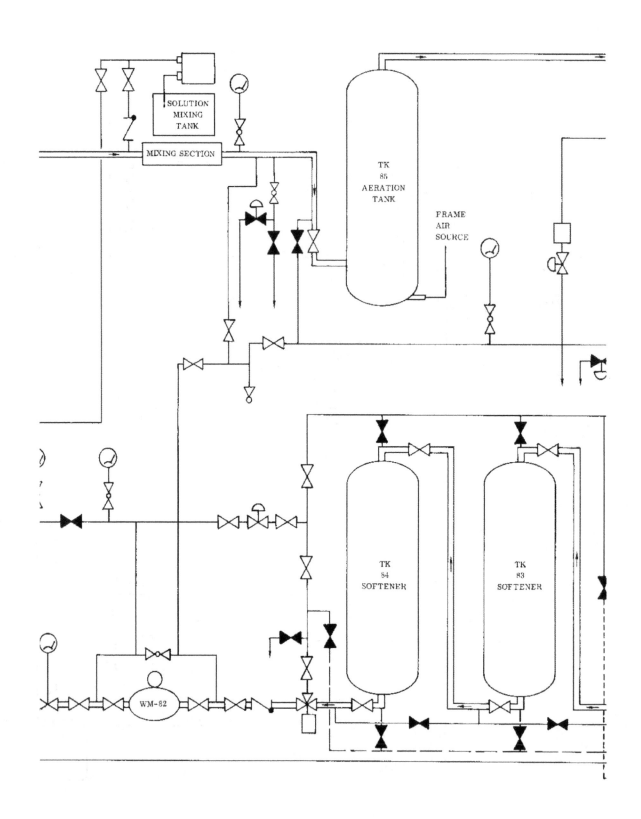

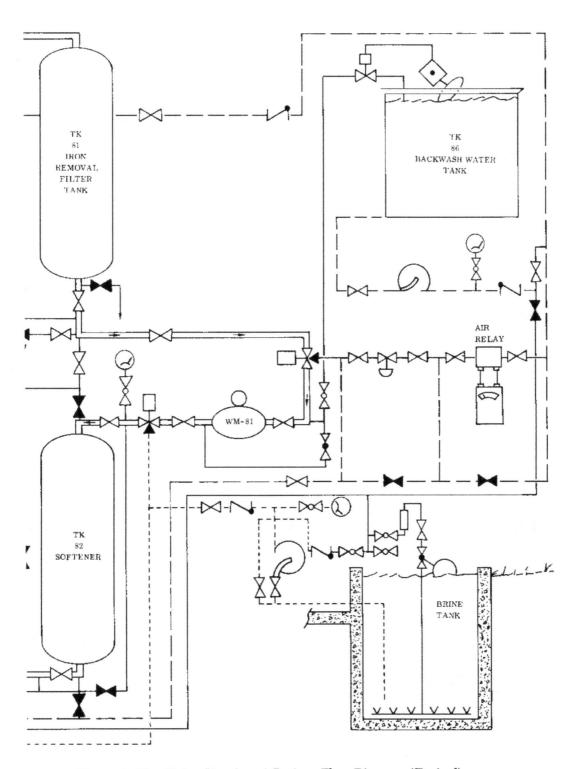

Figure 1-37.   Water Treatment System Flow Diagram (Typical)

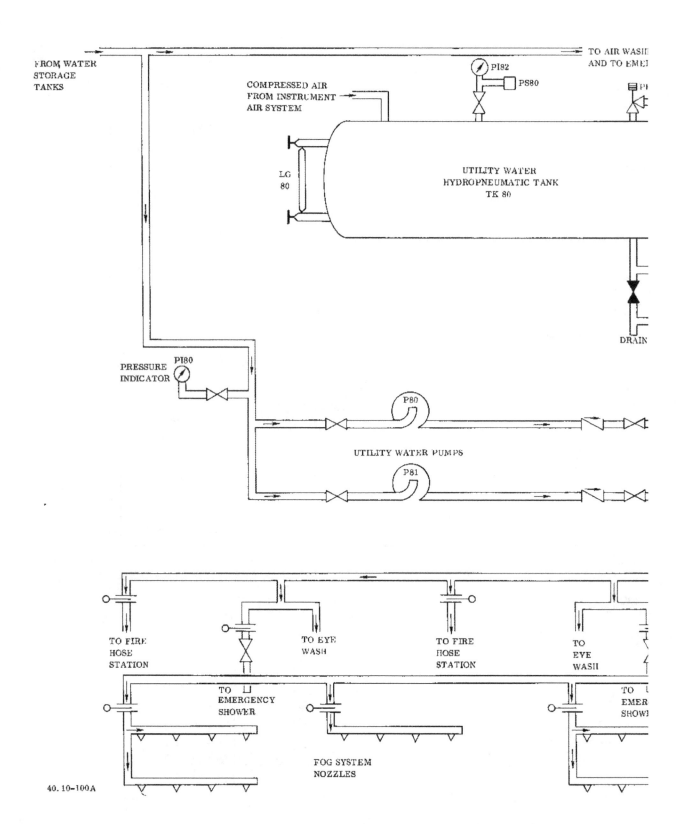

Figure 1-38. Utility Water

R, DUST COLLECTOR UNITS
GENCY WATER PUMP P-32

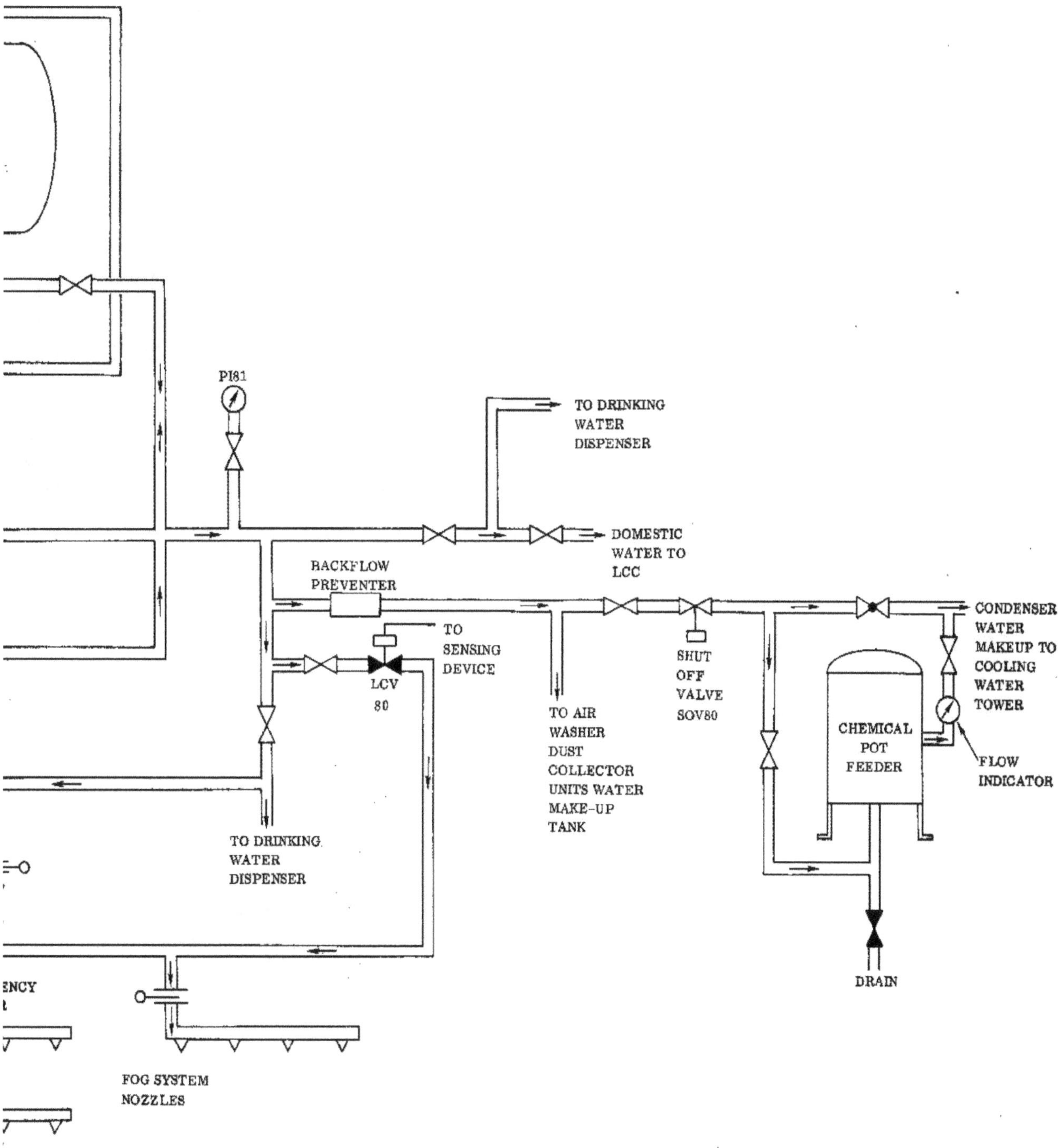

PI81
TO DRINKING WATER DISPENSER
DOMESTIC WATER TO LCC
BACKFLOW PREVENTER
TO SENSING DEVICE
LCV 80
TO AIR WASHER DUST COLLECTOR UNITS WATER MAKE-UP TANK
SHUT OFF VALVE SOV80
CONDENSER WATER MAKEUP TO COOLING WATER TOWER
CHEMICAL POT FEEDER
FLOW INDICATOR
TO DRINKING WATER DISPENSER
DRAIN
ENCY
FOG SYSTEM NOZZLES

ystem Flow Diagram (Typical)

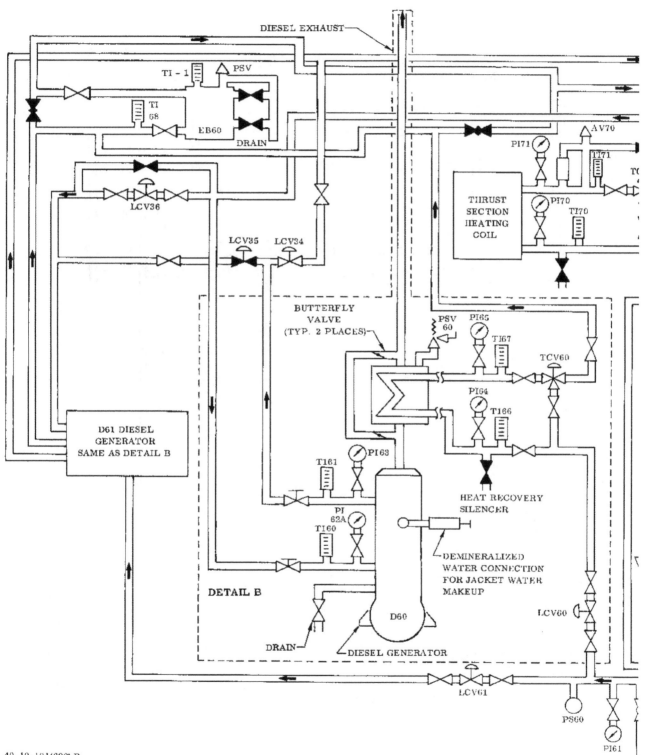

DIESEL EXHAUST

TI – 1
PSV
TI 68
EB60
DRAIN
LCV36
LCV35 LCV34

THRUST
SECTION
HEATING
COIL

PI71
AV70
TI71
PI70
TI70

BUTTERFLY
VALVE
(TYP. 2 PLACES)

PSV 60
PI65
TI67
TCV60
PI64
TI66

D61 DIESEL
GENERATOR
SAME AS DETAIL B

T161
PI63

HEAT RECOVERY
SILENCER

PI 62A
TI60

DEMINERALIZED
WATER CONNECTION
FOR JACKET WATER
MAKEUP

DETAIL B

LCV60

DRAIN
D60

DIESEL GENERATOR

LCV61

PS60

PI61

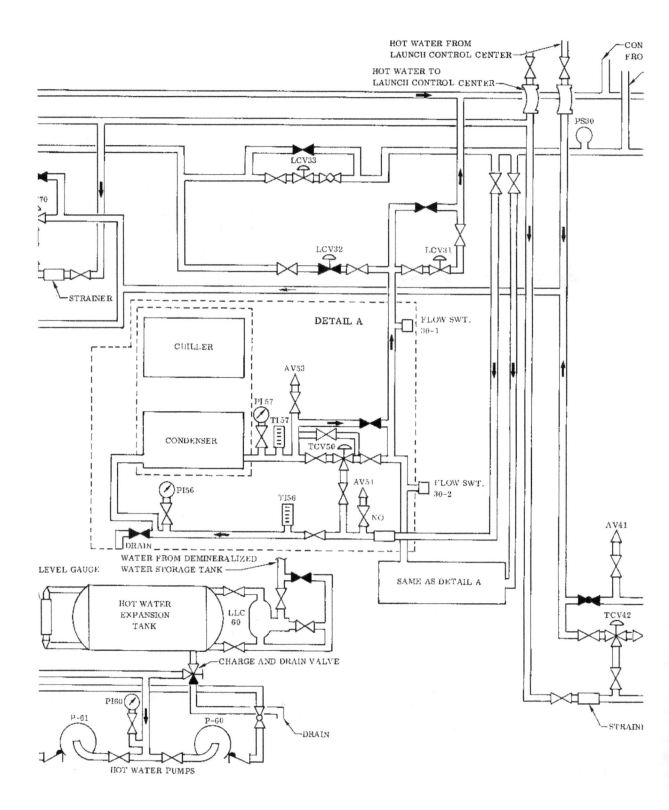

Figure 1-39. Condenser and Ho

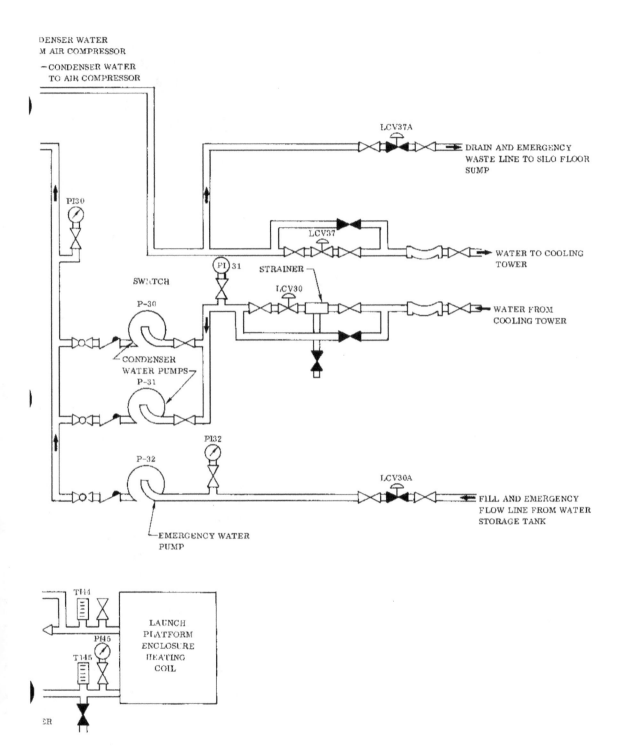

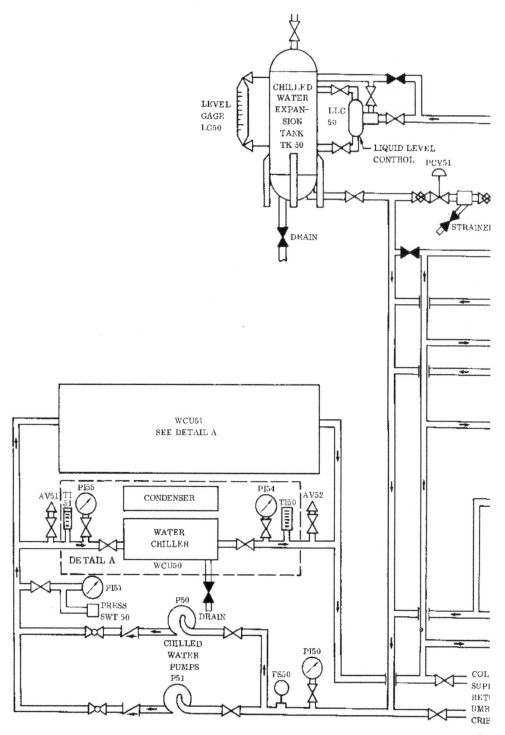

40.10-103(600)B

Figur

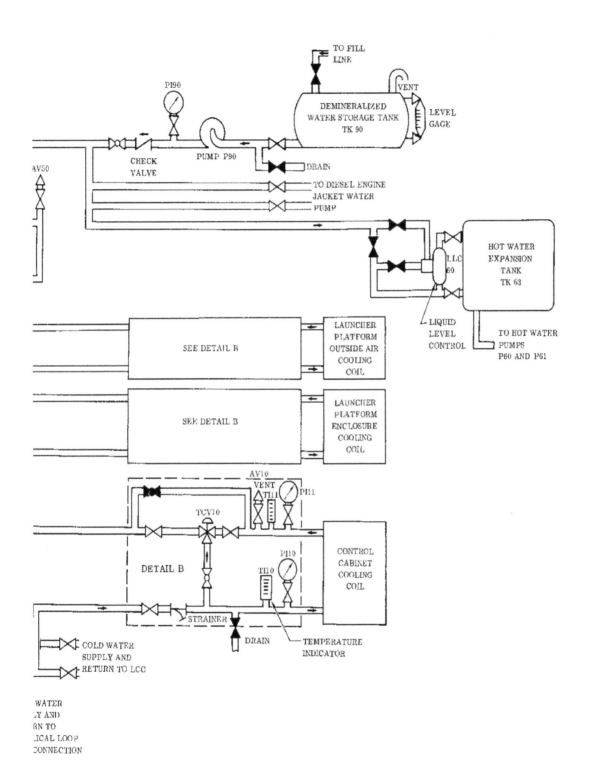

1-40. Demineralized And Chilled Water System Flow Diagram (Typical)

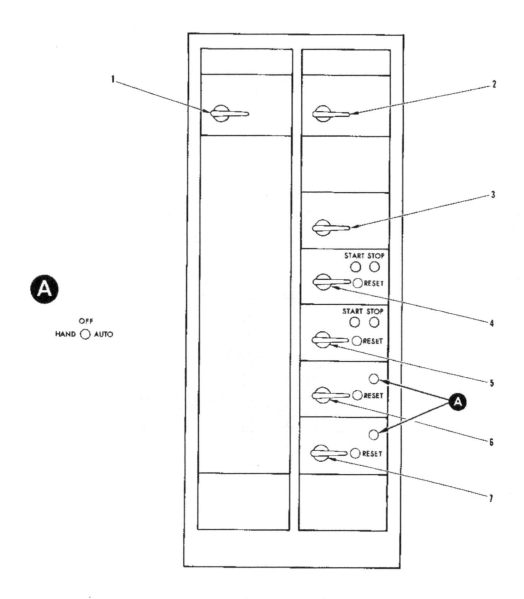

1. 480 VOLT CC DISCONNECT
2. 45 KVA TRANSFORMER
3. CONTROL TRANSF AND BREAKER
4. AC EXHAUST FAN E-1
5. AC CIRCULATION AND SUPPLY FAN S-1
6. SEWAGE PUMP NO. 1
7. SEWAGE PUMP NO. 2

40.10-40A

Figure 1-41. 480-Volt Control Center (OSTF-2)

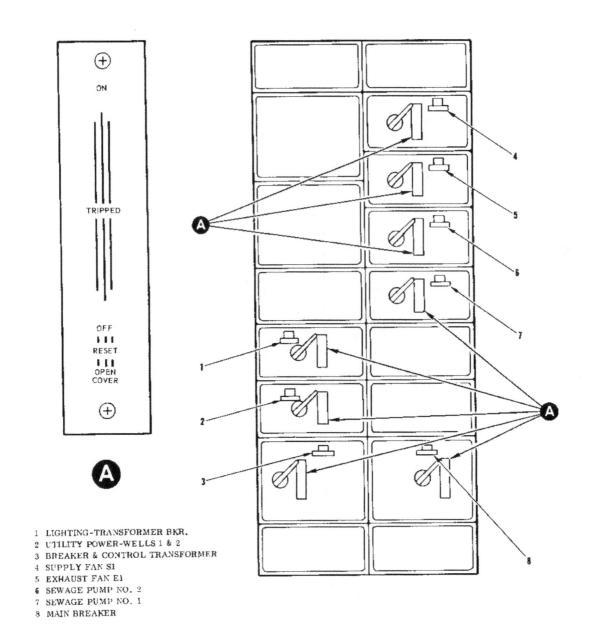

1 LIGHTING-TRANSFORMER BKR.
2 UTILITY POWER-WELLS 1 & 2
3 BREAKER & CONTROL TRANSFORMER
4 SUPPLY FAN S1
5 EXHAUST FAN E1
6 SEWAGE PUMP NO. 2
7 SEWAGE PUMP NO. 1
8 MAIN BREAKER

40.16-40(SAFB)B

Figure 1-42. 480-Volt Control Center (Typical)

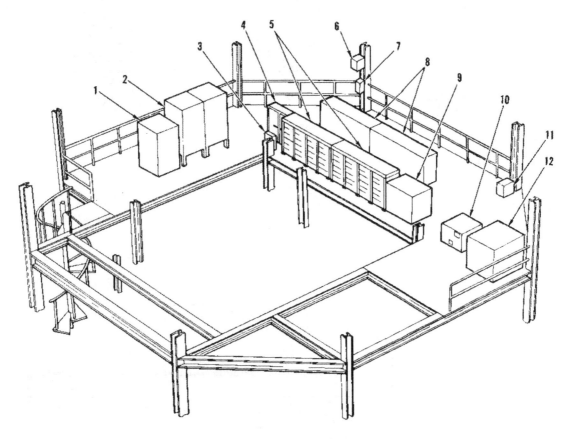

LEVEL 3

1 RE-ENTRY VEHICLE PRELAUNCH MONITOR
  AND CONTROL UNIT
2 COUNTDOWN GROUP
3 LIGHTING PANEL
4 FACILITIES INTERFACE CABINET
5 CONTROL-MONITOR GROUP
  1 OF 4 AND 2 OF 4
6 480-VOLT, 30-KVA TRANSFORMER

7 LAUNCH CONTROL POWER PANEL
8 CONTROL MONITOR GROUP 3 OF 4
   AND 4 OF 4
9 POWER SUPPLY - DISTRIBUTION SET
10 400 CYCLE SKID MOUNTED MOTOR-GENERATOR
   TYPE MD-2
11 DISTRIBUTION BOX
12 28 VDC BATTERY

40.10-126(600) B

Figure 1-43.   Silo Level Three Equipment Location

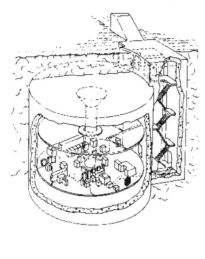

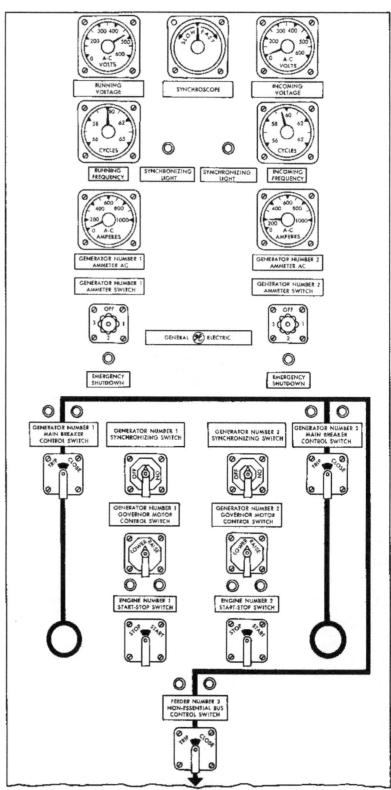

40.32-6A

Figure 1-44. Power Remote Control Panel (Typical)

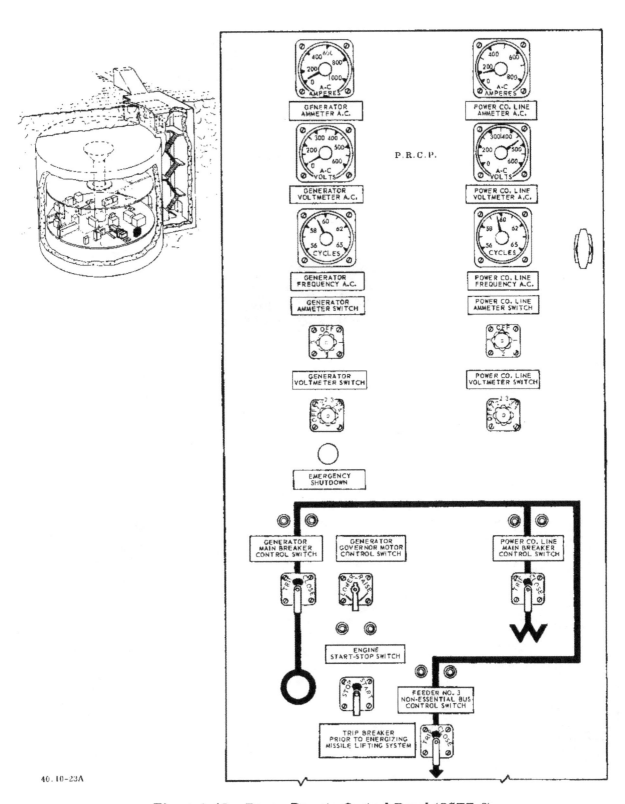

Figure 1-45.   Power Remote Control Panel (OSTF-2)

1 INPUT – AC VOLTMETER
2 ELAPSED TIME INDICATOR
3 OUTPUT – FREQUENCY METER
4 OUTPUT – AC VOLTMETER
5 OUTPUT – AC AMMETER
6 AMMETER PHASE SELECTOR OFF-A-B-C SWITCH
7 VOLTMETER PHASE SELECTOR OFF-A-B-C SWITCH
8 REGULATOR SENSING 1 PHASE - 3 PHASE SWITCH
9 ADJUST VOLTAGE RHEOSTAT
10 OUTPUT CONTACTOR ON-OFF PUSHBUTTON
11 MOTOR STARTER START - STOP PUSHBUTTON

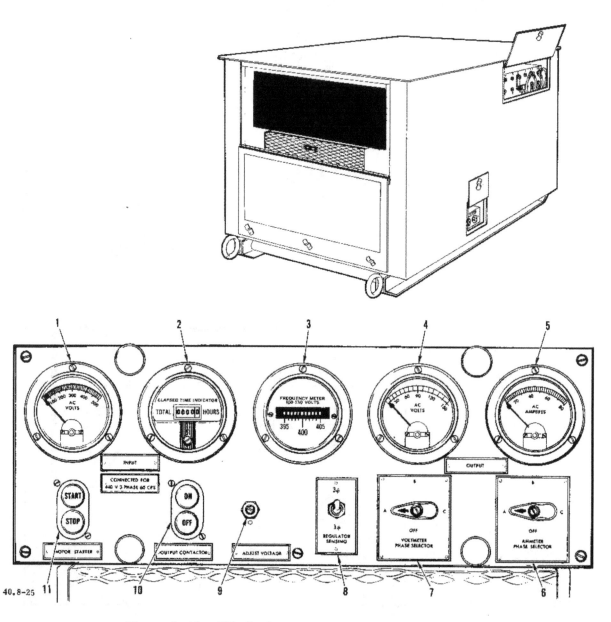

Figure 1-46. 400-Cycle Motor Generator Control Panel

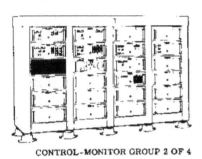

CONTROL-MONITOR GROUP 2 OF 4

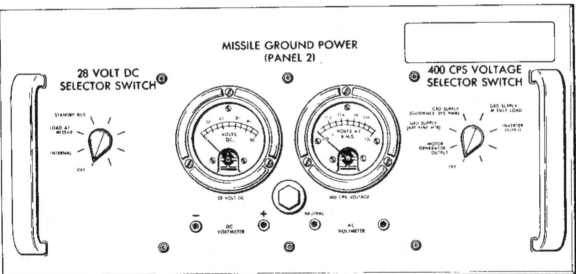

Figure 1-47. Missile Ground Power (Panel 2)

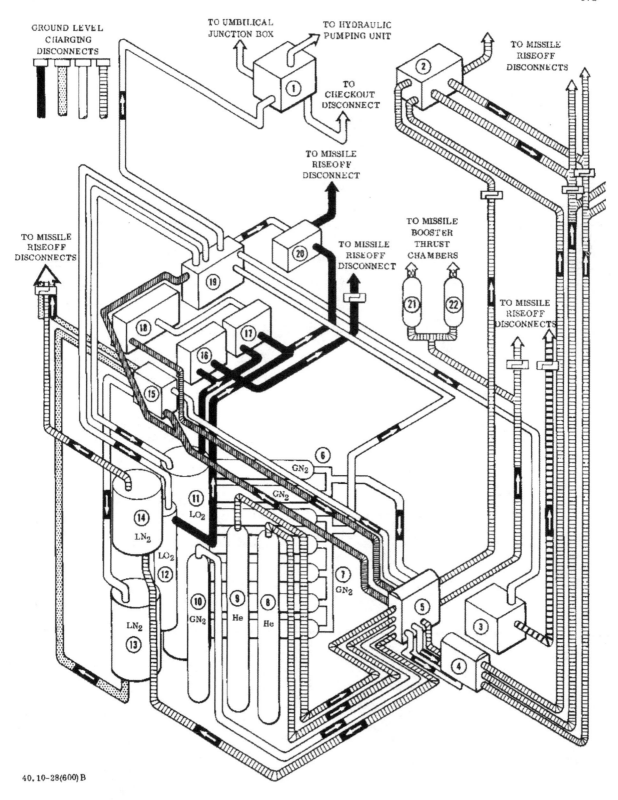

GROUND LEVEL CHARGING DISCONNECTS

TO UMBILICAL JUNCTION BOX

TO HYDRAULIC PUMPING UNIT

TO MISSILE RISEOFF DISCONNECTS

TO CHECKOUT DISCONNECT

TO MISSILE RISEOFF DISCONNECT

TO MISSILE BOOSTER THRUST CHAMBERS

TO MISSILE RISEOFF DISCONNECT

TO MISSILE RISEOFF DISCONNECTS

TO MISSILE RISEOFF DISCONNECTS

40.10-28(600)B

Figure 1-48. Silo Pneum

TO MISSILE
RISEOFF
DISCONNECTS

1 NITROGEN CONTROL UNIT
2 HELIUM CONTROL CHARGING
  UNIT
3 FUEL LOADING PREFAB
4 PRESSURE SYSTEM CONTROL
5 PNEUMATIC SYSTEM MANIFOLD
  REGULATOR
6 GASEOUS NITROGEN STORAGE
  TANKS
7 GASEOUS NITROGEN STORAGE
  TANKS
8 INFLIGHT HELIUM STORAGE
  TANK NO. 2
9 INFLIGHT HELIUM STORAGE
  TANK NO. 1
10 GASEOUS NITROGEN STORAGE
   TANK (GROUND PRESSURIZATION)
11 LIQUID OXYGEN STORAGE TANK
12 LIQUID OXYGEN TOPPING TANK
13 LIQUID NITROGEN STORAGE TANK
14 LIQUID NITROGEN/HELIUM HEAT
   EXCHANGER TANK
15 LIQUID NITROGEN PREFAB
16 LIQUID OXYGEN CONTROL PREFAB
17 LIQUID OXYGEN FILL PREFAB
18 INSTRUMENT AIR PREFAB
19 PRESSURIZATION PREFAB
20 LIQUID OXYGEN TOPPING CONTROL
   UNIT PREFAB
21 INERT FLUID ACCUMULATOR NO. 1
22 INERT FLUID ACCUMULATOR NO. 2

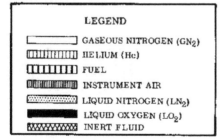

atics and Fluids

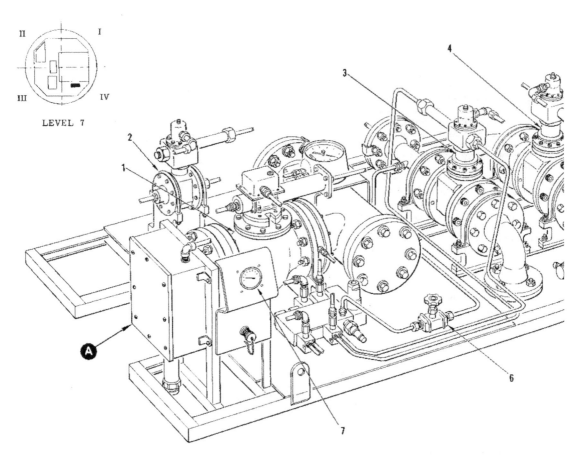

LEVEL 7

1 RAPID TOPPING VALVE L-
2 TOPPING CHILL VALVE L-
3 LINE DRAIN PRESSURIZATI
4 LINE VENT VALVE N-80
5 PLUG
6 MANUAL PURGE VALVE N-
7 GAGE PG-1

46.10-42(600)A

Figure 1-49 Liquid Oxygen Topping Cont

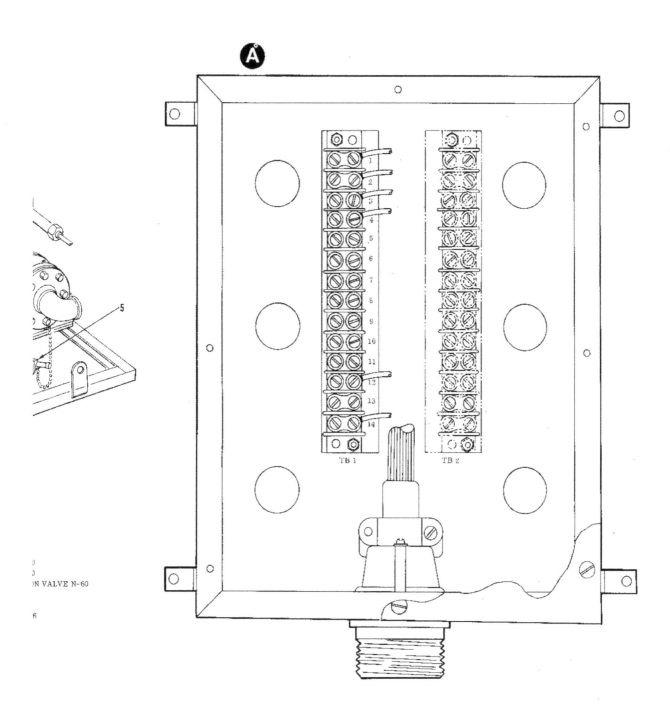

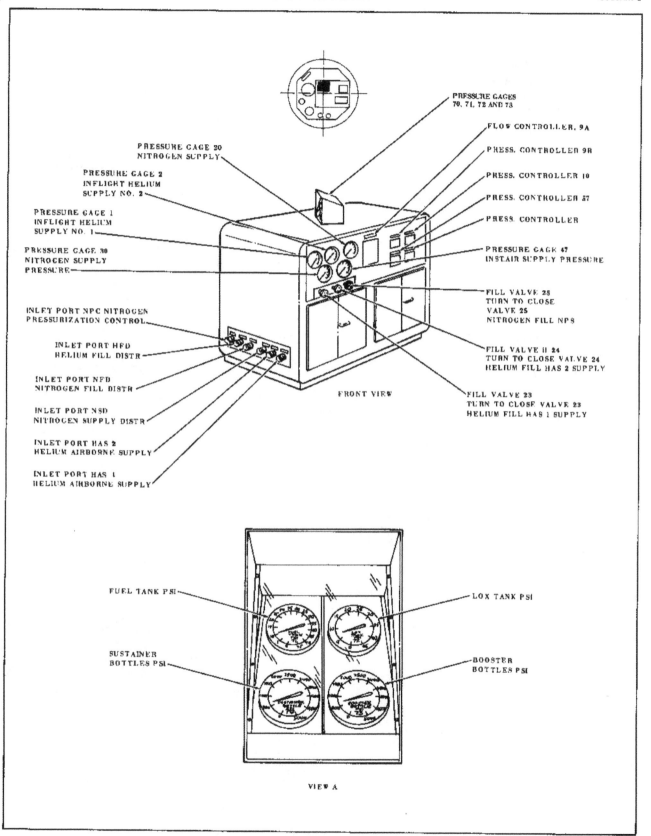

Figure 1-50. Pneumatic System Manifold Regulator

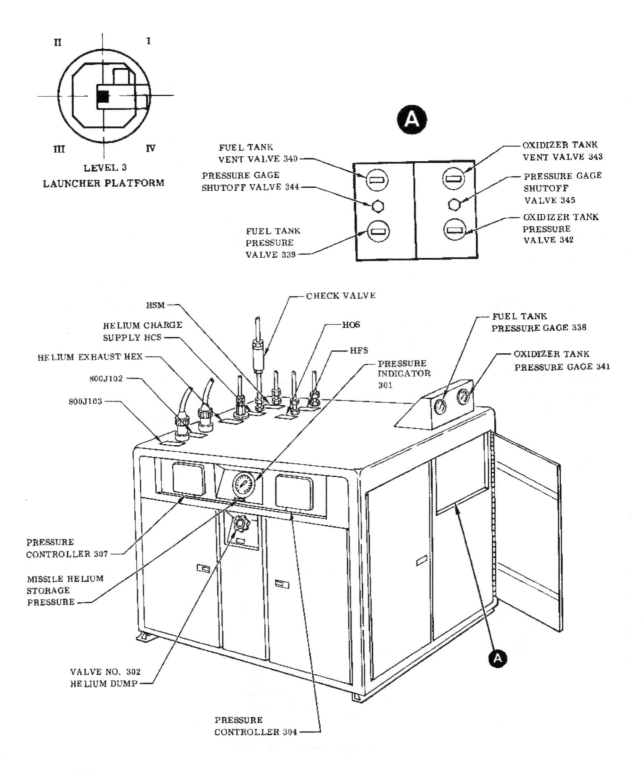

Figure 1-51. Helium Control Charging Unit

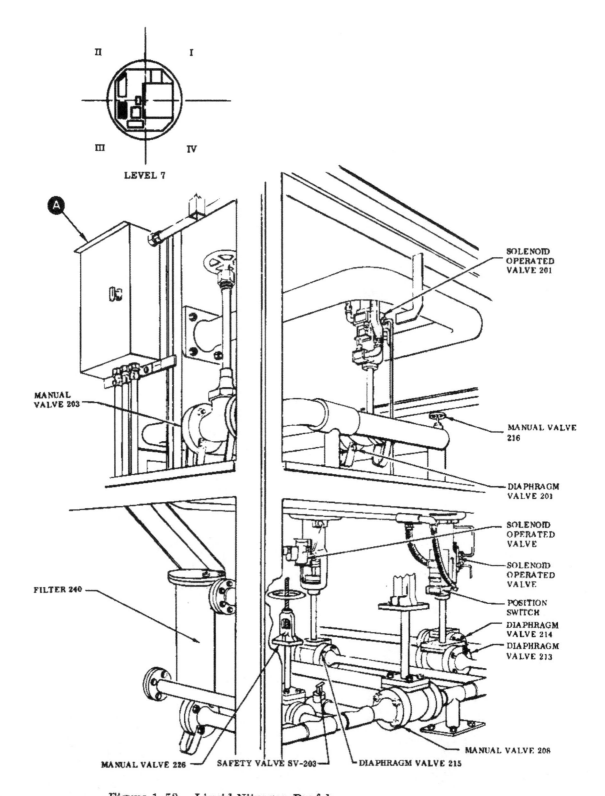

LEVEL 7

SOLENOID OPERATED VALVE 201

MANUAL VALVE 216

DIAPHRAGM VALVE 201

SOLENOID OPERATED VALVE

SOLENOID OPERATED VALVE

POSITION SWITCH

DIAPHRAGM VALVE 214

DIAPHRAGM VALVE 213

MANUAL VALVE 203

FILTER 240

MANUAL VALVE 226

SAFETY VALVE SV-203

DIAPHRAGM VALVE 215

MANUAL VALVE 208

40.10-44A

Figure 1-52.   Liquid Nitrogen Prefab

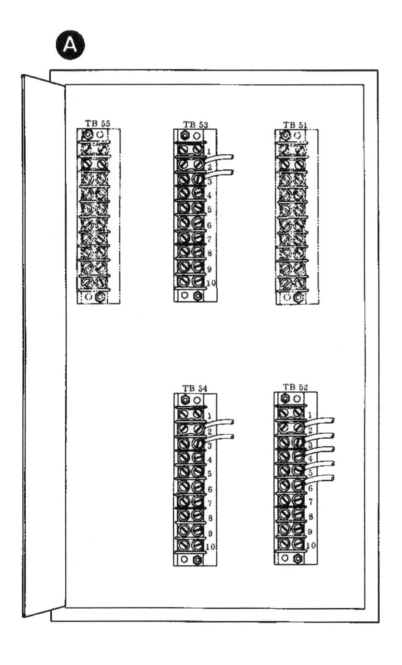

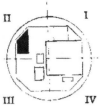

LEVEL 7

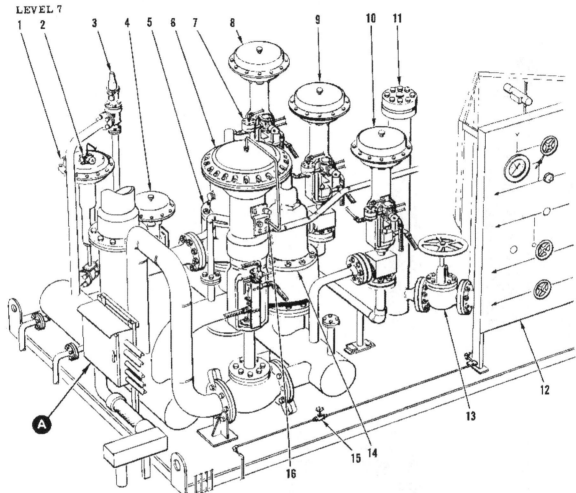

1 TOPPING TANK VENT VALVE N-4	9 STORAGE TANK PRESSURIZATION
2 TYPICAL SOLENOID VALVE	VALVE N-2
3 LIQUID OXYGEN TOPPING TANK	10 STORAGE TANK PRESSURIZATION
PRESSURIZATION RELIEF VALVE SV-5	VALVE N-1
4 TOPPING TANK PRESSURIZATION	11 PRESSURIZATION FILTER N-29
VALVE N-50	12 PRESSURIZATION PREFAB CONTROL
5 LIQUID OXYGEN STORAGE TANK	PANEL
PRESSURIZATION RELIEF VALVE SV-6A	13 LIQUID OXYGEN PRESSURIZATION
6 STORAGE TANK VENT VALVE N-5	SHUTOFF VALVE N-6
7 TYPICAL PRESSURE BOOSTER	14 LIQUID OXYGEN STORAGE TANK
8 STORAGE TANK PRESSURIZATION	PRESSURIZATION RELIEF VALVE SV-6B
VALVE N-3	15 INSTRUMENT AIR SHUTOFF VALVE

16 COUPLING NUT

40.10-43A

Figure 1-54. Pressurization Prefab

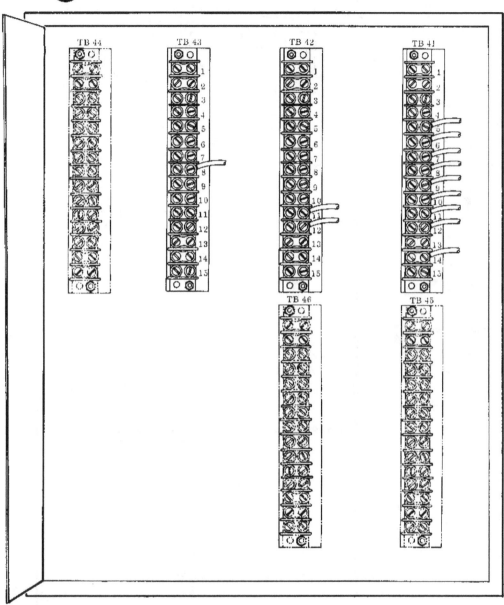

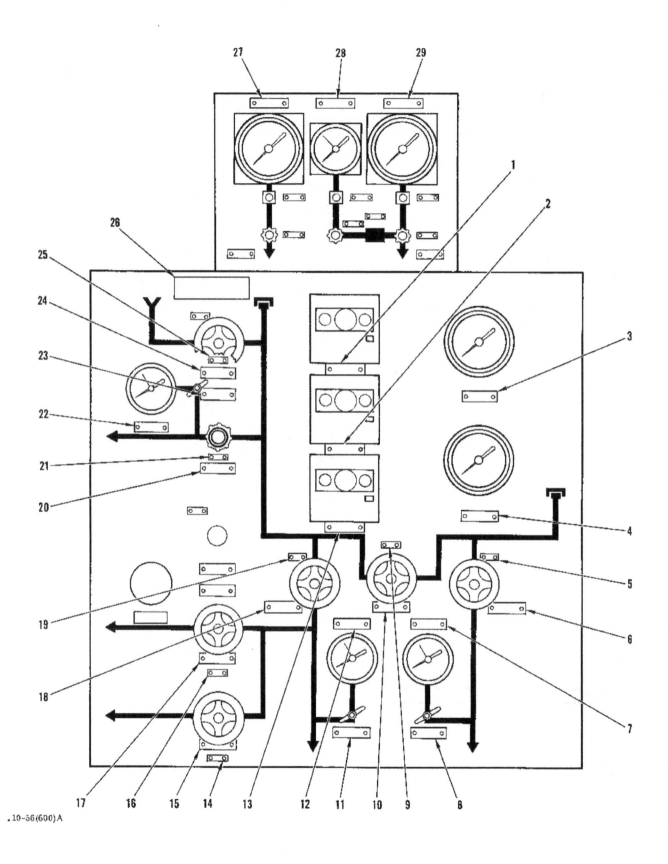

Figure 1-53 Pressurization Prefab Instrument Panel

.10-56(600)A

1 PRESSURE CONTROLLER
   COOLDOWN PRESSURE
2 PRESSURE CONTROLLER
   TRANSFER PRESSURE
3 LIQUID LEVEL LO_2 STORAGE
4 LIQUID LEVEL LO_2 TOPPING TANK
5 O-5
6 GO_2 STORAGE
7 VESSEL PRESSURE GO_2 STORAGE
8 PRESSURE GAGE SHUT OFF N-9
9 N-44
10 GO_2 STORAGE
11 PRESSURE GAGE SHUTOFF N-27
12 VESSEL PRESSURE LN_2 TRANSFER
   AND NCU SUPPLY
13 PRESSURE CONTROLLER
   TOPPING TRANSFER
14 N-48
15 LN_2 TRANSFER SUPPLY
16 N-40
17 GN_2 STORAGE NCU SUPPLY
18 GN_2 STORAGE
19 N-49
20 GN_2 STORAGE FUEL PREFAB
21 N-45
22 VESSEL PRESSURE GN_2 STORAGE
23 PRESSURE GAGE SHUTOFF N-19
24 SURFACE VENT GN_2 FILL
25 N-13
26 PRESSURIZATION PREFAB PANEL
27 PI-16 LOX TOPPING TRANSFER
28 PI-14 LOX STORAGE STANDBY OR CHILLDOWN
29 PI-15 LOX STORAGE TRANSFER

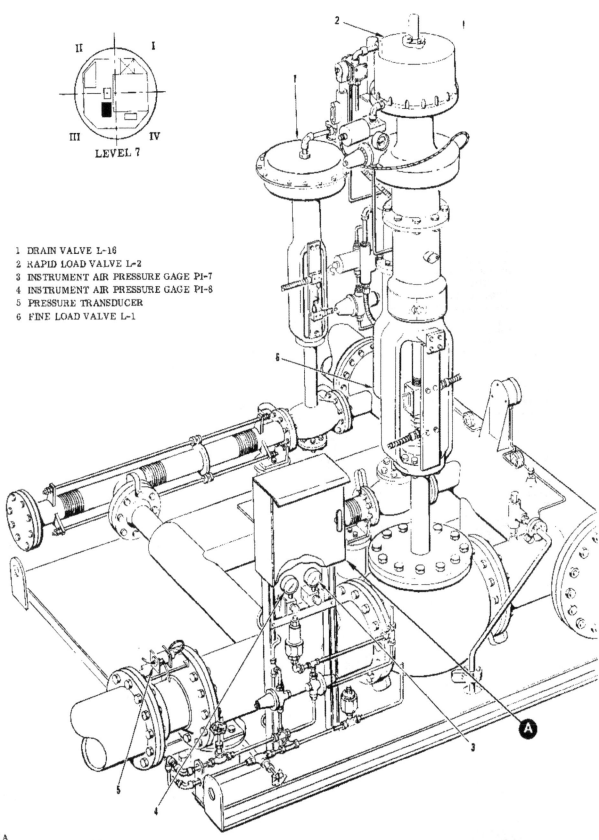

LEVEL 7

1 DRAIN VALVE L-16
2 RAPID LOAD VALVE L-2
3 INSTRUMENT AIR PRESSURE GAGE PI-7
4 INSTRUMENT AIR PRESSURE GAGE PI-8
5 PRESSURE TRANSDUCER
6 FINE LOAD VALVE L-1

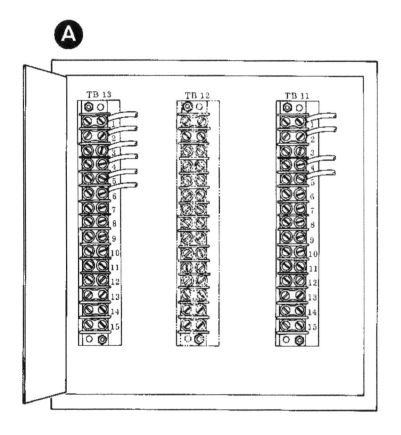

Figure 1-55. Liquid Oxygen Control Prefab

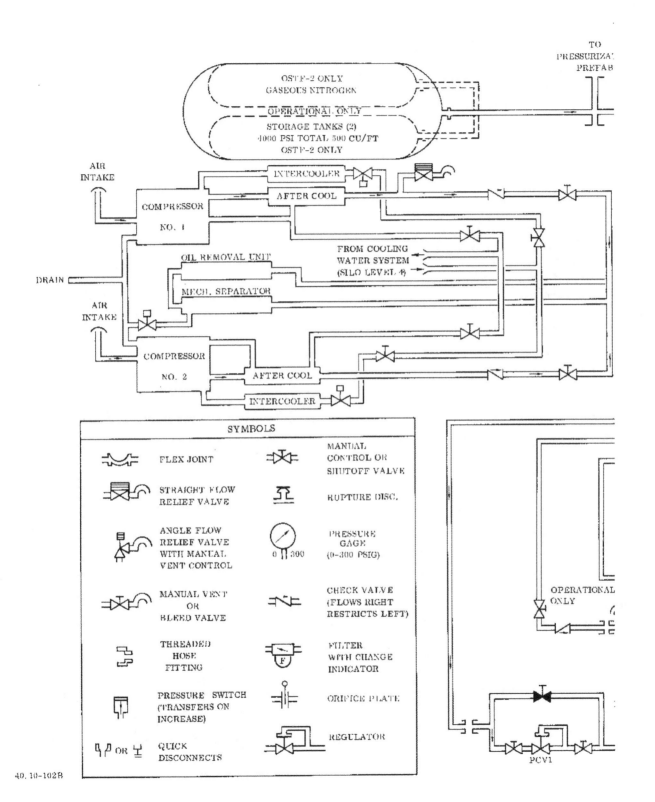

SYMBOLS

FLEX JOINT		MANUAL CONTROL OR SHUTOFF VALVE	
STRAIGHT FLOW RELIEF VALVE		RUPTURE DISC.	
ANGLE FLOW RELIEF VALVE WITH MANUAL VENT CONTROL		PRESSURE GAGE (0-300 PSIG)	0 || 300
MANUAL VENT OR BLEED VALVE		CHECK VALVE (FLOWS RIGHT RESTRICTS LEFT)	
THREADED HOSE FITTING		FILTER WITH CHANGE INDICATOR	
PRESSURE SWITCH (TRANSFERS ON INCREASE)		ORIFICE PLATE	
QUICK DISCONNECTS		REGULATOR	

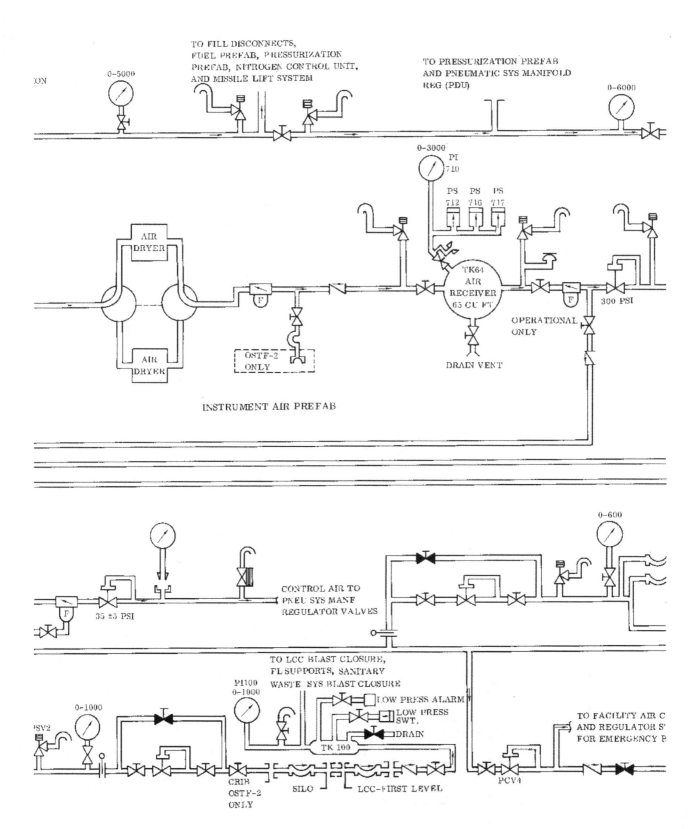

Figure 1-56. Instrument Air System Flow Diagram

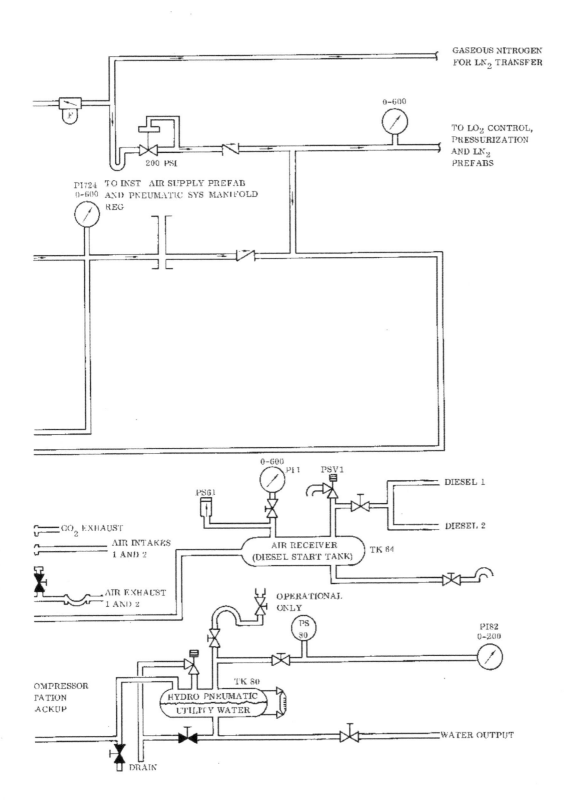

GASEOUS NITROGEN
FOR LN$_2$ TRANSFER

0-600

TO LO$_2$ CONTROL,
PRESSURIZATION
AND LN$_2$
PREFABS

200 PSI

PI724   TO INST AIR SUPPLY PREFAB
0-600   AND PNEUMATIC SYS MANIFOLD
REG

0-600   PI1   PSV1

PSG1   DIESEL 1

GO$_2$ EXHAUST   DIESEL 2

AIR INTAKES
1 AND 2

AIR RECEIVER   TK 64
(DIESEL START TANK)

AIR EXHAUST
1 AND 2

OPERATIONAL
ONLY

PS
30   PI82
                                   0-200

OMPRESSOR
TATION
ACKUP

TK 80

HYDRO PNEUMATIC
UTILITY WATER

WATER OUTPUT

DRAIN

n (Typical)

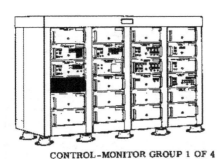

CONTROL-MONITOR GROUP 1 OF 4

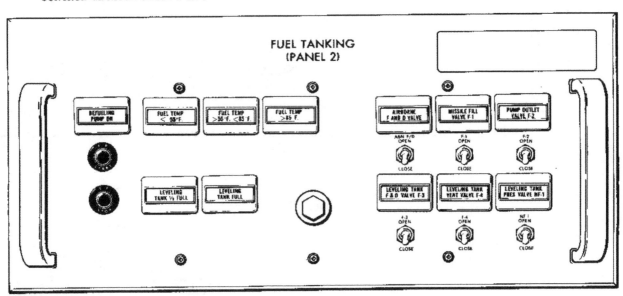

N40, 10-2

Figure 1-57.   Fuel Tanking (Panel 2)

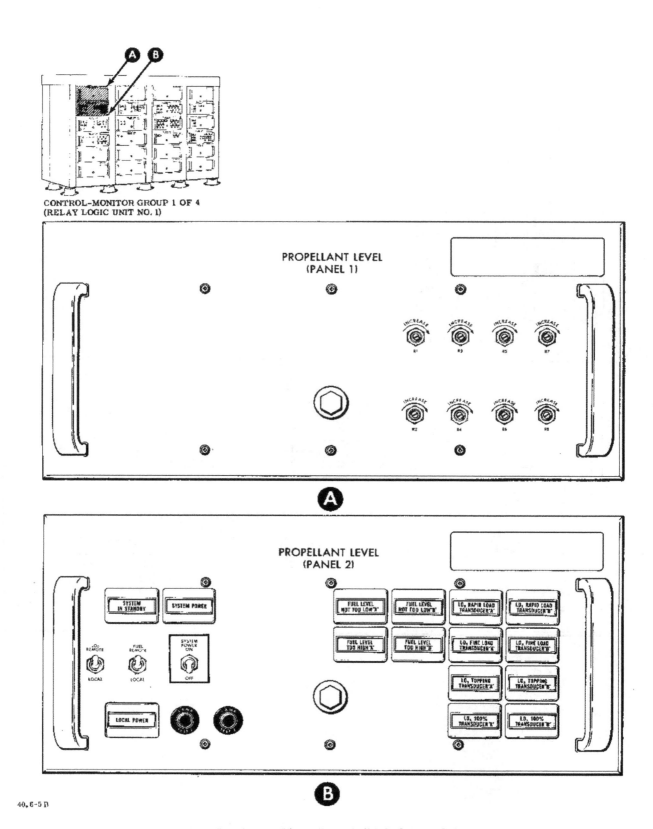

Figure 1-58. Propellant Level (Panel 1 and Panel 2)

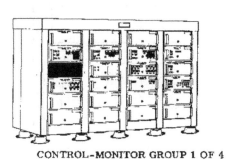

CONTROL-MONITOR GROUP 1 OF 4

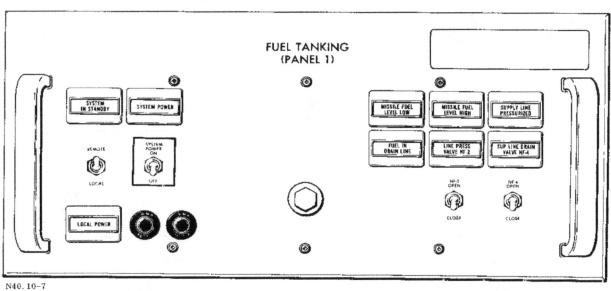

N40.10-7

Figure 1-59. Fuel Tanking (Panel 1)

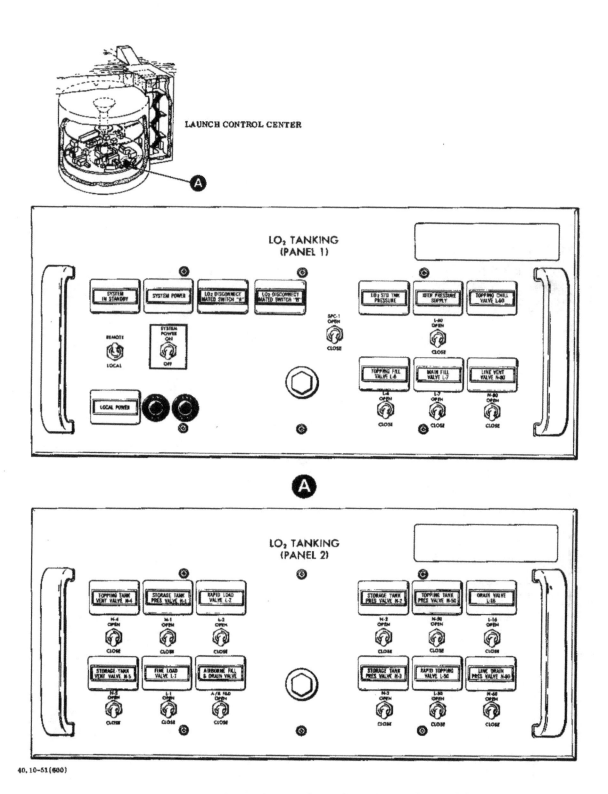

40.10-51(680)

Figure 1-60. Liquid Oxygen Tanking (Panel 1 and Panel 2)

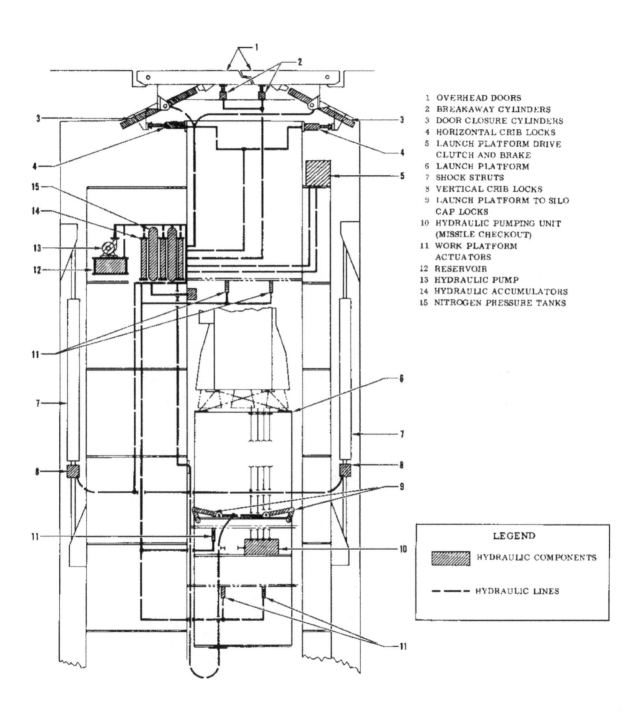

1 OVERHEAD DOORS
2 BREAKAWAY CYLINDERS
3 DOOR CLOSURE CYLINDERS
4 HORIZONTAL CRIB LOCKS
5 LAUNCH PLATFORM DRIVE
   CLUTCH AND BRAKE
6 LAUNCH PLATFORM
7 SHOCK STRUTS
8 VERTICAL CRIB LOCKS
9 LAUNCH PLATFORM TO SILO
   CAP LOCKS
10 HYDRAULIC PUMPING UNIT
   (MISSILE CHECKOUT)
11 WORK PLATFORM
   ACTUATORS
12 RESERVOIR
13 HYDRAULIC PUMP
14 HYDRAULIC ACCUMULATORS
15 NITROGEN PRESSURE TANKS

LEGEND

▨ HYDRAULIC COMPONENTS

— — — HYDRAULIC LINES

40.10-16 C

Figure 1-61. Silo Hydraulics

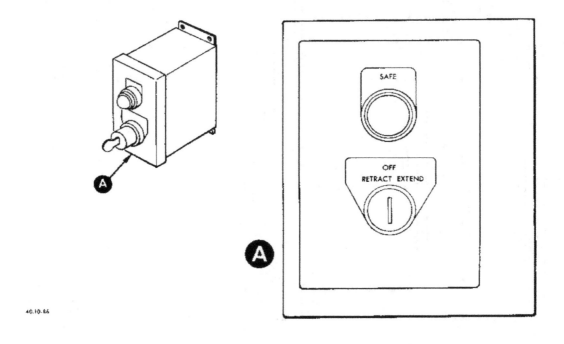

40.10-86

Figure 1-62. Work Platform Key Switch Control Panel (Typical)

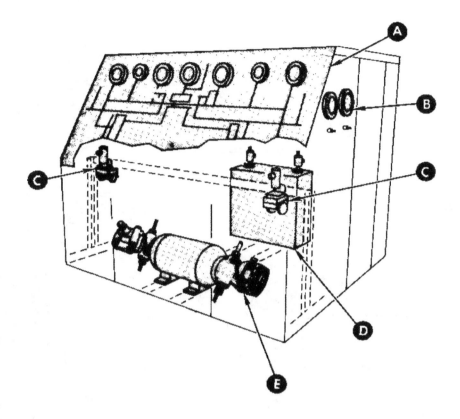

1 GAUGE HIGH PRESSURE SYSTEM (10)
2 GAUGE SHUTOFF (6)
3 HYDRAULIC FLUID TEMPERATURE-
   HIGH PRESSURE PUMP INLET (28)
4 GAUGE SHUTOFF (23)
5 GAUGE LOW PRESSURE SYSTEM (24)
6 GAUGE SHUTOFF (6A)
7 GAUGE RESERVOIR PRESSURE (46)
8 RESERVOIR PRESSURE BLEED
   VALVE (6B)
9 RESERVOIR PRESSURE REGULA-
   TOR (40)
10 GAUGE LOW PRESSURE SYSTEM (24)
11 GAUGE SHUTOFF (23)
12 HYDRAULIC FLUID TEMPERATURE-
   HIGH PRESSURE PUMP INLET (28)
13 GAUGE SHUTOFF (6)
14 GAUGE HIGH PRESSURE SYSTEM (10)

15 SECOND STAGE PRESSURE SHUTOFF
   VALVE (V-3)(84)
16 GAUGE EVACUATION CHAMBER
   PNEUMATIC PRESSURE (35) FIRST
   STAGE SYSTEM
17 GAUGE EVACUATION CHAMBER,
   PNEUMATIC PRESSURE (35)
   SECOND STAGE SYSTEM
18 OUTLET-EVACUATION CHAMBER
   CHARGING (90)
19 OUTLET-EVACUATION CHAMBER
   CHARGING (90)
20 HIGH PRESSURE RELIEF VALVE (12)
21 FLOWMETER SELECTOR VALVE (54)
22 FLOWMETER (20)
23 DISCONNECT RESERVOIR CONNECT
   EVACUATION CHAMBER (S6)
24 BYPASS SUSTAINER SYSTEM (V135)

Figure 1-63. Hydraulic Pumping Unit

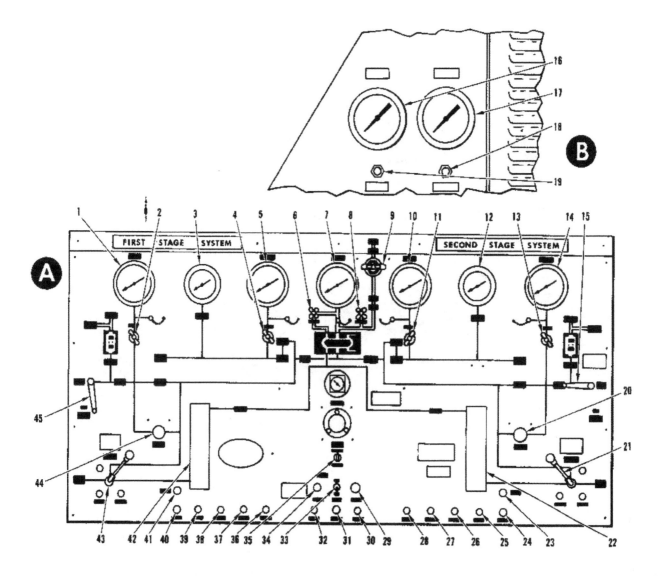

25 HIGH PRESSURE WITHIN LIMITS
26 PRESSURE DROP EXCESSIVE-HIGH
   PRESS. FILTER NO. 11
27 PRESSURE DROP EXCESSIVE-LOW
   PRESS. FILTER NO. 21
28 EVACUATING OIL
29 STOP UNIT (S4)
30 UNIT ON STANDBY STATUS (DS2)
31 PRESS TO TEST PANEL INDICATOR
   LIGHTS (S-10)
32 RESERVOIR OVERPRESSURIZED (DS1)
33 CONTROL POWER (S-8)
34 START UNIT (S3)
35 CONTROL STATUS (S1)
36 EVACUATING OIL
37 PRESSURE DROP EXCESSIVE-LOW
   PRESS. FILTER NO. 21
38 PRESSURE DROP EXCESSIVE-HIGH
   PRESS. FILTER NO. 11

39 HIGH PRESSURE WITHIN LIMITS
40 BYPASS BOOSTER SYSTEM (V13B)
41 DISCONNECT RESERVOIR CONNECT
   EVACUATION CHAMBER (S5)
42 FLOWMETER (20)
43 FLOWMETER SELECTOR VALVE (54)
44 HIGH PRESSURE RELIEF VALVE (12)
45 FIRST STAGE PRESSURE SHUTOFF
   VALVE (V1)(84)
46 ELECTRICAL CONNECTOR
47 BYPASS VALVE (13)
48 POWER SWITCH (CB-1)
49 RESET PUSHBUTTON
60 PRESSURE COMPENSATION CONTROL
51 VOLUME CONTROL
52 PUMP
53 PUMP MOTOR

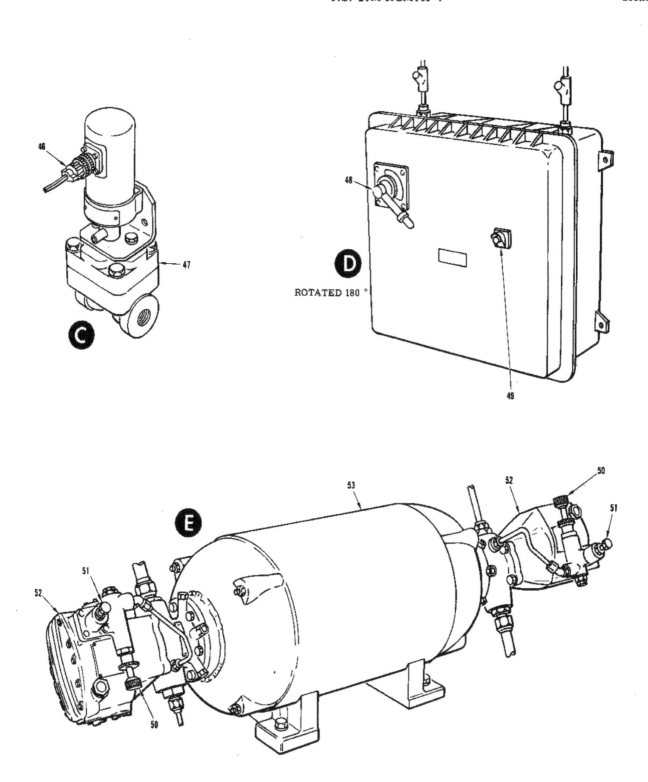

ROTATED 180 °

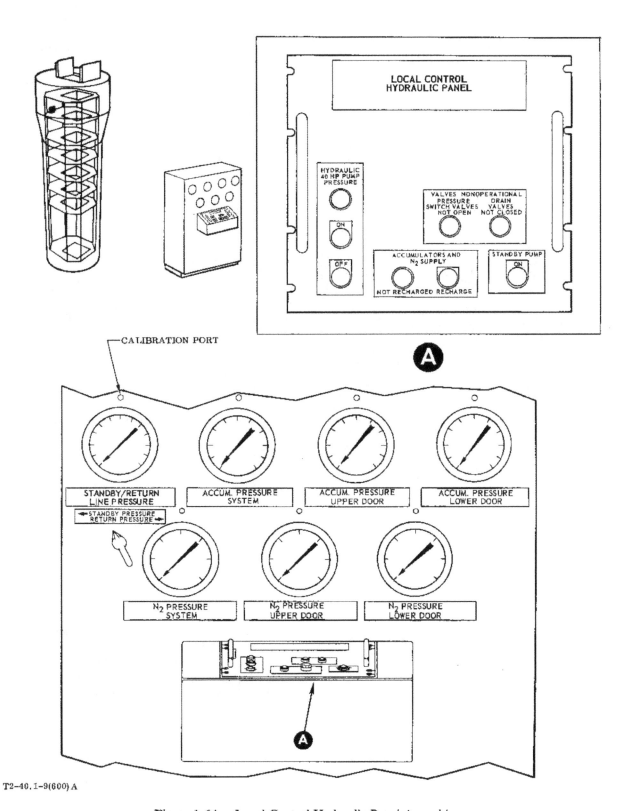

T2-40.1-9(600)A

Figure 1-64.   Local Control Hydraulic Panel Assembly

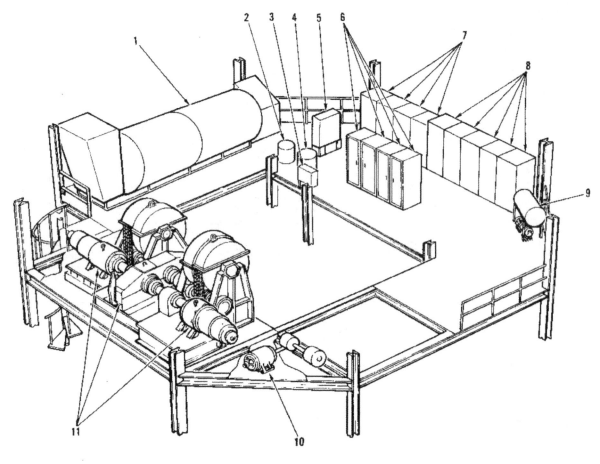

LEVEL 1

1 FRESH AIR DUST COLLECTOR, PUMP, AND
    WASHER
2 DUST COLLECTOR WATER MAKEUP TANK
3 OVERSPEED CONTROL BOX
4 CHILLED WATER EXPANSION TANK
5 INTERCONNECTING JUNCTION BOX
6 ELECTRICAL MISSILE LIFTING CONTROL
    SYSTEM

7 MISSILE LIFT SYSTEM MOTOR CONTROL CENTER
8 LAUNCH PLATFORM MISSILE LIFTING DRIVE
    ASSEMBLY CABINETS
9 DEMINERALIZED WATER STORAGE TANK AND
    PUMP P-90
10 FACILITY ELEVATOR DRIVE
11 MISSILE LIFTING LAUNCH PLATFORM DRIVE
    ASSEMBLY

40.10-60 (600) A

Figure 1-65. Silo Level 1 Equipment Location

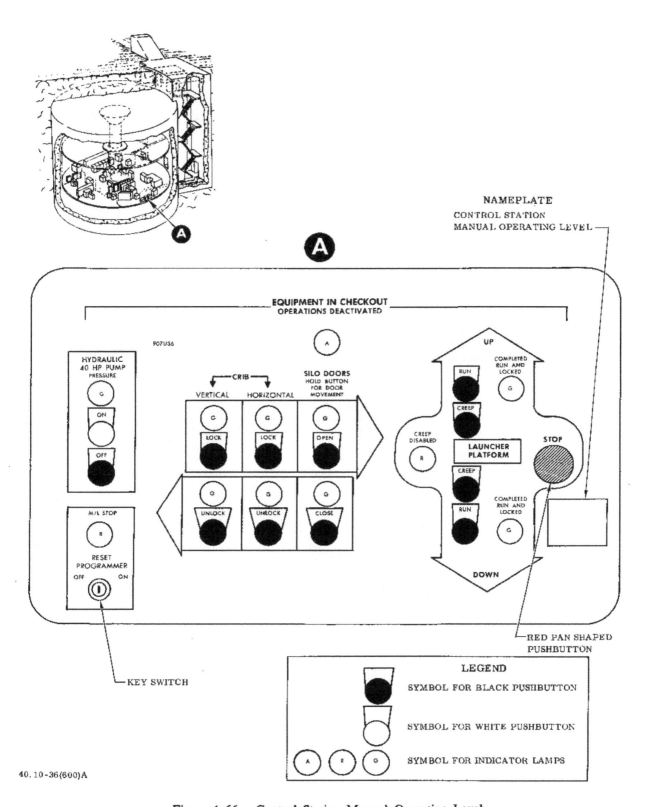

Figure 1-66.   Control Station Manual Operating Level

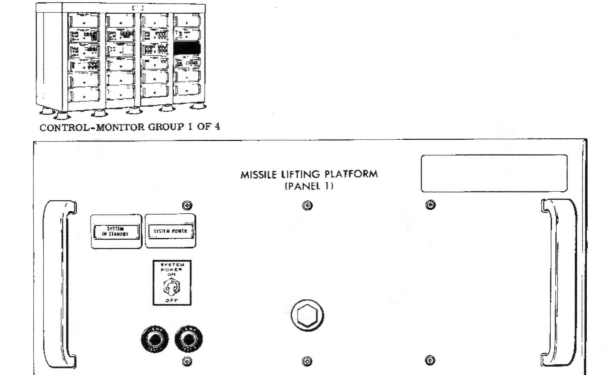

CONTROL-MONITOR GROUP 1 OF 4

MISSILE LIFTING PLATFORM
(PANEL 1)

N40. 10-6

Figure 1-67. Missile Lifting Platform (Panel 1)

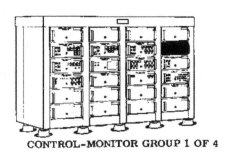

CONTROL-MONITOR GROUP 1 OF 4

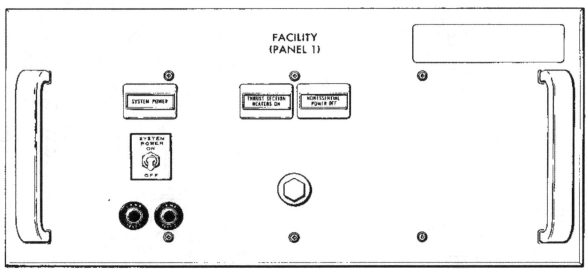

N40. 10-5

Figure 1-68.   Facility (Panel 1)

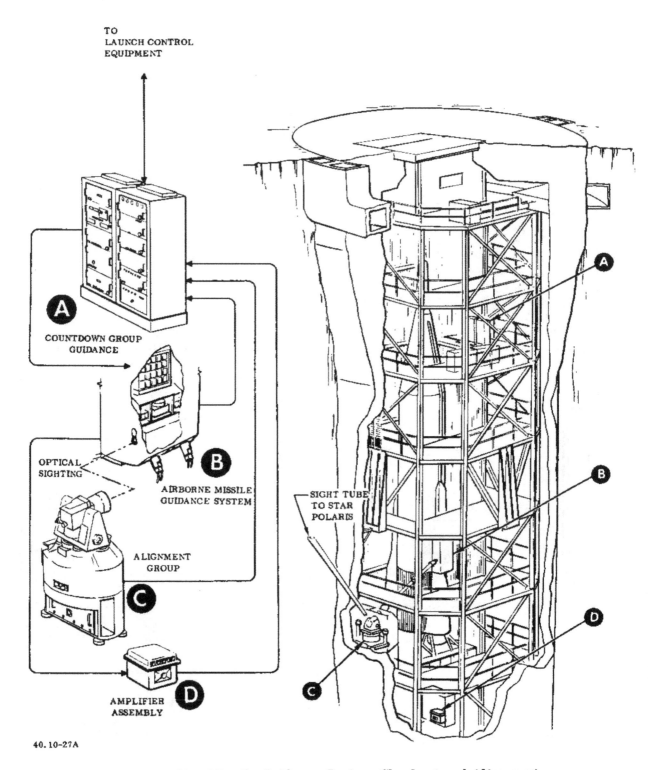

TO
LAUNCH CONTROL
EQUIPMENT

COUNTDOWN GROUP
GUIDANCE

OPTICAL
SIGHTING

AIRBORNE MISSILE
GUIDANCE SYSTEM

ALIGNMENT
GROUP

AMPLIFIER
ASSEMBLY

SIGHT TUBE
TO STAR
POLARIS

40.10-27A

Figure 1-69. Missile Guidance System Checkout and Alignment

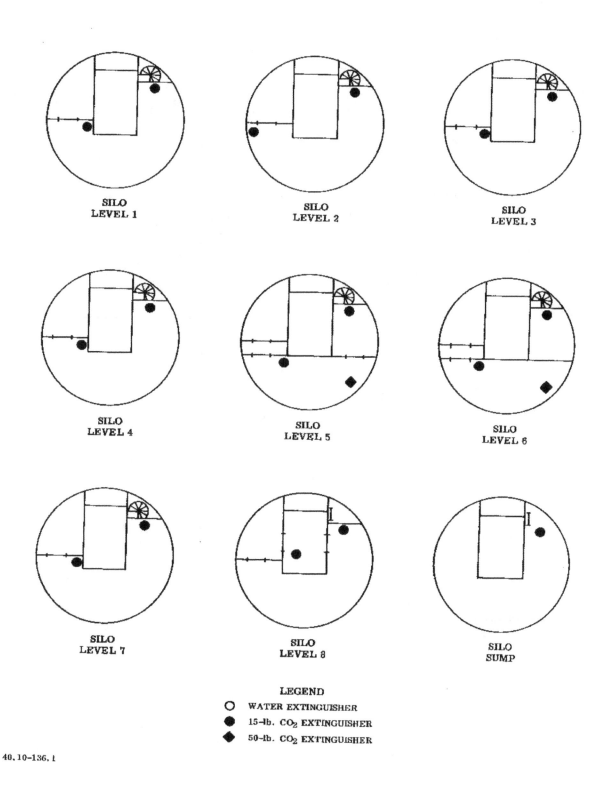

SILO
LEVEL 1

SILO
LEVEL 2

SILO
LEVEL 3

SILO
LEVEL 4

SILO
LEVEL 5

SILO
LEVEL 6

SILO
LEVEL 7

SILO
LEVEL 8

SILO
SUMP

LEGEND

○   WATER EXTINGUISHER
●   15-lb. CO_2 EXTINGUISHER
◆   50-lb. CO_2 EXTINGUISHER

40.10-136.1

Figure 1-70.   Portable Fire Extinguisher Location (Sheet 1 of 2)

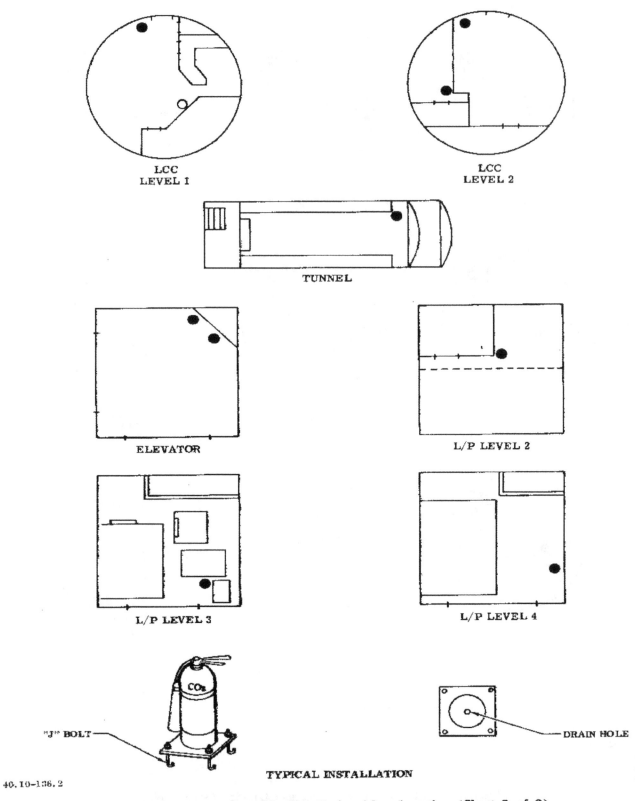

LCC
LEVEL 1

LCC
LEVEL 2

TUNNEL

ELEVATOR

L/P LEVEL 2

L/P LEVEL 3

L/P LEVEL 4

"J" BOLT

DRAIN HOLE

TYPICAL INSTALLATION

40.10-136.2

Figure 1-70. Portable Fire Extinguisher Location (Sheet 2 of 2)

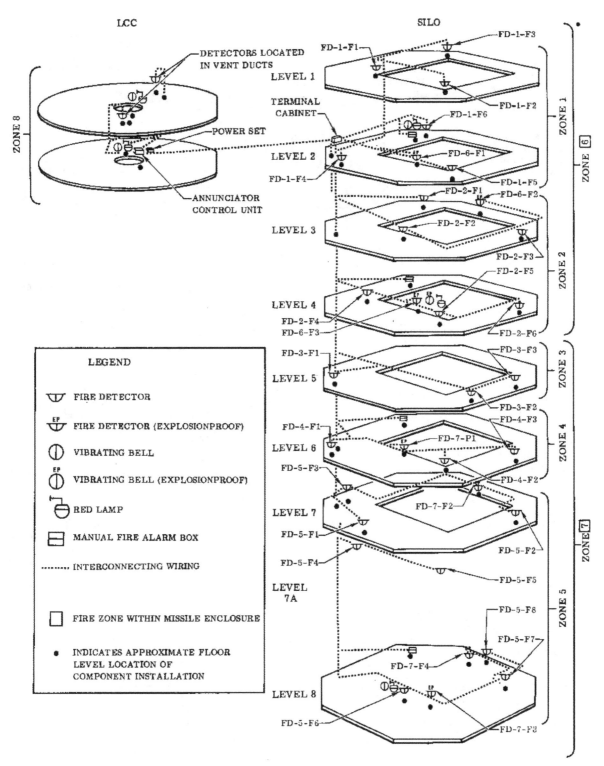

Figure 1-71. Fire Detection and Warning System Locations

40.10-71A

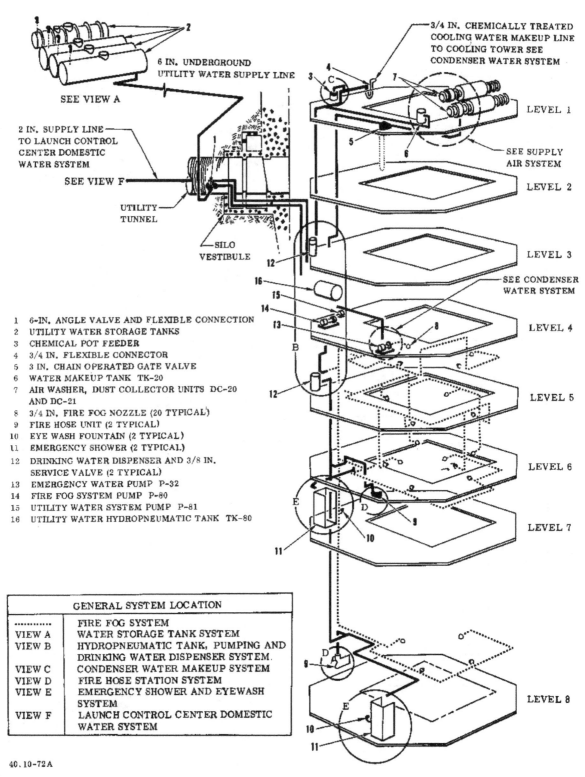

1   6-IN. ANGLE VALVE AND FLEXIBLE CONNECTION
2   UTILITY WATER STORAGE TANKS
3   CHEMICAL POT FEEDER
4   3/4 IN. FLEXIBLE CONNECTOR
5   3 IN. CHAIN OPERATED GATE VALVE
6   WATER MAKEUP TANK TK-20
7   AIR WASHER, DUST COLLECTOR UNITS DC-20
      AND DC-21
8   3/4 IN. FIRE FOG NOZZLE (20 TYPICAL)
9   FIRE HOSE UNIT (2 TYPICAL)
10   EYE WASH FOUNTAIN (2 TYPICAL)
11   EMERGENCY SHOWER (2 TYPICAL)
12   DRINKING WATER DISPENSER AND 3/8 IN.
      SERVICE VALVE (2 TYPICAL)
13   EMERGENCY WATER PUMP P-32
14   FIRE FOG SYSTEM PUMP P-80
15   UTILITY WATER SYSTEM PUMP P-81
16   UTILITY WATER HYDROPNEUMATIC TANK TK-80

GENERAL SYSTEM LOCATION	
··········	FIRE FOG SYSTEM
VIEW A	WATER STORAGE TANK SYSTEM
VIEW B	HYDROPNEUMATIC TANK, PUMPING AND DRINKING WATER DISPENSER SYSTEM
VIEW C	CONDENSER WATER MAKEUP SYSTEM
VIEW D	FIRE HOSE STATION SYSTEM
VIEW E	EMERGENCY SHOWER AND EYEWASH SYSTEM
VIEW F	LAUNCH CONTROL CENTER DOMESTIC WATER SYSTEM

40.10-72A

Figure 1-72. Fire Fog, Emergency Shower, Eyewash, and
Fire Hose Locations

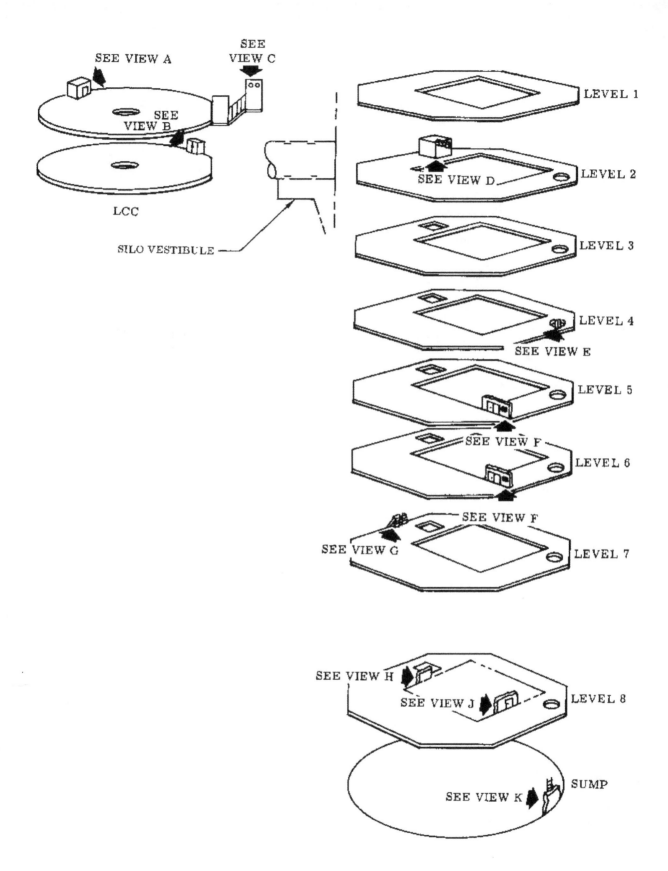

SEE VIEW A

SEE VIEW C

SEE VIEW B

LCC

SILO VESTIBULE

LEVEL 1

SEE VIEW D   LEVEL 2

LEVEL 3

LEVEL 4

SEE VIEW E

LEVEL 5

SEE VIEW F   LEVEL 6

SEE VIEW F

SEE VIEW G   LEVEL 7

SEE VIEW H

SEE VIEW J   LEVEL 8

SEE VIEW K   SUMP

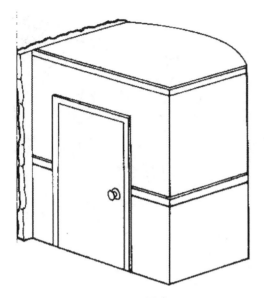

VIEW A
SAFETY EQUIPMENT
STORAGE CLOSET

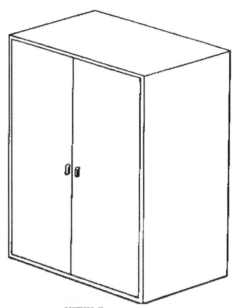

VIEW B
EMERGENCY RESCUE &
SAFETY EQUIPMENT CABINET

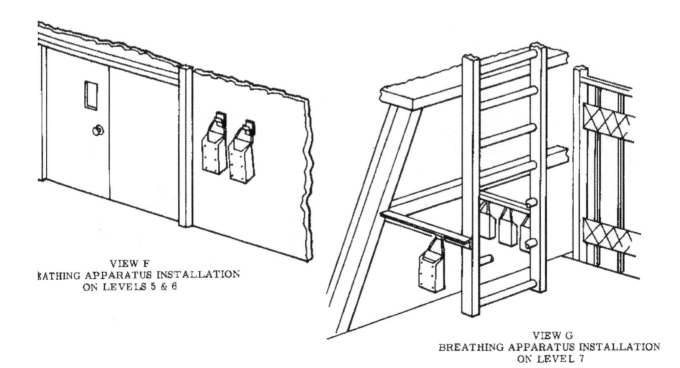

VIEW F
BREATHING APPARATUS INSTALLATION
ON LEVELS 5 & 6

VIEW G
BREATHING APPARATUS INSTALLATION
ON LEVEL 7

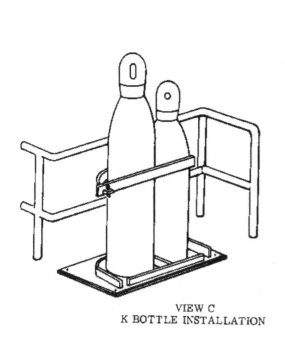

VIEW C
K BOTTLE INSTALLATION

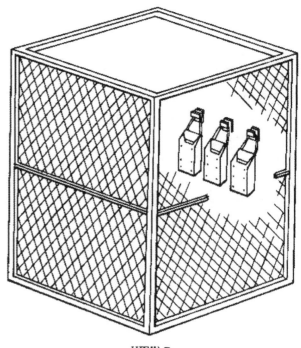

VIEW D
BREATHING APPARATUS INSTALLATION
IN ELEVATOR

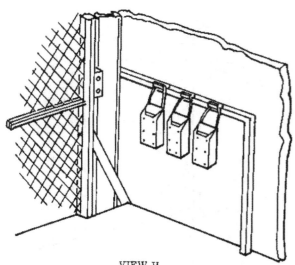

VIEW H
BREATHING APPARATUS INSTALLATION
ON LEVEL 8

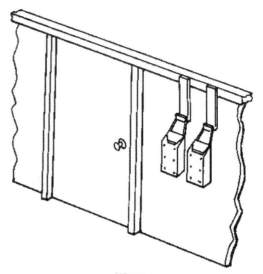

VIEW J
BREATHING APPARATUS INSTALLATION
ON LEVEL 8

Figure 1-73, Personnel Safety and

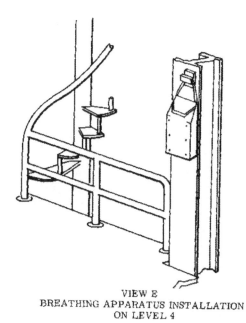

VIEW E
BREATHING APPARATUS INSTALLATION
ON LEVEL 4

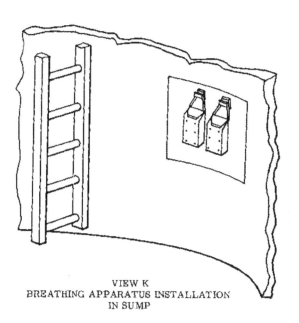

VIEW K
BREATHING APPARATUS INSTALLATION
IN SUMP

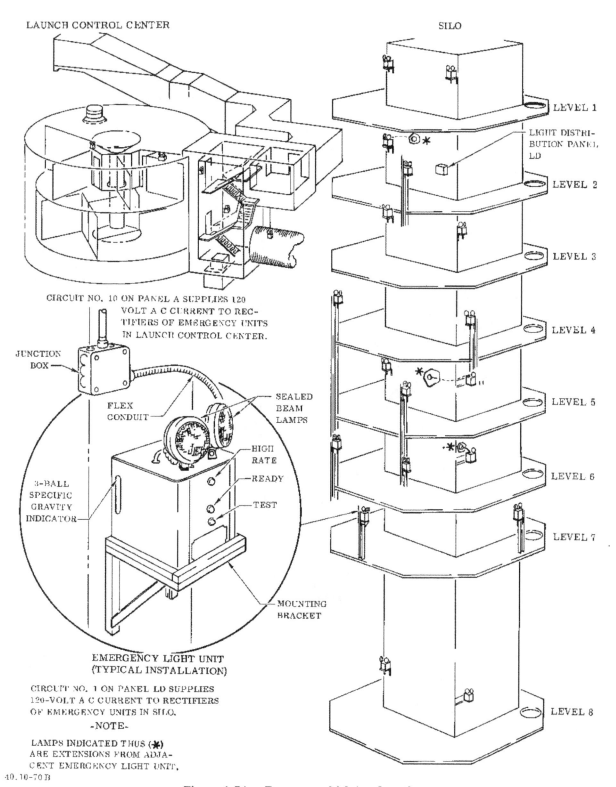

Figure 1-74.   Emergency Lighting Location

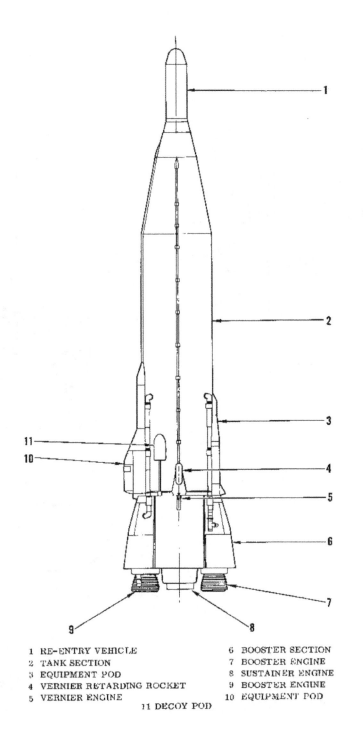

1 RE-ENTRY VEHICLE   6 BOOSTER SECTION
2 TANK SECTION   7 BOOSTER ENGINE
3 EQUIPMENT POD   8 SUSTAINER ENGINE
4 VERNIER RETARDING ROCKET 9 BOOSTER ENGINE
5 VERNIER ENGINE   10 EQUIPMENT POD
              11 DECOY POD

40. 10-4(600)B

Figure 1-75. HGM16F Strategic Missile

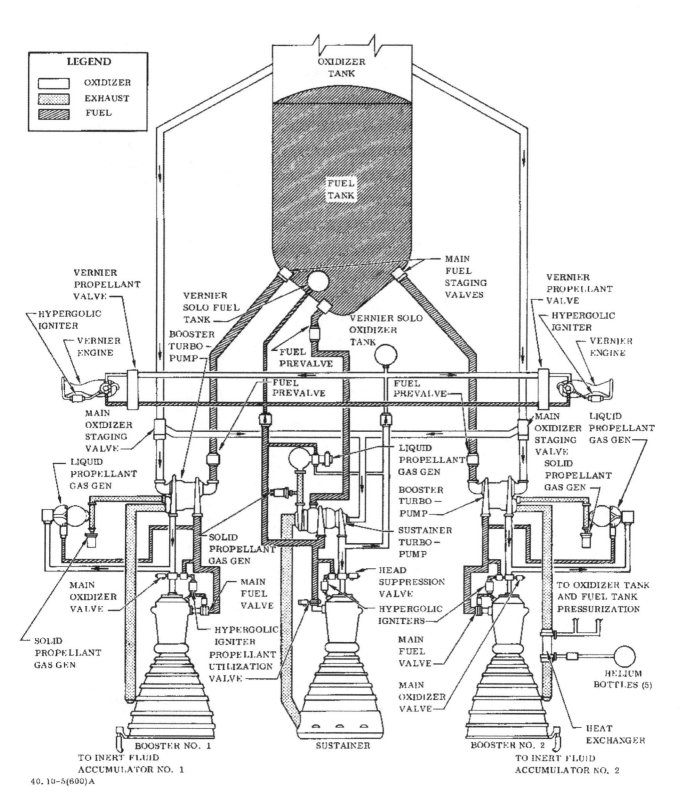

Figure 1-76. Missile Propulsion System

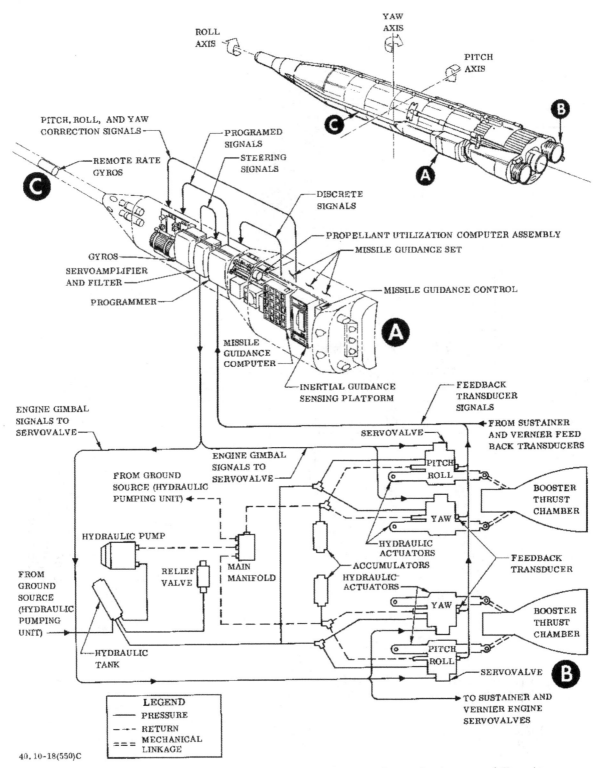

Figure 1-77. Missile Flight Control System, Guidance System, and Booster
Hydraulic Subsystem

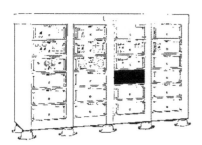

CONTROL-MONITOR GROUP 2 OF 4

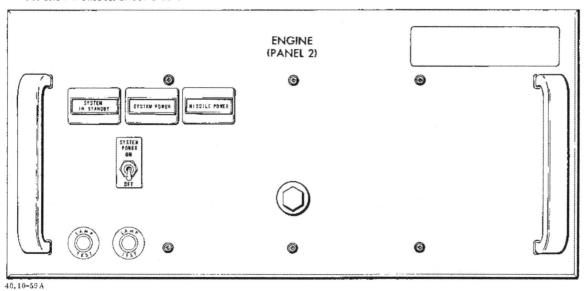

40.10-59 A

Figure 1-78. Engine (Panel 2)

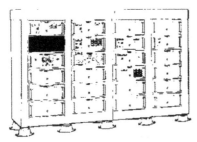

CONTROL-MONITOR GROUP 2 OF 4

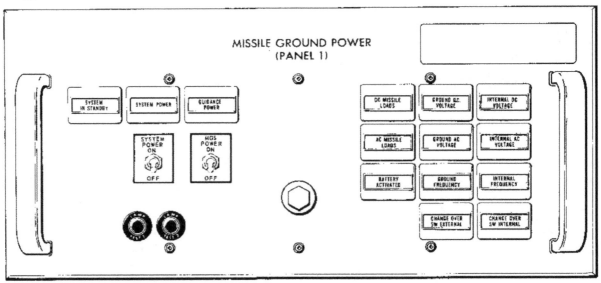

40.10-52 A

Figure 1-79. Missile Ground Power (Panel 1)

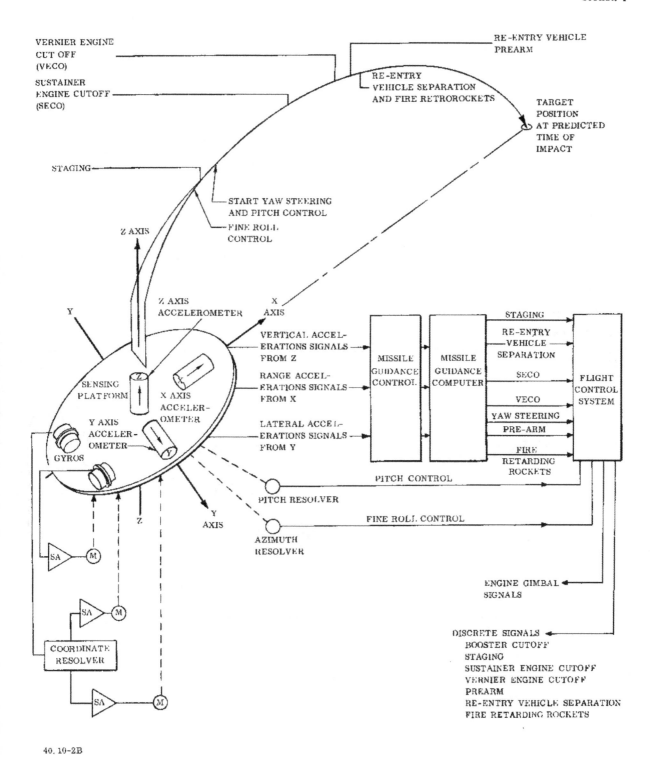

Figure 1-80. Inertial Guidance Subsystem

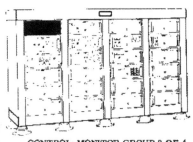

CONTROL-MONITOR GROUP 2 OF 4

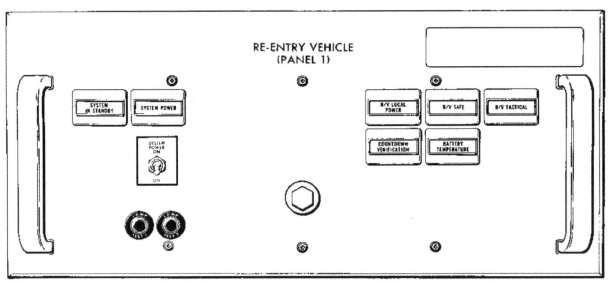

N40. 10-4

Figure 1-81.　Re-entry Vehicle (Panel 1)

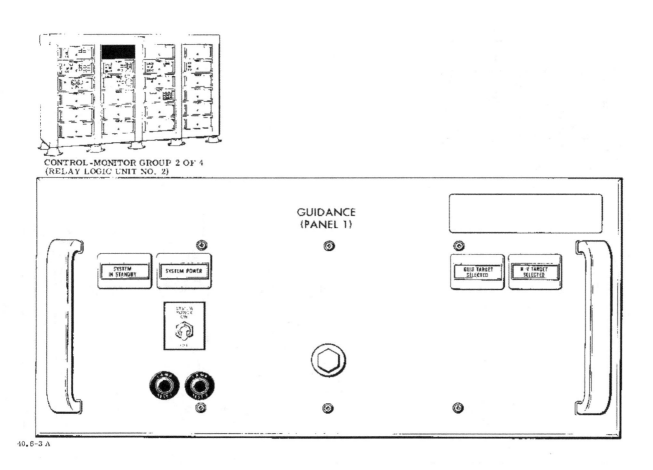

CONTROL-MONITOR GROUP 2 OF 4
(RELAY LOGIC UNIT NO. 2)

40.6-3 A

Figure 1-82.   Guidance (Panel 1)

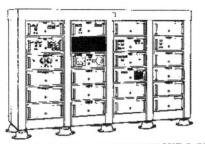

CONTROL-MONITOR GROUP 2 OF 4

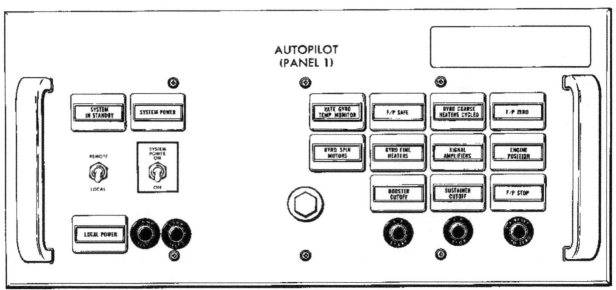

N40. 10-8

Figure 1-83.　Autopilot (Panel 1)

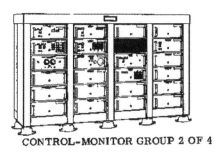

CONTROL-MONITOR GROUP 2 OF 4

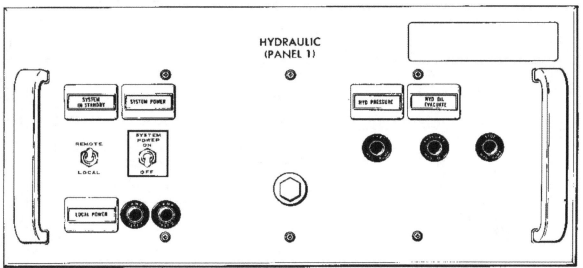

N40.10-9

Figure 1-84. Hydraulic (Panel 1)

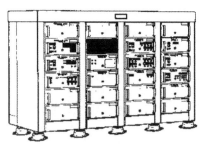

CONTROL-MONITOR GROUP 1 OF 4

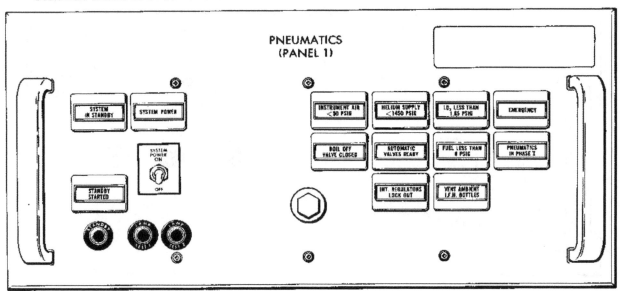

N40.10-3

Figure 1-85.   Pneumatics (Panel 1)

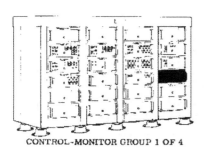

CONTROL-MONITOR GROUP 1 OF 4

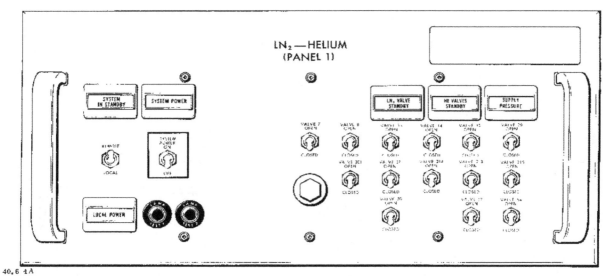

40.6 4A

Figure 1-86. Liquid Nitrogen–Helium (Panel 1)

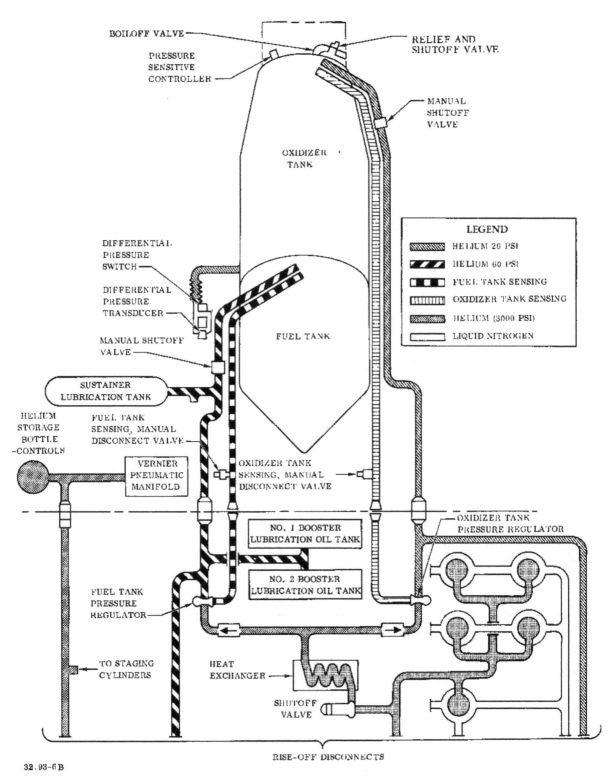

BOILOFF VALVE

PRESSURE
SENSITIVE
CONTROLLER

RELIEF AND
SHUTOFF VALVE

MANUAL
SHUTOFF
VALVE

OXIDIZER
TANK

LEGEND

HELIUM 26 PSI

HELIUM 60 PSI

FUEL TANK SENSING

OXIDIZER TANK SENSING

HELIUM (3000 PSI)

LIQUID NITROGEN

DIFFERENTIAL
PRESSURE
SWITCH

DIFFERENTIAL
PRESSURE
TRANSDUCER

FUEL TANK

MANUAL SHUTOFF
VALVE

SUSTAINER
LUBRICATION TANK

HELIUM
STORAGE
BOTTLE
-CONTROLS

FUEL TANK
SENSING, MANUAL
DISCONNECT VALVE

VERNIER
PNEUMATIC
MANIFOLD

OXIDIZER TANK
SENSING, MANUAL
DISCONNECT VALVE

OXIDIZER TANK
PRESSURE REGULATOR

NO. 1 BOOSTER
LUBRICATION OIL TANK

NO. 2 BOOSTER
LUBRICATION OIL TANK

FUEL TANK
PRESSURE
REGULATOR

TO STAGING
CYLINDERS

HEAT
EXCHANGER

SHUTOFF
VALVE

RISE-OFF DISCONNECTS

32.93-6 B

Figure 1-87. Missile Pneumatic System

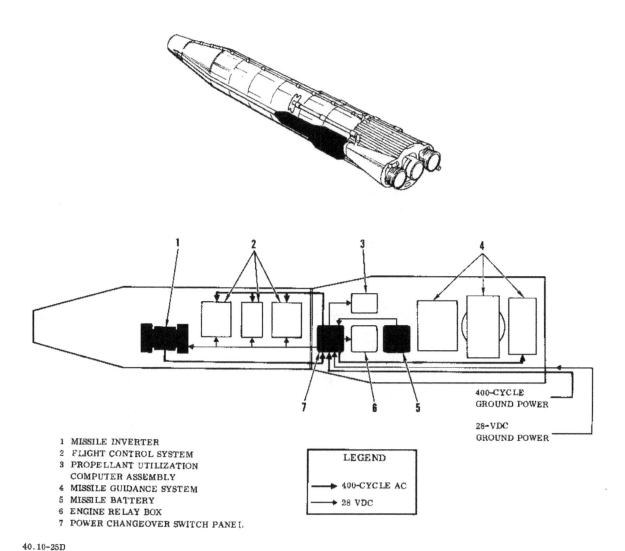

1 MISSILE INVERTER
2 FLIGHT CONTROL SYSTEM
3 PROPELLANT UTILIZATION
  COMPUTER ASSEMBLY
4 MISSILE GUIDANCE SYSTEM
5 MISSILE BATTERY
6 ENGINE RELAY BOX
7 POWER CHANGEOVER SWITCH PANEL

400-CYCLE
GROUND POWER

28-VDC
GROUND POWER

LEGEND

→ 400-CYCLE AC

→ 28 VDC

40.10-25D

Figure 1-88. Missile Electrical System

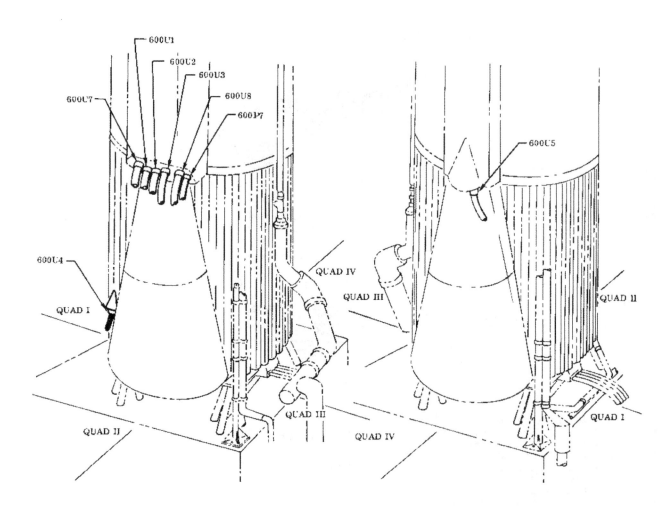

40.10-39A

Figure 1-89. Missile Umbilical Connections for Training Launch

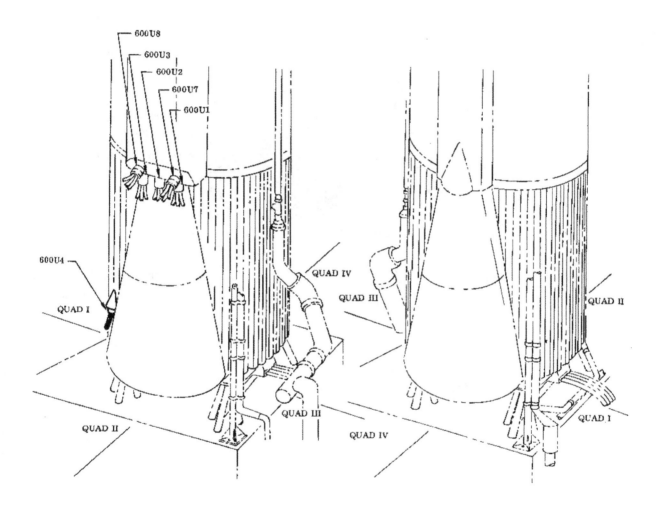

600U8
600U3
600U2
600U7
600U1

600U4

QUAD I

QUAD II

QUAD IV

QUAD III

QUAD III

QUAD IV

QUAD II

QUAD I

40.10-39(SAFB)B

Figure 1-90. Missile Umbilical Connections

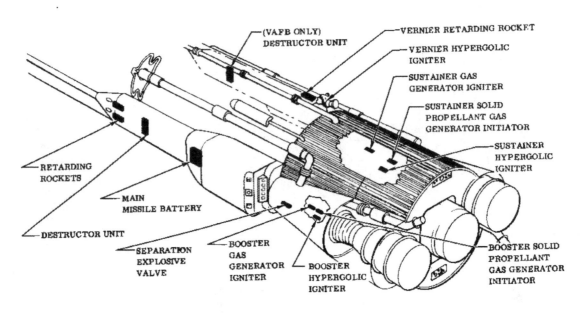

N40.10-17(600)A

Figure 1-91. Missile Explosive Assemblies

Changed 21 October 1964

1-172
(Pages 1-173 thru 1-178, Figures 1-92 thru 1-94 deleted).

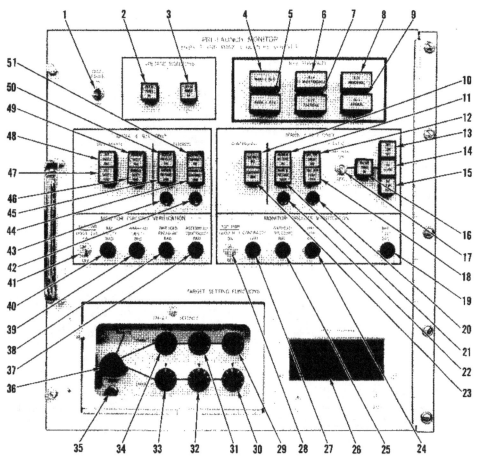

1 PANEL POWER TOGGLE SWITCH
2 28 VDC POWER ON INDICATOR (GREEN)
3 115 VAC POWER ON INDICATOR (GREEN)
4 MARK 3 R/V INDICATOR (WHITE)
5 MARK 4 R/V INDICATOR (WHITE)
6 R/V IN MAINTENANCE INDICATOR (WHITE)
7 R/V TACTICAL INDICATOR (GREEN)
8 R/V ABNORMAL INDICATOR (RED)
9 R/V NORMAL INDICATOR (GREEN)
10 CONTINUITY BAD INDICATOR (RED)
11 WARHEAD PRESSURE BAD INDICATOR (RED)
12 BAT. HTR CIRCUIT BAD INDICATOR (RED)
13 BAT. TEMP BAD INDICATOR (RED)
14 BAT. HEATING INDICATOR (WHITE)
15 BAT. TEMP GOOD INDICATOR (GREEN)
16 BAT. HTR POWER ON INDICATOR (GREEN)
17 BAT. HTR TOGGLE SWITCH
18 BAT. HTR CIRCUIT GOOD INDICATOR (GREEN)
19 BAT. TEMP BAD PUSHBUTTON SWITCH (RED)
20 BAT. HTR CIRCUIT MOMENTARY PUSHBUTTON SWITCH
21 WARHEAD PRESSURE GOOD INDICATOR (GREEN)
22 WARHEAD PRESSURE MOMENTARY PUSHBUTTON SWITCH
23 CONTINUITY GOOD INDICATOR (GREEN)
24 BAT. HTR BAD PUSHBUTTON SWITCH (RED)
25 WARHEAD PRESSURE BAD PUSHBUTTON SWITCH (RED)

26 TARGET POSITION INDICATOR
27 CONTINUITY BAD PUSHBUTTON SWITCH (RED)
28 TEST PWR/GOOD TEST TOGGLE SWITCH
29 TARGET A UNITS ROTARY SWITCH
30 TARGET B UNITS ROTARY SWITCH
31 TARGET A TENS ROTARY SWITCH
32 TARGET B TENS ROTARY SWITCH
33 TARGET B MODE ROTARY SWITCH
34 TARGET A MODE ROTARY SWITCH
35 RESET MOMENTARY PUSHBUTTON SWITCH
36 TARGET SELECT ROTARY SWITCH
37 ASSEMBLED CONTINUITY BAD PUSHBUTTON SWITCH (RED)
38 WARHEAD PRESSURE BAD PUSHBUTTON SWITCH (RED)
39 WARHEAD SAFETY BAD PUSHBUTTON SWITCH (RED)
40 A&F SAFETY BAD PUSHBUTTON SWITCH (RED)
41 TEST PWR/GOOD TEST TOGGLE SWITCH
42 ASSEMBLED CONTINUITY MOMENTARY PUSHBUTTON SWITCH
43 WARHEAD PRESSURE MOMENTARY PUSHBUTTON SWITCH
44 ASSEMBLED CONTINUITY GOOD INDICATOR (GREEN)
45 WARHEAD PRESSURE GOOD INDICATOR (GREEN)
46 WARHEAD SAFETY GOOD INDICATOR (GREEN)
47 A&F SAFETY GOOD INDICATOR (GREEN)
48 A&F SAFETY GAD INDICATOR (RED)
49 WARHEAD SAFETY BAD INDICATOR (RED)
50 WARHEAD PRESSURE BAD INDICATOR (RED)
51 ASSEMBLED CONTINUITY BAD INDICATOR (RED)

40.10-140

Figure 1-95. Re-entry Vehicle Pre-Launch Monitor Panel

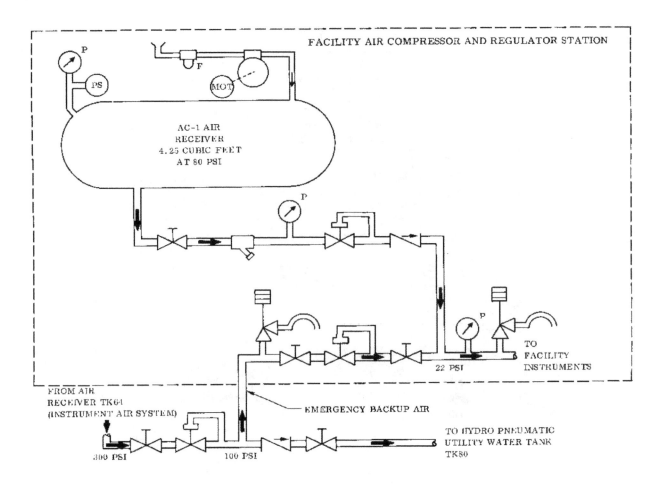

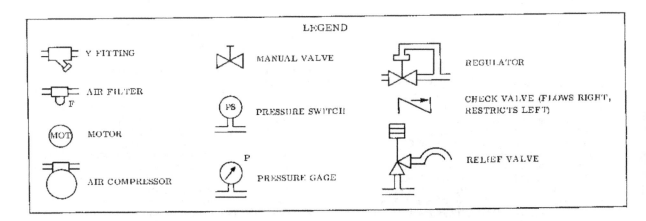

Figure 1-96. Facility Air Compressor and Regulator Station and Emergency
Backup System Flow Diagram (Squadron Complexes at VAFB and 550)

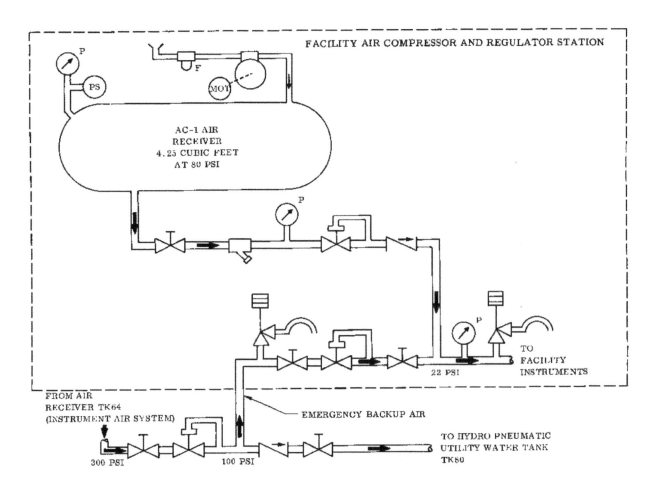

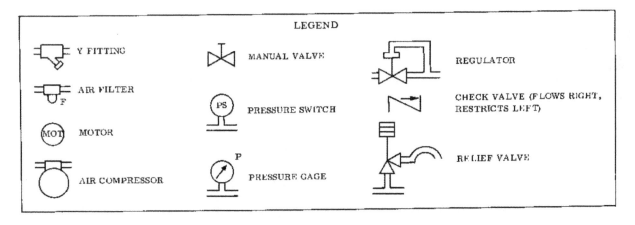

N 40. 10-12A

Figure 1-96. Facility Air Compressor and Regulator Station and Emergency
Backup System Flow Diagram (Squadron Complexes at VAFB and 550)

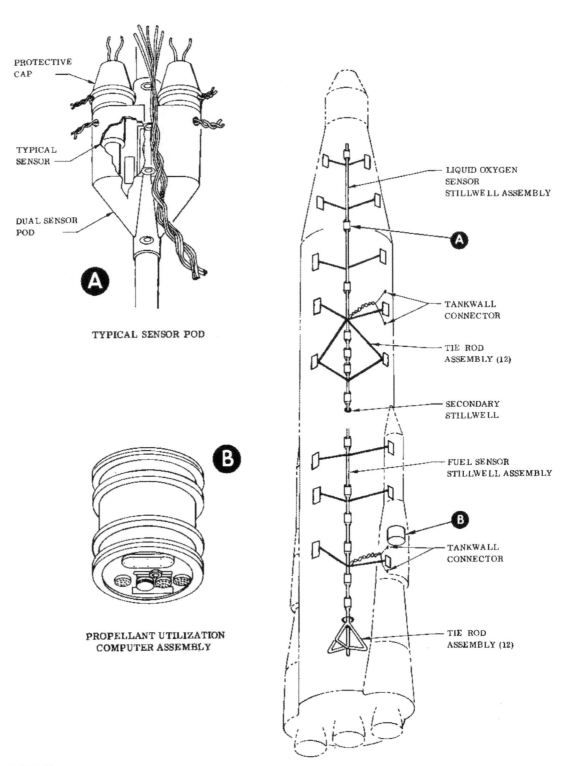

PROTECTIVE
CAP

TYPICAL
SENSOR

DUAL SENSOR
POD

A

TYPICAL SENSOR POD

B

PROPELLANT UTILIZATION
COMPUTER ASSEMBLY

LIQUID OXYGEN
SENSOR
STILLWELL ASSEMBLY

A

TANKWALL
CONNECTOR

TIE ROD
ASSEMBLY (12)

SECONDARY
STILLWELL

FUEL SENSOR
STILLWELL ASSEMBLY

B

TANKWALL
CONNECTOR

TIE ROD
ASSEMBLY (12)

N40.10-18

Figure 1-98.   Propellant Utilization System

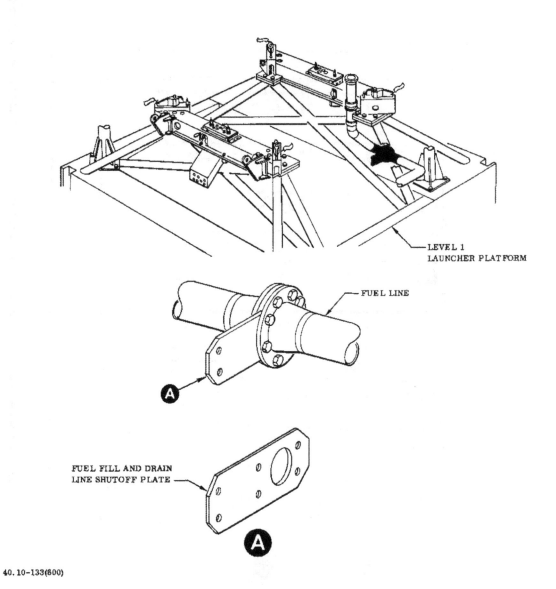

LEVEL 1
LAUNCHER PLATFORM

FUEL LINE

A

FUEL FILL AND DRAIN
LINE SHUTOFF PLATE

A

40.10-133(800)

Figure 1-99.   Fuel Fill-and-Drain Line Shutoff Plate (VAFB only)

LAUNCH CONTROL CONSOLE

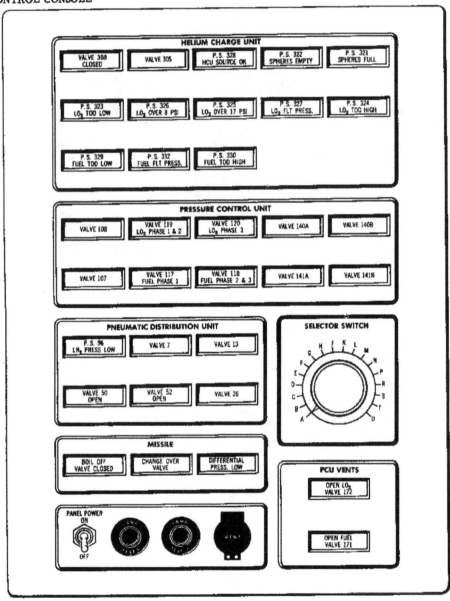

40.10-132(600)A

Figure 1-100.   Pneumatic Local Control Panel

Changed 15 April 1964

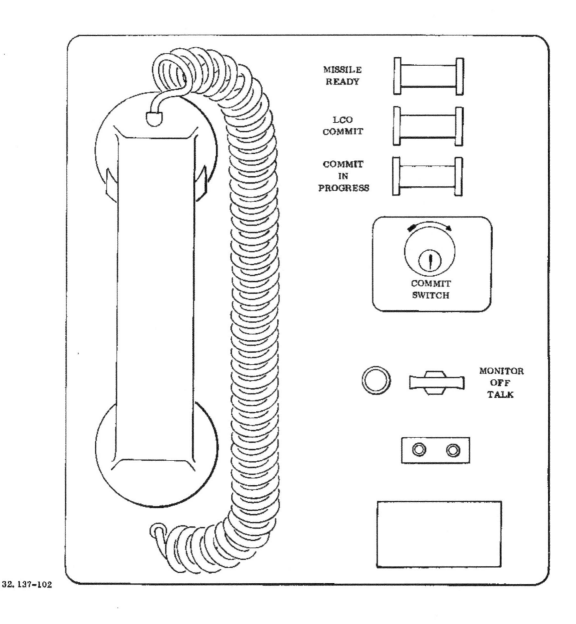

32.137-102

Figure 1-101. **ALCO COMM/CONTROL Panel**

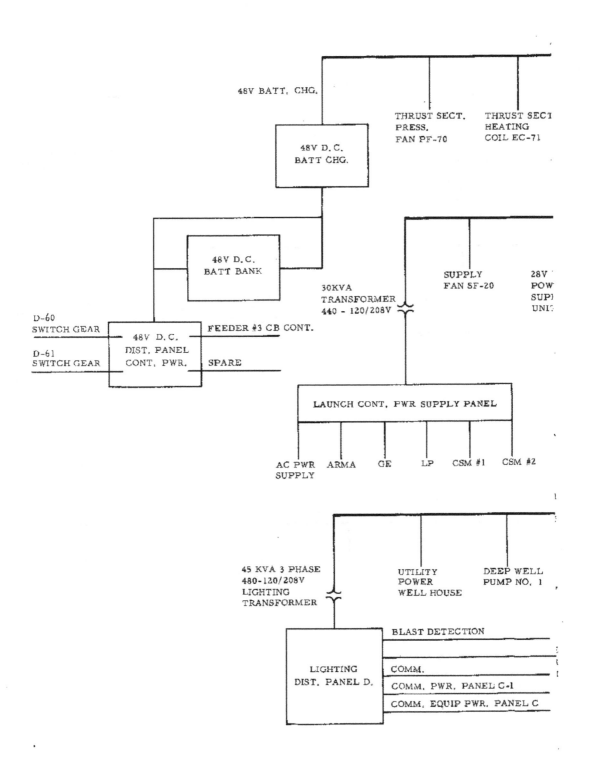

48V BATT. CHG.

THRUST SECT.
PRESS.
FAN PF-70

THRUST SECT
HEATING
COIL EC-71

48V D.C.
BATT CHG.

48V D.C.
BATT BANK

SUPPLY
FAN SF-20

28V
POW
SUP
UNIT

30KVA
TRANSFORMER
440 - 120/208V

D-60
SWITCH GEAR

D-61
SWITCH GEAR

48V D.C.
DIST. PANEL
CONT. PWR.

FEEDER #3 CB CONT.

SPARE

LAUNCH CONT. PWR SUPPLY PANEL

AC PWR
SUPPLY

ARMA

GE

LP

CSM #1

CSM #2

45 KVA 3 PHASE
480-120/208V
LIGHTING
TRANSFORMER

UTILITY
POWER
WELL HOUSE

DEEP WELL
PUMP NO. 1

LIGHTING
DIST. PANEL D.

BLAST DETECTION

COMM.

COMM. PWR. PANEL C-1

COMM. EQUIP PWR. PANEL C

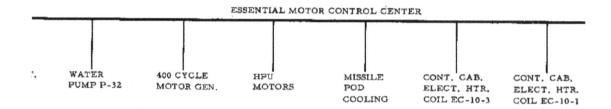

ESSENTIAL MOTOR CONTROL CENTER

| WATER PUMP P-32 | 400 CYCLE MOTOR GEN. | HPU MOTORS | MISSILE POD COOLING | CONT. CAB. ELECT. HTR. COIL EC-10-3 | CONT. CAB. ELECT. HTR. COIL EC-10-1 |

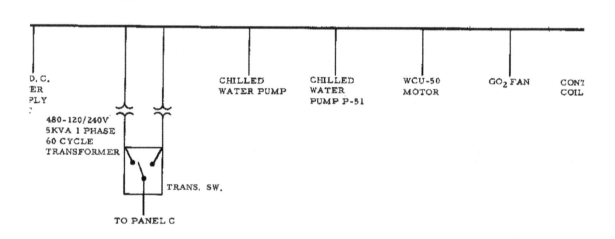

D. C.
ER
PLY

CHILLED WATER PUMP   CHILLED WATER PUMP P-51   WCU-50 MOTOR   GO$_2$ FAN   CONT COIL

480-120/240V
5KVA 1 PHASE
60 CYCLE
TRANSFORMER

TRANS. SW.

TO PANEL C

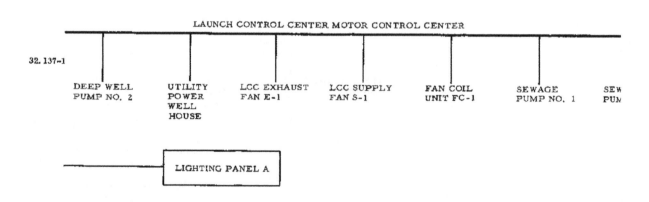

LAUNCH CONTROL CENTER MOTOR CONTROL CENTER

32.137-1

| DEEP WELL PUMP NO. 2 | UTILITY POWER WELL HOUSE | LCC EXHAUST FAN E-1 | LCC SUPPLY FAN S-1 | FAN COIL UNIT FC-1 | SEWAGE PUMP NO. 1 | SEW PUM |

LIGHTING PANEL A

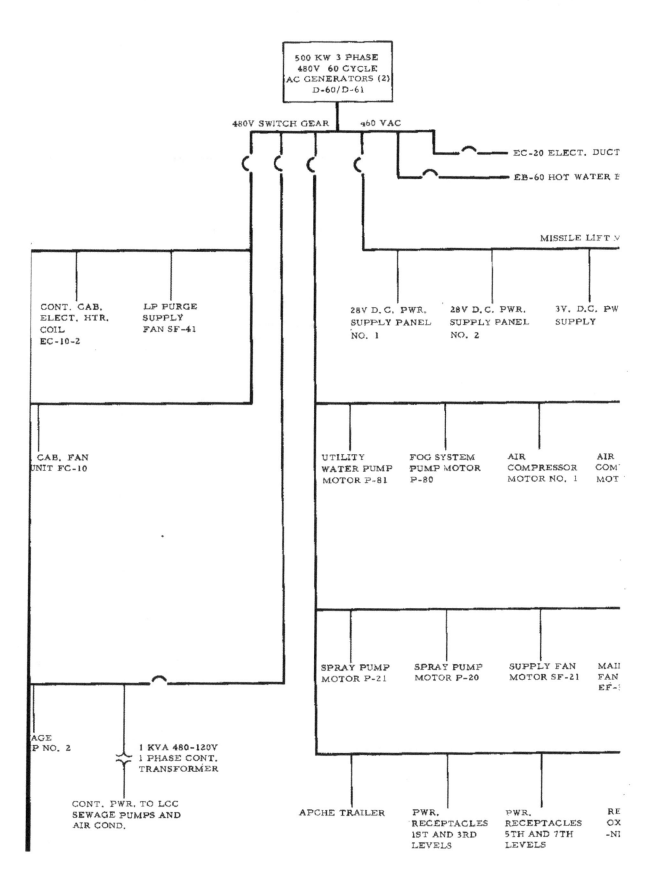

USE FOR REFERENCE OF CIRCUITRY TO MOTOR CONTROL CENTER ONLY.

NOT TO BE USED FOR CIRCUIT OR CIRCUIT BREAKER LOCATIONS.

HEATER

OILER

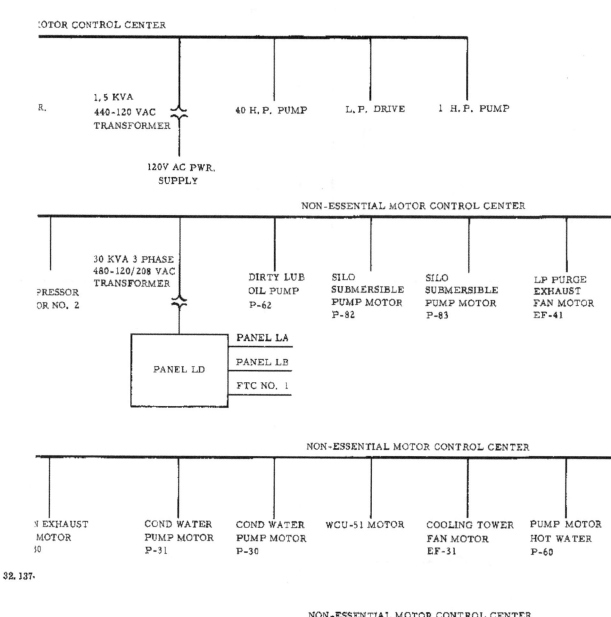

OTOR CONTROL CENTER

R.

1,5 KVA
440-120 VAC
TRANSFORMER

40 H.P. PUMP   L.P. DRIVE   1 H.P. PUMP

120V AC PWR.
SUPPLY

NON-ESSENTIAL MOTOR CONTROL CENTER

30 KVA 3 PHASE
480-120/208 VAC
TRANSFORMER

PRESSOR
OR NO. 2

DIRTY LUB
OIL PUMP
P-62

SILO
SUBMERSIBLE
PUMP MOTOR
P-82

SILO
SUBMERSIBLE
PUMP MOTOR
P-83

LP PURGE
EXHAUST
FAN MOTOR
EF-41

PANEL LA

PANEL LD

PANEL LB

FTC NO. 1

NON-ESSENTIAL MOTOR CONTROL CENTER

N EXHAUST
MOTOR
30

COND WATER
PUMP MOTOR
P-31

COND WATER
PUMP MOTOR
P-30

WCU-51 MOTOR

COOLING TOWER
FAN MOTOR
EF-31

PUMP MOTOR
HOT WATER
P-60

32.137-

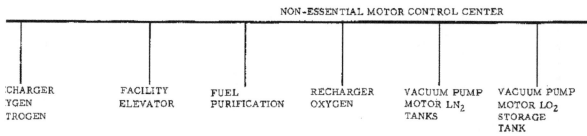

NON-ESSENTIAL MOTOR CONTROL CENTER

CHARGER
YGEN
TROGEN

FACILITY
ELEVATOR

FUEL
PURIFICATION

RECHARGER
OXYGEN

VACUUM PUMP
MOTOR LN$_2$
TANKS

VACUUM PUMP
MOTOR LO$_2$
STORAGE
TANK

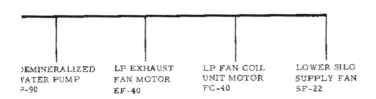

DEMINERALIZED
WATER PUMP
P-90

LP EXHAUST
FAN MOTOR
EF-40

LP FAN COIL
UNIT MOTOR
FC-40

LOWER SILO
SUPPLY FAN
SF-22

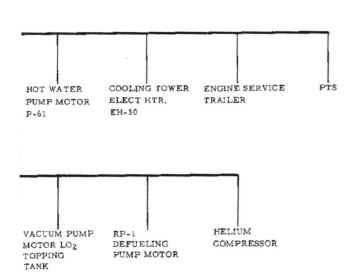

HOT WATER
PUMP MOTOR
P-61

COOLING TOWER
ELECT HTR.
EH-30

ENGINE SERVICE
TRAILER

PTS

VACUUM PUMP
MOTOR LO_2
TOPPING
TANK

RP-1
DEFUELING
PUMP MOTOR

HELIUM
COMPRESSOR

Figure 1-102   COMPLEX POWER DISTRIBUTION

SECTION II
RECEIPT-THROUGH-LAUNCH OPERATION PLAN

TABLE OF CONTENTS

LIST OF ILLUSTRATIONS

2-1. SCOPE

2-2. This section contains a receipt-through-launch operation plan and general block diagram flow chart for scheduled activities at the launch complex.

2-3. DESCRIPTION

2-4. The block diagram flow charts contained in this section, function as a managerial guide to be used by staff and missile combat crew personnel in planning scheduled maintenance, propellant loading exercises, and launches. Detailed functional flow charts for scheduled activities are contained in T.O. 21M-HGM16F-6 and T.O. 21M-HGM16F-3CL-1.

2-5. Figure 2-1, receipt-through-launch operation plan, illustrates the general overall flow of scheduled and unscheduled maintenance and training requirements in block diagram form. The receipt-through-launch operation plan presents general requirements from receipt of the missile from the factory, or missile assembly and maintenance shops (MAMS), through launch and postlaunch or return to MAMS for recycle. The requirements of the plan are performed using inspection requirements manuals, work cards, checklists, organizational maintenance manuals, and checkout manuals. Inspection requirements manuals provide the minimum requirements for scheduled maintenance inspections and scheduled replacement and calibration of equipment. The inspection requirements detail what equipment is inspected, the intervals at which such equipment is inspected, and what conditions are sought. An estimate of the number, types of personnel, and the time required to perform each inspection is included.

2-6. Figure 2-2 illustrates the general flow from EWO through scheduled activities to EWO. Reference to manuals containing detailed flow charts are shown as appropriate.

2-7. Functional flow charts for periodic inspections depict the systematic and sequential flow of work cards, checklist sections and (or) technical manuals. They are to be used in planning and performing preventive maintenance tasks, system calibrations, inspections and checkout of missile and aerospace ground equipment (AGE), starting from a known configuration.

2-8. Starting from a known configuration, ready state B functional flow charts depict the systematic and sequential flow of checklist sections to be performed to establish and verify the configuration of the missile, AGE and real property installed equipment (RPIE) for EWO readiness except for missile ordnance items and the re-entry vehicle.

2-9. Starting from a known configuration, ready state A functional flow charts depict the systematic and sequential flow of checklist sections to be performed to establish and verify the configuration of missile AGE and RPIE for EWO readiness.

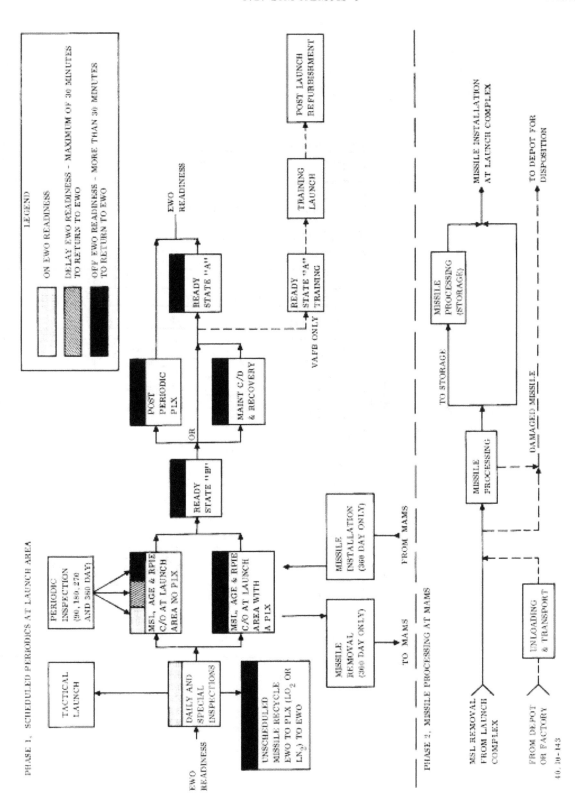

Figure 2-1. Receipt-Through-Launch Operation Plan Block Diagram

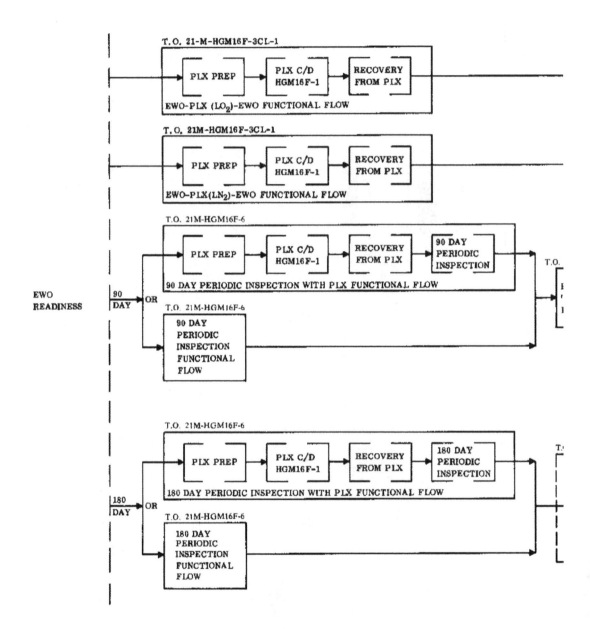

Changed 15 April 1964

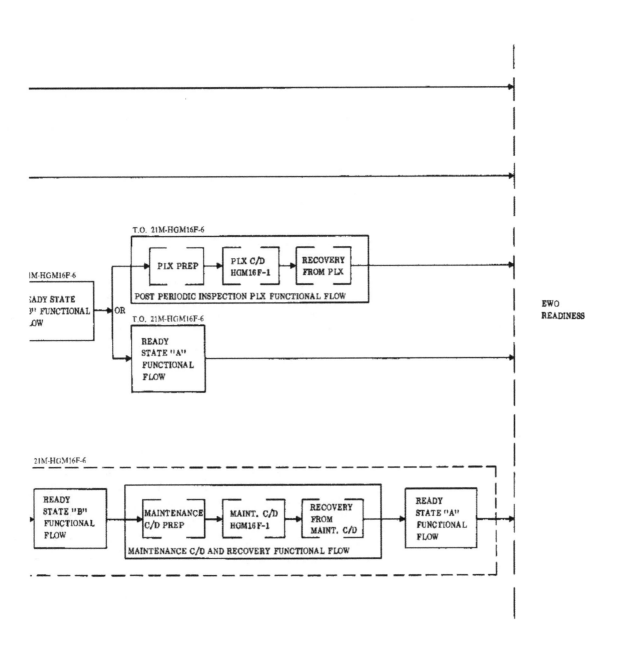

Figure 2-2.   From EWO Through Scheduled Activities to EWO Block Diagram   (Sheet 1 of 2)

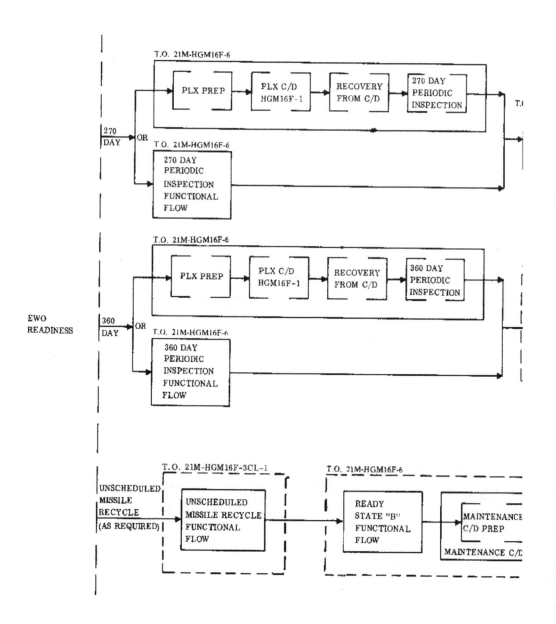

40.10-144.2

Changed 15 April 1964

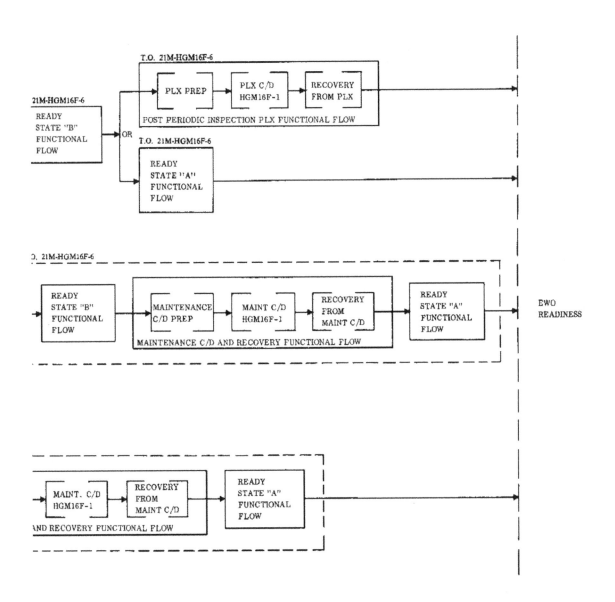

Figure 2-2. From EWO Through Scheduled Activities to EWO Block Diagram   (Sheet 2 of 2)

SECTION III

NORMAL OPERATING PROCEDURES

TABLE OF CONTENTS

LIST OF ILLUSTRATIONS

LIST OF TABLES

TABLE OF ASSOCIATED CHECKLISTS

T.O. 21M-HGM16F-1CL-1   Abbreviated Checklist, Normal Operating Procedures, Missile Combat Crew Commander (MCCC)

T.O. 21M-HGM16F-1CL-2   Abbreviated Checklist, Normal Operating Procedures, Deputy Missile Combat Crew Commander (DMCCC)

T.O. 21M-HGM16F-1CL-3   Abbreviated Checklist, Normal Operating Procedures, Electrical Power Production Technician (EPPT)

T.O. 21M-HGM16F-1CL-4   Abbreviated Checklist, Normal Operating Procedures Missile Facilities Technician (MFT)

T.O. 21M-HGM16F-1CL-5   Abbreviated Checklist, Normal Operating Procedures, Ballistic Missile Analyst Technician (BMAT)

3-1. SCOPE.

3-2. This section contains the normal operating procedures for the missile combat crew at the launch complex. Instructions and tables outling missile combat crew procedures, tasks, and required briefings, from complex acceptance through launch, are provided. The normal functional flow of missile combat crew duties during an alert tour is shown in figure 3-1. The procedures in this section are contained in abbreviated checklist form in T.O. 21M-HGM16F-1CL-1 through T.O. 21M-HGM16F-1CL-5 by sections as shown below:

a. Deleted

b. Section 2, COMPLEX SAFETY BRIEFING taken from table 3-2.

c. Section 3, MISSILE CREW CHANGE-OVER BRIEFING taken from table 3-3.

d. Section 4, EWO COMPLEX ACCEPTANCE PROCEDURES taken from table 3-4.

e. Section 5, MISSILE COMBAT CREW OPERATIONS BRIEFING taken from table 3-5.

f. Section 6, ACTIVITY COORDINATION BRIEFING taken from table 3-6.

g. Section 7, COMPLEX MONITORING taken from tables 7, and 7B.

h. Section 8, MISSILE CREW DAILY COMPLEX STATUS VERIFICATION taken from table 3-8.

i. Section 9, MGS ALIGNED, NOT-ALIGNED PROCEDURES taken from table 3-9.

j. Section 10, MISSILE LIFTING SYSTEM CORRELATION taken from table 3-10.

k. Section 11, DIESEL GENERATOR OPERATION AT PRCP (NA OSTF-2) taken from table 3-11.

l. Section 11A, DIESEL GENERATOR AND POWER CO LINE OPERATIONS AT PRCP (OSTF-2 ONLY) taken from table 3-11A.

m. Section 12, DIESEL GENERATOR OPERATION AT 480-VOLT SWITCH GEAR taken from table 3-12.

n. Section 12A, DIESEL GENERATOR AND POWER CO OPERATION AT 480-VOLT SWITCHGEAR (OSTF-2 ONLY) taken from table 3-12A.

o. Section 13, TV SYSTEM OPERATION taken from table 3-13.

p. Section 14, RE-ENTRY VEHICLE PRE-LAUNCH MONITOR TARGET SETTING PROCEDURES taken from table 3-14.

q. Section 15, PLX PREPARATION AND COORDINATION taken from table 3-15.

r. Section 16, COUNTDOWN taken from table 3-16.

s. Section 17, ABORT taken from table 3-17.

t.   Section 18, POSTLAUNCH ABORT (TAC-
TICAL) taken from table 3-18.

NOTE

Each step of an abbreviated checklist provided with a check-off space shall be checked
off after accomplishment. Individual crew
members performing tasks outlined in normal
operating procedures checklist sections 4,
7, 7B, and 8 shall enter their initials in the
check-off space in place of a check mark.
Abnormal indications shall be entered over
the step to which they apply. At crew changeover the relieving MCCC shall review checklist sections 7, 7B, and 8 performed by the
duty crew and section 4 performed by the
relieving crew prior to signing the exception
release. The MCCC shall review sections 4,
7, 7B, and 8 each time a section is accomplished. Initials shall remain on the checklists until crew changeover or until the checklist is to be reaccommplished.

3-3. COMPLEX ACCESS AND EXIT PRO-CEDURES.

WARNING

Except when directed, personnel shall not
approach within 1875 feet of the complex
area when the personnel warning light is
flashing red or the klaxon is sounding, or if a
loudspeaker announcement has been made
to clear the area. Exception to this warning
is authorized if communication with the
launch control center (LCC) cannot be
established, and after close surveillance of
the area, the approaching personnel can determine that no visible danger exists. At this
time one member of the approaching team
may proceed to the complex gate telephone
and make contact with the LCC for authority
to enter the complex.

a.   Vestibule blast doors shall be closed at all
times except:

(1)   During personnel entry or exit.

(2)   When the complex is off alert status during
heavy personnel traffic, and it is desired to
keep the doors open.

b.   All silo blast doors shall be closed:

(1)   During all countdowns and launches.

(2)   When servicing liquids and gases, except
diesel fuel to the soft storage tank and water
to the water system.

(3)   Whenever conditions exist in the silo which
required evacuation.

c.   At least one of the two blast doors nearest
the silo shall be closed at all times other than
listed in step b above, except during personnel entry or exit.

d.   Whenever conditions require the vestibule or
all silo doors to be closed, the blast door
interlock system shall be utilized. Only one
vestibule or silo door will be open at any one
time during entry or exit. Exception may be
made only if equipment installation or removal require opening of more than one door
at a time, or to expedite evacuation of the
LCC or silo.

3-4.   Authorization and procedures for entering
the complex shall be in accordance with SAC unit
security plans. The MCCC shall be notified in ad-
vance as to who will enter and why. Entry into the
complex shall be monitored closely by a crew member designated by the MCCC to act as complex controller. Monitoring shall be accomplished by observing TV monitor, gate control panel, and the facility
remote control panel (FRCP) light indications.
Normal entry into the missile complex shall only be
accomplished by the following procedures:

a.   Establish identity of personnel entering the
complex by employing local directives Communication contact shall be made by telephone located at the complex gate.

b.   After positive identity has been established,
entering personnel will be directed to proceed
directly to the outer door of the entrapment
area and to alert the LCC by depressing the
door buzzer signal button.

c.   The complex controller shall enable the complex gate to open after identification, and
shall ensure gate closure after entry by observing the gate control panel.

d.   Upon receiving the outer door buzzer signal,
the complex controller shall enable the outer

door to open by depressing NO. 1 entrapment door pushbutton on the gate control panel.

e. After entering personnel have entered the entrapment area and the outer door is secured, establish identity employing local directives.

f. After identification, personnel shall be directed to proceed to level 2 of the LCC and report to the duty MCCC. Instruct entering personnel that only one vestibule blast door shall be open at a time. Opening and closing of the vestibule doors shall be monitored on the FRCP to determine compliance with the above instructions and that the vestibule doors are properly closed.

3-5. Complex exit shall be accomplished in the following manner:

a. The MCCC or designated crew member shall ensure that the departing personnel have completed the tasks, duties, inspections forms, etc., which required their presence at the complex.

b. Departing personnel shall be briefed that only one door (entrapment or vestibule) shall be opened at any one time, to contact the LCC by telephone when in the entrapment area, and then to proceed directly to the complex gate. Upon arrival at the gate, contact the LCC on complex telephone for gate opening. After exiting the complex, departing personnel shall ensure that gate is properly closed and locked. They shall contact the LCC on gate telephone, ensure that LCC indications show gate is closed, and then request permission to depart the complex area. (Local procedures shall apply for enabling opening of gate until TV surveillance monitors are installed.)

c. Departure shall be monitored in the same manner as entry.

d. Deleted

3-6. MISSILE COMBAT CREW CHANGE-OVER PROCEDURES.

3-7. The MCCC of the crew being relieved shall conduct a formal briefing attended by duty and relief crews. This briefing shall be conducted upon arrival of the relieving crew. A current complex status chart shall be maintained at each complex. This chart shall indicate complex status, (equipment configuration that effects EWO or normal daily operation) EWO status, liquid and gas levels, and maintenance activities. It shall be used to aid in an expeditious and complete briefing. Procedures to be followed and items to be covered during this briefing are contained in Table 3-3. An abbreviated checklist condensed from this table is contained in T. O. 21M-HGM16F-1CL-1.

3-8. COMPLEX SAFETY BRIEFING.

3-9. The duty MCCC shall ensure that personnel arriving at the complex are briefed on all items and procedures as may be applicable in table 3-2, before departing the LCC for any destination within the complex. It is the responsibility of the MCCC to control personnel within the entire complex. The MCCC may designate a complex controller who shall be responsible for knowing the location of all personnel in the complex at all times. The complex safety briefing content may vary in detail depending on the experience level of personnel being briefed and is left to the discretion of the briefing officer. If numerous complex or silo entries and exits are to be made by the same personnel during one duty period the briefing need not be repeated. Visitors or other personnel entering the silo for the first time shall always be accompanied by an experienced person. Visitors shall be briefed on all items contained in table 3-2. Abbreviated checklists condensed from table 3-2 is contained in T.O. 21M-HGM16F-1CL-1 and T.O. 21M-HGM16F-1CL-2.

3-10. EWO COMPLEX ACCEPTANCE PROCEDURES.

3-11. Prior to accepting command of a complex the MCCC shall ensure that he has full understanding of the status of the complex, knows the location of portable equipment and material which may be used during his tour, and receipts for material as necessary. Prior to assuming complex command responsibilities, the MCCC shall accomplish items contained in table 3-4. Abbreviated checklists condensed from this table are contained in T. O. 21M-HGM16F-1CL-1.

3-12. MISSILE COMBAT CREW OPERATIONS BRIEFING.

3-13. The MCCC shall conduct a formal briefing to ensure proper crew coordination in event of actual EWO execution, "No-Notice" type training exercises, and scheduled training exercises. All combat crew members shall be present. This briefing shall be conducted immediately following completion of the positive control changeover procedures and prior to any other scheduled activities. For propellant loading exercise (PLX), the briefing shall be conducted at completion of preparation for PLX flow and prior to countdown execution. For LSR exercises, the briefing may be conducted at completion of LSR countdown preparation from EWO (paragraph 3-56) and prior to countdown execution if it has not been completed earlier in the duty tour. Items to be briefed and reviewed are provided in table 3-5. An abbreviated checklist condensed from this table is contained in T.O. 21M-HGM16F-1CL-1.

3-14. ACTIVITY COORDINATION BRIEFING.

3-15. The sector commander or his designated representative shall conduct a formal briefing prior to any operation or maintenance activity other than EWO launch operations. Lengthy activities such as an EWO-PLX-EWO flow may be briefed by individual tasks or groups of related tasks. All crew, maintenance, and support personnel involved in the activity shall be present. Briefing requirements are provided in table 3-6. Abbreviated checklists condensed from this table are contained in T.O. 21M-HGM16F-1CL-1 and T.O. 21M-HGM16F-1CL-2.

3-16. COMPLEX MONITORING PROCEDURES.

3-17. Continous monitoring of complex status, as reflected by consoles in the LCC, shall be accomplished by missile crew members at all times. In addition, periodic inspection or verification of specified equipment in the silo shall be performed during a duty tour regardless of complex status.

3-18. A list of items in the LCC requiring periodic inspection and verification is contained in table 3-7. Individual abbreviated checklist condensed from this table are contained in T.O. 21M-HGM16F-1CL-1 and T.O. 21M-HGM16F-1CL-2. Continuous monitoring of all items is required. In addition, a specific check of all items shall be accomplished each time there is a scheduled change in the two-man team, but not to exceed once every four hours. Any deviation from normal configuration will require immediate corrective action to retain or regain alert status.

3-19. A list of items in the silo requiring hourly inspection and verification is contained in table 3-7A. This check may be accomplished individually by either the EPPT, MFT, or BMAT. No checklist is required to support this table. The Daily Power Plant Operating Log, AF Form 1167, shall be completed in accordance with AFR 91-4A and instructions contained on the reverse side of the form.

3-20. A list of items in the silo requiring periodic inspection and verification is contained in table 3-7B. Individual abbreviated checklists condensed from this table are contained in T.O. 21M-HGM16F-1CL-3 through T.O. 21M-HGM16F-1CL-5. This check shall be accomplished a minimum of once every six hours. More frequent checks of individual items may be required when situations so dictate. Specific readings shall be recorded and logged on the Daily Complex Status Record, SAC 210a as source data for historical records. A sample format is shown in figure 3-2. Units may add to the form to meet local requirements.

3-21. MISSILE CREW DAILY COMPLEX STATUS VERIFICATION PROCEDURES.

3-22. A complex status verification and inspection shall be performed during each 24-hour tour of duty. A list of equipment to be checked daily is contained in table 3-8. Individual abbreviated checklists condensed from this table are contained in T.O. 21M-HGM16F-1CL-3 through T.O. 21M-HGM16F-1CL-5.

3-23. Complex status shall be verified at the earliest possible time during the alert tour to accomplish prompt verification of complex status. It shall be a scheduled item on the daily individual missile maintenance plan. The missile crew complex status checklist is designed to enable flexibility in the order in which silo levels and items on a level may be checked. The checklist shall be completed by a two-man team selected from the BMAT, MFT or EPPT. When the checklist is accomplished prior to a PLX, the safety representative may be an additional member of the team. In this instance, he shall be responsible for checking all complex safety equipment and safety conditions specified in table 3-8. Portions of the requirements in table 3-8 may be accomplished in conjunction with other EWO - PLX flow tasks, provided the equipment status of items checked will not be changed by completion of any unaccomplished tasks during the preparation flow.

3-24. PLACING MISSILE GUIDANCE SYSTEM IN ALIGNED STATUS.

3-25. Command policy directs that the missile guidance system (MGS) shall be placed in aligned status under certain conditions. Procedures for placing the guidance system in aligned status from a standby condition are provided in table 3-9, item 1. An abbreviated checklist condensed from this table is contained in T.O. 21M-HGM16F-1CL-5.

3-26. PLACING MISSILE GUIDANCE SYSTEM IN NOT-ALIGNED STATUS FROM ALIGNED STATUS.

3-27. When command policy directs the return to not-aligned status from the aligned status, procedures to follow are provided in table 3-9, item 2. An abbreviated checklist condensed from this table is contained in T.O. 21M-HGM16F-1CL-5.

3-28. MISSILE LIFTING SYSTEM (MLS) CORRELATION.

3-29. The MLS logic units shall be properly correlated prior to use. It is possible for correlation to be disturbed whenever the work platforms are raised or lowered or the control station manual operating level (CSMOL) is used to operate equipment in

the silo. Correlation shall be performed immediately prior to starting a PLX countdown and upon returning to EWO readiness condition. Procedures for correlation are contained in table 3-10. Abbreviated checklists condensed from this table are contained in T.O. 21M-HGM16F-1CL-4 and T.O. 21M-HGM16F-1CL-5.

3-30. DIESEL GENERATOR OPERATION.

3-31. Procedures for starting and paralleling the diesel generators from the power remote control panel (PRCP) are provided in table 3-11, item 1. An abbreviated checklist condensed from this table is contained in T.O. 21M-HGM16F-1CL-3. Prior to using procedures contained in table 3-11, the diesel generators shall have previously been paralleled from the 480-volt switch gear (table 3-12) (table 3-12A, OSTF-2 only), then shut down without disturbing the adjustment made during paralleling. The maximum time the diesel generators should be operated in parallel is 2-½ hours. Exception to this restriction may be made when power load is in excess of or reasonably expected to exceed, single generator limits. Diesel generators shall not be operated in parallel during standby, except momentarily for paralleling at the 480-volt switch gear, for parallel operational requirements, or if anticipated load on a single generator will exceed the optimum operating load of 500 KW (750 AMP).

3-32. Procedures for shutting down one paralleled diesel generator from the PRCP are provided in table 3-11, item 2. An abbreviated checklist condensed from this table is contained in T.O. 21M-HGM16F-1CL-3.

3-33. Procedures for transferring power from commercial power to the diesel generator and back on commercial power at OSTF-2 only, are provided in table 3-11A. An abbreviated checklist condensed from this table is contained in T.O. 21M-HGM16F-1CL-3.

3-34. A nontactical countdown shall not be initiated prior to generators being paralleled and placed on the line. Loss of one generator prior to MISSILE LIFT UP & LOCKED indicator AMBER requires abort except during a tactical countdown. If one generator fails after MISSILE LIFT UP & LOCKED indicator AMBER, continue the commit sequence. A tactical launch shall not be delayed for paralleling of generators, and may be initiated with one generator on the line. During single generator operation, the nonessential power bus should be de-energized when POWER INTERNAL indicator illuminates GREEN.

NOTE

At OSTF-2, a nontactical countdown shall be performed with the 500-KW diesel generator connected as the primary power source.

CAUTION

If cooling water tower is inoperative or destroyed, or EMERGENCY WATER PUMP P-32 ON indicator on FRCP is illuminated RED, do not start standby diesel generator unless a malfunction is indicated on the operating diesel generator. In this case, transfer power as quickly as possible and shut down faulty diesel generator. Failure to comply will cause the diesel engines to overheat.

3-35. Procedures for paralleling the diesel generators from the 480-volt switch gear at operational units are contained in table 3-12 (table 3-12A OSTF-2 only). Abbreviated checklists condensed from table 3-12 and 3-12A are contained in T.O. 21M-HGM16F-1CL-3.

3-36. **CLOSED CIRCUIT TELEVISION MONITOR.**

3-37. The permanently installed TV monitor located on level 2 of the LCC shall be in continuous operation to ensure monitoring capability at all times. During standby, the LIGHTING SYSTEM 1 INTENSITY switch shall be placed in the LOW position and the LIGHTING MAIN POWER switch turned OFF. During a tactical launch, the MFT shall place the INTENSITY switch to HIGH, the POWER SWITCH to ON, and position the monitor as directed by the MCCC. Procedures for operating the permanent TV monitor are provided in table 3-13, item 1. An abbreviated checklist condensed from table 3-13, item 1, is contained in T.O. 21M-HGM16F-1CL-4.

3-38. During a nontactical countdown three additional temporary TV monitors shall be positioned on level 2 of the LCC. These monitors shall be operated by the MFT or other crew member as directed by the MCCC. These units shall be used to observe the missile boiloff valve, launch platform movement, silo doors, missile frost level, electrical umbilical disconnects, LO_2 disconnects, etc, throughout the countdown and abort sequence. Procedures for operation of the four TV monitors are provided in table 3-13, item 2. An abbreviated checklist condensed from table 3-12, item 2, is contained in T.O. 21M-HGM16F-1CL-4.

3-39. **RE-ENTRY VEHICLE PRELAUNCH MONITOR TARGET SETTING PROCEDURES.**

3-40. Procedures for changing re-entry vehicle prelaunch monitor settings are presented in table 3-14. These procedures will be performed only when directed by proper authority to change target settings. The MCCC must be present while prelaunch monitor settings are being changed. T.O. 21M-HGM16F-1CL-1, section 14 contains the abbreviated checklist of table 3-14.

3-41. **BLAST DETECTION CABINET (NA-OSTF-2)**

NOTE

If the blast detection cabinet has been activated by detection of a nuclear blast or a malfunction, the system must be reset to enable detection of a second or further nuclear blast.

3-42. The blast detection cabinet, located on level 2 of the LCC shall be in continuous operation, providing 24-hour detection capability. Proper operation shall be observed when performing complex status verification (table 3-8). Procedures for resetting the blast detection system, after a blast has occurred or if a malfunction in the system has been corrected, are as follows:

a. At the blast detection cabinet:

(1) Position OUTPUT RELAY switch to DISCONNECT.

(2) Depress RCV 1 MANUAL TEST pushbutton.

(3) Depress ALARM RESET pushbutton.

(4) Position OUTPUT RELAY switch to CONNECT.

(5) Depress DETECTION MODE RESET pushbutton.

b. Verify that OPTICAL MODE indicator is illuminated and detector is cycling by observing channel indicators.

3-43. If an actual nuclear blast is detected during standby, do not open blast doors or silo doors until blast conditions (shock wave and effect) are over. If during countdown, do not start commit sequence until blast conditions are over.

3-44. SINGLE SIDEBAND RADIO (NA OSTF-2)

3-45. The UHF-HF single sideband radio shall be monitored at all times to ensure alternate communications availability. During standby condition, the selected receive-transmit capability will be indicated by POWER indicator illuminated GREEN.

a. Procedures for setting up the single sideband radio are as follows:

(1) Place HF POWER switch ON, observe POWER indicator illuminated.

(2) Adjust VOLUME control to desired level.

> **CAUTION**

Do not hold UHF POWER switch or TR SELECTOR switch in ON position more than five seconds.

(3) Place UHF POWER switch ON and release, observe POWER indicator illuminated.

(4) Turn TR SELECTOR switch to UHF position and release.

(5) Turn CHANNEL SELECTOR to desired frequency, adjust SQUELCH control until SQUELCH PUSH TO DISABLE indicator extinguishes.

(6) Turn TR SELECTOR switch to HF position and release.

(7) Select desired HF frequency, depress TUNING CYCLE PRESS TO START pushbutton, observe IN PROGRESS indicator illuminates and extinguishes, and COMPLETE indicator illuminates.

(8) Select appropriate 3-6 MC or 6-30 MC, PRESS TO SELECT pushbutton and observe indicator is illuminated.

(9) Depress UHF DISABLE, PUSH TO RE-SET pushbutton.

(10) Depress OPERATE MONITOR pushbutton and observe MONITOR indicator is illuminated.

b. To select UHF or HF, perform the following:

(1) Depress OPERATE MONITOR pushbutton.

(2) Turn TR SELECTOR to desired position.

(3) Observe OPERATE indicator is illuminated.

c. To turn single sideband radio off, place UHF and HF POWER switches to OFF position.

3-45A. HF HARDENED ANTENNA (NA 576G)

3-45B. The hardened HF antenna system is designed to provide an erectable-retractable, blast protected, shock isolated, emergency antenna to replace the soft HF antenna, used in normal operation, in the event of its failure. When the hardened antenna, housed in an underground steel and concrete structure, is actuated, the transmitter R-F power is switched from the soft to the hardened antenna. The ten-section, telescoping monopole antenna is automatically height-tuned to the transmitting frequency in the range of 2 to 30 megacycles. Antenna control provisions are local for maintenance only (antenna silo), and remote (from the LCC). The remote antenna control is mounted in the communications console located in the LCC.

3-45C. Whenever the soft antenna becomes inoperative, and command policy allows, the hardened antenna may be used.

a. The hardened antenna is activated as follows:

(1) Set the PWR switch on the antenna control to ON.

NOTE

If the SILO FAULT indicator on the antenna control illuminates, there is a malfunction in the hardened antenna facility. Correct the silo fault before proceeding further in operation cycle. Reference 31R2-4-189-2.

(2) Set the antenna selector switch on the antenna control to H.

(3) Turn the FREQUENCY control until the transmitting frequency is read under the scribe line of the MC indicator.

(4) Momentarily depress the ACTUATE switch on the antenna control. The IN MOTION indicator should illuminate. When the antenna has reached its selected height, the READY indicator will illuminate and the IN MOTION indicator will extinguish.

NOTE

If the READY indicator has not illuminated within 3 minutes after depressing the ACTUATE switch, and the IN MOTION indicator does not extinguish, a fault has occurred and the antenna has not positioned. Reference 31R2-4-189-2, Section V for troubleshooting procedures.

(5) To position the antenna for a new transmitting frequency, turn the FREQUENCY control until the new frequency is properly set, then repeat step (4).

CAUTION

To prevent damage to the antenna door seal, a visual check should be made before the antenna is lowered and the door closed to ensure that there is no debris on the door frame.

(6) To return to the soft antenna operation and to replace the hardened antenna in its protective environment, set the antenna selector switch on the antenna control to the S position. The antenna will automatically retract and the silo facility will close and lock the antenna silo door.

3-46. FIRE ALARM PANEL (ANNUNCIA-TOR) RESET.

NOTE

For fire detection systems, other than ANNUNCIATOR, refer to manufacturers operating instruction for reset procedures.

3-47. The fire alarm panel located on level 2 of the LCC is normally monitored for a green pilot light and a positive charge indication on the panel milliammeter. If a fire-detector head activates in the silo, it will cause a fire zone indicator to illuminate on the fire alarm panel. As long as the fire-detector head remains activated, the other fire detector heads within that zone cannot register a fire. The fire-detector heads in all other zones remain active and can cause a fire zone indicator to illuminate on the fire alarm panel. However, once reset, the fire alarm remains reset and cannot sound to announce a second fire indication. For fire detector systems manufactured by the Annunciator Corporation, the alarm capability can be restored to other zones by removing the plug-in type relay located immediately behind the illuminated fire zone indicator. To restore the system to normal perform the following steps:

a. Depress ALARM SILENCER and RESET pushbuttons located within the interior of the panel.

b. Replace triggered detection head if activation was due to excessive heat.

3-48. SECURITY SURVEILLANCE SYSTEM.

3-49. Procedures for operation of the security surveillance system will be provided when installation of the equipment is completed.

3-50. PRIMARY ALERTING SYSTEM (PAS) TAPE RECORDER.

3-51. The PAS tape recorder has been designed to continuously record any desired communication over any period of time up to 24 hours. A warning signal and automatic cutoff are incorporated in the recorder and activate when the end of the tape approaches the tape guide.

a. Tape replacement and operation procedures are as follows:

(1) Place OFF-VOLUME control to OFF.

(2) Position LOAD-RUN switch to LOAD.

(3) Remove old tape from recorder.

(4) Place new tape on recorder and thread through recorder head.

(5) Position red pointer on recorder to the time printed on the tape so as to correspond with the correct Zulu Time.

(6) Position LOAD-RUN switch to RUN.

(7) Position REC-STOP-PLAY switch to REC.

(8) Turn OFF-VOLUME control to ON and adjust until white dot is at 6 o'clock position.

b. Procedures to play back the recorder are as follows:

(1) Place OFF-VOLUME control to OFF.

(2) Position LOAD-RUN switch to LOAD.

(3) Wind tape backwards until red pointer points to time on tape that corresponds with the time of message receipt.

(4) Position REC-STOP-PLAY switch to PLAY.

(5) Position LOAD-RUN switch to RUN.

(6) Turn OFF-VOLUME control to ON and adjust volume as needed.

c. To return the recorder back to proper configuration for recording, proceed as follows:

(1) Place OFF-VOLUME control to OFF.

(2) Position LOAD-RUN switch to LOAD.

(3) Wind tape until red pointer points to the correct Zulu Time.

(4) Position LOAD-RUN switch to RUN.

(5) Position REC-STOP-PLAY switch to REC.

(6) Turn OFF-VOLUME control to ON and adjust white dot to 6 o'clock position.

d. Procedure to demagnetize the tape is as follows:

(1) Ensure that no watches or meters are near demagnetizer.

(2) Position and hold POWER switch to ON position.

(3) Ensure that the POWER indicator is illuminated.

(4) Place tape between poles and rotate three complete turns (approximately 15 seconds).

(5) Remove tape and release POWER switch.

3-52. ANNUNCIATOR CABINET (COMMUNICATION).

3-53. The annunciator cabinet located on LCC level 2, is used to continuously monitor the carrier bay communication equipment. If a malfunction occurs, a buzzer sounds and an indicator illuminates. To silence the annunciator alarm, depress ALMCO pushbutton. Listed below are the cabinet indicators and the communication circuits which are summarized by each indicator. When an indicator illuminates, perform a communications check on each circuit listed below the indicator and report outages to appropriate authorities.

a. Intersite alarm indicator RED:

(1) Direct line to command post.

(2) Direct line to alternate command post.

(3) All dial line circuits.

b. Launch maintenance nets 1, 2, 3 and 4.

c. Direct line NO. 1 fuse alarm indicator RED:

(1) All five explosion proof communication jacks.

(2) Direct line to APCHE telephone.

(3) Both direct lines to logic units.

(4) Direct line to command post.

(5) Direct line to alternate command post.

(6) Direct line to complex entry gate telephone.

d.   Direct line NO. 1 chain alarm indicator RED:

(1) All communications on the launch control console communications panel will be inoperative.

e.   PROTECTOR LAMP indicator AMBER:

(1) Nontactical radio at silo cap.

(2) Direct line to APCHE phase.

(3) Direct line to complex entry gate telephone.

(4) Direct line to explosionproof communications box at silo cap.

(5) Four maintenance communication boxes at silo cap.

f.   POWER MAJOR indicator RED:

(1) Any complex communications circuit could be faulty except for radio circuits.

g.   POWER MINOR indicator AMBER:

(1) Does not affect any complex communication circuit. Indicates communications battery charger fault.

3-54.   TACTICAL LAUNCH PREPARATIONS.

3-55.   Once a complex has been placed in ready state A, no special procedures are required prior to tactical launch. Completing normal alert tour checklists and briefings will verify the capability of immediate authorized launch.

(Paragraph 3-56 and 3-57 deleted.)

3-58.   PLX COUNTDOWN PREPARATIONS FROM EWO AND RETURN.

3-59.   From an EWO readiness condition, PLX countdowns with LO_2 or LN_2 may be performed by completing tasks in the applicable functional flow diagram contained in T.O. 21M-HGM16F-3CL-1. Table 3-9 shall be consulted for placing the MGS in aligned status. The functional flow defines and sequences the tasks required to perform a PLX, starting and ending with the launch complex in an EWO readiness status. The PLX may be performed with a fire-safe re-entry vehicle, no re-entry vehicle, or a training re-entry vehicle. If no re-entry vehicle is installed, an electrical circuit simulator and compensating ballast weights for the missile lifting system are used. A PLX shall not be initiated with a known malfunction or condition which would prevent completion of the exercise. The air conditioning, ventilation, GO_2 detection, television and fire fog system shall be in normal operating conditions. The MCCC or sector commander shall ensure that certain coordination and tasks are accomplished and support equipment placed in operations during PLX preparations. Table 3-15 provides the procedures to follow when notification is received to accomplish a PLX countdown. An abbreviated checklist condensed from table 3-15 is contained in T. O. 21M-HGM16F-1CL-1 and T. O. 21M-HGM16F-1CL-2. A minimum of four television sets located at the silo cap, silo level 3, 7 and 8 are required when performing a nontactical countdown. With necessary equipment available a camera on level 6A is required. During countdown the television monitors shall be operated by an individual fully knowledgeable of countdown and safety procedures.

3-60. TRAINING LAUNCH PREPARATION (VAFB).

3-61. From an EWO readiness condition, a training launch (CTL) may be executed after completion of tasks in the applicable functional flow diagrams contained in T.O. 21M-HGM16F-3CL-1. Prior to participating in a training launch, missile crews must be thoroughly briefed on VAFB operational peculiarities, equipment, complex configuration, and safety procedures. Table 3-15 shall be used in conjunction with the training launch flow.

3-62. COUNTDOWN AND ABORT PROCEDURES.

3-63. Countdown procedures are provided in table 3-16. Table 3-17 provides for the abort sequence. Table 3-18 delineates postlaunch procedures after a tactical launch. T.O. 21M-HGM16F-3CL-1 contains procedures to follow after completing a nontactical launch. Missile combat crew must be completely familiar with these tables before performing a countdown using the abbreviated checklists. The tables provide step-by step actions, the basic logic of countdown and abort indications, and necessary action or references to be employed in the event of malfunctions. In the event a malfunction or emergency is not covered, the MCCC must use his best judgment, based on experience and system knowledge, to determine proper action to place the missile in a safe condition. Appropriate agencies shall be contacted at the earliest opportunity for assistance and guidance.

3-64. Countdown, abort, or postlaunch procedures shall be conducted by missile crew members using individual abbreviated checklists condensed from tables 3-16, 3-17, and 3-18. The MCCC shall direct the overall operation and control crew actions and coordination. Crew members shall monitor their assigned equipment and be ready to detect, announce, and recommend corrective actions in the event of an abnormal indication. Extraneous conversation shall be kept at a minimum. It is imparative that verbal direction or announcements be made in a clear, audible manner.

3-65. Abbreviated checklists are designed to enable crew members to monitor or react to progress of the countdown or abort sequences, through announcements or observation of indicators. As each checklist step is accomplished, it shall be checked in the space provided. Required action or emergency checklist reference for various sequential malfunction indications appear either in the left-hand column of the countdown or abort checklists or in notes preceeding the step. TAC and N/T appearing in the left-hand column denote tactical countdown (TAC) and nontactical countdown (N/T). For example, if N/T ⊚ and TAC 14 appear for a malfunction action or emergency checklist reference, it denotes that during a nontactical countdown, the MCCC shall abort and if during a tactical countdown, the MCCC shall perform emergency checklist section 14. The symbol ⊚ denotes abort. If no prefix (TAC or N/T) appear, the reference or action applies to both tactical and nontactical countdowns. Required reactions to nonsequential emergencies are contained in appropriate sections of the emergency checklists. When tactical troubleshooting checklists are referenced, troubleshooting shall be delayed until the countdown sequence stops. When an abort is directed in the left-hand column, the MCCC shall ensure that crew members are prepared to follow the abort sequence and without delay for accomplishment of malfunction analysis, initiate the abort sequence. Where action or reference is not reflected, the abnormal indication will either have no effect on the sequence or will automatically stop the sequence. With no reference in the left-hand column and an eventual abort situation has developed, initiation of abort should be delayed during a tactical launch until malfunction analysis has been completed as outlined in column 3 of appropriate table of section V. During a PLX or training launch, malfunction analysis shall be completed only to the point silo entry is necessary prior to initiating abort.

3-66. Deleted

3-67. Responsibility for aborting a countdown rests sololy with the MCCC. It is the responsibility of all other crew members to assist and make recommendations to the MCCC. Missile crew members shall note and record abnormal indications and significant related data as practicable, to assist in analysis and troubleshooting.

3-68. COMMUNICATION PROCEDURES.

3-69. Standard communication procedures are essential for safe and reliable complex operations. Communications within the complex shall normally be by public address, direct line or maintenance conference network. Communications to stations not within the complex shall be by direct line, dial line, nontactical radio or order wire.

3-70. COMMUNICATION DISCIPLINE.

3-71. Clear and precise language and slightly slower, well modulated speech shall be used in all inter-phone and radio communications. Proper communication identifiers, commands, reports and action responses should be used during all complex operations. Communications shall be limited to that which is pertinent to complex operations on all circuits except the dial phones. The proper procedure for testing the operation of headsets is to announce, "Testing 1, 2, 3, test out". Blowing in the microphone is considered poor communication practice.

3-72. Missile combat crew members will use the following communication identifiers:

a.	MCCC —	"COMMANDER"
b.	DMCCC —	"DEPUTY"
c.	BMAT —	"A-1"
d.	MFT —	"M-1"
e.	EPPT —	"L-1"

3-73. Additional communication identifiers such as "A-2", "M-2" and "L-2" should be temporarily assigned to additional crews and members of the relief crews while they are on the complex. Non-crew members may be assigned call signs identifiers or addressed by rank and last name. Area identifiers may also be used for initial contact calls. The word "ROGER" shall be used when acknowledging the understanding of instructions or commands.

3-74. Initiation and termination of inter-phone communications on the maintenance conference network and maintenance safety network shall follow the general format listed in steps a through d.

a. Calls on an activated line: State communications identifier of persons or station being addressed, state your communications identifier and location, give brief reason for call. For example, "Commander, M-1, L-1 in position at PSC ready to compare pressure".

b. Public address system calls: State communications identifier of persons or locations being called, state action to be taken or information to be conversed. For example, "L-1 call A-1 on net two".

c. Answering an inter-phone call or signal: State your communication identifier and location. For example, "A-1, level 2".

d. Terminating inter-phone communications: State your communications identifier, then "Releasing net". For example, "M-1 releasing net".

3-75. Most inter-phone communications are made up of commands, responses to action commands, and reporting of conditions or indications. Commands may be for equipment activation or for a report of conditions or indications. Commands requiring action such as depressing pushbuttons, positioning switches and moving valves shall require naming of the item and the position required. For example, "N-5 switch OPEN". Commands for conditions or indications should only state the name of item or indicator which is being examined. For example, "N-5 indicator (wait for response)".

3-76. The response to an action command should include the entire command as stated followed by a short pause prior to action accomplishment. The pause allows time for correction by the commander if the verbal response was incorrect. For example, "N-5 switch ON"..... pause..... then position the N-5 switch to the ON position. Since there are many SYSTEM POWER RESPONDER P O W E R and LOCAL-REMOTE switches, the panel title should proceed the name of the switch to prevent confusion in the response. This is accomplished by reading the panel title, then the switch title before taking action. For example, "Propellant Level panel 2, SYSTEM POWER switch ON"..... pause..... then position the switch to the ON position.

3-77. Reports of indications or conditions are short, direct statements of observed condition. For example, "GREEN, OPEN, SECURE, NORMAL, ILLUMINATED, etc.". Normally, numerical designators on valves shall be expressed by the most commonly used term. For example, 60 CYCLES, 3000 PSI, etc. When necessary to emphasize, clarify, or avoid confussion, numerical designators on valves shall be expressed or identified as single digits. For example, three-nine-eight-point five.

3-78. Commands to install jumpers between two given points shall be completely understood and separate responses must be given for each end of the jumper. The responses shall be the same as to action commands but with a slightly longer pause. The pause allows time for corretction by the commander if the item was wrong. Command: "Connect jumper between terminal board 5, pin 6 and terminal board CHARLIE (C), pin 1". Response: "Jumper to terminal board 5, pin 6, ... pause. . . . . connect, and state "Connected"; "Jumper to terminal board CHARLIE (C), pin 1", ... pause. . . . . connect, and state "Connected".

3-79. During countdown and abort, commands, responses, or announcements shall be made as indicated in the appropriate checklists. Abnormal indications shall be announced in a manner that all crew members are able to hear and understand. For example; " STORAGE AREA OXYGEN 25% INDICATOR RED, ABORT REQUIRED" or "VALVE N-4 INDICATOR NOT AMBER, ABORT REQUIRED".

3-80. COMMUNICATION SYSTEM.

3-81. A detailed explanation of the complex communications system and its individual unit operation is contained in section I. Calls to the LCC may be initiated at any of the maintenance communications panels on any one of the four networks, or at any maintenance safety network jacks. Calls from the LCC or from one station to another must be initiated by a public address announcement or a prior agreement. Contact with the LCC shall be established immediately upon reaching a work area to ensure that the MCCC has knowledge of the location of all personnel in the complex.

3-82. The location of personnel may be confirmed at any time by performing a personnel location check using the public address and maintenance conference net 2. The personnel location check shall be conducted by announcing over the public address system, "This is a personnel location check. All personnel report your location to launch control using net 2". Net 2 shall then be monitored until all personnel have reported.

3-83. If call is received on a circuit of higher priority than the one currently in use, the LCC monitor shall announce his communications identifier and state that he is leaving the line (or net) momentarily. For example, "This is the commander leaving the line (or net) momentarily". Then answer the higher priority call. Upon returning to the original line (or net), announce, "This is the commander back on the line (or net)". Answering priority for the communication circuits in the LCC are as follows:

a. Maintenance conference net 1.

b. Command post direct line.

c. Alternate command post direct line.

d. Order wire.

e. All other circuits.

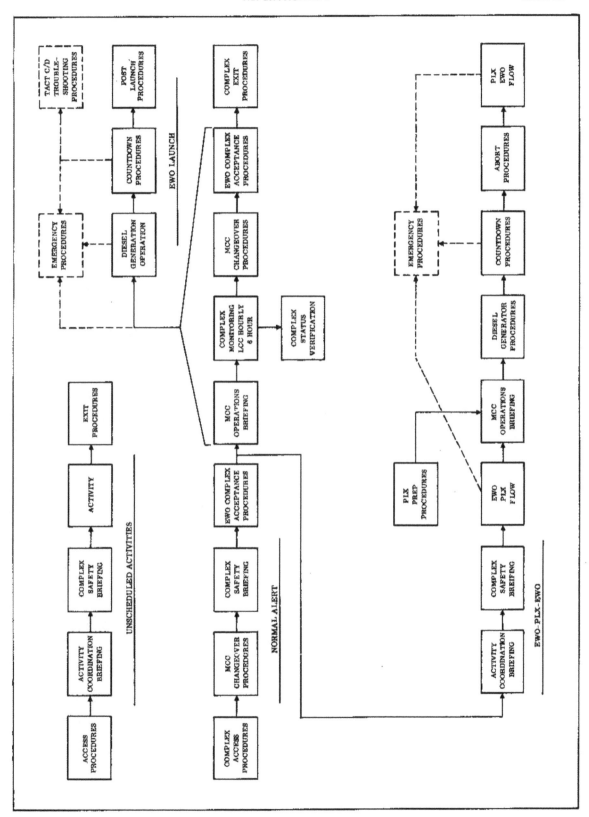

Figure 3-1. Normal Operating Procedures Function Flow Diagram

DAILY COMPLEX STATUS RECORD		COMPLEX NUMBER 551-08		DATE 29 FEB 1964			
RECORDED BY A2C JONES							

	ITEM / TIME	1000	1600	2200	0400			
COMPLEX MONITORING – SIX HOUR REQUIREMENT	POD AIR I-1							
	" " I-2							
	" " I-4							
	" " I-5							
	" " I-7	LIST READINGS IN THESE COLUMNS					LIST ADDITIONAL READINGS IN THESE COLUMNS AS REQUIRED	
	" " I-8							
	INST AIR COMP 1 OIL							
	" " COMP 2 OIL							
	" " HOUR METER 1							
	" " HOUR METER 2							
	WCU 50 TI-50							
	" " SUCTION PRESS							
	" " DISCHARGE PRESS							
	WCU 51 TI-52							
	" " SUCTION PRESS							
	" " DISCHARGE PRESS							
COMPLEX STATUS VERIFICATION	TK-90							
	PI-993							
	LN2 PREFAB LLI-220							
	" " LLI-221							
	PRESS PREFAB PI-1							
	" " PI-3	LIST READINGS IN THIS COLUMN	USE THIS COLUMN AS REQUIRED		LIST ADDITIONAL ITEMS IN THESE COLUMNS		LIST READINGS IN THIS COLUMN	USE THIS COLUMN AS REQUIRED
	" " PI-4							
	" " LLI-1							
	" " LLI-2							
	FUEL PREFAB LLI-3							
	PSMR PI-1							
	" PI-2							
	" PI-20							
	" PI-30							
	HPU RESERVOIR							
	DIESEL STORAGE							
	WATER STORAGE							

40.10-139

Figure 3-2.   Daily Complex Status Record

Table 3-2. Complex Safety Briefing

STEP	REQUIREMENT	ACTION
1	BRIEFING INTRODUCTION (official visitors only) a. Name, rank and duty title.	Briefing officer will state his name, rank, and duty title.
2	COMPLEX STATUS (official visitors only)	Brief complex EWO status, silo and missile status from complex status charts. Brief tasks in progress and number of personnel involved.
3	EWO ACTIONS	Brief, whenever a PAS message is broadcast, all personnel except crew members shall proceed from level 2 of LCC to level 1 and remain there until recalled. Brief that the 1000-cycle tone shall be activated when an EWO message is received. Crew members shall report to level 2 and prepare for countdown. If personnel are in silo they shall evacuate and report to complex controller in level 2 of the LCC. If PAS message is in process upon arrival at the LCC, they shall remain outside until termination of message, then check in with the complex controller and proceed to level 1 of the LCC.
4	LAUNCH CONTROL CENTER DEPARTURE	Determine why personnel desire to leave the LCC to establish which briefing items shall require emphasis.
5	BREATHING APPARATUS a. Operation proficiency	Ensure that personnel have been instructed and are proficient in the use of all emergency breathing apparatus used in the complex. If not, ensure that instructions are given and that proficiency is demonstrated prior to silo entry. Brief as to when and why apparatus will be used.
	b. Location and status	Brief as to the locations and number of breathing apparatus located in areas which persons shall proceed. Any abnormal number or location caused by repair or malfunction shall be briefed.
	c. Limitation	Brief that approximately 5 to 7 minutes of breathing air is available under normal breathing rate when apparatus is charged above 1800 PSI (Scott) or 2100 PSI (MSA).
	d. When used	Brief that emergency breathing apparatus shall be used whenever abnormal oxygen content is suspected or announced over public address system and by personnel located below silo level 6 when evacuation signal is sounded.
6	FIRE FIGHTING EQUIPMENT a. Operation proficiency	Insure proficiency or brief personnel on the operation of portable extinguishers, fire hoses, and fire alarm boxes.
	b. Location and status	Brief the location of portable extinguishers, fire hoses, and fire alarm boxes in the area to which personnel shall proceed.

Table 3-2. Complex Safety Briefing (CONT)

STEP	REQUIREMENT	ACTION
7	CLOTHING AND EQUIPMENT	
	a. Hard hats	Brief that hard hats shall be worn at all times in the silo. They shall be secured by a chin strap whenever entry is made to the missile enclosure area (MEA).
	b. Rings, watches, lighters or matches and loose articles	Brief that: Rings shall be removed or taped to prevent catching and causing injury; watches with metal bands shall be removed to prevent possible injury due to electric shock; lighters and matches shall not be carried into the silo to prevent accidental or careless ignition; loose articles, such as pencils, wrenches, etc., which may be dropped or caught on equipment shall be placed inside trouser pocket.
	c. Designated smoking areas	Brief that designated smoking areas are levels 1 and 2 of LCC. (Smoking area on silo cap is 125 feet back from center of cap.)
8	SAFETY CAUTIONS	
	a. Stairways, ladders, and platforms	Brief that stairways shall be used one step at a time and handrails shall be used at all times. Extreme caution must be taken when using ladders and platforms.
	b. Equipment handling	Brief that no switches, valves, or other equipment shall be operated without authority and that the LCC will be notified prior to operation of equipment which may affect complex status. Ensure all switches that are turned off to de-energize equipment are properly tagged.
	c. High-pressure lines, super-cooled equipment, high voltages, and toxic agents	Brief that high-pressure lines, super-cooled equipment, and high voltages are present throughout the silo and are dangerous; that caution must be exercised by personnel when in the vicinity of this equipment; toxic agents are used for cleaning agents and inhalation should be avoided.
	d. Buddy system (SAFETY)	Brief that in the interest of personnel safety, a minimum of two people shall normally occupy any one level of the silo whenever possible.
	e. Blast doors (NA OSTF-2)	Brief that blast doors are to be opened and closed one at a time.
9	ELEVATOR FACILITY	
	a. Use	Access to launcher platform and safety platform is from the elevator. Elevator should be retained at these levels. Brief that the elevator shall be maintained at the lowest level occupied if duties require personnel on level 7 or 8. If the elevator is used for access to other levels during this time personnel on levels 7 or 8 shall ensure the immediate return of the elevator to the lowest level occupied. The elevator should normally be left in a configuration where it may be used upon signal from other silo levels.

Table 3-2. Complex Safety Briefing (CONT)

STEP	REQUIREMENT	ACTION
10	EMERGENCY EVACUATION	
	a. Signals	Brief that personnel shall be alerted to evacuate the silo by the following signals:
		a. Public address system.
		b. Fire alarm.
		c. Silo lights turned on and off at 1-second intervals.
		d. 1000-cycle tone.
		All personnel shall evacuate the silo upon receipt of any one of the signals. Whenever possible, personnel should return all equipment to a safe configuration prior to evacuating.
		Brief that personnel shall be alerted to evacuate the silo cap by the sounding of the warning horn and activation of the personnel warning light.
	b. Procedure	Upon receipt of an evacuation signal all persons on silo levels 1 through 6 shall evacuate via the stairway and tunnel to LCC level 2. All personnel on levels 7 and 8 shall immediately don emergency breathing equipment and use the facility elevator to leave the silo. The elevator operator shall visually check the elevator entrance at each level enroute to level 2 and will stop at each entrance to the missile enclosure area (MEA) to determine if any personnel are in need of assistance. Personnel in the MEA should make themselves plainly visible. All personnel shall report to the complex controller on LCC level 2. When personnel on silo cap are alerted to evacuate, non-crew members shall proceed to the fall back area; crew members shall proceed to LCC level 2. Evacuation of the LCC, if required, shall be by way of the stairway or by use of the escape hatch, located on level 1 of the LCC, as directed.
11	CIRCULATION PROCEDURES	
	a. Control	Brief that personnel shall report to the complex controller prior to entering the silo and immediately upon return to the LCC. They shall report to the LCC via interphone when reaching and departing a work location. Brief personnel on "no lone" zone procedures.
	b. Communication	Assign communication net to be used. Instruct on proper INTERCOM procedures. Brief that net 1 is reserved for immediate action calls to the LCC. Brief the location of explosion proof and non-explosion proof head sets and status of communication panels by exception. Brief if communication with the LCC on the assigned maintenance net is lost, personnel in the silo will immediately attempt to re-establish communications using an alternate net, explosion proof jack, and/or spare head sets. If communication cannot be re-established and the task being performed is not critical, return to the LCC.

Table 3-2. Complex Safety Briefing (CONT)

STEP	REQUIREMENT	ACTION
12	EMERGENCIES a. Fire	Brief personnel to report fires immediately to the LCC, giving the nature and location. If contact cannot be established with the LCC, activate any one of the manual fire alarms located on silo levels 2, 4, 6, or 8, and combat the fire. If fire is out of control, notify the LCC if possible and evacuate.
13	FIRST AID FACILITIES a. Showers and eye wash	Brief, when applicable, that emergency showers and eye wash fountains are located on levels 7 and 8. These showers are operated with a quick opening, push-type valve that remains open until manually closed.
	b. First aid supplies	Brief that a limited supply of first aid items are located in the LCC and to report any injury immediately to the LCC.
14	HAZARDOUS CONDITIONS	Brief on known hazardous condition due to the present status of the complex, which may not be a normal condition.

Table 3-3. Missile Crew Changeover Briefing

STEP	REQUIREMENT	ACTION
1	CREW PERSONNEL a. Required personnel	Ensure all crew members are present.
	b. Personal equipment	Ensure crew members are adequately dressed and possess hard hats.
2	WEATHER a. Present	Brief on present and forecasted weather which may have effect on complex activities.
	b. Forecast	If there is no significant weather, this item need only be noted.
3	COMPLEX STATUS a. Current security	Brief on the current security conditions.
	b. EWO	Brief on DEFCON condition, ready state of the complex, and target situation.
	c. Checklists	Brief on checklists accomplished.
	d. Missile	Brief on status of complex equipment. Items which affect operations, safety, or EWO status shall be covered.
	e. Aerospace ground equipment (AGE)	
	f. Real property installed equipment (RPIE)	
	g. Power production	

Table 3-3. Missile Crew Changeover Briefing (CONT)

STEP	REQUIREMENT	ACTION
4	COMPLEX STATUS FORMS	Relief crew shall review complex forms with duty crew.
5	MAINTENANCE a. Accomplished b. In progress c. Scheduled	Brief on what maintenance was accomplished during the duty tour, maintenance that is in progress at time of crew changeover, and that which is scheduled.
6	SAFETY a. Breathing apparatus	Brief on the location and status of all breathing apparatus by exception (missing items).
	b. Fire extinguishers	Brief on status and location of portable fire extinguishers by exception (missing items).
	c. General safety	Brief on new bulletins, standard operating procedures (SOP), or activities affecting safety of personnel or equipment.
7	MISCELLANEOUS a. Special activities	Brief on known activities scheduled for the tour. Visitors, operational readiness inspection (ORI), target change, etc.
	b. Technical data	Brief on changes received or new material.
	c. Communication headsets	Brief on location of headsets within the silo and on silo cap, by exception (missing items).

Table 3-4. EWO Complex Acceptance Procedures

STEP	REQUIREMENT	ACTION
1	COMPLEX SAFETY BRIEFING	Conduct a complex safety briefing using table 3-2 or section 2 of T.O. 21M-HGM16F-1CL-1.
2	CONSOLE MANNING	Establish console manning assignments during changeover. Normally the duty DMCCC and relief BMAT shall be assigned to the launch control console.
3	DMCCC and BMAT TO SILO LEVEL 3	Direct relief DMCCC and duty BMAT to proceed to silo level 3 and establish contact with LCC.
4	SAFETY EQUIPMENT CLOSET INVENTORY	Direct EPPT to inventory equipment contained in safety equipment closet.

Table 3-4. EWO Complex Acceptance Procedures (CONT)

STEP	REQUIREMENT	ACTION
5	EMERGENCY AND PORTABLE MATERIAL CHECK	Direct MFT to verify operation and proper location of the following material on level 2 of LCC: 30-minute type breathing apparatus (5), combustible gas indicator, portable oxygen detector, emergency tool kit containing authorized tools and jumper wires, PSM-6 multimeter, keys, and safety equipment cabinet sealed.
6	OPERATION MANUAL AND CHECKLIST	Direct relief BMAT to ensure that current operation manual and associated checklists are available and in proper location.
7	LAUNCH ENABLE SYSTEM (LES) CONTROL BOXES	Verify that all LES control boxes are locked and seals are secure.
8	START C/D PUSHBUTTON	Verify that seal is secured.
9	COMMIT START KEY COVER	Verify that seal is secured.
10	LAUNCH CONTROL CONSOLE REAR DOORS	Verify that seal is secured.
10A	ALCO COMM/CONTROL PANEL	Verify that door is sealed and secured.
11	PRELAUNCH MONITOR (PLM) TARGET SETTINGS	Verify with DMCCC, using appropriate documents, that settings are correct.
12	TARGET SETTING COVER	Verify with DMCCC that cover is locked.
13	PLM ACCESS PANELS	Verify with DMCCC that front, rear, top, side and 28-volt access panels are closed and sealed.
14	CONSTANT BOARDS	Verify with DMCCC that appropriate constant boards are secure and sealed in guidance countdown group.
15	GUIDANCE ALIGNMENT	Verify with DMCCC that guidance is aligned in accordance with command policy. The MGS CHECKOUT COMPLETE indicator must be green if the system is not in aligned status. ALIGNMENT COMPLETE indicator must be green if system is in aligned status.
16	CONTROL-MONITOR GROUP PANELS (104)	Verify with DMCCC that all panel seals on control—monitor group 1, 2, 3 and 4 of 4 are secure.
17	CONTROL-MONITOR GROUP REAR ACCESS DOORS	Verify with DMCCC that all doors of control—monitor group 1, 2, 3, and 4 of 4 are closed and sealed.
18	CROSS RANGE AND DOWN RANGE SETTINGS	Verify correct settings using current target printout sheet.
19	FAST REACTION MESSAGE MATERIAL	Insure that all required checklists and copy formats are available, current, and in proper location.

Table 3-4. EWO Complex Acceptance Procedures (CONT)

STEP	REQUIREMENT	ACTION
20	CLASSIFIED DOCUMENT INVENTORY	Complete inventory and receipt for material as required.
21	SAFE AND LOCK COMBINATIONS	Verify with duty MCCC and DMCCC appropriate combination of safes and locks.
22	SAFETY EQUIPMENT CLOSET INVENTORY	Verify with EPPT that inventory has been completed and all equipment was present.
23	BREATHING APPARATUS	Verfiy with MFT that equipment or material in steps 23 through 29 have been checked, are operational, and in proper location.
24	COMBUSTIBLE GAS INDICATOR	
25	PORTABLE OXYGEN DETECTOR	
26	TOOL KIT	
27	PSM-6 MULTIMETER	
28	KEYS	
29	SAFETY EQUIPMENT CABINET	
29A	CREW CHECKLISTS	Review sections 7, 7B and 8 of duty crew and section 4 of relief crew.
30	EXCEPTIONAL RELEASE	Sign appropriate block of Form 207.
31	POSITIVE CONTROL CHANGEOVER	MCCC and DMCCC (duty and relief) complete changeover procedure in accordance with appropriate directives.

Changed 21 October 1964

Table 3-5. Missile Combat Crew Operations Briefing

STEP	REQUIREMENT	ACTION
1	CONSOLE ASSIGNMENTS	Establish console manning schedule for the tour of duty. Launch control console manning shall be accomplished by a two-man team composed of the MCCC and DMCCC or designated NCO. One member of the team shall be disignated as complex controllor. In addition to initial fast reaction message requirements, he shall be responsible for monitoring missile tank pressures, complex communications, access control, complex control, maintaining the complex log, control of keys, etc. He will assign single sideband monitoring responsibilities. Adherence to the two-man concept shall be stressed.
2	MOVEMENT OF OFFICERS	Brief on adherence to the two-officer movement restrictions and emphasize limitations imposed by existing directives.
3	CURRENT CALL SIGNS	Review current call signs used in PAS transmissions. Emphasize command post and complex call signs.
4	MESSAGES REQUIRING REACTION	Identify only those messages requiring initiation of the 1000-cycle tone. Special instructions shall be given for PLX and simulated countdown exercises when applicable.
5	EWO OR OPERATIONAL EXERCISE REACTION	Brief what each individual's reaction shall be when 1000-cycle tone sounds. Special emphasis shall be placed on paralleling diesels, securing from maintenance activities, clearing the silo, etc.
6	COUNTDOWN POSITIONS AND RESPONSIBILITIES	Reviewed by each crew member as outlined in the amplified countdown table.
7	NOTES, CAUTIONS, AND WARNING IN COUNTDOWN AND ABORT CHECKLISTS	Discuss each note, caution, and warning in countdown and commit checklists. Identify individual responsibilities and crew coordination for compliance.
8	COUNTDOWN PROCEDURES	Review applicable sections of abbreviated checklists. Cover communications, timing, and coordination requirements. If desired, a simulated talk-through countdown may be accomplished.
9	DAILY TRAINING/MAINTENANCE SCHEDULE	Establish training/maintenance to be accomplished during the duty shift (not applicable to PLX or LSR brifings).

T.O. 21M-HGM16F-1

Table 3-6. Activity Coordination Briefing

STEP	REQUIREMENT	ACTION
1	REQUIRED PERSONNEL	Verify all crew, maintenance, and support personnel involved in activity are present.
2	ACTIVITY AUTHORIZATION	Verify that job request or valid operational requirement exists.
3	PERSONNEL QUALIFICATIONS	Verify that personnel involved are qualified to perform assigned tasks.
4	CHECKLISTS AND TECHNICAL DATA	Verify that necessary checklists and technical data are available and current. Brief on proper use.
5	BREAK OF INSPECTION (BOI) REQUIREMENTS	Verify that approval is obtained for required BOI seals.
6	TOOLS AND EQUIPMENT	Verify that proper tools and equipment are available.
7	SEQUENCE OF EVENTS	Review proper sequence of action and procedures from checklists, technical data, and flow charts.
8	INDIVIDUAL RESPONSI-BILITIES	Brief each individual on his assigned responsibilities.
9	COORDINATION	Brief on the coordination necessary to perform the task. All actions shall be coordinated through the LCC.
10	COMMUNICATIONS	Brief on communication requirements, net assignments, and procedures.
11	BACKOUT PROCEDURES	Brief on proper backout procedures in case of emergency or EWO requirements.
12	EMERGENCIES	Brief on any potential emergencies or hazards which may be expected and action to be taken.
13	CONDITIONS AFFECTING THE TASK	Brief on weather, equipment status, configuration changes, or other existing conditions that may have an effect on the activity.
14	CLEAN-UP	Brief that all equipment shall be secured, all areas cleaned, and loose items removed when task is completed.
15	DOCUMENTATION	Brief that all forms shall be properly completed, inspection seals replaced, and clearance obtained from MCCC before departing the complex.
16	COMPLEX SAFETY BRIEFING	Brief all items in table 3-2 that are applicable in the area that work is to be performed.

Table 3-7. Complex Monitoring — LCC Requirements

STEP	EQUIPMENT OR INDICATION	CONDITION OR POSITION
	NOTE If alarm sounds on launch control console during standby, immediate attention should be given to the PRESSURE MODE indicator. When this malfunction appears on the console, immediate action is required. ✶ ⬛ **WARNING** Do not depress ABORT pushbutton unless a countdown is in progress. Failure to comply may result in inadvertent operation of certain equipment associated with the missile lifting system. **NOTE** If missile guidance system is to be placed in aligned status as directed by command policy, direct BMAT to place guidance system in aligned status using checklist T.O. 21M-HGM16F-1CL-5. **NOTE** If missile guidance system is in aligned status and command policy directs placing it in the not-aligned status, direct BMAT to place guidance system in not-aligned status using checklist T.O. 21M-HGM16F-1CL-5.	
1	AUTHENTICATION AND DECODING MATERIAL	Check currency and location.
2	READY FOR COUNTDOWN INDICATOR	Verify that indicator is illuminated green.
3	TARGET SELECTION	Verify current target selection and spotting box settings.
4	PAS TAPE RECORDER	Check that tape is feeding and recorder is operating.
5	SINGLE SIDEBAND RADIO (NA OSTF-2)	Check operation and log.
6	CSMOL PROGRAMMER KEY	Check key to be in the proper position.
7	RUNNING FREQUENCY METER (PRCP)	Check for 60 CPS.
8	RUNNING VOLTAGE METER (PRCP)	Check for 460 (+2, -8) volts.

Table 3-7. Complex Monitoring — LCC Requirements (CONT)

STEP	EQUIPMENT OR INDICATION	CONDITION OR POSITION
9	GENERATOR AMMETER AC METER (PRCP)	Check operating generator in positions 1, 2, and 3, for balanced load. Reading will vary depending on operation being performed but all phases should register approximately the same load.
10	SILO AIR INTAKE CLOSURES OPEN INDICATOR (FRCP)	Check that indicators in steps 10 through 13 are illuminated green.
11	STARTING AIR RECEIVER NORMAL PRESS INDICATOR (FRCP)	
12	SILO AIR EXHAUST CLOSURES OPEN INDICATOR (FRCP)	
13	LCC VESTIBULE BLAST DOOR (EXTERIOR AND INTERIOR) CLOSED INDICATORS (FRCP) (NA OSTF-2)	
13A	EMERGENCY WATER PUMP P-32 ON INDICATOR (FRCP) (NA OSTF-2)	Verify ON indicator is extinguished.
14	SILO BLAST DOOR CLOSED INDICATOR (NA OSTF-2)	Verify silo blast doors are closed as required. If both silo blast doors are closed, observe SILO BLAST DOORS CLOSED indicator GREEN.
15	ENTRY GATE AND ENTRAPMENT DOORS (NA OSTF-2)	Verify that complex entry gate and both entrapment doors are closed by observing the three GREEN indicators on gate and door control panel to be illuminated GREEN.
16	LO$_2$ TANKING (PANEL 1 AND PANEL 2)	Verify valve control switches N-4 and N-5 are in OPEN position and all other valve control switches in CLOSED position.
17	COMPLEX STATUS CHARTS	Check currency and completeness.

Table 3-7A. Complex Monitoring — Hourly Requirements

STEP	EQUIPMENT OR INDICATOR	CONDITION OR POSITION
		NOTE This table shall be used only for reference in filling out AF Form 1167.
	OPERATING DIESEL GENERATOR	
1	JACKET WATER TEMPERATURE IN GAUGE	Check for reading of 150 to 165°F. Reading will vary depending on load and condenser water temperature.
2	JACKET WATER TEMPERATURE OUT GAUGE	Check for reading of 165 to 180°F. Reading will vary depending on load and jacket water inlet temperature.
3	LUBE OIL TEMPERATURE IN GAUGE	Check for reading of 155 to 165°F. Reading will vary depending on load and condenser water temperature.
4	LUBE OIL TEMPERATURE OUT GAUGE	Check for reading of 170 to 180°F. Reading will vary depending on load and temperature of lube oil inlet temperature.
5	CYLINDER EXHAUST TEMPERATURE PYROMETER	Check for reading of 940°F maximum. Readings shall be taken from position 1 through 8 on the pyrometer. Readings will vary depending on load. **NOTE** The temperature should not exceed 40°F between any two cylinders with approximately 400-KW load on the generator.
6	TURBOCHARGER EXHAUST TEMPERATURE PYROMETER	Check for reading of 800°F maximum. Reading shall be taken from position number 9 on the pyrometer. Reading will vary depending on load.
7	ROOM TEMPERATURE	Check generator room temperature. Reading shall be taken from a thermometer within the generator room.
8	LUBE OIL FILTER PRESSURE OUT GAUGE	Check for reading of 29 to 34 PSI.
9	FUEL OIL FILTER PRESSURE OUT GAUGE	Check for reading of 15 to 20 PSI. If operating off the soft storage tank (selftopping) then reading could be as high as 50 PSI depending on the amount of fuel in soft storage tank.
10	AIR MANIFOLD PRESSURE GAUGE	Check for reading of 0 to 9 PSI. This reading will vary depending on the load.

Table 3-7A. Complex Monitoring — Hourly Requirements (CONT)

STEP	EQUIPMENT OR INDICATOR	CONDITION OR POSITION
11	TURBOCHARGER LUBE OIL PRESSURE GAUGE	Check for reading of 29 to 34 PSI. This reading shall be taken from the lube oil pressure gauge on engine panel.
12	STARTING AIR PRESSURE GAUGE	Check for reading of 250 to 300 PSI.
13	TURBOCHARGER VIBRATION	Check degree of vibration by touching unit.
14	GOVERNOR OIL	The governor oil level should normally be between the low and high mark on the light gauge.
15	CRANKCASE LUBE OIL SUMP	The oil level should be between the normal and full mark on dip stick.
16	FUEL OIL DAY TANK	The fuel level in the sight gauge should be full at all times if operating from soft storage tank. If operating from day tank only, the fuel level should never be allowed to get below ½-full.
17	KW METER (480-VOLT SWITCHGEAR)	This reading will vary depending on load.
18	VOLTAGE METER (480-VOLT SWITCHGEAR)	Check for reading of 460 (+2, -8) VOLTS.
19	AMMETER, PHASE 1, 2, AND 3 (480-VOLT SWITCHGEAR)	The reading will vary depending on load.
20	POWER FACTOR METER (480-VOLT SWITCHGEAR)	This figure will vary depending on type of load. If operating generators are in parallel, both power factor meters should have the same reading.

Table 3-7B. Complex Monitoring — 6-Hour Requirements

STEP	EQUIPMENT OR INDICATOR	CONDITION OR POSITION
	NOTE Certain gauge readings may differ with each individual missile complex. Local balance, adjustment, and calibration procedures shall establish acceptable ranges. These ranges shall be appropriately marked on the gauge and be considered as normal indications.	
	POD AIR CONDITIONER	
1	PRIMARY COOLING COIL INLET TEMPERATURE INDICATOR I-1	Verify indicator indicates 44 to 50°F Record reading.
2	CHILLED WATER RETURN TEMPERATURE INDICATOR I-2	Verify indicator indicates 50 to 75°F. Record reading.
3	SUPPLY AIR TEMPERATURE INDICATOR I-4	Verify indicator indicates 43 to 49°F. Record reading.
4	SUPPLY AIR PRESS. INDICATOR I-5	Verify indicator indicates 32 to 39 inches of water. Record reading.
5	REFRIG. SUCT. PRESSURE INDICATOR I-6	Verify indicator indicates normal.
6	REFRIG. DISCH. PRESSURE INDICATOR I-7	Verify indicator indicates 170 to 240 PSI. Record reading.
7	COMPR. OIL PRESSURE INDICATOR I-8	Verify indicator indicates between a minimum of 15 PSI and a maximum of 40 PSI higher than REFRIG. SUCT. PRESSURE indicator I-6. Record reading.
8	REGEN-PROCESS PRESS. DIFF. MANOMETER I-10	Verify manometer indicates 0.1 to 0.4 inches of water.
	NOTE Manometer I-12 should always indicate a reading of 0.1 to 0.4 higher than Manometer I-11 to prevent hot air from passing into the processing side of the pod air conditioner.	
9	PROCESS PRESS. DEHUMIDIFIER MANOMETER I-11	Verify manometer indicates 0.37 to 0.8 inches of water (0.6 to 1.1 inches of water VAFB only).
10	REGEN. PRESS DEHUMIDIFIER MANOMETER I-12	Verify manometer indicates 0.5 to 1.1 inches of water (0.7 to 1.2 inches of water VAFB only).

Table 3-7B.   Complex Monitoring — 6 Hour Requirements (CONT)

STEP	EQUIPMENT OR INDICATOR	CONDITION OR POSITION
11	POD AIR CONDITIONER (CONT) AFTER-FILTER PRESS. DROP MANOMETER I-13	Verify manometer indicates 0.6 inches of water maximum. Replace filter if above 0.6 inches of water.
	NOTE Service recorders as required, using manufacturer's recorder ink. Fill temperature pen with blue ink and dew point and air flow pens with red ink.	
12	RECORDER PC1	Verify dew point is being maintained 0 to 26°F and dry bulb temperature 43 to 49°F.
13	RECORDER PC2	Verify minimum chart indication of 4.0 (+0.2) major chart divisions. At VAFB only, minimum chart indication of 5.6 (+0.2) major chart divisions.
14	REFRIGERANT SIGHT GAUGE (RA-3)	Verify no bubbles in liquid refrigerant.
15	REFRIGERANT COMPRESSOR	Check for proper oil level (¼ to ½-full).
	INSTRUMENT AIR PREFAB **NOTE** Both air compressors, if serviceable, shall be checked for the following items:	**NOTE** If PI-710 indicates less than 1200 PSI, both air compressors shall be operating. To check oil level, if both air compressors are operating, number 1 shall be stopped, checked, and restarted before number 2 is stopped and checked. The air compressors can be stopped or started by placing the HAND/OFF/AUTO switch in the appropriate position. Unless directed, the HAND/OFF/AUTO switch shall be left in the AUTO position.
1	AIR COMPRESSOR CRANK-CASE	Check for proper oil level. Record oil consumption.
2	HOUR METER NUMBER	Record readings.
3	OIL PRESSURE GAUGE	30 to 50 PSI (45 PSI is desired).
4	FIRST STAGE GAUGE	48 to 52 PSI.

Table 3-7B.   Complex Monitoring — 6 Hour Requirements (CONT)

STEP	EQUIPMENT OR INDICATOR	CONDITION OR POSITION
5	SECOND STAGE GAUGE	235 to 265 PSI.
6	THIRD STAGE GAUGE	750 to 850 PSI.
7	FOURTH STAGE GAUGE	Verify gauge indicates same as PI-710 gauge reading.
		NOTE The air dryers are cycled by an automatic 4-way valve at 8-hour intervals, causing the green indicator to cycle to the air dryer in use. A timer initiates a 5-hour heating period (red indicator illuminated) and a 3-hour cooling period (red indicator extinguished) on the air dryer not in use.
8	AIR DRYERS	Verify proper operation and cycling.
	WATER CHILLER UNIT WCU50 **NOTE** Pressure and temperature indications in the chilled water system (WCU50 and 51) must be kept within specifications, as abnormal pressures or temperatures will have an adverse effect on the air conditioning system throughout the silo. The following indicators shall be verified.	
1	CHILLED WATER INLET TEMP GAUGE TI-51	Check for normal indication (less than 60°F).
2	CHILLED WATER INLET PRESS. GAUGE PI-55	Check for normal indication (35 to 45 PSI).
3	CHILLED WATER OUT PRESS. GAUGE PI-54	Check for normal indication (30 to 40 PSI).
4	CHILLED WATER OUT TEMP GAUGE TI-50	Check for normal indication (43 to 50°F).
5	EVAPORATOR PRESSURE GAUGE	Evaporator pressure 50 to 65 PSI. Record reading.

Changed 21 October 1964

Table 3-7B. Complex Monitoring - 6 Hour Requirements (CONT)

STEP	EQUIPMENT OR INDICATOR	CONDITION OR POSITION
	WATER CHILLER UNIT WCU50 (CONT)	
6	OIL PRESSURE GAUGE	60 to 70 PSI higher than evaporator pressure.
7	OIL SIGHT GAUGE	1/2-full if compressor is running, 7/8-full if compressor is stopped.
8	CONDENSER PRESS. GAUGE	Check for normal indication (condenser pressure 170 to 250 PSI).
9	LIQUID LINE SIGHT GAUGE	No bubbles in line.
10	3-INCH AND 6-INCH Y STRAINERS	Blow down Y strainers in condenser water line (frequency determined by local conditions).
11	CONTROL CIRCUIT INDICATOR	Illuminated.
12	ANNUNCIATION DISABLE SWITCH	Verify switch is positioned to on/off whichever enables the FRCP annunciator system (site peculiar).
	WATER CHILLER UNIT WCU51	
1	CHILLED WATER INLET TEMP GAUGE TI-53	Check for normal indication (less than 60°F).
2	CHILLED WATER INLET PRESS. GAUGE PI-53	Check for normal indication (35 to 45 PSI).
3	CHILLED WATER OUT PRESS. GAUGE PI-52	Check for normal indication (30 to 40 PSI).
4	CHILLED WATER OUT TEMP GAUGE TI-52	Check for normal indications (43 to 50°F).
5	EVAPORATOR PRESSURE GAUGE	Evaporator pressure 50 to 65 PSI. Record reading.
6	OIL PRESSURE GAUGE	60 to 70 PSI higher than evaporator pressure.
7	OIL SIGHT GAUGE	1/2-full if compressor is running, 7/8-full if compressor is stopped.
8	CONDENSER PRESS. GAUGE	Check for normal indication (condenser pressure 170 to 250 PSI).
9	LIQUID LINE SIGHT GAUGE	No bubbles in line.

Changed 21 October 1964

Table 3-7B.   Complex Monitoring — 6 Hour Requirements (CONT)

STEP	EQUIPMENT OR INDICATOR	CONDITION OR POSITION
	WATER CHILLER UNIT WCU51 (CONT)	
10	3-INCH Y STRAINER	Blow down Y strainer in condenser water line (frequency determined by local conditions).
11	CONTROL CIRCUIT INDICATOR	Illuminated.
12	ANNUNCIATION DISABLE SWITCH	Verify switch is positioned to on/off whichever enables the FRCP annunciator system (site peculiar).

Table 3-8.   Missile Crew Daily Complex Status Verification Procedures

STEP	EQUIPMENT OR INDICATOR	CONDITION OR POSITION
	NOTE The following five items are general procedures which are to be performed when specified. a. General condition shall include, where applicable, inspection of an area and major items of equipment for cleanness corrosion, damage, spills, and leaks, to include all liquilds and gases, burned out light bulbs, and broken BOI seals. Doors, panels, and equipment drawers should be closed and secured to ensure proper ventilation and cleanness. Operating equipment will be checked for leaks, overheating, unusual noise, wear, adjustments, and for abnormal indications on all associated gauges. b. Emergency lighting units shall be lamp tested to ensure bulbs are serviceable. Lighting unit batteries shall be checked for proper electrolyte level and for fully-charged condition. The unit case should not indicate damage or corrosion. c. Fire extinguishers shall be inspected for availability, unbroken seals, proper condition, and proper stowage. Manual fire alarm boxes, located on silo levels 2, 4, 6, and 8 shall be properly stowed and inspected for serviceability and obvious damage. d. One non-explosive proof headset shall be placed on each silo level (1 through 8) and on the facility elevator. In addition, silo levels 7 and 8, sump area, and silo cap area shall each have one explosive proof headset. Headset cords shall be connected to the communications panel at all times. The initial call to the LCC shall constitute a communications panel check. The public address system shall also be checked during initial call to the LCC, by requesting a PA announcement. e. Breathing apparatus shall be inspected for cleanness, damage, serviceability, and proper pressure (Scott 1500 to 1980 PSI) (MSA 1800 to 2250 PSI). Flashlights attached to the units shall be checked for proper operation. Each unit shall be located and stowed in accordance with this manual.	

Table 3-8. Missile Crew Daily Complex Status Verification Procedures (CONT)

STEP	EQUIPMENT OR INDICATOR	CONDITION OR POSITION
	NOTE Certain gauge readings may differ with each individual missile complex. Local balance, adjustment, and calibration procedures shall establish acceptable ranges. These ranges shall be appropriately marked on the gauge and shall be considered normal indications. **NOTE** When a condition is found to exist that is in variance with specified indications, it shall be immediately reported to the MCCC prior to initiating corrective action. Fluid and gas levels shall be recorded and checked against requirements specified in section VI.	
1	LCC LEVEL 1 MECHANICAL EQUIPMENT ROOM	
	a. SUPPLY FAN S-1	Check for general condition. (See general procedure a.)
	b. EXHAUST FAN E-1	Check for general condition. (See general procedure a.)
	c. SUPPORT CYLINDER	Check for general condition. (See general procedure a.)
	d. CBR FILTER MANO-METER (IF INSTALLED)	Verify manometer indicates less than 3.5.
	e. EQUIPMENT ROOM	Check for general condition. (See general procedure a.)
2	READY ROOM	
	a. AIR RECEIVER TANK	Drain condensation from tank by cracking drain valve until no visible moisture is present.
	b. PRESSURE GAUGE PI-100	Verify PI-100, located on air receiver, indicates normal pressure.
	c. PRESSURE GAUGE PI-103	Verify PI-103, located on air receiver, indicates normal pressure.
	d. AIR RECEIVER	Check for general condition. (See general procedure a.)
	e. SUPPORT CYLINDER	Check for general condition. (See general procedure a.)
	f. LCC ESCAPE HATCH	Visually inspect hatch to be free of obstructions and inspect handle for seal.
	g. FIRE EXTINGUISHERS (2)	Check for general condition of both extinguishers. (See general procedure c.)

Table 3-8. Missile Crew Daily Complex Status Verification Procedures (CONT)

STEP	EQUIPMENT OR INDICATOR	CONDITION OR POSITION
2 (CONT)	READY ROOM (CONT)	
	h. EMERGENCY LIGHTING UNITS (2)	Check both units. (See general procedure b.)
	i. READY ROOM	Check for general condition. (See general procedure a.)
3	UTILITY ROOM	
	a. SUPPORT CYLINDER	Check for general condition. (See general procedure a.)
	b. UTILITY ROOM	Check for general condition. (See general procedure a.)
4	KITCHEN	
	a. SUPPORT CYLINDER	Check for general condition. (See general procedure a.)
	b. KITCHEN	Check for general condition. (See general procedure a.)
5	LATRINE	
	a. LATRINE AREA	Check for general condition. (See general procedure a.)
6	STAIRWELL	
	a. EMERGENCY LIGHTING UNITS (2)	Check both units. (See general procedure b.)
	b. K BOTTLES (2)	Check both bottles for pressure. One bottle shall have a minimum pressure of 1800 PSI.
	c. SEWAGE PUMPS	Check both automatic and manual operation and general condition. Return pumps to automatic.
	d. STAIRWELL	Check for general condition. (See general procedure a.)
	LCC LEVEL 2	
1	COMMUNICATION ROOM	
	a. FAN COIL UNIT	Verify fan coil unit is operating and check general condition. (See general procedure a.)
	b. 480-VOLT CONTROL CENTER CIRCUIT BREAKERS	All circuit breakers should normally be ON.

Table 3-8. Missile Crew Daily Complex Status Verification Procedures (CONT)

STEP	EQUIPMENT OR INDICATOR	CONDITION OR POSITION
1 (CONT)	LCC LEVEL 2 (CONT)	
	c. FIRE EXTINGUISHER	Check condition. (See general procedure c.)
	d. COMMUNICATION ROOM	Check for general condition. (See general procedure a.)
2	COMMUNICATION BATTERY ROOM	
	a. BATTERY BANK	Check for proper electrolyte level and general condition. (See general procedure a.)
	b. BATTERY CHARGER AMMETER	Verify ammeter indicates less than 20 AMP.
	c. AIR COMPRESSOR TANK (WHERE INSTALLED)	Drain condensation from tank by cracking drain valve until no visible moisture is present.
	d. AIR COMPRESSOR OIL LEVEL	Check for proper oil level as indication on dip stick.
	e. BATTERY ROOM	Check for general condition. (See general procedure a.)
3	LAUNCH CONTROL ROOM	
	a. BLAST DETECTION CABINET	Verify blast detection cabinet OUTPUT RELAY disconnect switch is in CONNECT position and that OPTICAL MODE indicator is illuminated. Ensure detector is cycling.
	b. FIRE ALARM PANEL	Check that all zone and trouble indicators are extinguished. A positive charge indication should be indicated on meter. Batteries shall be checked for charge and proper electrolyte level.
	c. EMERGENEY LIGHTING UNIT	Check for proper condition. (See general procedure b.)
	d. FIRE EXTINGUISHER	(See general procedure c.)
	e. LAUNCH CONTROL ROOM	Check for general condition. (See general procedure a.)

Table 3-8. Missile Crew Daily Complex Status Verification Procedures (CONT)

STEP	EQUIPMENT OR INDICATOR	CONDITION OR POSITION
4	FACILITIES REMOTE CONTROL PANEL INDICATORS	
	NOTE If any indicator not listed is illuminated, an abnormal condition is probable.	
	a. LAMP TEST	Accomplish lamp test and verify all warning panel indicators illuminate.
	b. WARNING PANEL	Verify all indicators are extinguished.
	c. DIESEL GENERATOR D-60 OR D-61 ON	Verify one or both (D-60 and D-61) ON indicators are illuminated GREEN.
	d. LAUNCH PLATFORM EXHAUST FAN ON	GREEN.
	e. LAUNCH PLATFORM DAMPERS OPEN	RED.
	f. GASEOUS OXYGEN VENT CLOSED	RED (extinguished for PLX).
	g. LCC SEWER VENT OPEN	GREEN.
	h. GRADE ENTRY DOOR (S) CLOSED	GREEN.
	i. GASEOUS OXYGEN VENT OPEN	GREEN (PLX only).
	j. LAUNCH PLATFORM FAN COIL UNIT ON	GREEN.
	k. SILO AIR INTAKE CLOSURES OPEN	GREEN.
	l. LCC AIR INTAKE OPEN	GREEN.
	m. LCC VESTIBULE EXTERIOR BLAST DOOR CLOSED (NA OSTF-2)	GREEN.
	n. STARTING AIR RECEIVER NORMAL PRESS.	GREEN.

Table 3-8.   Missile Crew Daily Complex Status Verification Procedures (CONT)

STEP	EQUIPMENT OR INDICATOR	CONDITION OR POSITION
4 (CONT)	FACILITIES REMOTE CONTROL PANEL INDICATORS (CONT)	
	o. SILO AIR EXHAUST CLOSURES OPEN	GREEN.
	p. LCC AIR EXHAUST OPEN	GREEN.
	q. LCC VESTIBULE INTERIOR BLAST DOOR CLOSED (NA OSTF-2)	GREEN.
	r. SILO BLAST DOOR CLOSED (NA OSTF-2)	GREEN.
	s. RPI AND FIRE FOG SYSTEM DAMPERS CLOSED	GREEN.
	t. LCC STAIRWELL AIR EXHAUST OPEN (NA OSTF-2)	GREEN.
	u. LCC ESCAPE HATCH DOOR CLOSED	GREEN.
	v. SPRAY PUMP NORMAL (OSTF-2 ONLY)	GREEN.
	w. SETTLING TANK NORMAL (OSTF-2 ONLY)	GREEN.
5	TELEVISION MONITORS	
	a. LIGHTING SYSTEM MAIN POWER SWITCH	Positioned to ON.
	b. LIGHTING SYSTEM MAIN POWER INDICATOR	Observe indicator is illuminated.
	NOTE Allow 2 minutes warmup between steps b and c.	
	c. LIGHTING SYSTEM INTENSITY SWITCH	Position to HIGH.
	d. CONTROLS FOR PICTURE	Adjusted for clearest picture.

Table 3-8.   Missile Crew Daily Complex Status Verification Procedures (CONT)

STEP	EQUIPMENT OR INDICATOR	CONDITION OR POSITION
5 (CONT)	TELEVISION MONITORS (CONT)	
	NOTE To place system in standby, perform steps e and f.	
	e. LIGHTING SYSTEM 1 INTENSITY SWITCH	Positioned to LOW.
	f. LIGHTING SYSTEM MAIN POWER SWITCH	Positioned to OFF.
6	PRCP (NA OSTF-2)	
	a. FEEDER NUMBER 3 NON-ESSENTIAL BUS CONTROL SWITCH IND	Observe RED indicator is illuminated.
	b. RUNNING FREQUENCY METER	Verify meter indicates 60 CPS.
	c. RUNNING VOLTAGE METER	Verify meter indicates 460 (+2, -8) volts.
	OPERATING GENERATOR	
	a. GENERATOR MAIN BREAKER CONTROL SWITCH INDICATOR	Observe RED indicator is illuminated.
	b. ENGINE START-STOP SWITCH INDICATOR	Observe RED indicator is illuminated.
	c. AMMETER PHASE 1, 2 AND 3	Verify AMMETER AC indicates approximately the same in each phase. This indication will vary depending on load.
	ALTERNATE GENERATOR	
	a. GENERATOR MAIN BREAKER CONTROL SWITCH INDICATOR	Observe GREEN indicator is illuminated.
	b. ENGINE START-STOP SWITCH INDICATOR	Observe GREEN indicator is illuminated.

Table 3-8.   Missile Crew Daily Complex Status Verification Procedures (CONT)

STEP	EQUIPMENT OR INDICATOR	CONDITION OR POSITION
7	PRCP (OSTF-2 ONLY)	
	IF 500-KW GENERATOR OUTPUT IS CONNECTED AS PRIMARY POWER	
	a. ENGINE START-STOP SWITCH INDICATOR	Observe indicator is illuminated RED.
	b. GENERATOR MAIN BREAKER CONTROL SWITCH INDICATOR	Observe indicator is illuminated RED.
	c. POWER CO. LINE MAIN BREAKER CONTROL SWITCH INDICATOR	Observe indicator is illuminated GREEN.
	d. GENERATOR FREQUENCY A.C. METER	Verify meter indicates 60 CPS.
	e. POWER CO. LINE FREQUENCY A.C. METER	Verify meter indicates 60 CPS.
	f. GENERATOR VOLTMETER A.C. METER	Verify GENERATOR VOLTMETER SWITCH is in 1-2, 2-3, or 3-4 position and that meter indicates 460 (+2, -8) volts.
	g. POWER CO. LINE VOLTMETER A.C. METER	Verify POWER CO. LINE VOLTMETER SWITCH is in 1-2, 2-3, and 3-4 position and that meter indicates 460 (+2, -8) volts.
	h. GENERATOR AMMETER A.C. METER	Verify GENERATOR AMMETER SWITCH is in 1, 2, or 3 position and that ammeter indicates approximately 300 AMP.
	i. POWER CO. LINE AMMETER A.C. METER	Verify POWER CO. LINE AMMETER SWITCH is in 1, 2, or 3 position and that ammeter indicates zero AMP.
	j. FEEDER NO 3. NONESSENTIAL BUS CONTROL SWITCH IND	Observe indicator is illuminated RED.
	IF POWER CO. IS CONNECTED AS PRIMARY POWER (OSTF-2 ONLY)	
	a. ENGINE START-STOP SWITCH INDICATOR	Observe indicator is illuminated GREEN.

Table 3-8. Missile Crew Daily Complex Status Verification Procedures (CONT)

STEP	EQUIPMENT OR INDICATOR	CONDITION OR POSITION
7 (CONT)	IF POWER CO. IS CONNECTED AS PRIMARY POWER (OSTF-2 ONLY) (CONT)	
	b. GENERATOR MAIN BREAKER CONTROL SWITCH INDICATOR	Observe indicator is illuminated GREEN.
	c. POWER CO. LINE MAIN BREAKER CONTROL SWITCH INDICATOR	Observe indicator is illuminated RED.
	d. POWER CO. LINE FREQUENCY A.C. METER	Verify meter indicates 60 CPS.
	e. POWER CO. LINE VOLT METER A.C. METER	Verify POWER CO. LINE VOLTMETER SWITCH is in 1-2, 2-3, or 3-4 position and that meter indicates 460 (+2, -8) volts.
	f. POWER CO. LINE AMMETER A.C. METER	Verify POWER CO. LINE AMMETER SWITCH is in 1, 2, or 3 position and that ammeter indicates approximately 300 AMP.
	g. FEEDER NO. 3 NON-ESSENTIAL BUS CONTROL SWITCH IND	Observe indicator is illuminated RED.
1	UTILITY TUNNEL (NA OSTF-2)	
	a. EMERGENCY LIGHTING UNITS (2)	(See general procedure b.)
	b. FIRE EXTINGUISHER	(See general procedure c.)
	c. TUNNEL	Check for general condition. (See general procedure a.)
1	SILO LEVEL 1, GENERAL	
	a. COMMUNICATION PANEL	(See general procedure d.)
	b. SAFETY PLATFORM	Check that platform is properly latched and that chain is attached.

Table 3-8. Missile Crew Daily Complex Status Verification Procedures (CONT)

STEP	EQUIPMENT OR INDICATOR	CONDITION OR POSITION
1 (CONT)	SILO LEVEL 1, GENERAL (CONT)	
	c. SAFETY PLATFORM HINGE PINS	Check that hinge pins are properly installed.
	d. DEMINERALIZED WATER TANK TK-90 LEVEL SIGHT GAUGE	Check water level sight gauge (½-full, minimum). Record reading.
	e. CHEMICAL POT FEEDER FLOW INDICATOR	Check flow indicator for water flow. Flow indicator will not indicate flow if no makeup water is flowing to the cooling tower.
	f. CHILLED WATER EX-PANSION TANK TK-50 LEVEL SIGHT GAUGE	Check water level sight gauge (½-full, nominal).
	g. AIR CONDITIONING MAKEUP TANK TK-20 LEVEL SIGHT GAUGE	Check water level sight gauge (½-full, nominal).
	h. SUPPLY FAN SF-20	Check for general condition. (See general procedure a.)
	i. SUPPLY FAN SF-21	Check for general condition. (See general procedure a.)
	j. SPRAY PUMP PRESSURE GAUGE PI-21	Check gauge for normal indication.
	k. SPRAY PUMP PRESSURE GAUGE PI-20	Check gauge for normal indication.
	l. SPRAY NOZZLE PRESS. GAUGE PI-21A (IF INSTALLED)	Check gauge for normal indication.
	m. LP OVERSPEED CON-TROL PANEL	Perform lamp test.
	n. LP OVERSPEED INDICATOR	Observe indicator is extinguished.
	o. LP DRIVE CABINET 1 DC REFERENCE POWER SUPPLY ON INDICATOR	Verify indicator is illuminated.
	p. (Deleted)	
	q. MLS DRIVE ASSEMBLY	Check for general condition. (See general procedure a.)

Table 3-8.   Missile Crew Daily Complex Status Verification Procedures (CONT)

STEP	EQUIPMENT OR INDICATOR	CONDITION OR POSITION
1 (CONT)	SILO LEVEL 1 GENERAL (CONT)	
	r. LOW SPEED MOTOR SHAFT	Rotate shaft by hand to ensure shaft is free and coupling is disengaged.
	s. Deleted	
	t. FIRE EXTINGUISHERS (2)	(See general procedure c.)
	u. EMERGENCY LIGHTING UNITS (2)	(See general procedure b.)
	v. LEVEL 1	Check for general condition. (See general procedure a.)
1	SILO LEVEL 2 GENERAL	
	a. COMMUNICATION PANEL	(See general procedure d.)
	b. SILO EXHAUST SYSTEM EF-30	Check for general condition. (See general procedure a.)
	c. MANUAL FIRE ALARM BOX	(See general procedure c.)
	d. FACILITY ELEVATOR	Check for general condition. (See general procedure a.)
	e. BREATHING APPARATUS ON ELEVATOR (3)	(See general procedure e.)
	f. FIRE EXTINGUISHERS ON ELEVATOR (2)	(See general procedure c.)
	g. FIRE EXTINGUISHERS (2)	(See general procedure c.)
	h. EMERGENCY LIGHTING UNITS (2)	(See general procedure b.)
	i. ESSENTIAL MOTOR CONTROL CENTER (EMCC)	Check for general condition. (See general procedure a.)
	j. ALL CIRCUIT BREAKERS ON EMCC	All circuit breakers shall normally be ON.

Table 3-8.   Missile Crew Daily Complex Status Verification Procedures (CONT)

STEP	EQUIPMENT OR INDICATOR	CONDITION OR POSITION
1 (CONT)	SILO LEVEL 2 (CONT)	
	k. MISSILE ENCLOSURE AIR CONDITIONER FC-40	Check for general condition. (See general procedure a.)
	l. HOT WATER SUPPLY GAUGE TI-45	Verify gauge indicates normal temperature.
	m. SAND SETTLING TANK DRAIN VALVE	Open valve momentarily to drain sediment.
	n. SILO HYDRAULIC SYSTEM EQUIPMENT	Check for general condition. (See general procedure a.)
	o. UTILITY AIR COMPRESSOR (IF INSTALLED)	Check for general condition. (See general procedure a.)
	p. UTILITY AIR COMPRESSOR OIL LEVEL	Check oil for proper level.
	q. AIR COMPRESSOR TANK	Open tank drain until moisture is drained.
	r. UTILITY AIR COMPRESSOR RECEIVER GAUGE	Check for normal indication.
	s. BLAST CLOSURE SYSTEM AIR PRESSURE GAUGE PI-3	Check gauge for normal pressure.
	t. LEVEL 2	Check for general condition. (See general procedure a.)
2	NONESSENTIAL MOTOR CONTROL CENTER (NEMCC), FACILITIES TERMINAL CABINET (FTC) NO. 2	
	a. PF-70 ON PUSHBUTTON	Depress PF-70 ON pushbutton.
	b. PF-70 ON INDICATOR	Observe PF-70 ON indicator is illuminated RED.
	c. EC-71 ON INDICATOR	Observe EC-71 ON indicator is illuminated RED.
	d. PUMP P-80 HAND-OFF-AUTO SWITCH	Verify switch is in AUTO position.
	e. PUMP AND FAN INDICATORS	Verify pump and fan indicators match the pump and fan selections made.

Table 3-8.   Missile Crew Daily Complex Status Verification Procedures (CONT)

STEP	EQUIPMENT OR INDICATOR	CONDITION OR POSITION
2 (CONT)	NONESSENTIAL MOTOR CON-TROL CENTER (NEMCC), FACILITIES TERMINAL CAB-INET (FTC) NO. 2 (CONT)	
	f. PF-70 OFF PUSHBUTTON	Depress PF-70 OFF pushbutton.
	g. PF-70 ON INDICATOR	Observe PF-70 ON indicator is extinguished.
	h. EC-71 ON INDICATOR	Observe EC-71 ON indicator is extinguished.
	i. ALTERNATE PUMPS AND FANS	Position all selector switches to select the standby unit. Wait 30 seconds, then verify that running indicator matches the unit selected. Selector switches shall be left in this position.
		Chilled Water Pumps P/50/51 must be restarted after selector switch has been positioned to the alternate pump.

Table 3-8. Missile Crew Daily Complex Status Verification Procedures (CONT)

STEP	EQUIPMENT OR INDICATOR	CONDITION OR POSITION
2 (CONT)	NONESSENTIAL MOTOR CONTROL CENTER (NEMCC), FACILITIES TERMINAL CABINET (FTC) NO. 2 (CONT)	
	j. COOLING TOWER EXHAUST FAN EF-31 RUN INDICATOR	Observe indicator is illuminated if outside air temperature is above 40°F or water temperature is above 90°F.
	k. NEMCC	Check for general conditions. (See general procedure a.)
3	LOCAL CONTROL HYDRAULIC PANEL	
	a. STANDBY/RETURN LINE PRESSURE GAUGE	17 to 27 PSI. (Selector valve in RETURN PRESSURE position.)
	aA. PRESSURE SELECTOR VALVE	Position to STANDBY PRESSURE.
	aB. STANDBY/RETURN LINE PRESSURE GAUGE	Increased to 90 to 220 PSI.
	aC. PRESSURE SELECTOR VALVE	Position to RETURN PRESSURE.
	CAUTION Gauge readings on items c, d, and e shall be taken with the STANDBY PUMP ON pushbutton depressed. If pressures do not stabilize at 280 to 320 PSI, release ON pushbutton immediately. Failure to comply may result in damage to standby pump and require fill and bleed operations.	
	b. STANDBY PUMP ON PUSHBUTTON	Depressed and held.
	c. ACCUM. PRESSURE SYSTEM GAUGE	Increased and stabilized at 280 to 320 PSI.
	d. ACCUM. PRESSURE UPPER DOOR GAUGE	Increased and stabilized at 280 to 320 PSI.

Table 3-8. Missile Crew Daily Complex Status Verification Procedures (CONT)

STEP	EQUIPMENT OR INDICATOR	CONDITION OR POSITION
3 (CONT)	LOCAL CONTROL HYDRAU-LIC PANEL (CONT)	
	e. ACCUM. PRESSURE LOWER DOOR GAUGE	Increased and stabilized at 280 to 320 PSI.
	f. STANDBY PUMP ON PUSHBUTTON	Released.
	g. N$_2$ PRESSURE SYSTEM GAUGE	Verify gauge indicates 3250 PSI (minimum).
	h. N$_2$ PRESSURE UPPER DOOR GAUGE	Verify gauge indicates 3250 PSI (minimum).
	i. N$_2$ PRESSURE LOWER DOOR GAUGE	Verify gauge indicates 3250 PSI (minimum).
	j. PRESS-TO-TEST INDICATORS	Perform lamp test of all press-to-test indicators on control panel.
	k. ACCUMULATOR AND N$_2$ SUPPLY NOT RE-CHARGED INDICATOR	Verify indicator is extinguished.
	l. FILTERS FR-115, FR-116 AND FR-102	Check filters for no-clogged indication. (Red flags not extended.)
	m. STANDBY ACCUMULA-TOR GAUGE GA 986	Verify gauge indicates 85 to 95 PSI.
	n. RESERVOIR LEVEL	Verify fluid level is above minimum operating level and below maximum drain level on sight glass. (With silo overhead doors closed and work platforms retracted.)
	SILO LEVEL 3	
1	GENERAL	
	a. COMMUNICATION PANEL	(See general procedure d.)
	b. ELECTRONIC CABINET AIR CONDITIONING SYSTEM	Check for general condition. (See general procedure a.)
	c. ELECTRONIC CABINET AIR CONDITIONING AIR FLOW	Verify air flow is being supplied to cabinets by observing ducting.

Table 3-8.   Missile Crew Daily Complex Status Verification Procedures (CONT)

STEP	EQUIPMENT OR INDICATOR	CONDITION OR POSITION
1 (CONT)	SILO LEVEL 3 (CONT)	
	d. FIRE EXTINGUISHERS (2)	(See general procedure c.)
	e. EMERGENCY LIGHTING UNITS (2)	(See general procedure b.)
	f. LEVEL 3	Check for general condition. (See general procedure a.)
2	400-CPS MOTOR GENERATOR	
	a. GENERAL CONDITION	(See general procedure a.)
3	28-VOLT DC POWER DISTRIBUTION SET	(See general procedure a.)
	a. GENERAL CONDITION	(See general procedure a.)
4	CONTROL-MONITOR GROUP 1 OF 4, PROPELLANT LEVEL (PANED 2)	**CAUTION** When propellant level panel FUEL REMOTE-LOCAL switch is in LOCAL position and FUEL LEVEL TOO HIGH "A" and (or) "B" indicators are illuminated, immediate action shall be initiated to adjust missile fuel level. Failure to comply may result in fuel contamination of pressure lines and PSC. Notify LCC prior to performing fuel level check. Any abnormal indication shall be investigated and corrected. If both FUEL LEVEL NOT TOO LOW "A" and "B" indicators are extinguished an unknown fuel level exists. A tactical countdown may be initiated if only FUEL LEVEL NOT TOO LOW "B" indicator is extinguished.
	a. FUEL REMOTE-LOCAL SWITCH	Position switch to LOCAL.
	b. FUEL LEVEL NOT TOO LOW "A" INDICATOR	Observe indicator is illuminated GREEN.
	c. FUEL LEVEL NOT TOO LOW "B" INDICATOR	Observe indicator is illuminated GREEN.
	d. FUEL LEVEL TOO HIGH "A" INDICATOR	Observe indicator is extinguished.

Table 3-8. Missile Crew Daily Complex Status Verification Procedures (CONT)

STEP	EQUIPMENT OR INDICATOR	CONDITION OR POSITION
4 (CONT)	CONTROL-MONITOR GROUP 1 OR 4, PROPELLANT LEVEL (PANEL 2) (CONT)	
	e. FUEL LEVEL TOO HIGH "B" INDICATOR	Observe indicator is extinguished.
5	FUEL TANKING (PANEL 1)	
	a. MISSILE FUEL LEVEL LOW INDICATOR	Observe indicator is extinguished.
	b. MISSILE FUEL LEVEL HIGH INDICATOR	Observe indicator is extinguished.
6	PROPELLANT LEVEL (PANEL 2)	
	a. FUEL REMOTE-LOCAL SWITCH	Position switch to REMOTE.
7	CONTROL-MONITOR GROUP 2 OF 4, MISSILE GROUND POWER (PANEL 1)	**NOTE** When missile battery simulator is connected and activated, BATTERY ACTIVATED indicator will be illuminated GREEN.
	a. BATTERY ACTIVATED INDICATOR	Observe indicator is extinguished.
8	MISSILE GROUND POWER (PANEL 2)	
	a. 28 VOLT DC SELECTOR SWITCH	Position selector switch to STANDBY BUS.
	b. DC VOLTMETER	Verify voltmeter indicates 29.5 ($+1.5$) volts.
	c. 28 VOLT DC SELECTOR SWITCH	Position selector switch to OFF.
	d. 400 CPS VOLTAGE SELECTOR SWITCH	Position selector switch to MOTOR GENERATOR OUTPUT.
	e. AC VOLTMETER	Verify voltmeter indicates 116.7 ($+1.4$) volts.
	f. 400 CPS VOLTAGE SELECTOR SWITCH	Position selector switch to OFF.

Table 3-8.   Missile Crew Daily Complex Status Verification Procedures (CONT)

STEP	EQUIPMENT OR INDICATOR	CONDITION OR POSITION
	SILO LEVEL 4	
1	GENERAL	
	a. COMMUNICATION PANEL	(See general procedure d.)
	b. MANUAL FIRE ALARM BOX	(See general procedure c.)
	c. EMERGENCY LIGHTING UNITS (2)	(See general procedure b.)
	d. FIRE EXTINGUISHERS (2)	(See general procedure c.)
	e. BREATHING APPARATUS	(See general procedure e.)
	f. CHILLED WATER SYSTEM	Check for general condition. (See general procedure a.)
	g. LEVEL 4	Check for general condition. (See general procedure a.)
2	WATER SYSTEM PUMPS	
	a. PUMPS P-30 AND P-31	Check for proper gland drip and ensure pumps are not locked out.
	b. PUMP P-32	Check for proper gland drip and ensure pumps are not locked out.
	c. PUMPS P-50 AND P-51	Check for proper gland drip and ensure pumps are not locked out.
	d. PUMPS P-60 AND P-61	Check for proper gland drip and ensure pumps are not locked out.
	e. PUMPS P-80 AND P-81	Check for proper gland drip and ensure pumps are not locked out.
3	UTILITY WATER SYSTEM	
	a. PRESSURE GAUGE PI-82	Verify PI-82 on tank TK-80 indicates normal pressure.
	b. LEVEL GAUGE LG-80	Verify LG-80 on tank TK-80 indicates ½-full, minimum.
	c. UTILITY WATER SYSTEM	Check for general condition. (See general procedure a.)

Table 3-8. Missile Crew Daily Complex Status Verification Procedures (CONT)

STEP	EQUIPMENT OR INDICATOR	CONDITION OR POSITION
4	HOT WATER SYSTEM	
	a. LEVEL GAUGE LG-63	Verify level gauge LG-63 on tank TK-63 indicates ½-full, nominal.
	b. PRESSURE GAUGE PI-60	Verify PI-60 indicates normal pressure.
	c. PRESSURE GAUGE PI-61	Verify PI-61 indicates normal pressure.
	d. HOT WATER SYSTEM	Check for general condition. (See general procedure a.)
5	CONDENSER WATER SYSTEM	
	a. INLET PRESSURE GAUGE PI-56	Verify PI-56 indicates normal pressures.
	b. INLET PRESSURE GAUGE PI-58	Verify PI-58 indicates normal pressure.
	c. INLET TEMPERATURE GAUGE TI-54	Verify TI-54 indicates normal temperature.
	d. INLET TEMPERATURE GAUGE TI-56	Verify TI-56 indicates normal temperature.
	e. OUTLET PRESSURE GAUGE PI-57	Verify PI-57 indicates normal pressure (unit operating).
	f. OUTLET PRESSURE GAUGE PI-59	Verify PI-59 indicates normal pressure (unit operating).
	g. OUTLET TEMPERATURE GAUGE TI-57	Verify TI-57 indicates normal temperature (unit operating).
	h. OUTLET TEMPERATURE GAUGE TI-55	Verify TI-55 indicates normal temperature (unit operating).
	i. OUTLET PRESSURE GAUGE PI-30	Verify PI-30 indicates normal pressure.
	j. INLET PRESSURE GAUGE PI-31	Verify PI-31 indicates normal pressure.
	k. INLET PRESSURE GAUGE PI-32	Verify PI-32 indicates normal pressure.
	l. CONDENSER WATER SYSTEM	Check for general condition. (See general procedure a.)

Table 3-8.   Missile Crew Daily Complex Status Verification Procedures (CONT)

STEP	EQUIPMENT OR INDICATOR	CONDITION OR POSITION
6	WATER CHILLER UNIT WCU-50 CAPACITY CONTROL PANEL	
	a. SELECTOR SWITCH	Position selector switch to alternate unit.
	SILO LEVEL 5	
1	480-VOLT SWITCHGEAR AREA	
	a. COMMUNICATION PANEL	(See general procedure d.)
	b. FIRE EXTINGUISHER	(See general procedure c.)
	c. EMERGENCY LIGHTING UNIT	(See general procedure b.)
	d. 480-VOLT SWITCHGEAR AREA	Check for general condition. (See general procedure a.)
2	480-VOLT SWITCHGEAR	
	a. GROUND INDICATORS	Verify brilliance of indicators are equal and that no indicators are extinguished.
	b. NEUTRAL RELAY	Check seal-in unit for target drop.
	NOTE At OSTF-2, perform either step 3 or 4 as applicable.	
3	OPERATING GENERATOR PANEL	
	a. GENERATOR MAIN BREAKER CONTROL SWITCH INDICATOR	Observe RED indicator is illuminated.
	b. GENERATOR SYNCHRO-NIZING SWITCH (NA OSTF-2)	Verify synchronizing switch handle is installed and positioned to OFF.
	c. GENERATOR VOLTAGE REGULATOR TRANSFER SWITCH	Verify switch is positioned to AUTOMATIC.

T.O. 21M-HGM16F-1

Table 3-8. Missile Crew Daily Complex Status Verification Procedures (CONT)

STEP	EQUIPMENT OR INDICATOR	CONDITION OR POSITION
3 (CONT)	OPERATING GENERATOR PANEL (CONT)	
	d. OVERCURRENT RELAYS (3)	Check seal-in unit for target drop.
	e. ANTI MONITORING RELAY (NA OSTF-2)	Check seal-in unit for target drop.
4	ALTERNATE GENERATOR PANEL	
	a. GENERATOR MAIN BREAKER CONTROL SWITCH	Observe GREEN indicator is illuminated.
	b. GENERATOR VOLTAGE REGULATOR TRANSFER SWITCH	Verify switch is positioned to AUTOMATIC.
	c. OVERCURRENT RELAY (3)	Check seal-in unit for target drop.
	d. ANTI-MONITORING RELAY (NA OSTF-2)	Check seal-in unit for target drop.
5	SUPPLEMENTARY PANEL	
	a. ELECTRIC HOT WATER BOILER EB-60 CIRCUIT BREAKER	Verify circuit breaker is in ON position.
	b. ELECTRIC DUCT HEATER EC-20 CIRCUIT BREAKER (NA VAFB)	Verify circuit breaker is in ON position.
6	NO. 1 GENERATOR ROOM	
	a. COMMUNICATION PANEL	(See general procedure d.)
	b. BREATHING APPARATUS (2)	(See general procedure e.)
	c. EMERGENCY LIGHTING UNITS (2)	(See general procedure c.)
	d. CLEAN LUBE OIL STORAGE TANK LIQUID LEVEL SIGHT GAUGE	Check liquid level sight gauge for sufficient level ($\frac{1}{3}$-full, minimum).

Table 3-8. Missile Crew Daily Complex Status Verification Procedures (CONT)

STEP	EQUIPMENT OR INDICATOR	CONDITION OR POSITION
6 (CONT)	NO. 1 GENERATOR ROOM (CONT)	
	e. DIRTY LUBE OIL STORAGE TANK LIQUID LEVEL SIGHT GAUGE	Verify liquid level sight gauge indicates less than ¾-full.
	f. DIESEL FUEL OIL DAY TANK	Drain approximately ½-gallon fuel oil to remove sediment and condensation.
	g. FUEL DRIP RETURN PUMP SWITCH (OSTF-2 ONLY)	Verify switch is in ON position.
	h. ELECTRIC HOT WATER BOILER EB-60 SWITCH (OSTF-2 ONLY)	Verify switch is in ON position.
	i. ELECTRICAL HOT WATER BOILER EB-60 POWER ON INDICATOR (OSTF-2 ONLY)	Observe POWER ON indicator is illuminated.
	j. EB-60 OUTLET WATER TEMPERATURE GAUGE TI-1 (OSTF-2 ONLY)	Verify gauge indicates normal.
	k. EB-60 INLET WATER TEMPERATURE GAUGE TI-68 (OSTF-2 ONLY)	Verify gauge indicates normal.
	l. STARTING AIR RECEIVER TANK (OSTF-2 ONLY)	Drain condensation from receiver.
	m. CONDENSER WATER INLET PRESSURE GAUGE PI-62A	Check for normal pressure.
	n. CONDENSER WATER INLET TEMPERATURE GAUGE TI-60	Check for normal temperature.
	o. CONDENSER WATER OUTLET PRESSURE GAUGE PI-63	Check for normal pressure.
	p. CONDENSER WATER OUTLET TEMPERATURE GAUGE TI-61	Check for normal temperature.

Table 3-8.   Missile Crew Daily Complex Status Verification Procedures (CONT)

STEP	EQUIPMENT OR INDICATOR	CONDITION OR POSITION
6 (CONT)	NO. 1 GENERATOR ROOM (CONT)	
	q. FIRE FOG SYSTEM SHUT-OFF VALVE	Verify manual valve, located upstream of LCV-80, is open.
	r. FIRE EXTINGUISHERS (2)	(See general procedure c.)
	s. GENERATOR ROOM	Check for general condition. (See general procedure a.)

NOTE

Perform step 7 only if D-60 is in standby.
Perform step 8 only if D-60 is operating.

STEP	EQUIPMENT OR INDICATOR	CONDITION OR POSITION
7	STANDBY DIESEL GENERATOR	

WARNING

MASTER CONTROL SWITCH shall be in OFF position before performing any checks or inspections of any parts subject to move if engine is accidently started.

STEP	EQUIPMENT OR INDICATOR	CONDITION OR POSITION
	a. MASTER CONTROL SWITCH	Position switch to OFF.
	b. DIESEL LUBE OIL LEVEL	Verify lube oil level is between normal and full as indicated on dip stick.
	c. FUEL OIL RETURN LINE VALVE	Verify fuel oil return line valve is open.
	d. GENERATOR FIELD BREAKER	Verify generator field breaker is in ON position.
	e. GENERATOR PEDESTAL BEARING OIL CUP	Verify generator pedestal bearing oil cup is full.
	f. CONDENSER WATER INLET VALVE	Verify condenser water inlet valve is open.
	g. AIR INTAKE FILTER OIL LEVEL	Check air intake filter oil level is up to mark. Do not overfill.

Table 3-8. Missile Crew Daily Complex Status Verification Procedures (CONT)

STEP	EQUIPMENT OR INDICATOR	CONDITION OR POSITION
7 (CONT)	STANDBY DIESEL GENERATOR (CONT)	
	h. JACKET WATER EXPANSION TANK LIQUID LEVEL GAUGE	Verify liquid level gauge is ½-full, maximum.
	i. CONDENSER WATER OUTLET VALVE	Verify condenser water outlet valve is open.
	j. FUEL OIL INLET LINE VALVE	Verify fuel oil inlet line valve is open.
	k. STARTING AIR VALVE	Verify starting air valve is open.
	l. GOVERNOR OIL LEVEL	Verify governor oil level is above low mark and not above full mark.
	m. GOVERNOR SPEED DROOP CONTROL	Verify governor speed droop is set at 3.
	n. GOVERNOR PERCENT LOAD CONTROL	Verify governor percent load is set at MAXIMUM FUEL.
	o. INDICATOR (SNIFTER) VALVES (8)	Verify indicator (snifter) valve (8) are in open position.
	p. AIR START PLUNGER	Depress air-start plunger on utility Air Regulator momentarily and allow engine to make at least two revolutions. Observing for condensation discharge.
	q. INDICATOR (SNIFTER) VALVES (8)	Verify indicator (snifter) valves (8) are in closed position.
	r. WATER PUMP BELTS	Check water pump belts for wear and tension.
	s. EXCITER DRIVE BELTS	Check exciter drive belts for wear and tension.
	t. MASTER CONTROL SWITCH	Position switch to ON.
	u. ALARM TEST PUSHBUTTON	Establish contact with LCC, then depress ALARM TEST pushbutton.
	v. ALL INDICATOR AND ALARM	Observe all indicators are illuminated and that alarm sounds, then release ALARM TEST pushbutton.
	w. GENERATOR	Check for general condition. (See general procedure a.)

Table 3-8. Missile Crew Daily Complex Status Verification Procedures (CONT)

STEP	EQUIPMENT OR INDICATOR	CONDITION OR POSITION
8	OPERATING DIESEL GENERATOR	
	a. GOVERNOR SPEED DROOP CONTROL	Verify governor speed droop is set at 3.
	b. GOVERNOR PERCENT LOAD CONTROL	Verify governor percent load is set at MAXIMUM FUEL.
	c. JACKET WATER EXPANSION TANK LIQUID LEVEL GAUGE	Verify liquid level gauge indicates ½-full, maximum.
	d. GENERATOR PEDESTAL BEARING OIL CUP	Verify generator pedestal bearing oil is full.
	e. GENERATOR	Check for general condition. (See general procedure a.)
1	SILO LEVEL 6 NO. 2 GENERATOR ROOM	
	a. COMMUNICATION PANEL	(See general procedure d.)
	b. BREATHING APPARATUS (2)	(See general procedure e.)
	c. FIRE EXTINGUISHERS (2)	(See general procedure c.)
	d. EMERGENCY LIGHTING UNITS (2)	(See general procedure b.)
	e. ELECTRIC HOT WATER BOILER EB-60 (NA OSTF-2)	Verify electric hot water boiler ON switch is ON.
	f. ELECTRIC HOT WATER EB-60 POWER ON INDICATOR (NA OSTF-2)	Observe POWER ON indicator is illuminated.
	g. EB-60 OUTLET WATER TEMPERATURE GAUGE TI-1 (NA OSTF-2)	Verify gauge indicates normal.
	h. EB-60 INLET WATER TEMPERATURE GAUGE TI-68 (NA OSTF-2)	Verify gauge indicates normal.

Table 3-8.   Missile Crew Daily Complex Status Verification Procedures (CONT)

STEP	EQUIPMENT OR INDICATOR	CONDITION OR POSITION
1 (CONT)	NO. 2 GENERATOR ROOM (CONT)	
	i. FUEL DRIP RETURN PUMP (NA OSTF-2)	Verify fuel drip return pump switch is ON.
	j. STARTING AIR RECEIVER TANK TK-64 (NA OSTF-2)	Blow down air tank for condensation.
	k. CONDENSER WATER INLET PRESSURE GAUGE PI-66 (NA OSTF-2)	Check for normal pressure.
	l. CONDENSER WATER INLET TEMPERATURE GAUGE TI-62 (NA OSTF-2)	Check for normal temperature.
	m. CONDENSER WATER OUTLET PRESSURE GAUGE PI-67 (NA OSTF-2)	Check for normal pressure.
	n. CONDENSER WATER OUTLET TEMPERATURE GAUGE TI-63 (NA OSTF-2)	Check for normal temperature.
	o. GENERATOR ROOM	Check for general condition. (See general procedure a.)
		NOTE If D-61 is in standby, perform step 7 under silo level 5. If D-61 is operating, perform step 8 under silo level 5.
2	48 VOLT BATTERY CHARGER AREA	
	a. EMERGENCY LIGHTING UNIT	(See general procedure b.)
	b. FIRE EXTINGUISHER	(See general procedure c.)
	c. BATTERY BANK ELECTROLYTE	Check electrolyte for proper level.
	d. BATTERY CHARGE INDICATORS	Check indicators for proper charge level.

Table 3-8.   Missile Crew Daily Complex Status Verification Procedures (CONT)

STEP	EQUIPMENT OR INDICATOR	CONDITION OR POSITION
2 (CONT)	48 VOLT BATTERY CHARGER AREA (CONT)	
	e. BATTERY CHARGER VOLTMETER	Check voltmeter for normal indication.
	f. BATTERY CHARGER AMMETER	Check ammeter for normal indication.
	g. BATTERY CHARGER GROUND INDICATORS	Check ground indicator lights for dull glow (Brilliant glow indicates a ground.)
	h. TEST SWITCH (TFR CHARGER ONLY)	Check test switch (alarm failure) is in normal position.
	i. BATTERY CHARGER AREA	Check for general condition. (See general procedure a.)
1	SILO LEVEL 7 GENERAL	
	a. COMMUNICATION PANEL	(See general procedure d.)
	b. BREATHING APPARATUS (4)	(See general procedure e.)
	c. FIRE HOSE STATION	(See general procedure c.)
	d. EMERGENCY LIGHTING UNITS (2)	(See general procedure b.)
	e. SUPPLY FAN SF-22	Check for general condition. (See general procedure a.)
	f. LEVEL 7	Check for general condition. (See general procedure a.)
2	PRESSURIZATION PREFAB	
	a. FILTER N-29	Verify mechnical indicator indicates a noclogged condition (no red flag).
	b. PRESSURE GAUGE PI-1	Record pressure.
	c. PRESSURE GAUGE PI-3	Record pressure.
	d. PRESSURE GAUGE PI-4	Record pressure.

Table 3-8.   Missile Crew Daily Complex Status Verification Procedures (CONT)

STEP	EQUIPMENT OR INDICATOR	CONDITION OR POSITION
2 (CONT)	PRESSURIZATION PREFAB (CONT)	
	e. LIQUID LEVEL INDICA-TOR LLI-1	Record level.
	f. LIQUID LEVEL INDICA-TOR LLI-2	Record level.
	g. PRESSURIZATION VALVE N-6	Verify valve N-6 is open.
	h. CONTROLLER PC-1	Verify controller is set at 25 PSI for LO_2 (20 PSI for LN_2).
	i. CONTROLLER PC-2	Verify controller is set at 135 PSI for LO_2 (100 PSI for LN_2).
	j. CONTROLLER PC-3	Verify controller is set at 150 PSI for LO_2 (110 PSI for LN_2).
	jA. LO_2 STORAGE STAND-BY OR CHILLDOWN PRESS. GAUGE PI-14	Verify pressure gauge indicates 2.4 (+1) PSI.
	k. PRESSURIZATION PRE-FAB	Check for general condition. (See general procedure a.)
3	LO_2 STORAGE TANK VACUUM PUMP	
	a. MICRON GAUGE	Check gauge for 150 microns, nominal.
		WARNING
		Vacuum pump oil is toxic. Use extreme caution not to inhale fumes or come in contact with oil.
	b. Deleted	
	c. OIL LEVEL	Verify oil level sight glass indicates ¼-full, minimum.
	d. VACUUM PUMP	Check for general condition. (See general procedure a.)
4	LO_2 CONTROL PREFAB	
	a. PRESSURE GAUGE PI-8	Verify PI-8 indicates 35 PSI.
	b. PRESSURE GAUGE PI-7	Verify PI-7 indicates 250 PSI.
	c. L-15 FILTER DIFFEREN-TIAL PRESSURE GAUGE	Ensure gauge is set at zero.

Table 3-8. Missile Crew Daily Complex Status Verification Procedures (CONT)

STEP	EQUIPMENT OR INDICATOR	CONDITION OR POSITION
4 (CONT)	LO₂ CONTROL PREFAB (CONT)	
	d. LO₂ CONTROL PREFAB	Check for general condition. (See general procedure a.)
5	INSTRUMENT AIR PREFAB	
	a. TANK TK-64	Drain condensation from TK-64 by opening drain valve until no moisture is evident.
	b. INSTRUMENT AIR PRE-FAB	Check for general condition. (See general procedure.)
6	GASEOUS OXYGEN DETECTOR	
	a. TROUBLE INDICATOR	Observe TROUBLE indicator is extinguished.
	b. STATION INDICATOR	Observe STATION indicators are cycling.
	c. OXYGEN PERCENT METER	Verify oxygen percent meter indicates 20.8%.
	d. SPAN GAS PRESS GAUGE (MSA ONLY)	Check for greater than 300 PSI. Record pressure.
7	DIESEL FUEL VAPOR DETECTOR	
	a. NORMAL POWER INDICATOR	Observe NORMAL POWER indicator is illuminated.
	b. EXPLOSIVE LEVEL METER	Verify explosive level meter is indicating 0% LEL.
	c. TROUBLE INDICATOR	Observe TROUBLE indicator is extinguished.
8	LIQUID NITROGEN PREFAB	**NOTE** Record vessel levels (or pressures). Readings shall be compared with values specified in table 6-1, for replenishment requirements and EWO or PLX minimums. Request re-servicing as required.
	a. LIQUID LEVEL INDICATOR LLI-220	Record level.
	b. LIQUID LEVEL INDICATOR LLI-221	Record level.

Table 3-8.   Missile Crew Daily Complex Status Verification Procedures (CONT)

STEP	EQUIPMENT OR INDICATOR	CONDITION OR POSITION
8 (CONT)	LIQUID NITROGEN PREFAB (CONT)	**NOTE** Manual valves are of the softseat type. When checking valves to be in the closed position, apply only sufficient pressure to ensure valves are closed.
	c. MANUAL VALVES 203, 216, 208 AND 226	Verify valves are closed.
	d. FILTER N-230	Check for no-clogged indication (no red flag).
	e. PREFAB	Check for general condition. (See general procedure a.)
9	LO₂ TOPPING CONTROL UNIT PREFAB	
	a. L-51 FILTER DIFFERENTIAL PRESSURE GAUGE	Verify gauge is set at zero.
	b. PRESSURE GAUGE PG-1	Verify gauge indicates 1000 PSI.
	c. PREFAB	Check for general condition. (See general procedure a.)
	SILO LEVEL 8	
1	GENERAL	
	a. COMMUNICATION PANEL	(See general procedure d.)
	b. BREATHING APPARATUS (5)	(See general procedure e.)
	c. FIRE HOSE STATION	(See general procedure c.)
	d. MANUAL FIRE ALARM BOX	(See general procedure c.)
	e. EMERGENCY LIGHTING UNITS (2)	(See general procedure b.)
	f. FIRE EXTINGUISHERS (2)	(See general procedure c.)
	g. VACUUM PUMP RUN INDICATORS	Verify RUN indicators (3) on vacuum pump local control station are illuminated.
	h. LEVEL 8	Check for general condition. (See general procedure a.)

Table 3-8. Missile Crew Daily Complex Status Verification Procedures (CONT)

STEP	EQUIPMENT OR INDICATOR	CONDITION OR POSITION
2	LN$_2$/HELIUM AND LO$_2$ TOPPING TANK VACUUM PUMPS	**NOTE** Check both vacuum pumps for the following conditions:
	a. MICRON GAUGE	Check micron gauge for 150 microns nominal.
		WARNING Vacuum pump oil is toxic. Use extreme caution not to inhale fumes or come in contact with oil.
	b. Deleted	
	c. OIL LEVEL SIGHT GAUGE	Verify oil level sight gauge indicates ¼-full, minimum.
	d. VACUUM PUMP	Check for general condition. (See general procedure a.)
3	THRUST SECTION HEATERS ASSEMBLY	
	a. PRESSURE GAUGE PI-70	Check for normal indication.
	b. TEMPERATURE GAUGE TI-70	Check for normal indication.
	c. PRESSURE GAUGE PI-71	Check for normal indication.
	d. TEMPERATURE GAUGE TI-71	Check for normal indication.
4	MISSILE ENCLOSURE AREA (MEA) SILO LEVEL 8	
	a. LN$_2$ EVAPORATOR TANK	Check for general condition. (See general procedure a.)
	b. UMBILICAL LOOP	Check for general condition. (See general procedure a.)
	c. LAUNCHER PLATFORM (LP) SHEAVES AND CABLES	Check for general condition. (See general procedure a.)
	d. HOT AND COLD DISCONNECTS	Check for general condition. (See general procedure a.)
	e. LP (BOTTOM)	Check for general condition. (See general procedure a.)

Table 3-8.   Missile Crew Daily Complex Status Verification Procedures (CONT)

STEP	EQUIPMENT OR INDICATOR	CONDITION OR POSITION
5	PRESSURE SYSTEM CONTROL (PSC)	
	a. MISSILE FUEL TANK PRESSURE GAUGE PI-129	Verify pressure (11.9 to 13.0 PSI) with LCC.
	b. MISSILE LO₂ TANK PRESSURE GAUGE PI-130	Verify pressure (3.4 to 4.2 PSI) with LCC.
	c. PSC	Check for general condition. (See general procedure a.)
6	FUEL LEVELING PREFAB	
	a. PRESSURE GAUGE PI-9	Verify PI-9 indicates 750 (+40) PSI.
	b. LIQUID LEVEL INDICATOR LLI-3	Record level.
	c. PREFAB	Check for general condition. (See general procedure a.)
7	PNEUMATIC SYSTEM MANIFOLD REGULATOR (PSMR)	
	a. PRESSURE GAUGE PI-1, AIRBORNE HELIUM SUPPLY NO. 1	Record pressure
	b. PRESSURE GAUGE PI-2, AIRBORNE HELIUM SUPPLY NO. 2	Record pressure
	c. PRESSURE GAUGE PI-20, NITROGEN SUPPLY	Record pressure
	d. PRESSURE GAUGE PI-30, NITROGEN SUPPLY PRESSURE	Record pressure
	e. PRESSURE GAUGE PI-47, INST AIR SUPPLY PRESS.	Verify PI-47 indicates 300 (+25) PSI.
	eA. FUEL TANK PRESSURE GAUGE PI-71	Verify pressure indication with PSC gauge PI-129.
	eB. LO₂ TANK PRESSURE GAUGE PI-72	Verify pressure indication with PSC gauge PI-130.
	f. PSMR	Check for general condition. (See general procedure a.)

Table 3-8.   Missile Crew Daily Complex Status Verification Procedures (CONT)

STEP	EQUIPMENT OR INDICATOR	CONDITION OR POSITION
1	**SUMP AREA** GENERAL a. COMMUNICATION PANEL	(See general procedure d.)
	b. BREATHING APPARATUS (2)	(See general procedure e.)
	c. EXHAUST FAN EF-40	Check for general condition. (See general procedure a.)
	d. EXHAUST FAN EF-41 START PUSHBUTTON	Depress EF-41 START pushbutton.
	e. EXHAUST FAN EF-41	Check for general condition. (See general procedure a.)
	f. LP COUNTERWEIGHT AREA	Verify counterweight area is free of obstructions.
	g. SUMP PUMPS (2)	Position pump switch to MANUAL and check pumps for general condition. Ensure pumps are switched back to AUTO after completion of check.
	h. FIRE EXTINGUISHER	(See general procedure c.)
	i SUMP AREA	Check for general condition. (See general procedure a.)
1	**LAUNCHER PLATFORM LEVEL 4** GENERAL a. GUIDE ROLLERS AND SHEAVES	Check guide rollers and sheaves to be free of obstructions.
	b. FIRE EXTINGUISHER	(See general procedure c.)
	bA. INERT FLUID IN-JECTION MODULE INLET VALVE	Open and safetied for launch. Closed and safetied for PLX.
	bB. INERT FLUID IN-JECTION MODULE VENT VALVE	Closed and safetied.
	c. LP LEVEL 4	Check for general condition. (See general procedure a.)
1	**LAUNCHER PLATFORM LEVEL 3** GENERAL a. COMMUNICATION PANEL	(See general procedure d.)
	b. FIRE EXTINGUISHER	(See general procedure c.)

Table 3-8. Missile Crew Daily Complex Status Verification Procedures (CONT)

STEP	EQUIPMENT OR INDICATOR	CONDITION OR POSITION
1 (CONT)	LAUNCHER PLATFORM LEVEL 3 (CONT)	
	c. HELIUM CONTROL CHARGING UNIT (HCU)	Check for general condition. (See general procedure a.)
	cA. FUEL TANK PRESS. GAUGE 338	Verify indication with launch control console FUEL TANK pressure gauge.
	cB. OXIDIZER TANK PRESS. GAUGE 341	Verify indication with launch control console LO$_2$ TANK pressure gauge.
	d. NITROGEN CONTROL UNIT (NCU)	Check for general condition. (See general procedure a.)
	e. LP LEVEL 3	Check for general condition. (See general procedure a.)
2	HYDRAULIC PUMPING UNIT (HPU)	
	a. HPU RESERVOIR GAUGE	Verify gauge indicates between 12 gallons (minimum) and 16 gallons (maximum).
	b. FIRST AND SECOND STAGE PRESSURE SHUT-OFF VALVES	Verify valves are in open position and safetied.
	c. FIRST AND SECOND STAGE EVACUATION CHAMBER PRESSURE GAUGES	Verify gauges indicate 500 (+50) PSI.
	d. HPU	Check for general condition. (See general procedure a.)
1	LAUNCHER PLATFORM LEVEL 2 GENERAL	
	a. COMMUNICATION PANEL	(See general procedure d.)
	b. FIRE EXTINGUISHER	(See general procedure c.)
	c. MAIN LOCKS AND GUIDE ROLLERS	Check for obstructions.
	d. UMBILICAL JUNCTION BOX CONNECTORS AND PLUGS	Check for caps to be properly installed and that plugs are covered.
	e. LP LEVEL 2	Check for general condition. (See general procedure a.)

Table 3-8. Missile Crew Daily Complex Status Verification Procedures (CONT)

STEP	EQUIPMENT OR INDICATOR	CONDITION OR POSITION
	LAUNCHER PLATFORM LEVEL 1	
1	GENERAL	
	a. UMBILICAL LANYARDS	Check lanyards for proper slack.
	b. SIGHT TUBE	Ensure sight tube is installed and secured.
	c. FLAME DEFLECTOR	Check for general condition. (See general procedure a.)
	d. MEA WALLS	Check walls for protruding obstructions.
	e. ENGINE HUMIDITY INDICATORS (5)	**NOTE** If humidity indicator is over 30%, the desiccant must be changed. If the indicator is 40%, a flush and purge operation is required. Check indicator to be indicating less than 30%.
	f. MISSILE	Check for general condition. (See general procedure a.)
	g. LP LEVEL 1	Check for general condition. (See general procedure a.)
	COMPLEX GRADE LEVEL	
1	GENERAL	
	a. COMMUNICATION PANEL	(See general procedure d.)
	b. INTAKE AND EXHAUST AIR VENTS	Check vents to be free of obstructions.
	c. FILL AND VENT SHAFT	Check vent shaft to be free of debris and contamination.
	d. ELECTRICAL STUBUPS	Check stubups for cleanness and damage. Ensure stubups are properly capped.
	e. FIRE EXTINGUISHER	(See general procedure c.)
	f. SECURITY LIGHTS	Check for proper operation and damage.
	g. SILO DOOR RAIN COVERS	Check covers for damage and proper installation.
	h. DIESEL FUEL OIL SOFT STORAGE TANK	Check level of storage tank and record level.
	i. PERIMETER FENCE AND GATE	Check fence and gate for damage.

            Changed 21 October 1964

Table 3-8.   Missile Crew Daily Complex Status Verification Procedures (CONT)

STEP	EQUIPMENT OR INDICATOR	CONDITION OR POSITION
2	COOLING TOWER	
	a. FAN MOTOR	Check for general condition. (See general procedure a.)
	b. WATER LEVEL	Check water level to be approximately ½-inch below overflow pipe.
	c. FLOAT CONTROL VALVE	Check float control valve for free operation. (Test by hand.)
	d. SCREENS	Check screens for proper installation and to be free of obstructions.
	e. MANUAL BYPASS VALVE	**Ensure manual bypass valve for TCV 30 is closed.**
	f. COOLING TOWER BASIN	Check for general condition. (See general procedure a.)
	g. ELECTRIC HEATERS	Verify electric heaters are operating (if temperature is below 40°F).
3	WATER TREATMENT PLANT (IF INSTALLED)	
	NOTE Perform all steps that are applicable to the type of unit installed.	
	a. BUILDING STRUCTURE	Check for general condition. (See general procedure a.)
	b. BRINE MEASURING TANK (SALT TANK)	Verify salt level is approximately 20 inches from top of tank.
	c. WATER LEVEL	Verify water level is approximately 1 inch below overflow.
	d. WATER PUMP	Check water pump operation by selecting HAND position, then AUTO position.
	e. HYPOCHLORINATOR	Verify operation, (pump must be on when well pump is on.)
	f. CHLORINE RESIDUAL AND HARDNESS OF WATER	Perform chlorine reading at water treatment plant and at one drinking fountain in silo. Chlorine readings should be 0.2 to 0.6 parts per million.
	g. SOLEMATIC VALVES	Check valves for proper position.
	h. BUILDING HEATERS	Verify that building heaters are on if during cold weather.
	i. HARDNESS MONITOR	Record reading from monitor.
	j. WATER METER	Check water meter from proper flow (approximately 15 GPM).

Table 3-9. MGS Aligned, Not-Aligned Procedures

STEP	EQUIPMENT OR INDICATOR	CONDITION OR POSITION
	ITEM 1. ALIGNED PRO-CEDURES COUNTDOWN GROUP	**NOTE** Verify target select switch position on count-down group is the same as target selected on LCOC.
1	CONTROL SELECTOR SWITCH	Position control selector switch to LOCAL-AUTO.
2	READINESS STATUS SWITCH	Position switch to ALIGNED.
3	MGS CHECKOUT COMPLETE INDICATOR	Observe indicator is extinguished.
4	MGS CHECKOUT INDICATOR	Observe indicator is illuminated.
5	ALIGNMENT COMPLETE INDICATOR	Observe indicator is illuminated 6 to 13 minutes after per-forming step 2.
6	CONTROL SELECTOR SWITCH	Position switch to REMOTE-AUTO.
	ITEM 2. NOT-ALIGNED PRO-CEDURES, COUNTDOWN GROUP	**NOTE** The following steps shall be performed to return the missile guidance system to the not-aligned status from an aligned status.
1	ALIGNED COMPLETE INDICATOR	Observe that indicator is illuminated.
2	CONTROL SELECTOR SWITCH	Position switch to LOCAL-AUTO.
3	READINESS STATUS SWITCH	Position switch to NOT-ALIGNED.
4	RETURN TO READINESS PUSHBUTTON	Momentarily depress pushbutton.
5	ALIGNMENT COMPLETE INDICATOR	Observe indicator is extinguished
6	MGS CHECKOUT COMPLETE INDICATOR	Observe indicator is illuminated.
7	CONTROL SELECTOR SWITCH	Position switch to LOCAL-MANUAL.
8	STEPPING SWITCHES ON CHASSIS 1A2A2, 1A1A5 and 1A2A4	Observe all stepping switches are in zero position.
9	CONTROL SELECTOR SWITCH	Position switch to REMOTE-AUTO.

Table 3-10.   Missile Lifting System Correlation

STEP	EQUIPMENT OR INDICATOR	CONDITION OR POSITION
	CSMOL, LCC LEVEL 2	
1	RESET PROGRAMMER KEY SWITCH	Position key switch ON.
2	HYDRAULIC 40 HP PUMP ON INDICATOR	Observe indicator is extinguished.
3	HYDRAULIC 40 HP PUMP PRESSURE INDICATOR	Observe indicator is extinguished.
	SILO LEVEL 1 MLS LOGIC UNIT 1A4A2	
1	DIRECTORY SWITCH	Verify switch is in position 1.
2	TEST SELECTOR SWITCH	Verify switch is in position 1.
3	LAUNCHER STATUS AND TEST START PUSHBUTTONS	Momentarily depress both pushbuttons simultaneously.
4	LAUNCHER STATUS INDICATOR	Observe indicator is illuminated green.
	CSMOL, LCC LEVEL 2	
1	RESET PROGRAMMER KEY SWITCH	Position key switch OFF.

Table 3-11.   Diesel General Operations at Power Remote Control Panel (NA OSTF-2)

STEP	EQUIPMENT OR INDICATOR	CONDITION OR POSITION
	NOTE The diesel generators must have previously been paralleled at the 480-volt switch-gear, then shutdown without disturbing the adjustments made during paralleling. **NOTE** The diesel engine shall be run for a period of at least 30 minutes after starting to allow accessories to reach normal operating temperature before shutdown.	
	ITEM 1 STARTING AND PAR-ALLELING ALTERNATE GEN-ERATOR	
1	**GENERATOR AMMETER**	Position **GENERATOR AMMETER SWITCH** to **1, 2,** or **3.**
2	**SYNCHRONIZING SWITCH**	Insert **SYNCHRONIZING SWITCH** handle in **SYNCHRONIZING SWITCH** of generator to be started and position switch handle to **ON.** **NOTE** If SYNCHRONIZING SWITCH is not in the ON position, INCOMING VOLTAGE INCOMING FREQUENCY, and SYN-CHROSCOPE meters will not operate and GENERATOR MAIN BREAKER CONTROL SWITCH cannot be closed.
3	**ENGINE START-STOP SWITCH INDICATOR**	Observe **ENGINE START-STOP SWITCH GREEN** indicator for generator to be started is illuminated.
4	**ENGINE START-STOP SWITCH**	Position **ENGINE START-STOP SWITCH** of generator to be started to **START** and hold momentarily. Observe **GREEN ENGINE START-STOP SWITCH** indicator is extinguished.
5	**ENGINE START-STOP SWITCH INDICATOR**	Observe **ENGINE START-STOP SWITCH RED** indicator illuminates when **ENGINE START-STOP SWITCH** is released.
6	**INCOMING VOLTAGE METER**	Monitor **INCOMING VOLTAGE** meter until voltage has stabilized at 460 (+2, -8) volts. **NOTE** SYNCHROSCOPE will not operate unless incoming frequency is between 56 and 64 CPS.
7	**INCOME FREQUENCY METER**	Observe **INCOMING FREQUENCY** meter indicates 60 CPS. If necessary, adjust incoming frequency with **GOVERNOR MOTOR CONTROL SWITCH.**

Table 3-11. Diesel General Operations at Power Remote Control Panel (NA OSTF-2) (CONT)

STEP	EQUIPMENT OR INDICATOR	CONDITION OR POSITION
	ITEM 1. STARTING AND PAR-ALLELING ALTERNATE GEN-ERATOR (CONT)	
8	SYNCHROSCOPE	Operate started generator GOVERNOR MOTOR CONTROL SWITCH so that SYNCHROSCOPE rotates slowly clockwise. **NOTE** When incoming GENERATOR MAIN BREAKER CONTROL SWITCH is closed, incoming AMMETER AC indication will increase and running generator AMMETER AC indication will decrease.
9	GENERATOR MAIN BREAKER CONTROL SWITCH	When SYNCHROSCOPE hand indicates 11:55 (clock position), position started generator MAIN BREAKER CONTROL SWITCH to CLOSE.
10	GENERATOR MAIN BREAKER CONTROL SWITCH INDICATOR	Observe both GENERATOR MAIN BREAKER CONTROL SWITCH RED indicators are illuminated and both GREEN indicators are extinguished.
11	BOTH AMMETER AC METERS	Observe AMMETER AC meters indicate same current. If both ammeters do not indicate the same current, operate one GOVERNOR MOTOR CONTROL SWITCH until both AMMETER AC meters indicate the same current, retaining 60 CPS. **NOTE** A variation of 50 AMP between ammeters shall require adjustment.
12	SYNCHRONIZING SWITCH	Position started generated SYNCHRONIZING SWITCH to OFF.
13	RUNNING VOLTAGE METER	Observe RUNNING VOLTAGE meter indicates 460 (+2, -8) volts.
14	RUNNING FREQUENCY METER	Observe RUNNING FREQUENCY meter indicates 60 CPS. If necessary, operate one or both GOVERNOR MOTOR CONTROL SWITCH and correct frequency. **NOTE** If one diesel generator fails during parallel operation, immediate action is required by EPPT.

Table 3-11.   Diesel General Operations at Power Remote Control Panel (NA OSTF-2) (CONT)

STEP	EQUIPMENT OR INDICATOR	CONDITION OR POSITION
	ITEM 2. PARALLELED GENERATOR SHUT DOWN	
1	GENERATOR MAIN BREAKER CONTROL SWITCH	Position GENERATOR (NO. 1 or NO. 2) MAIN BREAKER CONTROL SWITCH to TRIP.
2	GENERATOR MAIN BREAKER CONTROL SWITCH INDICATOR	Observe GENERATOR MAIN BREAKER CONTROL SWITCH GREEN indicator is illuminated and RED indicator is extinguished.
		NOTE To cool engines, allow engine to run for approximately 5 minutes if not an emergency shutdown.
3	ENGINE START-STOP SWITCH	Position ENGINE START-STOP SWITCH to STOP and hold momentarily.
4	ENGINE START-STOP SWITCH INDICATOR	Observe ENGINE START-STOP SWITCH RED indicator extinguishes and GREEN indicator illuminates when switch is released.

Table 3-11A.   Diesel Generator and Power CO. Line Operation at Power Remote Control Panel (OSTF-2 only)

STEP	EQUIPMENT OR INDICATOR	CONDITION OR POSITION
	NOTE The diesel generator must have previously been started and voltage adjusted, then shut down without disturbing the adjustments, prior to transferring from POWER CO. line power to diesel generator power from the PRCP.	
	CAUTION Prior to closing GENERATOR MAIN BREAKER CONTROL SWITCH, POWER CO. LINE MAIN BREAKER CONTROL SWITCH must be tripped.	
	ITEM 1. POWER CO. LINE POWER TO DIESEL GENERATOR POWER **NOTE** The following steps shall be performed when transferring from POWER CO. line power to diesel generator power.	

Table 3-11A.   Diesel Generator and Power CO. Line Operation at Power Remote
Control Panel (OSTF-2 only) (CONT)

STEP	EQUIPMENT OR INDICATOR	CONDITION OR POSITION
	ITEM 1. POWER CO. LINE POWER TO DIESEL GENER-ATOR POWER (CONT)	
1	GENERATOR AMMETER SWITCH	Position GENERATOR AMMETER SWITCH to 1, 2, or 3.
2	GENERATOR VOLTMETER SWITCH	Position GENERATOR VOLTMETER SWITCH to 1-2, 2-3, or 3-4.
3	ENGINE START-STOP SWITCH INDICATOR	Observe ENGINE START-STOP SWITCH GREEN indicator is illuminated.
		NOTE
		Diesel engine shall be run for a period of at least 30 minutes after starting to allow accessories to reach normal operating temperature before shutdown.
4	ENGINE START-STOP SWITCH	Position ENGINE START-STOP SWITCH to START and hold momentarily. Observe GREEN ENGINE START-STOP SWITCH indicator is extinguished.
5	ENGINE START-STOP SWITCH INDICATOR	Observe ENGINE START-STOP SWITCH RED indicator illuminates when switch is released.
6	GENERATOR VOLTMETER A. C. METER	Monitor GENERATOR VOLTMETER, A.C. meter until voltage has stabilized at 460 (+2, -8) volts.
7	GENERATOR FREQUENCY A. C. METER	Observe GENERATOR FREQUENCY A.C. meter indicates 60 CPS.
		NOTE
		If necessary, adjust generator frequency with GOVERNOR MOTOR CONTROL SWITCH.
8	FEEDER NO. 3 NON-ESSENTIAL BUS CONTROL SWITCH	Position FEEDER NO. 3 NON-ESSENTIAL BUS CONTROL SWITCH to TRIP.
9	FEEDER NO. 3 NON-ESSENTIAL BUS CONTROL SWITCH INDICATOR	Observe FEEDER NO. 3 NON-ESSENTIAL BUS CONTROL switch RED indicator is extinguished and GREEN indicator is illuminated.
		CAUTION
		Prior to closing GENERATOR MAIN BREAKER CONTROL SWITCH, POWER CO. LINE MAIN BREAKER CONTROL SWITCH shall be tripped.

Table 3-11A.   Diesel Generator and Power CO. Line Operation at Power Remote
Control Panel (OSTF-2 only) (CONT)

STEP	EQUIPMENT OR INDICATOR	CONDITION OR POSITION
	ITEM 1. POWER CO. LINE POWER TO DIESEL GENER- ATOR POWER (CONT)	
10	POWER CO. LINE MAIN BREAKER CONTROL SWITCH	Position POWER CO. LINE MAIN BREAKER CONTROL SWITCH to TRIP.
11	POWER CO. LINE MAIN BREAKER CONTROL SWITCH INDICATOR	Observe POWER CO. LINE MAIN BREAKER CONTROL SWITCH RED indicator is extinguished and GREEN indicator is illuminated.
12	GENERATOR MAIN BREAKER CONTROL SWITCH	Position GENERATOR MAIN BREAKER CONTROL SWITCH to CLOSE.
13	GENERATOR MAIN BREAKER CONTROL SWITCH INDICATOR	Observe GENERATOR MAIN BREAKER CONTROL SWITCH GREEN indicator is extinguished and RED indicator is illuminated.
14	FEEDER NO. 3 NON-ESSENTIAL BUS CONTROL SWITCH	Position FEEDER NO. 3 NON-ESSENTIAL BUS CONTROL SWITCH to CLOSE.
15	FEEDER NO. 3 NON-ESSENTIAL BUS CONTROL SWITCH INDICATORS	Observe FEEDER NO. 3 NON-ESSENTIAL BUS CONTROL SWITCH GREEN indicator is extinguished and RED indicator is illuminated.
16	GENERATOR FREQUENCY A. C. METER	Observe GENERATOR FREQUENCY A.C. meter indicates 60 CPS.
17	GENERATOR VOLTMETER A. C. METER	Observe GENERATOR VOLTMETER A.C. meter indicates 460 (+2, -8) volts.
18	COMPLEX ELECTRICAL SYSTEM	Reset of complex electrical systems shall be accomplished after power is transferred to diesel generator. Emergency checklist procedures (section 7) for complex electrical reset shall be used.
	ITEM 2. DIESEL GENERATOR POWER TO POWER CO. LINE POWER	
	NOTE The following steps shall be performed when transferring from generator power to POWER CO. line power.	
1	POWER CO. LINE AMMETER SWITCH	Position POWER CO. LINE AMMETER SWITCH to 1, 2, or 3.
2	POWER CO. LINE VOLTMETER SWITCH	Position POWER CO. LINE VOLTMETER SWITCH to 1-2, 2-3, or 3-4.

Table 3-11A.　Diesel Generator and Power CO. Line Operation at Power Remote
Control Panel (OSTF-2 only) (CONT)

STEP	EQUIPMENT OR INDICATOR	CONDITION OR POSITION
	ITEM 2. DIESEL GENERATOR POWER TO POWER CO. LINE POWER (CONT)	
3	POWER CO. LINE VOLTMETER A. C. METER	Observe POWER CO. LINE VOLTMETER A.C. meter indicates 460 (+2, -8) volts.
4	POWER CO. LINE FREQUENCY A. C. METER	Observe POWER CO. LINE FREQUENCY A.C. meter indicates 60 CPS.
5	FEEDER NO. 3 NON-ESSENTIAL BUS CONTROL SWITCH	Position FEEDER NO. 3 NON-ESSENTIAL BUS CONTROL SWITCH to TRIP.
6	FEEDER NO. 3 NON-ESSENTIAL BUS CONTROL SWITCH INDICATORS	Observe FEEDER NO. 3 NON-ESSENTIAL BUS CONTROL SWITCH RED indicator is extinguished and GREEN indicator is illuminated.

> **CAUTION**
>
> Prior to closing POWER CO. LINE MAIN BREAKER CONTROL SWITCH, GENERATOR MAIN BREAKER CONTROL SWITCH must be tripped.

STEP	EQUIPMENT OR INDICATOR	CONDITION OR POSITION
7	GENERATOR MAIN BREAKER CONTROL SWITCH	Position GENERATOR MAIN BREAKER CONTROL SWITCH to TRIP.
8	GENERATOR MAIN BREAKER CONTROL SWITCH INDICATORS	Observe GENERATOR MAIN BREAKER CONTROL SWITCH RED indicator is extinguished and GREEN indicator is illuminated.
9	POWER CO. LINE MAIN BREAKER CONTROL SWITCH	Position POWER CO. LINE MAIN BREAKER CONTROL SWITCH to CLOSE.
10	POWER CO. LINE MAIN BREAKER CONTROL SWITCH INDICATORS	Observe POWER CO. LINE MAIN BREAKER CONTROL SWITCH GREEN indicator is extinguished and RED indicator is illuminated.
11	FEEDER NO. 3 NON-ESSENTIAL BUS CONTROL SWITCH	Position FEEDER NO. 3 NON-ESSENTIAL BUS CONTROL SWITCH to CLOSE.
12	FEEDER NO. 3 NON-ESSENTIAL BUS CONTROL SWITCH INDICATORS	Observe FEEDER NO. 3 NON-ESSENTIAL BUS CONTROL SWITCH GREEN indicator is extinguished and RED indicator is illuminated.

NOTE

Allow engine to run for approximately 5 minutes to cool, if not an emergency shutdown.

Table 3-11A. Diesel Generator and Power CO. Line Operation at Power Remote
Control Panel (OSTF-2 only) (CONT)

STEP	EQUIPMENT OR INDICATOR	CONDITION OR POSITION
	ITEM 2. DIESEL GENERATOR POWER TO POWER CO. LINE POWER (CONT)	
13	ENGINE START-STOP SWITCH	Position ENGINE START-STOP SWITCH to STOP and hold momentarily.
14	ENGINE START-STOP SWITCH INDICATORS	Observe ENGINE START-STOP SWITCH RED indicator extinguishes and GREEN indicator illuminates when switch is released.
15	COMPLEX ELECTRICAL SYSTEMS	Reset of complex electrical systems shall be accomplished after power is transferred from diesel generator to POWER CO. line. Emergency checklist procedures (section 7) for complex electrical reset shall be used.

Table 3-12. Diesel Generator Operations at 480-Volt Switchgear (NA OSTF-2)

STEP	EQUIPMENT OR INDICATOR	CONDITION OR POSITION
	NOTE This table is to be accomplished a minimum of every seven days or after power maintenance has been performed. Steps 1 through 12 of item 1 need not be accomplished if no maintenance has been performed and missile crew daily complex status verification has been accomplished.	
	ITEM 1. STARTING AND PARALLELING ALTERNATE GENERATOR	
	ALTERNATE DIESEL GENERATOR (D-60 OR D-61)	
1	DIESEL LUBE OIL	Verify lube oil level is between normal and full on dip stick.
2	FUEL OIL RETURN LINE VALVES	Verify fuel oil return line valves are open.
3	GENERATOR FIELD BREAKER	Verify generator field breaker is in closed position.
4	GENERATOR PEDESTAL BEARING OIL CUP	Verify generator pedestal bearing oil cup is full.

Table 3-12.   Diesel Generator Operations at 480-Volt Switchgear (NA OSTF-2) (CONT)

STEP	EQUIPMENT OR INDICATOR	CONDITION OR POSITION
	ITEM 1. STARTING AND PAR-ALLELING ALTERNATE GEN-ERATOR (CONT)	
5	CONDENSER WATER INLET VALVE	Verify condenser water inlet valve is open.
6	AIR INTAKE FILTER OIL LEVEL	Verify air intake filter oil level is up to mark. (Do not over-fill.)
7	JACKET WATER EXPAN-SION TANK LIQUID LEVEL GAUGE	Verify liquid level gauge is ½-full, maximum.
8	CONDENSER WATER OUT-LET VALVE	Verify condenser water outlet valve is open.
9	FUEL OIL INLET LINE VALVE	Verify fuel oil inlet line valve is open.
10	STARTING AIR VALVE	Verify starting air valve is open.
11	GOVERNOR OIL	Verify governor oil level is above low mark and not above full mark.
12	GOVERNOR SPEED DROOP	Verify governor speed droop is set at 3.
		NOTE Steps 13 through 17 need not be accomplished if diesel jacket water temperature is above 120°F.
13	GOVENOR PERCENT LOAD CONTROL	Verify governor percent load control is set at minimum fuel.
14	INDICATOR (SNIFTER) VALVES (8)	Verify indicator (snifter) valves (8) are in open position.
15	MASTER CONTROL SWITCH	Position MASTER CONTROL switch to OFF.
16	AIR-START PLUNGER	Depress air-start plunger on utility air regulator momentarily and allow engine to make at least two revolutions.
17	INDICATOR (SNIFTER) VALVES (8)	Verify indicator (snifter) valves (8) are in closed position.
18	GOVERNOR PERCENT LOAD CONTROL	Verify governor percent load control is set at maximum fuel.
19	MASTER CONTROL SWITCH	Position MASTER CONTROL switch to ON.

Table 3-12. Diesel Generator Operations at 480-Volt Switchgear (NA OSTF-2) (CONT)

STEP	EQUIPMENT OR INDICATOR	CONDITION OR POSITION
	ITEM 1. STARTING AND PAR-ALLELING ALTERNATE GEN-ERATOR (CONT)	
		NOTE Diesel engine shall be run for a period of at least 30 minutes after starting to allow accessories to reach normal operating temperature before shutdown.
20	START PUSHBUTTON	Depress START pushbutton momentarily. Engine alarm will sound until lube oil pressure is 13 to 17 PSI.
21	LUBE OIL PRESSURE GAUGE	Observe that lube oil pressure gauge indicates 29 to 34 PSI.
22	FUEL OIL PRESSURE GAUGE	Observe gauge indicates 15 to 26 PSI day-tank operation, or 30 to 65 PSI constant topping operation from soft storage tank.
23	TACHOMETER	Observe TACHOMETER indicates 720 RPM. Adjust speed with governor speed control knob if necessary.
	ALTERNATE GENERATOR PANEL (480-VOLT SWITCH GEAR)	**NOTE** Allow engine to run until jacket water temperature outlet gauge indicates not less than 120°F.
1	GENERATOR AMMETER SWITCH	Place alternate generator ammeter switch in the same position (1, 2, or 3) as operating generator ammeter switch.
2	SYNCHRONIZING SWITCH	Insert SYNCHRONIZING SWITCH handle in SYNCHRONIZING SWITCH of alternate generator and position to ON.
3	VOLTAGE REGULATOR TRANSFER SWITCH	Position VOLTAGE REGULATOR TRANSFER switch (MANUAL-AUTOMATIC) to MANUAL.
4	INCOMING FREQUENCY METER	Observe INCOMING FREQUENCY meter indicates 60 CPS. **NOTE** Operate GOVERNOR MOTOR CONTROL switch to correct frequency if necessary.
5	INCOMING VOLTMETER	Observe INCOMING VOLTMETER indicates 465 volts. If voltage is incorrect adjust EXCITER FIELD RHEOSTAT until voltage is correct.

Table 3-12.   Diesel Generator Operations at 480-Volt Switchgear (NA OSTF-2) (CONT)

STEP	EQUIPMENT OR INDICATOR	CONDITION OR POSITION
	ITEM 1. STARTING AND PAR-ALLELING ALTERNATE GEN-ERATOR (CONT)	
6	VOLTAGE REGULATOR TRANSFER SWITCH	Position VOLTAGE REGULATOR TRANSFER switch (MANUAL-AUTOMATIC) to AUTOMATIC.
7	INCOMING VOLTMETER	Observe INCOMING VOLTMETER indicates (460 +2, -8) (same voltage as RUNNING VOLTMETER). If voltage is incorrect, adjust VOLTAGE ADJUSTMENT RHEOSTAT until voltage is equal to running voltage.
8	SYNCHROSCOPE	Operate started GENERATOR GOVERNOR M O T O R CONTROL switch so that SYNCHROSCOPE hand rotates slowly clockwise. **NOTE** Observe RUNNING GENERATOR AM-METER to determine amount of load incoming generator should pick up when placed on the line. Incoming generator should pick up one-half of the load.
9	GENERATOR MAIN BREAKER CONTROL SWITCH	When SYNCHROSCOPE hand indicates 11:55 (clock position), position started GENERATOR MAIN BREAKER CONTROL SWITCH to CLOSE.
10	BOTH GENERATOR MAIN BREAKER CONTROL SWITCH INDICATORS	Observe both GENERATOR MAIN BREAKER C O N-TROL SWITCH RED indicators are illuminated and that both GREEN indicators are extinguished.
11	BOTH GENERATOR AMMETERS ·	If both AMMETERS do not indicate the same current, operate one GENERATOR GOVERNOR MOTOR CONTROL switch until both AMMETERS indicate the same current, retaining 60 cycles. **NOTE** The SYNCRONIZING S W I T C H shall remain in OFF position to enable remote paralleling of generators.
12	SYNCHRONIZING SWITCH	Position started generator SYNCHRONIZING SWITCH to OFF.
13	RUNNING FREQUENCY METER	Observe RUNNING FREQUENCY meter indicates 60 CPS. If necessary operate one or both GOVERNOR MOTOR CONTROL switches to correct frequency.

Table 3-12.   Diesel Generator Operations at 480-Volt Switchgear (NA OSTF-2) (CONT)

STEP	EQUIPMENT OR INDICATOR	CONDITION OR POSITION
	ITEM 1. STARTING AND PARALLELING ALTERNATE GENERATOR (CONT)	**NOTE** With high voltage, lower voltage on generator with a lagging power factor. With low voltage, raise voltage on generator with a leading power factor.
14	RUNNING VOLTMETER	Observe RUNNING VOLTMETER indicates 460 volts. If voltage is incorrect adjust VOLTAGE ADJUSTMENT RHEOSTAT until voltage is correct.
15	BOTH POWER FACTOR METERS	Both POWER FACTOR meters should indicate the same. If readings are not the same adjust voltage on generators until both POWER FACTOR meters have equal readings while retaining 460 volts.
	ITEM 2. PARALLELED GENERATOR SHUTDOWN	
1	GENERATOR MAIN BREAKER CONTROL SWITCH	Trip GENERATOR (D-60 or D-61) MAIN BREAKER CONTROL SWITCH.
2	GENERATOR MAIN BREAKER CONTROL SWITCH INDICATOR	Observe GENERATOR MAIN BREAKER CONTROL SWITCH RED indicator is extinguished and GREEN indicator is illuminated. **NOTE** To cool engine, allow engine to run for approximately 5 minutes if not an emergency shutdown.
	ENGINE PANEL (D-60 OR D-61)	
1	MASTER CONTROL SWITCH	Position MASTER CONTROL switch to OFF. **NOTE** MASTER CONTROL switch must be in ON position to start diesel generator either locally or remotely.
2	MASTER CONTROL SWITCH	Position MASTER CONTROL switch to ON.

Table 3-12A. Diesel Generator and Power CO. Line Operations at 480-Volt
Switchgear (OSTF-2 only)

STEP	EQUIPMENT OR INDICATOR	CONDITION OR POSITION
	ITEM 1. POWER CO. LINE POWER TO DIESEL GENERATOR POWER	
	NOTE The following steps shall be performed when starting diesel generator and transferring from POWER CO. line power to diesel generator power.	
	NOTE Step 1 through 12 need not be accomplished if no maintenance has been performed and missile crew daily complex status verification has been accomplished.	
	DIESEL GENERATOR, (LEVEL 5)	
1	DIESEL LUBE OIL	Verify lube oil level is between normal and full on dip stick.
2	FUEL OIL RETURN LINE VALVES	Verify fuel oil return line valves are open.
3	GENERATOR FIELD BREAKER	Verify GENERATOR FIELD BREAKER is in CLOSED position.
4	GENERATOR PEDESTAL BEARING OIL CUP	Verify generator pedestal bearing oil cup is full.
5	CONDENSER WATER INLET VALVE	Verify condenser water inlet valve is open.
6	AIR INTAKE FILTER OIL	Verify air intake filter oil level is up to mark. (Do not overfill.)
7	JACKET WATER EXPANSION TANK LIQUID LEVEL GAUGE	Verify liquid level gauge is ½-full (MAX).
8	CONDENSER WATER OUTLET VALVE	Verify condenser water outlet valve is open.
9	FUEL OIL INLET LINE VALVE	Verify fuel oil inlet line valve is open.

Table 3-12A. Diesel Generator and Power CO. Line Operations at 480-Volt
Switchgear (OSTF-2 only) (CONT)

STEP	EQUIPMENT OR INDICATOR	CONDITION OR POSITION
	ITEM 1. POWER CO. LINE POWER TO DIESEL GENER-ATOR POWER (CONT)	
10	STARTING AIR VALVE	Verify starting air valve is open.
11	GOVERNOR OIL LEVEL	Verify governor oil level is above low mark and not above full mark.
12	GOVERNOR SPEED DROOP	Verify governor speed droop is set at 3.
		NOTE Steps 13 through 17 need not be accomplished if diesel jacket water temperature is above 120°F.
13	GOVERNOR PERCENT LOAD CONTROL	Verify governor percent load control is set at minimum fuel.
14	INDICATOR (SNIFTER) VALVES (8)	Verify indicator (snifter) valves (8) are in OPEN position.
15	MASTER CONTROL SWITCH	Position MASTER CONTROL switch to OFF.
16	AIR-START PLUNGER	Depress air-start plunger on utility air regulator momentarily and allow engine to make at least two revolutions.
17	INDICATOR (SNIFTER) VALVES (8)	Verify indicator (snifter) valves (8) are in the CLOSED position.
18	GOVERNOR PERCENT LOAD	Verify governor percent load is set at maximum fuel.
19	FUEL DRIP RETURN PUMP SWITCH	Verify fuel drip return pump switch is ON.
20	MASTER CONTROL SWITCH	Position MASTER CONTROL SWITCH to ON.
		NOTE Diesel engine shall be run for a period of at least 30 minutes after starting to allow accessories to reach normal operating temperature before shutdown.
21	START PUSHBUTTON	Depress START pushbutton momentarily. Engine alarm will sound until lube oil pressure is 13 to 17 PSI.
22	LUBE OIL PRESSURE GAUGE	Observe lube oil pressure gauge indicates 29 to 34 PSI.

Table 3-12A.   Diesel Generator and Power CO. Line Operations at 480-Volt
Switchgear (OSTF-2 only) (CONT)

STEP	EQUIPMENT OR INDICATOR	CONDITION OR POSITION
	ITEM 1. POWER CO. LINE POWER TO DIESEL GENER-ATOR POWER (CONT)	
23	FUEL OIL PRESSURE GAUGE	Observe fuel oil pressure gauge indicates 15 to 26 PSI day tank operation, or 30 to 65 PSI constant topping operation from soft storage tank.
24	TACHOMETER	Observe tachometer indicates 720 RPM. Adjust speed with GOVERNOR SPEED CONTROL knob if necessary.
		NOTE
		Allow engine to run until jacket water temperature outlet gauge indicates not less than 120°F.
	GENERATOR PANEL 480-VOLT SWITCHGEAR	
1	GENERATOR AMMETER SWITCH	Position GENERATOR AMMETER switch to 1, 2, or 3.
2	GENERATOR VOLTMETER SWITCH	Position GENERATOR VOLTMETER switch to 1-2, 2-3, or 3-4.
3	VOLTAGE REGULATOR TRANSFER SWITCH	Position VOLTAGE REGULATOR TRANSFER switch (MANUAL-AUTOMATIC) to MANUAL.
4	GENERATOR FREQUENCY METER	Observe GENERATOR FREQUENCY meter indicates 60 CPS.
		NOTE
		Operate GOVERNOR MOTOR CONTROL switch to correct frequency if necessary.
5	GENERATOR VOLTMETER	Observe GENERATOR VOLTMETER indicates 465 volts. If voltage is incorrect adjust EXCITER FIELD RHEOSTAT until voltage is correct.
6	VOLTAGE REGULATOR TRANSFER SWITCH	Position VOLTAGE REGULATOR TRANSFER switch (MANUAL-AUTOMATIC) to AUTOMATIC.
7	GENERATOR VOLTMETER	Observe GENERATOR VOLTMETER indicates 460 volts. If voltage is incorrect, adjust VOLTAGE ADJUSTING RHEOSTAT until voltage is correct.
8	FEEDER NO. 3 NON-ESSEN-TIAL BUS BREAKER TRIP PUSHBUTTON	Depress FEEDER NO. 3 NON-ESSENTIAL BUS BREAKER TRIP pushbutton momentarily.

Table 3-12A. Diesel Generator and Power CO. Line Operations at 480-Volt
Switchgear (OSTF-2 only) (CONT)

STEP	EQUIPMENT OR INDICATOR	CONDITION OR POSITION
	ITEM 1. POWER CO. LINE POWER TO DIESEL GENER-ATOR POWER (CONT)	
9	FEEDER NO. 3 NON-ESSEN-TIAL BUS BREAKER INDICATORS	Observe FEEDER NO. 3 NON-ESSENTIAL BUS BREAKER RED indicator is extinguished and GREEN indicator is illuminated.
	POWER CO. LINE PANEL	
1	POWER CO. LINE MAIN BREAKER CONTROL SWITCH	Trip POWER CO. LINE MAIN BREAKER CONTROL SWITCH.
2	POWER CO. LINE MAIN BREAKER CONTROL SWITCH INDICATORS	Observe POWER CO. LINE MAIN BREAKER CONTROL SWITCH RED indicator is extinguished and GREEN indicator is illuminated.
		CAUTION Prior to closing GENERATOR MAIN BREAKER CONTROL SWITCH, POWER CO. LINE MAIN BREAKER CONTROL SWITCH shall be tripped.
3	GENERATOR MAIN BREAKER CONTROL SWITCH	Position GENERATOR MAIN BREAKER CONTROL SWITCH to CLOSE.
4	GENERATOR MAIN BREAKER CONTROL SWITCH INDICATORS	Observe GENERATOR MAIN BREAKER CONTROL SWITCH GREEN indicator is extinguished and RED indicator is illuminated.
		NOTE Prior to performing next step ensure a person is stationed at vestibule of silo entrance to restart silo lighting after nonessential bus is energized.
5	FEEDER NO. 3 NON-ESSEN-TIAL BUS BREAKER CLOSE PUSHBUTTON	Depress FEEDER NO. 3 NON-ESSENTIAL BUS BREAKER CLOSE pushbutton momentarily.
6	FEEDER NO. 3 NON-ESSEN-TIAL BUS BREAKER INDICATOR	Observe FEEDER NO. 3 NON-ESSENTIAL BUS BREAKER GREEN indicator is extinguished and RED indicator is illuminated.
7	GENERATOR FREQUENCY METER	Observe GENERATOR FREQUENCY METER indicates 60 CPS.

Table 3-12A.   Diesel Generator and Power CO. Line Operations at 480-Volt
Switchgear (OSTF-2 only) (CONT)

STEP	EQUIPMENT OR INDICATOR	CONDITION OR POSITION
	ITEM 1. POWER CO. LINE POWER TO DIESEL GENER-ATOR POWER (CONT)	
8	GENERATOR VOLTMETER	Observe GENERATOR VOLTMETER indicates 460 volts. If voltage is incorrect, adjust VOLTAGE ADJUSTING RHEOSTAT until voltage is correct.
9	COMPLEX ELECTRICAL SYSTEMS	Reset of complex electrical systems shall be accomplished after power is transferred from POWER CO. line to diesel generator. Emergency checklist procedures (section 1) for complex electrical reset shall be used.
	ITEM 2. DIESEL GENERATOR POWER TO POWER CO. LINE POWER **NOTE** The following steps shall be performed when tranferring from diesel generator power to POWER CO. line power. POWER CO. LINE PANEL 480-VOLT SWITCHGEAR	
1	POWER CO. LINE AMMETER SWITCH	Position POWER CO. LINE AMMETER switch to 1, 2, or 3.
2	POWER CO. LINE VOLT-METER SWITCH	Position POWER CO. LINE VOLTMETER switch to 1-2, 2-3, or 3-4.
3	POWER CO. LINE VOLT-METER	Observe POWER CO. LINE VOLTMETER indicates 460 (+2 -8) volts.
4	FEEDER NO. 3 NON-ESSEN-TIAL BUS BREAKER TRIP PUSHBUTTON	Depress FEEDER NO. 3 NON-ESSENTIAL BUS BREAKER TRIP pushbutton momentarily.
5	FEEDER NO. 3 NON-ESSEN-TIAL BUS BREAKER INDICATORS	Observe FEEDER NO. 3 NON-ESSENTIAL BUS BREAKER RED indicator is extinguished and GREEN indicator is illuminated.
6	GENERATOR MAIN BREAKER CONTROL SWITCH	Trip GENERATOR MAIN BREAKER CONTROL SWITCH.
7	GENERATOR MAIN BREAKER CONTROL SWITCH INDICATORS	Observe GENERATOR MAIN BREAKER CONTROL SWITCH RED indicator is extinguished and GREEN indicator is illuminated.

Table 3-12A.   Diesel Generator and Power CO. Line Operations at 480-Volt
Switchgear (OSTF-2 only) (CONT)

STEP	EQUIPMENT OR INDICATOR	CONDITION OR POSITION
	ITEM 2. DIESEL GENERATOR POWER TO POWER CO. LINE POWER (CONT)	
		CAUTION Prior to closing POWER CO. LINE MAIN BREAKER CONTROL SWITCH, GENERATOR MAIN BREAKER CONTROL SWITCH shall be tripped.
8	POWER CO. LINE MAIN BREAKER CONTROL SWITCH	Position POWER CO. LINE MAIN BREAKER CONTROL SWITCH to CLOSE.
9	POWER CO. LINE MAIN BREAKER CONTROL SWITCH INDICATORS	Observe POWER CO. LINE MAIN BREAKER CONTROL SWITCH GREEN indicator is extinguished and RED indicator is illuminated.
		NOTE Prior to performing next step ensure a person is stationed at vestibule of silo entrance to restart silo lighting after non-essential bus is energized.
10	FEEDER NO. 3 NON-ESSENTIAL BUS BREAKER CLOSE PUSHBUTTON	Depress FEEDER NO. 3 NON-ESSENTIAL BUS BREAKER CLOSE pushbutton momentarily.
11	FEEDER NO. 3 NON-ESSENTIAL BUS BREAKER INDICATORS	Observe FEEDER NO. 3 NON-ESSENTIAL BUS BREAKER GREEN indicator is extinguished and RED indicator is illuminated.
		NOTE To cool engine, allow engine to run for approximately 5 minutes if not on emergency shutdown.
	GENERATOR PANEL	
1	MASTER CONTROL SWITCH	Position MASTER CONTROL switch to OFF.
		NOTE The MASTER CONTROL SWITCH must be in ON position to start diesel generator either locally or remotely.
2	MASTER CONTROL SWITCH	Position MASTER CONTROL SWITCH to ON.
3	COMPLEX ELECTRICAL SYSTEMS	Reset of complex electrical systems shall be accomplished after power is transferred from diesel generator power to POWER CO. line power. Emergency checklist procedures (Section 7) for complex electrical reset shall be used.

Table 3-13. TV System Operation (NA OSTF-2)

STEP	CONTROL OR INDICATOR	CONDITION OR POSITION
	ITEM 1. PERMANENT-TV MONITOR OPERATION TURN-ON PROCEDURE **CAUTION** Ensure all toggle switches are in OFF or LOW position except sync generator power switch, which must be in ON position.	
1	CABINET POWER KEY SWITCH	Position key switch to ON.
2	CABINET MAIN POWER IN-DICATOR	Observe indicator is illuminated.
3	SYNC GENERATOR POWER INDICATOR	Observe indicator is illuminated.
4	LIGHTING CONTROL MAIN POWER SWITCH	Position switch to ON.
5	LIGHTING CONTROL MAIN POWER INDICATOR	Observe indicator is illuminated.
6	LIGHTING SYSTEM 1 POWER SWITCH	Position power switch to ON.
7	LIGHTING SYSTEM 1 POWER INDICATOR	Observe indicator is illuminated.
8	MONITOR BRIGHTNESS CONTROL	Rotate control fully counter-clockwise.
9	MONITOR CONTRAST CONTROL	Rotate control fully counter-clockwise.
10	MONITOR POWER SWITCH	ON.
	NOTE Approximately 1 minute delay is required for monitor voltage to reach operating level.	
11	MONITOR CONTRAST CONTROL	Rotate control 75% clockwise.

Table 3-13. TV System Operation (NA OSTF-2) (CONT)

STEP	CONTROL OR INDICATOR	CONDITION OR POSITION
	ITEM 1. PERMANENT-TV MONITOR OPERATION (CONT)	
	TURN-ON PROCEDURE (CONT)	
12	MONITOR BRIGHTNESS CONTROL	Rotate control clockwise until trace of brightness appears on screen.
13	CAMERA BEAM CONTROL	Rotate control fully counter-clockwise.
14	CAMERA TARGET CONTROL	Position selector switch to AUTO.
15	CAMERA POWER SWITCH	Position power switch to ON.
16	CAMERA POWER INDICATOR	Observe indicator is illuminated.
	CAUTION	
	Ensure 2 minutes have elapsed since performing step 6 before positioning I N T E N S I T Y switch to HIGH.	
17	LIGHTING SYSTEM 1 INTENSITY SWITCH	Position switch to HIGH.
18	LIGHTING SYSTEM 1 INTENSITY INDICATOR	Observe indicator is illuminated.
19	CAMERA BEAM CONTROL	Rotate control slowly clockwise until scene highlights discharge.
20	CAMERA IRIS	Set iris to minimum open position that provides the best but not the brightest picture lighting.
21	CAMERA FOCUS	Adjust focus for NEAR or FAR.
22	MONITOR CONTRAST CONTROL	Adjust control for best picture.
23	MONITOR BRIGHTNESS CONTROL	Adjust control for best picture.

Table 3-13.   TV System Operation (NA OSTF-2) (CONT)

STEP	CONTROL OR INDICATOR	CONDITION OR POSITION
	ITEM 1. PERMANENT-TV MONITOR OPERATION (CONT)	
	TURN-ON PROCEDURE (CONT)	
24	CAMERA PAN, TILT, AND ZOOM CONTROLS	Adjust controls for desired scene.
	NOTE The TV system may be turned off or placed in a standby status after the best picture is obtained. Standby status is accomplished by turning floodlights off.	
	STANDBY PROCEDURE	
1	LIGHTING SYSTEM 1 INTENSITY SWITCH	Position switch to LOW.
2	LIGHTING SYSTEM MAIN POWER SWITCH	Position switch to OFF.
	FLOODLIGHT TURN-ON PROCEDURE FROM STAND-BY FOR OPERATIONAL USE	
1	LIGHTING SYSTEM MAIN POWER SWITCH	Position switch to ON.
2	LIGHTING SYSTEM MAIN POWER INDICATOR	Observe indicator is illuminated.
	CAUTION Allow 2 minute warmup before positioning INTENSITY switch to HIGH.	
3	LIGHTING SYSTEM 1 INTENSITY SWITCH	Position switch to HIGH.
4	LIGHTING SYSTEM 1 INTENSITY INDICATOR	Observe indicator is illuminated.

Table 3-13.   TV System Operation (NA OSTF-2) (CONT)

STEP	CONTROL OR INDICATOR	CONDITION OR POSITION
	ITEM 1. PERMANENT-TV MONITOR OPERATION (CONT)	
	FLOODLIGHT TURN-ON PROCEDURE (CONT)	
	NOTE Minor adjustment of monitor contrast and brightness; camera focus, iris, pan, tilt; and zoom controls may be made to obtain best picture of desired scene.	
	TURN-OFF PROCEDURE	
1	LIGHTING SYSTEM 1 INTENSITY SWITCH	Position switch to LOW.
2	CAMERA TILT	Tilt camera to point down approximately 45 degrees.
3	CAMERA IRIS	Close iris completely.
4	LIGHTING SYSTEM 1 POWER SWITCH	Position switch to OFF.
5	LIGHTING SYSTEM MAIN POWER SWITCH	Position switch to OFF.
6	CAMERA BEAM CONTROL	Rotate control fully counter-clockwise.
7	CAMERA POWER SWITCH	Position switch to OFF.
8	MONITOR CONTRAST CONTROL	Rotate control fully counter-clockwise.
9	MONITOR BRIGHTNESS CONTROL	Rotate control fully counter-clockwise.
10	MONITOR POWER SWITCH	Position switch to OFF.
11	CABINET POWER KEY	Position key to OFF.
12	ALL INDICATORS	Extinguished.

Table 3-13. TV System Operation (NA OSTF-2) (CONT)

STEP	CONTROL OR INDICATOR	CONDITION OR POSITION
	ITEM 2. FOUR-TV MONITOR OPERATION	
	TURN-ON PROCEDURE	
	CAUTION	
	Ensure all toggle switches are in OFF or LOW position, except Snyc Generator 1 and/or 2 power switches which must be in ON position.	
	CABINET U202	
1	CABINET POWER KEY	Position key switch to ON.
2	CABINET POWER IND	Observe indicator is ILLUMINATED.
3	SNYC GENERATOR POWER INDICATOR	Observe indicator is ILLUMINATED.
	CABINET U204	
1	CABINET POWER KEY	Position key switch to ON.
2	CABINET POWER IND	Observe indicator is ILLUMINATED.
3	SYNC GENERATOR POWER INDICATOR	Observe indicator is ILLUMINATED.
	CABINET U202	
1	SYNC CHANGEOVER SW	Turn to operating SYNC Generator.
	CABINET U201	
1	CABINET POWER KEY	Position key switch to ON.
2	CABINET POWER IND	Observe indicator is ILLUMINATED.
	CABINET U203	
1	CABINET POWER KEY	Position key switch to ON.
2	CABINET POWER IND	Observe indicator is ILLUMINATED.

Table 3-13. TV System Operation (NA OSTF-2) (CONT)

STEP	CONTROL OR INDICATOR	CONDITION OR POSITION
	ITEM 2. FOUR-TV MONITOR OPERATION (CONT)	
	TURN-ON PROCEDURE (CONT)	
	CABINET U204	
1	LIGHTING SYSTEM MAIN POWER SWITCH	Position switch to ON.
2	LIGHTING SYSTEM MAIN POWER INDICATOR	Observe indicator is ILLUMINATED.
3	LIGHTING SYSTEM 1 POWER SWITCH	Position switch to ON.
	NOTE Allow 2 minutes warmup before positioning INTENSITY switch to HIGH.	
	ON EACH CABINET TO BE OPERATED	
1	MONITOR BRIGHTNESS CONTROL	Rotate control fully counter-clockwise.
2	MONITOR CONTRAST CONTROL	Rotate control fully counter-clockwise.
3	MONITOR POWER SWITCH	Position switch to ON.
	NOTE Delay approximately 1 minute to allow monitor voltage to reach operating level.	
4	MONITOR CONTRAST CONTROL	Rotate control 75% clockwise.
5	MONITOR BRIGHTNESS CONTROL	Rotate control clockwise until a trace of brightness appears on screen.
6	CAMERA BEAM CONTROL	Rotate control fully counter-clockwise.
7	CAMERA TARGET CONTROL	Position switch to AUTO.

Table 3-13. TV System Operation (NA OSTF-2) (CONT)

STEP	CONTROL OR INDICATOR	CONDITION OR POSITION
	ITEM 2. FOUR-TV MONITOR OPERATION (CONT)	
	TURN-ON PROCEDURE (CONT)	
8	CAMERA POWER SWITCH	Position switch to ON.
9	CAMERA POWER IND	Observe indicator is ILLUMINATED.
	CABINET U204	
1	LIGHTING SYSTEM 1 INTENSITY SWITCH	Position switch to HIGH.
2	LIGHTING SYSTEM 1 INTENSITY INDICATOR	Observe indicator is ILLUMINATED.
	ON EACH CABINET TO BE OPERATED	
1	CAMERA BEAM CONTROL	Rotate control slowly clockwise until scene highlights discharge.
2	CAMERA IRIS	Set iris to minimum open position that provides the best but not the brightest picture lighting.
3	CAMERA FOCUS	Adjust focus for best picture.
4	MONITOR CONTRAST CONTROL	Adjust control for best picture.
5	MONITOR BRIGHTNESS CONTROL	Adjust control for best picture.
6	CAMERA PAN, TILT, AND ZOOM CONTROLS	Adjust controls for desired scene.
	TURN-OFF PROCEDURES	
	CABINET U204	
1	LIGHTING SYSTEM 1 INTENSITY SWITCH	Position switch to LOW.
	ON EACH CABINET	
1	CAMERA TILT	Tilt camera to point down approximately 45 degrees.
2	CAMERA IRIS	Close iris completely.

Table 3-13. TV System Operation (NA OSTF-2) (CONT)

STEP	CONTROL OR INDICATOR	CONDITION OR POSITION
	ITEM 2. FOUR-TV MONITOR OPERATION (CONT)	
	TURN-OFF PROCEDURE (CONT)	
	CABINET U204	
1	LIGHTING SYSTEM 1 POWER SWITCH	Position switch to OFF.
2	LIGHTING SYSTEM MAIN POWER SWITCH	Position switch to OFF.
	ON EACH CABINET	
1	CAMERA BEAM CONTROL	Rotate control fully counter-clockwise.
2	CAMERA POWER SWITCH	Position switch to OFF.
3	MONITOR POWER SWITCH	Position switch to OFF.
	CAUTION Ensure all camera power switches are off before positioning cabinet key switches to OFF.	
4	CABINET POWER KEY SWITCHES	Position key switches to OFF.

Table 3-14. Re-Entry Vehicle Prelaunch Monitor Target Setting Procedures

STEP	CONTROL OR INDICATOR	CONDITION OR POSITION
	RE-ENTRY VEHICLE PRE-LAUNCH MONITOR PANEL. SILO LEVEL 3	
1	MARK 4 R/V INDICATOR	Observe indicator is illuminated WHITE.
2	R/V TACTICAL INDICATOR	Observe indicator illuminated GREEN.
3	R/V NORMAL INDICATOR	Observe indicator is illuminated GREEN.

Table 3-14. Re-Entry Vehicle Prelaunch Monitor Target Setting Procedures (CONT)

STEP	CONTROL OR INDICATOR	CONDITION OR POSITION
	RE-ENTRY VEHICLE PRE-LAUNCH MONITOR PANEL SILO LEVEL 3 (CONT)	
4	PANEL POWER TOGGLE SWITCH	Position switch to ON.
5	28 VDC POWER ON IND	Observe indicator is illuminated GREEN.
6	115 VAC POWER ON IND	Observe indicator is illuminated GREEN.
7	A & F SAFETY GOOD IND	Observe indicator is illuminated GREEN.
8	WARHEAD SAFETY GOOD INDICATOR	Observe indicator is illuminated GREEN.
9	TARGET POSITION IND	Observe indicator is illuminated.
10	TEST PWR/GOOD SWITCH	Verify switch is positioned to OFF.
11	TARGET A AND B ROTARY SWITCH	Position switches to desired target settings.
12	TARGET SELECT ROTARY SWITCH	Position rotary switch to TARGET B.
13	RESET PUSHBUTTON	Depressed momentarily.
14	R/V TACTICAL INDICATOR	Observe indicator is extinguished.
15	R/V IN MAINTENANCE INDICATOR	Observe indicator is illuminated WHITE.
16	TARGET POSITION IND	Verify indicator resets to HOMING, HOMING, HOMING.

NOTE

If target position indicator fails to count to TARGET B settings, depress RESET pushbutton.

17	TARGET POSITION IND	Verify indicator steps to TARGET B setting.
18	TARGET SELECT ROTARY SWITCH	Position rotary switch to TARGET A.
19	RESET PUSHBUTTON	Depressed momentarily.
20	TARGET POSITION IND	Verify indicator reset to HOMING, HOMING, HOMING.

Table 3-14.   Re-Entry Vehicle Prelaunch Monitor Target Setting Procedures (CONT)

STEP	CONTROL OR INDICATOR	CONDITION OR POSITION
		NOTE If TARGET POSITION indicator fails to count to TARGET A setting, depress RE-SET pushbutton.
21	TARGET POSITION IND	Verify indicator steps to TARGET A settings.
22	TARGET SELECT ROTARY SWITCH	Position switch to REMOTE.
23	R/V IN MAINTENANCE IND	Observe indicator is extinguished.
24	R/V TACTICAL INDICATOR	Observe indicator is GREEN.
		NOTE Within 15 seconds, TARGET POSITION indicator should step to settings for target selected on launch control console.
25	TARGET POSITION IND	Verify that indicator steps to proper target selected settings (A or B).
26	PANEL POWER SWITCH	Position switch to OFF.
27	MARK 4 R/V INDICATOR	Observe indicator illuminates WHITE.
28	R/V TACTICAL INDICATOR	Observe indicator illuminates GREEN.
29	R/V NORMAL INDICATOR	Observe indicator illuminates GREEN.
30	ALL OTHER INDICATORS	Observe all other indicators are extinguished.

Table 3-15. PLX Preparation and Coordination Procedures

STEP	REQUIREMENTS	ACTION
	NOTE The MCCC or sector commander shall ensure that activity briefings are conducted as required. **NOTE** This procedure is to be used in conjunction with the applicable flow in the preparation of the missile complex for a PLX, maintenance countdown, or training launch. Only those steps identified by a double asterisk need be accomplished or verified prior to the second of two successive PLX'S.	
1	SUPPORT ORGANIZATIONS a. WING COMMAND POST b. SAFETY OFFICER c. FIRE DEPARTMENT d. HOSPITAL e. LES OFFICER f. UNIT MAINTENANCE g. MUNITIONS MAINTE-NANCE SQUADRON	Notified. Upon notification to prepare complex for a PLX, the MCCC shall verify with job control that all supporting organizations have been notified of scheduled PLX.
2	SUPPORT EQUIPMENT a. (Deleted) b. MISSILE ENCLOSURE PURGE UNIT (MEPU) c. TELEVISION MONITORS d. ANEMOMETER e. BREATHING APPARATUS f. TAPE RECORDER g. FALL BACK AREA TELEPHONE	Dispatched or on hand. Verify with job control or on site that required support equipment is dispatched or on hand. A tape recorder to record countdown action is optional.

Table 3-15. PLX Preparation and Coordination Procedures (CONT)

STEP	REQUIREMENTS	ACTION
		NOTE In conjunction with performing tasks to prepare complex for countdown, complete or verify steps 3 thru 12. Steps need not be completed in sequence.
3	TV MONITORS	Installed and operating.
4	ANEMOMETER	On hand and operating.
**5	GO$_2$ VENT BLAST CLOSURE MANUAL VALVE	OPEN-BYPASS position.
**6	GO$_2$ VENT FAN OPERATION	Checked by depressing local GO$_2$ VENT FAN TEST pushbutton.
7	LO$_2$ SAMPLE	Obtain a liquid oxygen sample as required by T.O. 42B-6-1-1.
**8	FALL-BACK AREA TELEPHONE	Installed.
**9	TAPE RECORDER	Checked for operation.
10	PORTABLE WATER PUMPS	Checked for operation if available.
11	FILL AND VENT SHAFT	Inspected and cleaned. Thoroughly inspect fill and vent shaft for foreign matter. Remove all matter, especially that of hydrocarbon base. During both inspections and cleaning procedures, oxygen analyzers, self-contained breathing apparatus, and the buddy system shall be used. Ensure lifelines are immediately available during operation.
		NOTE Upon completion of tasks to prepare the complex for countdown, complete or verify the following steps. Steps need not be completed in sequence.
12	MISSILE CREW DAILY COMPLEX STATUS VERIFICATION CHECKLIST	Completed. As areas are cleared during the preparation flow, portions of the verification checklist may be completed.
**13	FLUID AND GASES	Verify that fluid and gas levels are sufficient to comply with the minimum requirements as shown in table 6-1.
**14	GUIDANCE ALIGNED	Accomplished. Place guidance in the aligned status. Before the second successive PLX, guidance will be removed from the aligned status and then realigned. Refer to table 3-9.

Table 3-15. PLX Preparation and Coordination Procedures (CONT)

STEP	REQUIREMENTS	ACTION
**15	MLS CORRELATION	Accomplished. Perform MLS correlation in accordance with checklist provided.
**15a	THERMAL OVERLOAD RELAYS CHECKED	Accomplished. Proceed to Missile lift system drive cabinet No. 2 and ensure relays OL-1, OL-2 and BMOL are not tripped. If relays are tripped refer to applicable tech data.
**16	AFR 122-50 ACTIONS	Verified. Removal of missile ordnance, locking of fuel prevalve locks and fuel line fuel flow checks shall be verified as required.
**17	LO$_2$ TOPPING QUICK DISCONNECT PURGE	Accomplished just prior to clearing silo for each PLX. Blow clean dry gaseous nitrogen in the annular space between the male and female portions of the LO$_2$ topping quick disconnect valve to be sure valve is free of moisture or water.
**18	COMPLEX STATUS FORMS	Reviewed. Review and ensure that complex forms are correct and reflect proper configuration for countdown. Countdown shall not be initiated with a known malfunction which would prevent completion of the exercise. The air conditioning system, ventilation system, GO$_2$ purge system, GO$_2$ detection system, fire fog system and television system shall be verified as being in normal operating condition during the accomplishment of missile crew daily complex status verification.
19	SUPPORT PERSONNEL	Duties assigned and briefed. Ensure that Support Personnel are qualified and briefed on assigned duties and responsibilities during the countdown and abort including foreseeable emergencies.
**20	NONESSENTIAL PERSONNEL	Evacuated. Evacuate nonessential personnel from the area as soon as their tasks are completed.
**21	ENTRAPMENT DOOR KEY	In place. The outer entrapment door key shall be positioned in the lock prior to all training or maintenance PLX.
**22	SAFETY INSPECTION	Completed. Direct performance of safety inspection as soon as silo and area is clear.
**23	LAUNCH ENABLE SYSTEM (LES)	Enabled. Enable LES in accordance with current directives.
**24	PERSONNEL BRIEFING	Completed. Brief and ensure that all personnel authorized to be in the LCC understand use of the breathing apparatus, evacuation procedures, assigned positions, limitations, limited nonessential conversation and smoking restriction during countdown.
**25	CREW OPERATION BRIEFING	Completed in accordance with briefing checklist provided.
**26	WEATHER	Checked to ensure that wind and weather restrictions are within allowable limits for countdown. Compare site weather with nomogram (section VI) to determine action required in event of boiloff valve failure during abort.

Table 3-15.   PLX Preparation and Coordination Procedures (CONT)

STEP	REQUIREMENTS	ACTION
**27	SECTOR COMMAND	Ready. Verify and coordinate with sector command that complex and personnel are ready for countdown.
**28	SAFETY MONITOR	Ready. Verify and coordinate with safety monitor that complex is ready for countdown.
**29	MEPU	Start generator set and ensure MEPU is retracted if a PLX or maintenance countdown is to be performed. For a training launch the MEPU shall be positioned at the fallback area.
**30	TAPE RECORDER	ON.
**31	MAINTENANCE CONTROL	Contact maintenance control and verify that no outstanding discrepancies exist on complex AGE, RPIE, or missile which will prohibit the completion of the PLX or create a hazardous condition, and that all mandatory conditions for conducting a PLX have been accomplished.
**32	COMMAND POST	Notify command post that complex is ready for countdown, establish a time hack, and standby for authority to initiate countdown.

Table 3-16.   Amplified Countdown Procedures

STEP	CREW POS	REQUIREMENT
1		**NOTE** Normal indications shall be announced as indicated in this table. Abnormal or emergency indications shall be announced in all instances and recommendations made by responsible crew member. **NOTE** If guidance is on memory, target change cannot be accomplished. **NOTE** Emergency actions required for emergency or abnormal indications displayed on other than countdown, commit, and abort patches are contained in tables 4-2, 4-3, and 4-4. **NOTE** When necessary, position of the boiloff valve can be determined by panning level 3 TV camera vertically across the MEA. If B-2 pod transition fairing is above camera level, the boiloff valve is above the silo cap. **NOTE** The DMCCC shall announce countdown progression time at one-minute intervals. (One minute in countdown, etc.)

Table 3-16. Amplified Countdown Procedures (CONT)

STEP	CREW POS	REQUIREMENT
1 (CONT)		**NOTE** During countdown or abort, if a malfunction occurs which requires that the LO$_2$ Tanking Panels be placed in LOCAL, the READY FOR COUNTDOWN, LO$_2$ LINE FILLED, RAPID LO$_2$ LOAD, FINE LO$_2$ LOAD, LO$_2$ COMMIT and LO$_2$ DRAIN COMPLETE indicators on the LCC will be extinguished. The LO$_2$ AND FUEL indicator will illuminate red. The above indications will return to normal when the LO$_2$ Tanking Panels are returned to REMOTE. **NOTE** During a countdown, abort may be initiated by depressing ABORT pushbutton anytime prior to MISSILE LIFT UP & LOCKED indicator GREEN. After MISSILE LIFT UP & LOCKED indicator illuminates GREEN, it is not possible to initiate an abort sequence until one of the following has occurred: a. ABORT indicator illuminated RED, which will occur if guidance does not go inertial in one second, or if engine cutoff signal is received and the missile is not away in five seconds plus the time it took for guidance to go inertial. b. Missile is away and engine cutoff signal is received. In this case, ABORT indicator will remain extinguished. c. A partial abort is automatically initiated 15 seconds after MISSILE LIFT UP & LOCKED indicator illuminates GREEN. At this time, missile power will change to external, autopilot programmer will return to safe, and commit lockup summary will drop out. **CAUTION** Do not make radio transmissions after countdown start. **CAUTION** If a malfunction requires the LO$_2$ Tanking Panel to be placed in LOCAL, ensure that N-5 and N-4 valve switches are in the OPEN and all other valve switches are in the CLOSED position. **WARNING** Prior to start of countdown, personnel shall be alerted and the silo shall be evacuated.

T.O. 21M-HGM16F-1

Table 3-16. Amplified Countdown Procedures (CONT)

STEP	CREW POS	REQUIREMENT
1 (CONT)		When 1000-cycle tone sounds, all crew members shall report to launch control center immediately.
	MCCC	a. Take position on left side of launch control console. b. Accomplish controller fast reaction checklist. c. Ensure countdown and emergency checklists and stopwatches are available. d. Break seal on START C/D pushbutton.
	DMCCC	a. Take position on right side of launch control console. b. Accomplish controller fast reaction checklist. c. Ensure countdown and emergency checklists and stopwatches are available. d. Position personnel warning light switch on FRCP to ON.
	BMAT	a. Lamp test LO₂ TANKING (PANEL 1 and PANEL 2) and launch control console. b. Take position behind MCCC. c. Monitor pressurization until MCCC and DMCCC have completed controller fast reaction checklist.
	MFT	a. Ensure all personnel are clear of silo and that blast doors and vent valve 600 are closed. b. Ensure RESET PROGRAMMER key is inserted in CSMOL and that key is in OFF position. c. Ensure TV monitor is on, INTENSITY switch is in LOW position and LIGHTING SYSTEM MAIN POWER switch is on. d. Take position in front of FRCP. e. Ensure FRCP indicators do not indicate a condition which will prevent countdown.

Table 3-16. Amplified Countdown Procedures (CONT)

STEP	CREW POS	REQUIREMENT
	EPPT	a. *Take position in front of PRCP.* b. *Parallel diesel generators.* **NOTE** A tactical countdown shall not be delayed until generators are paralleled and may be completed on one generator. **NOTE** DIESEL GENERATOR (D-60 or D-61) OVERSPEED, LOW LUBE OIL PRESS, HI TEMP indicator on FRCP will illuminate RED for less than eight seconds when alternate generator is started.
2	MCCC	Verify all crew members and equipment are ready for countdown. Crew members shall respond in the following manner and sequence, to MCCC announcement, "CREW REPORT".
	DMCCC	"DEPUTY READY".
	BMAT	"A-1 READY".
	MFT	"M-1 READY".
	EPPT	**NOTE** EPPT shall respond in accordance with step a, b, or c, depending on condition of generators. a. "L-1 READY, GENERATOR PARALLELED". b. "L-1 READY, GENERATOR BEING PARALLELED". c. "L-1 READY, COUNTDOWN ON ONE GENERATOR".
3	MCCC	When ready for countdown: a. Announce, "COUNTDOWN START ON MY MARK-MARK" and depress START C/D pushbutton at MARK announcement.

Table 3-16. Amplified Countdown Procedures (CONT)

STEP	CREW POS	REQUIREMENT
3 (CONT)	DMCCC	a. Start stopwatch. b. Log Zulu Time of countdown start.
	BMAT	Monitor launch control console.
	MFT	Monitor FRCP.
	EPPT	Monitor PRCP.
4	MCCC	LN₂ LOAD indicator AMBER. Indicator illuminates AMBER when liquid nitrogen valve 214 has opened and LN₂ load is not complete. If indicator fails to illuminate AMBER at start of countdown, abort during a nontactical countdown. During a tactical countdown, see table 4-11. Announce, "LN₂ LOAD AMBER".
	DMCCC	Monitor tank pressures (phase I). LO₂ tank pressure 3.4 - 4.2 PSI. Differential pressure greater than 5 PSI. FUEL tank pressure 11.9 - 13.0 PSI.
	BMAT	Observe the following launch control console indicators, announce any abnormal indications, and advise MCCC if abort is required. (Time at which abort is initiated shall be at MCCC discretion. Refer to section V for malfunction procedures.): a. MISSILE POWER indicator AMBER to GREEN. Indicator will illuminate AMBER when countdown bus is energized and will illuminate GREEN after AC and DC buses in missile are energized from ground power system. b. HEATERS ON indicator AMBER to GREEN. Indicator will illuminate AMBER when countdown bus is energized and illuminate GREEN when engine valve heaters are energized. c. MISSILE BATTERY ACTIVATED indicator AMBER. Indicator illuminates amber when battery activate signal is sent and a 2-minute timer is started to allow the battery sufficient time to generate a full load carrying capability. d. GUIDANCE READY indicator AMBER. Indicator will illuminate amber when guidance countdown is initiated. Test of airborne computer and calibration of accelerometers begin.

Changed 15 April 1964

Table 3-16. Amplified Countdown Procedures (CONT)

STEP	CREW POS	REQUIREMENT
4 (CONT)	BMAT (CONT)	e. R/V BATTERY TEMPERATURE indicator GREEN. Indicator will illuminate green when start countdown signal to prelaunch monitor is received. f. AUTOPILOT ON indicator AMBER. Indicator illuminates AMBER when 400-cycle power is applied to autopilot system for gyro spin motor operation. A 4-minute timer is started to prevent initiation of autopilot test until GYROS reach proper operating speed. g. HYDRAULIC PRESSURE indicator AMBER. Indicator illuminates AMBER after power is applied to autopilot system and booster and sustainer hydraulic pressures are not between 1750 and 2250 PSI.
	MFT	Observe the following FRCP indications: a. GASEOUS OXYGEN VENT OPEN indicator GREEN and GASEOUS OXYGEN VENT CLOSED indicator extinguished. If vent fails to open during a tactical countdown, countdown may continue. If vent closes during a nontactical countdown prior to commit start, abort is required. b. GASEOUS OXYGEN VENT FAN ON indicator illuminated. If indicator fails to illuminate during a nontactical countdown, abort is required. If during a tactical countdown, countdown may continue. c. RP1 AND FIRE FOG SYSTEM DAMPERS CLOSED indicator extinguished. If damper fails to open, countdown may continue. Position INTENSITY switch on TV monitor to HIGH.
	EPPT	Monitor PRCP.
5		PNEUMATICS IN PHASE II indicator AMBER. Indicator will illuminate AMBER after a 5-second period for propellant level control unit (PLCU) warmup and fuel level check and if pneumatics not in phase II. If the PNEUMATICS IN PHASE II indicator fails to illuminate AMBER, but the fuel tank pressure increases normally, continue countdown.
	MCCC	Announce, "PNEUMATICS IN PHASE II AMBER". Acknowledge DMCCC announcement, "FUEL TANK PRESSURE RISING NORMALLY".

Table 3-16.   Amplified Countdown Procedures (CONT)

STEP	CREW POS	REQUIREMENT
5 (CONT)	DMCCC	a. Acknowledge MCCC announcement, "PNEUMATICS IN PHASE II AMBER". b. Observe fuel tank pressure starts to rise at a steady even rate. With a normal rate of increase, fuel pressure tank should reach 53.0 PSI in approximately 20 seconds. If an erratic rate of increase is observed during a nontactical countdown, abort is required. During a tactical launch, countdown may be continued, however, close observation of pressures is necessary. **CAUTION** When PRESSURE MODE indicator is illuminated GREEN and an unscheduled rapid change of pressures not within limits is observed, depress EMERGENCY pushbutton immediately. Manually correct missile tank pressures with appropriate RAISE or LOWER pushbutton, then return to automatic control by depressing AUTOMATIC pushbutton. c. Announce, "FUEL TANK PRESSURE RISING NORMALLY". d. Continue to observe missile tank pressures.
	BMAT	Monitor launch control console.
	MFT	Monitor FRCP.
	EPPT	Monitor PRCP.
6		HYDRAULIC PRESSURE indicator GREEN. Indicator will illuminate GREEN when power is applied to the autopilot system and booster and sustainer hydraulic pressure are between 1750 and 2250 PSI. If indicator fails to illuminate GREEN and remain GREEN within 30 seconds after illuminating AMBER abort is required during a nontactical countdown. During a tactical countdown, see table 4-12.
	MCCC	Observe HYDRAULIC PRESSURE indicator GREEN.
	DMCCC	Monitor fuel tank pressure increasing to phase II pressure.
	MFT	Monitor FRCP.
	EPPT	Monitor PRCP.

Table 3-16. Amplified Countdown Procedures (CONT)

STEP	CREW POS	REQUIREMENT
7	MCCC	**CAUTION** If PNEUMATICS IN PHASE II indicator is not GREEN prior to HELIUM LOAD indicator AMBER, depress ABORT pushbutton immediately. PNEUMATICS IN PHASE II indicator GREEN. Indicator will illuminate GREEN when fuel tank pressure is between 53.0 and 67.5 PSI. **CAUTION** If AIRBORNE FILL & DRAIN VALVE indicator fails to illuminate GREEN within 10 seconds after PNEUMATICS IN PHASE II indicator illuminates GREEN, depress ABORT pushbutton immediately. Failure to comply may result in damage to load lines or possible loss of missile. a. Announce, "MARK-PNEUMATICS IN PHASE II GREEN". b. Start stopwatch and count seconds aloud until AIRBORNE FILL & DRAIN VALVE indicator is GREEN or until 10 seconds have elapsed. Continue timing for LO_2 LINE FILLED indicator GREEN or RAPID LO_2 LOAD indicator AMBER within 45 seconds.
	DMCCC	Observe missile tank pressures stabilized at phase II pressures. (LO_2 tank pressure 3.4 to 8.0 PSI, differential pressure greater than 5 PSI, fuel tank pressure 62.5 to 63.9 PSI).
	BMAT	a. Acknowledge MCCC announcement, "MARK - PNEUMATICS IN PHASE II GREEN" and take position at LO_2 tanking panels. b. Observe AIRBORNE FILL & DRAIN VALVE indicator on LO_2 TANKING (PANEL 2) illuminates GREEN. c. Announce, "AIRBORNE FILL & DRAIN VALVE GREEN". d. Observe STORAGE TANK VENT VALVE N-5 indicator on LO_2 TANKING (PANEL 2) is illuminated AMBER.
	MFT	Monitor FRCP.
	EPPT	Monitor PRCP.

Table 3-16. Amplified Countdown Procedures (CONT)

STEP	CREW POS	REQUIREMENT
8	MCCC	LO₂ LINE FILLED indicator AMBER. Indicator will illuminate AMBER after pneumatics ready for chilldown, LO₂ storage tank vent valve N-5 closed, and line chilldown not complete (40-second timer not picked up). If indicator fails to illuminate AMBER and STORAGE TANK VENT VALVE N-5 indicator is illuminated AMBER, continue countdown and observe LO₂ LINE FILLED indicator illuminates GREEN 40 seconds after PNEUMATICS IN PHASE II GREEN. If STORAGE TANK VENT VALVE N-5 indicator is not illuminated AMBER, abort a nontactical countdown or see table 4-13, item 1 if during a tactical countdown. Observe LO₂ LINE FILLED indicator AMBER. Acknowledge announcements: a. BMAT: "AIRBORNE FILL & DRAIN VALVE GREEN". b. DMCCC: "PRESSURES STABILIZED AT PHASE II". c. BMAT: "LO₂ TANKING PANELS NORMAL".
	DMCCC	Announce. "PRESSURES STABLIZED AT PHASE II", after BMAT has announced "AIRBORNE FILL & DRAIN VALVE GREEN".
	BMAT	**NOTE** The following is a general summarization of the events occuring during the LO₂ loading sequence. The LO₂ load sequence is started by the pneumatics ready for chilldown signal. LO₂ LINE FILLED indicator illuminates AMBER, a 40-second timer starts, N-5 and N-4 valves close, and L-1, L-2, L-60, N-50, N-1, and airborne fill-and-drain valves open. Opening of valve N-1 pressurizes the LO₂ storage tank to approximately 25 PSI, forcing LO₂ through the loading system into the missile and chilling down the loading lines. At the expiration of the 40-second timer, LO₂ LINE FILLED indicator illuminates GREEN, RAPID LO₂ LOAD indicator illuminates AMBER, and valve N-2 opens. Opening of valve N-2 allows the transfer pressure to the LO₂ storage tank to increase to approximately 135 PSI, which rapid loads the missile at approximately 5500 GPM. When the 95% sensor in the missile becomes wet, RAPID LO₂ LOAD indicator illuminates GREEN, FINE LO₂ LOAD indicator illuminates AMBER L-2 closes, and L-50 opens. When the 99% sensor in the missile becomes wet, FINE LO₂ LOAD indicator illuminates GREEN, L-1, airborne fill-and-drain valve, N-1, N-2, and

Table 3-16.　Amplified Countdown Procedures (CONT)

STEP	CREW POS	REQUIREMENT
8 (CONT)		N-3 (if open) close. Valves L-16 and N-60 open and a 90-second line drain timer starts. At the same time, topping continues through the booster engine turbopump by way of L-50 and L-60. When the 99.25% sensor in the missile becomes wet L-50 closes and 15 seconds later L-60 closes.
		When the 90-second timer runs out L-16 and N-60 close and the LO_2 loading lines are vented through N-80, which opens at this time for 40 seconds. As the LO_2 in the missile boils off, successive topping cycles through L-60 will continue. The LO_2 READY indicator illuminates GREEN 50 seconds after FINE LO_2 LOAD indicator is illuminated GREEN.
	BMAT	Observe the following:
		a.　TOPPING TANK VENT VALVE N-4 indicator AMBER. If indicator fails to illuminate AMBER, abort is required during a nontactical countdown, or see table 4-13, item 2 if during a tactical countdown.
		b.　STORGAE TANK PRESS VALVE N-1 indicator GREEN. The indicator may return to AMBER after illuminating GREEN until RAPID LO_2 LOAD indicator illuminates AMBER. If indicator fails to illuminate GREEN, at least momentarily, prior to RAPID LO_2 LOAD indicator AMBER, during a nontactical countdown abort is required. Continue a tactical countdown, however, close observation of LO_2 loading is required.
		c.　RAPID LOAD VALVE L-2 indicator GREEN. If indicator fails to illuminate GREEN during a nontactical countdown, abort is required. Tactical countdown may be continued. Rapid LO_2 loading may be accomplished through fine load valve L-1. Maximum allowable hold time will be reduced 2-1/2 minutes for each minute of rapid load in excess of 3-1/2 minutes, based on a minimum of 2650 gallons of LO_2 in the topping tank.
		d.　FINE LOAD VALVE L-1 indicator GREEN. Fine load valve L-1 must have been open to enable airborne fill-and-drain valve to open. If FINE LOAD VALVE L-1 indicator fails to illuminate GREEN, during a nontactical countdown abort.
		e.　TOPPING TANK PRESS. VALVE N-50 indicator GREEN. If indicator fails to illuminate GREEN, abort a nontactical countdown, or see table 4-13, item 3 if during a tactical countdown.
		f.　TOPPING CHILL VALVE L-60 indicator GREEN. If indicator fails to illuminate GREEN, abort is required during a nontactical countdown, or see table 4-13, item 4 if during a tactical countdown.
		Announce, "LO_2 TANKING PANELS NORMAL".
	MFT	Monitor FRCP.
	EPPT	Monitor PRCP.

Table 3-16. Amplified Countdown Procedures (CONT)

STEP	CREW POS	REQUIREMENT
9		LO₂ LINE FILLED indicator GREEN. Indicator illuminates GREEN upon expiration of 40-second line chilldown timer. If indicator fails to illuminate GREEN within 45 seconds (40 seconds normal) after PNEUMATICS IN PHASE II indicator GREEN, continue countdown only if RAPID LO₂ LOAD indicator illuminates AMBER.
	MCCC	a. Observe LO₂ LINE FILLED indicator GREEN. b. Reset stopwatch.
	DMCCC	Monitor missile fuel pressure for phase II.
	BMAT	Monitor LO₂ TANKING (PANEL 1 and PANEL 2).
	MFT	Monitor FRCP.
	EPPT	Monitor PRCP.
10		RAPID LO₂ LOAD indicator AMBER. Indicator will illuminate AMBER when 95% sensor is dry and LO₂ load signal is present. If indicator fails to illuminate AMBER within 45 seconds (40 seconds normal) after PNEUMATICS IN PHASE II indicator GREEN, abort is required.
	MCCC	a. Announce. "MARK - RAPID LO₂ LOAD AMBER". b. Start stopwatch to time LO₂ rapid load sequence. c. Acknowledge BMAT announcement. "LO₂ TANKING PANELS NORMAL". **CAUTION** If the RAPID LO₂ LOAD and FINE LO₂ LOAD indicators illuminate GREEN simultaneously, abort is required since a double LO₂ sensor failure has occurred.
	DMCCC	Monitor phase II pressures.

Table 3-16. Amplified Countdown Procedures (CONT)

STEP	CREW POS	REQUIREMENT
10 (CONT)		Acknowledge MCCC announcement, "MARK - RAPID LO$_2$ LOAD AMBER". Observe LO$_2$ TANKING (PANEL 1 and PANEL 2) for thhe following indications:
	BMAT	a. LO$_2$ STG TNK PRESSURE indicator AMBER. If indicator fails to illuminate AMBER shortly after RAPID LO$_2$ LOAD indicator illuminates AMBER, abort a nontactical countdown immediately. This condition indicates a possible pressure switch failure which could cause back-filling the missile with LO$_2$ if an abort is initiated after LO$_2$ has been loaded and the storage tank has not been vented.

b. STORAGE TANK PRESS. VALVE N-1 and STORAGE TANK PRESS. VALVE N-2 indicators GREEN. Valve N-2 indicator may cycle from GREEN to AMBER until the ullage space in the LO$_2$ storage tank is large enough the require valve N-2 to remain open. If valve N-1, N-2, and N-3 indicators remain AMBER after RAPID LO$_2$ LOAD indicator illuminates AMBER, an abort is required.

c. STORAGE TANK PRESS. VALVE N-3 indicator GREEN or AMBER. Valve N-3 indicator will illuminate GREEN only if the LO$_2$ transfer pressure supply falls below 2500 PSI.

Announce, "LO$_2$ TANKING PANELS NORMAL". |
	MFT	Monitor FRCP.
	EPPT	Monitor PRCP.
11		HELIUM LOAD indicator AMBER. Indicator will illuminate AMBER when 2-minute helium load delay timer times out, abort external signal is not present, and pressure in the shrouded helium spheres is less than 2950 PSI. If indicator fails to illuminate AMBER, abort is required during a nontactical countdown. If during a tactical countdown, see table 4-14.
	MCCC	Observe HELIUM LOAD indicator AMBER.
	DMCCC	Monitor phase 11 pressures.
	BMAT	Monitor LO$_2$ TANKING (PANEL 1 and PANEL 2)
	MFT	Monitor FRCP.
	EPPT	Monitor PRCP.

Table 3-16. Amplified Countdown Procedures (CONT)

STEP	CREW POS	REQUIREMENT
12		MISSILE BAT. ACTIVATED indicator GREEN. Indicator will illuminate GREEN after 2-minute battery sensing timer has picked up and battery output is within specifications. If indicator fails to illuminate GREEN, abort is required. Refer to section V for malfunction procedures.
	MCCC	Observe MISSILE BAT. ACTIVATED indicator GREEN.
	DMCCC	Monitor phase II pressures.
	BMAT	Monitor LO$_2$ TANKING (PANEL 1 and PANEL 2).
	MFT	Monitor FRCP.
	EPPT	Monitor PRCP.
13		ENG & MISSILE POWER READY indicator GREEN. Indicator will illuminate GREEN when the following conditions exist: a. Missile AC and DC loads ready. b. Engine valve heaters on. c. (Deleted) d. Missile battery activated and output voltage within tolerance. e. Engines are ready. If indicator fails to illuminate GREEN, abort is required. Refer to section V for malfunction procedures.
	MCCC	a. Announce, "ENGINE AND MISSILE POWER READY GREEN". b. Verify validity of launch order.
	DMCCC	Verify validity of launch order.
	BMAT	Monitor LO$_2$ TANKING (PANEL 1 and PANEL 2).
	MFT	Monitor FRCP.
	EPPT	Monitor PRCP.

Table 3-16. Amplified Countdown Procedures (CONT)

STEP	CREW POS	REQUIREMENT
14		LN$_2$ LOAD indicator GREEN. Indicator will illuminate GREEN, provided rapid load valve 214 has opened, 3-minute LN$_2$ load timer has timed out, and transfer pressure is greater than 75 PSI. If indicator fails to illuminate GREEN during a tactical countdown, see table 4-11. During a nontactical countdown abort is required. Refer to section V for malfunction procedures.
	MCCC	Announce, "LN$_2$ LOAD GREEN".
	DMCCC	Monitor phase II pressures.
	BMAT	Stand by to position REMOTE LOCAL switch on LO$_2$ TANKING (PANEL 1) to LOCAL if required.
	MFT	Monitor FRCP.
	EPPT	Monitor PRCP.
15		AUTOPILOT ON indicator GREEN. Indicator will illuminate GREEN provided that 400-cycle, 115 VAC, 3-phase power is available at the missile and the 4-minute autopilot test delay timer is timed out. If indicator fails to illuminate GREEN, abort is required. Refer to section V for malfunction procedures.
	MCCC	Observe AUTOPILOT ON indicator GREEN.
	DMCCC	Monitor phase II pressures.
	BMAT	Monitor LO$_2$ TANKING (PANEL 1 and PANEL 2).
	MFT	Monitor FRCP.
	EPPT	Monitor PRCP.
16		AUTOPILOT TEST indicator AMBER. Indicator will illuminate AMBER when the 4-minute autopilot test delay timer is timed out, hydraulic pressure is between 1750 and 2250 PSI, and the 90-second autopilot test timer has

Table 3-16. Amplified Countdown Procedures (CONT)

STEP	CREW POS	REQUIREMENT
16 (CONT)		not timed out. If the indicator fails to illuminate AMBER, abort is required. Refer to section V for malfunction procedures.
	MCCC	Observe AUTOPILOT TEST indicator AMBER.
	DMCCC	Monitor phase II pressures.
	BMAT	Monitor LO$_2$ TANKING (PANEL 1 and PANEL 2).
	MFT	Monitor FRCP.
	EPPT	Monitor PRCP.
17		┌─────────────┐ │ **CAUTION** │ └─────────────┘ During a nontactical countdown, if L-2 valve indicator fails to illuminate AMBER within 8 seconds after RAPID LO$_2$ LOAD indicator GREEN, immediately position REMOTE LOCAL switch on LO$_2$ TANKING (PANEL 1) to LOCAL and initiate abort. After LO$_2$ STG TNK PRESSURE indicator illuminates GREEN, position REMOTE LOCAL switch to REMOTE. LO$_2$ tank pressure must be carefully monitored during LO$_2$ drain because of the increase of drain flow if valve L-2 is open. If LO$_2$ tank pressure decreases to 2.0 PSI, immediately position REMOTE LOCAL switch to LOCAL. Return REMOTE LOCAL switch to REMOTE when LO$_2$ tank pressure stabilizes. During a tactical countdown, if valve L-2 fails to close, allow countdown to continue. Do not initiate commit start until an automatic topping sequence is complete, as indicated by L-60 valve indicator cycling from AMBER to GREEN and back to AMBER. RAPID LO$_2$ LOAD indicator GREEN. Indicator will illuminate GREEN when the 95% sensor is wet and LO$_2$ load signal is still present. RAPID LO$_2$ LOAD indicator should illuminate GREEN approximately 3 minutes after illuminating AMBER. In a nontactical countdown, abort will be initiated

Table 3-16. Amplified Countdown Procedures (CONT)

STEP	CREW POS	REQUIREMENT
17 (CONT)		if RAPID LO_2 LOAD indicator fails to illuminate GREEN 6 minutes after illuminating AMBER. In a tactical countdown, with the LO_2 tanking panels normal, LO_2 loading may be continued longer than 6 minutes.
	MCCC	a. Announce, "MARK - RAPID LO_2 LOAD GREEN".
		b. Stop and restart stopwatch, count aloud up to 8 seconds or until RAPID LOAD VALVE L-2 indicator AMBER. Continue timing for FINE LO_2 LOAD indicator GREEN in approximately 30 seconds.
	DMCCC	Monitor phase II pressures.
	BMAT	a. Acknowledge MCCC announcement, "MARK - RAPID LO_2 LOAD GREEN".
		b. Announce, "L-2 AMBER". RAPID LOAD VALVE L-2 indicator should illuminate AMBER within 8 seconds after RAPID LO_2 LOAD indicator illuminates GREEN. If indicator fails to illuminate AMBER, see caution above. A tactical countdown shall be allowed to continue. Commit shall be delayed until a topping cycle has been completed. If L-2 fails to close, the FINE LO_2 LOAD indicator will illuminate GREEN sooner than normal. At FINE LO_2 LOAD indicator GREEN, fine load valve L-1 will close, then the airborne fill and drain valve will close, stopping the rapid and fine load sequence. The airborne fill-and-drain valve is capable of stopping a full flow rapid load.
		c. Observe RAPID TOPPING VALVE L-50 indicator GREEN. If indicator fails to illuminate GREEN, during a training launch, abort is required due to marginal chilldown of the turbopumps during commit sequence.
		d. Continue to stand by REMOTE LOCAL switch on LO_2 TANKING (PANEL 1).
	MFT	Monitor FRCP.
	EPPT	Monitor PRCP.
18		FINE LO_2 LOAD indicator AMBER. Indicator will illuminate AMBER when the 95% sensor is wet, the 99% sensor is not wet, and the LO_2 load signal is present. If FINE LO_2 LOAD indicator fails to illuminate AMBER and LO_2 tanking panels are normal, continue countdown for approximately 30 seconds to observe if FINE LO_2 LOAD indicator GREEN.

Table 3-16. Amplified Countdown Procedures (CONT)

STEP	CREW POS	REQUIREMENT
18 (CONT)	MCCC	a. Observe FINE LO$_2$ LOAD indicator AMBER. b. Acknowledge BMAT announcement, "L-2 AMBER".
	DMCCC	Monitor phase II pressures.
	BMAT	Standby at LO$_2$ TANKING (PANEL 1 and PANEL 2).
	MFT	Monitor FRCP.
	EPPT	Monitor PRCP.
19	MCCC	┌─────────────────┐ **CAUTION** └─────────────────┘ If FINE LOAD VALVE L-1 indicator fails to illuminate AMBER within 5 seconds after FINE LO$_2$ LOAD indicator illuminates GREEN, position REMOTE LOCAL switch on LO$_2$ TANKING (PANEL 1) to LOCAL and start abort. Return REMOTE LOCAL switch on LO$_2$ TANKING (PANEL 1) to REMOTE when LO$_2$ STG TNK PRESSURE indicator illuminates GREEN and monitor abort sequence. Failure to comply may result in missile tank overfill and subsequent damage. FINE LO$_2$ LOAD indicator GREEN. Indicator will illuminate GREEN when 95% and 99% sensors are wet and LO$_2$ load signal is present. If indicator fails to illuminate GREEN approximately 30 seconds after RAPID LO$_2$ LOAD indicator has illuminated GREEN, abort is required. a. Announce, MARK - FINE LO$_2$ LOAD GREEN". b. Start stopwatch and count aloud up to 5 seconds or until FINE LOAD VALVE L-1 indicator AMBER. Continue timing until 120 seconds have elapsed. Announce 20 seconds, for timing RAPID TOPPING VALVE L-50 indicator to illuminate AMBER and 120 seconds for timing completion of LO$_2$ line drain. c. Acknowledge BMAT announcement, "L-1 AMBER".
	DMCCC	Monitor phase II pressures.

Table 3-16. Amplified Countdown Procedures (CONT)

STEP	CREW POS	REQUIREMENT
19 (CONT)	BMAT	a. Acknowledge MCCC announcement, "MARK - FINE LO$_2$ LOAD GREEN". b. Announce, "L-1 AMBER". See caution above. c. Observe the following: (1) AIRBORNE FILL & DRAIN VALVE indicator AMBER. If indicator fails to illuminate AMBER, the LO$_2$ line drain sequence will not start. The LO$_2$ READY indicator will fail to illuminate GREEN and abort will be required. (2) DRAIN VALVE L-16 indicator GREEN. (3) LINE DRAIN PRES VALVE N-60 indicator GREEN. During a tactical countdown, if indicators for valves L-16 and N-60 fail to illuminate GREEN and AIRBORNE FILL & DRAIN VALVE indicator is illuminated AMBER, LO$_2$ line drain shall be accomplished by: (a) Positioning REMOTE LOCAL switch on LO$_2$ TANKING (PANEL 1) to LOCAL. (b) Positioning L-1 valve switch to OPEN after LO$_2$ STG TNK PRESSURE indicator illuminates GREEN. (c) Waiting 2 minutes, then returing L-1 valve switch to CLOSE. (d) Positioning REMOTE LOCAL switch to REMOTE. Commit sequence can now be initiated. If indicators for valves L-16 and N-60 fail to illuminate GREEN during a nontactical countdown, abort is required. (4) RAPID TOPPING VALVE L-50 indicator AMBER. If indicator fails to illuminate AMBER within 20 seconds after FINE LO$_2$ LOAD indicator illuminates GREEN, an immediate abort is required.
	MFT	Monitor FRCP.
	EPPT	Monitor PRCP.
20		LO$_2$ READY indicator GREEN. Indicator will illuminate GREEN when 99.25% sensor has been wet, airborne fill-and-drain valve is closed, and the 50-second commit delay timer has timed out. If FINE LO$_2$ LOAD indicator illuminates AMBER and remains AMBER after LO$_2$ READY indicator

Table 3-16. Amplified Countdown Procedures (CONT)

STEP	CREW POS	REQUIREMENT
20 (CONT)		has illuminated GREEN, commit sequence must be started within 10 minutes or abort is required as a malfunction in the LO_2 topping system is indicated. If LO_2 READY indicator fails to illuminate GREEN, abort is required. Refer to section V for malfunction procedures.
	MCCC	Announce "LO_2 READY GREEN".
	DMCCC	Monitor phase II pressures.
	BMAT	Monitor LO_2 TANKING (PANEL 1 and PANEL 2).
	MFT	Monitor FRCP.
	EPPT	Monitor PRCP.
21		AUTOPILOT TEST indicator GREEN. Indicator will illuminate GREEN when the 90-second autopilot test timer times out and an autopilot fail signal is not present. If indicator fails to illuminate GREEN, abort is required. Refer to section V for malfunction procedures.
	MCCC	Observe AUTOPILOT TEST indicator GREEN.
	DMCCC	Monitor phase II pressures.
	BMAT	Monitor LO_2 TANKING (PANEL 1 and PANEL 2).
	MFT	Monitor FRCP.
	EPPT	Monitor PRCP.
22		HELIUM LOAD indicator GREEN. Indicator will illuminate GREEN when missile shrouded spheres reach a pressure greater than 2950 PSI. If indicator fails to illuminate GREEN within 8 minutes after being AMBER during a tactical countdown, see table 4-14. During a nontactical countdown, abort is required.
	MCCC	Observe HELIUM LOAD indicator GREEN.

Table 3-16. Amplified Countdown Procedures (CONT)

STEP	CREW POS	REQUIREMENT
22 (CONT)	DMCCC	Monitor phase II pressures.
	BMAT	Monitor LO₂ TANKING (PANEL 1 and PANEL 2).
	MFT	Monitor FRCP.
	EPPT	Monitor PRCP.
23		HYD PNEU & LN₂ - HE READY indicator will illuminate GREEN when all of the following conditions exists: a. Pneumatics in phase II. b. Helium control charging unit greater than 4500 PSI. c. Hydraulic pressure within limits. d. Helium load complete. e. LN₂ load complete. If HYD PNEU & LN₂ - HE READY indicator fails to illuminate GREEN during a tactical countdown, see table 4-15. During a nontactical countdown, abort is required. Refer to section V for malfunction procedures.
	MCCC	Announce, " HYD PNEU & LN₂ - HE READY GREEN"
	DMCCC	a. Monitor phase II pressures. b. After controllers fast reaction checklist is completed and after being relieved at launch control console by BMAT, depress ALCO COMM pushbutton and proceed to ALCO COMM/CONTROL panel.
	BMAT	┌─────────────────┐ │ CAUTION │ └─────────────────┘ If DRAIN VALVE L-16 or LINE DRAIN PRESS VALVE N-60 has failed to illuminate AMBER within 120 seconds from FINE LO₂ LOAD GREEN, immediately position REMOTE LOCAL switch on LO₂ TANKING (PANEL 1) to LOCAL. After LO₂ STG TNK PRESSURE indicator illuminates GREEN, return REMOTE LOCAL switch to REMOTE. ABORT a nontactical countdown. Commit a tactical countdown.

Table 3-16. Amplified Countdown Procedures (CONT)

STEP	CREW POS	REQUIREMENT
23 (CONT)	BMAT (CONT)	Observe the following within 120 seconds from beginning of LO$_2$ line drain. (FINE LO$_2$ LOAD indicator GREEN). a. DRAIN VALVE L-16 indicator AMBER. If indicator fails to illuminate AMBER, see caution above. b. LINE DRAIN PRES VALVE N-60 indicator AMBER. If indicator fails to illuminate AMBER. See caution above. STORAGE TANK VENT VALVE N-5 indicator GREEN or EXTINGUISHED within 30 seconds after DRAIN VALVE L-16 indicator AMBER. If indicator remains AMBER, abort a nontactical countdown. d. LINE VENT VALVE N-80 indicator GREEN. Indicator should remain GREEN for approximately 40 seconds. If indicator fails to illuminate GREEN during a nontactical countdown, abort is required. After observing LO$_2$ TANKING (PANEL 1 and PANEL 2) for proper line drain indications BMAT shall relieve DMCCC at launch control console to monitor missile tank pressures and report countdown timing at 1-minute intervals.
	MFT	Monitor FRCP.
	EPPT	Monitor PRCP.
24		GUIDANCE READY indicator GREEN. Indicator will illuminate GREEN when guidance countdown is complete. If indicator fails to illuminate GREEN within 13 minutes from start countdown the GUIDANCE FAIL indicator will illuminate RED, see table 4-2, item 10.
	MCCC	Observe GUIDANCE READY indicator GREEN.
	DMCCC	When relieved by BMAT proceed to ALCO COMM/CONTROL panel, open panel, insert key, position COMM switch to TALK, monitor countdown, and stand by for MISSILE READY indicator to illuminate GREEN.
	BMAT	Monitor phase II pressures.
	MFT	Monitor FRCP.
	EPPT	Monitor PRCP.

Table 3-16. Amplified Countdown Procedures (CONT)

STEP	CREW POS	REQUIREMENT
25		FLIGHT CONTROL & R/V READY indicator GREEN. Indicator will illuminate GREEN when the following conditions exist: a. Autopilot ready. b. Guidance ready. c. R/V ready. d. Target selected. If FLIGHT CONTROL & R/V READY indicator fails to illuminate GREEN, abort is required. Refer to section V for malfunction procedures.
	MCCC	Announce, "FLIGHT CONTROL & R/V READY GREEN".
	DMCCC	Stand by at ALCO COMM/CONTROL panel for MISSILE READY indicator to illuminate GREEN.
	BMAT	Monitor phase II pressures.
	MFT	Monitor FRCP.
	EPPT	Monitor PRCP.
26		READY FOR COMMIT indicator GREEN. Indicator will illuminate GREEN when the following conditions exist: a. Engine and missile power ready. b. Flight control and R/V ready. c. HYD-PNEU and LN_2-HE ready. d. LO_2 ready. e. Missile lifting system in standby. If READY FOR COMMIT indicator fails to illuminate GREEN, abort is required. Refer to section V for malfunction procedures.
	MCCC	a. Announce, "READY FOR COMMIT GREEN." b. Acknowledge EPPT announcement of power condition. c. Acknowledge deputy announcement, "DEPUTY READY".

Table 3-16. Amplified Countdown Procedures (CONT)

STEP	CREW POS	REQUIREMENT
26 (CONT)	DMCCC	MISSILE READY indicator will illuminate GREEN when READY for COMMIT indicator on launch control console illuminates GREEN. a. Observe MISSILE READY indicator GREEN. b. Acknowledge, "DEPUTY READY" when MCCC announces "READY FOR COMMIT GREEN".
	BMAT	Monitor phase II pressures.
	MFT	Monitor FRCP.
	EPPT	a. Acknowledge MCCC announcement. "READY FOR COMMIT GREEN." b. Announce power condition (single or parallel generator operation.)
27		**NOTE** If only one generator is operating, trip FEEDER NUMBER 3 NON-ESSENTIAL BUS CONTROL SWITCH when POWER INTERNAL indicator illuminates GREEN. **NOTE** During commit sequence, UTILITY WATER PRESSURE and SILO WATER CHILLER UNITS MALFUNCTION indicators may illuminate RED. This is a normal condition if FEEDER NUMBER 3 NON-ESSENTIAL BUS CONTROL SWITCH GREEN indicator is illuminated. **CAUTION** If wind velocity or wind gust velocity exceeds maximum allowable anemometer reading measured at a distance of 10 feet above ground, do not start commit sequence except for tactical launch. (Refer to classified supplement to this manual.) Failure to comply may result in structural damage to missile. **WARNING** For training launches only, missile flight safety system instrumentation and range safety shall be ready for commit sequence to start. **NOTE** During missile commit sequence abort may be accomplished by depressing ABORT pushbutton anytime prior to MISSILE LIFT UP & LOCKED indicator illuminating GREEN or after ABORT indicator is illuminated red.

Table 3-16. Amplified Countdown Procedures (CONT)

STEP	CREW POS	REQUIREMENT
27 (CONT)	MCCC	**CAUTION** If a nuclear blast is detected during countdown, do not start commit sequence until blast conditions (shock wave and effect) are over. a. Announce, "M-1, CLOSE LCC BLAST CLOSURES". b. Complete controller fast reaction checklist if launch order was received when in a hold configuration. c. Acknowledge MFT announcement, "BLAST CLOSURES CLOSED". d. Break COMMIT START key cover seal.
	DMCCC	Monitor ALCO COMM/CONTROL panel.
	BMAT	Monitor phase II pressures.
	MFT	a. Acknowledge MCCC announcement, "M-1, CLOSE LCC BLAST CLOSURES". b. Depress LCC BLAST CLOSURES MANUAL OPERATION CLOSE pushbutton. c. Observe the following indications: (1) LCC AIR INTAKE CLOSED indicator RED. (2) LCC AIR EXHAUST CLOSED indicator RED. (3) LCC STAIRWELL AIR EXHAUST CLOSED indicator RED (NA OSTF-2). **NOTE** If the above indications are abnormal, continue countdown. d. Announce, "BLAST CLOSURES CLOSED". e. Take position at CSMOL.
	EPPT	Monitor PRCP.

Table 3-16. Amplified Countdown Procedures (CONT)

STEP	CREW POS	REQUIREMENT
28	MCCC	Announce, "COMMIT START ON MY MARK...MARK" and rotate COMMIT START key.
	DMCCC	The ALCO COMM/CONTROL panel COMMIT SWITCH key must be rotated fully clockwise within 3 seconds after either LCO COMMIT indicator illuminates GREEN or MCCC "MARK" announcement. Rotate ALCO COMM/CONTROL panel COMMIT SWITCH key.
	BMAT	Monitor missile tank pressures. Reset and start stopwatch. Announce commit timing at 1-minute intervals.
	MFT	Monitor FRCP.
	EPPT	Monitor PRCP.
29		LAUNCH ENABLED indicator GREEN. Indicator will illuminate GREEN if launch has been enabled at the command post and alternate command post, and if the ALCO COMM/CONTROL panel COMMIT SWITCH key is rotated within 3 seconds after launch control console COMMIT START key is rotated.
	MCCC	Observe LAUNCH ENABLED indicator GREEN.
	DMCCC	Monitor ALCO COMM/CONTROL panel.
	BMAT	Monitor missile tank pressures.
	MFT	Monitor FRCP.
	EPPT	Monitor PRCP.
30		POWER INTERNAL indicator AMBER, then GREEN. Indicator will illuminate AMBER after commit start and before changeover switch internal. This condition may remain for 60 seconds if the missile inverter has been running for 5 minutes before commit start. Indicator illuminates GREEN if internal AC power is within tolerance and the power changeover switch is internal. If POWER INTERNAL indicator remains AMBER and fails to illuminate GREEN, abort is required.

Changed 21 October 1964

Table 3-16. Amplified Countdown Procedures (CONT)

STEP	CREW POS	REQUIREMENT
30 (CONT)		**NOTE** If POWER INTERNAL indicator illuminates AMBER, then extinguishes, and LO_2 COMMIT, PNEUMATICS INTERNAL, and GUIDANCE COMMIT indicators illuminate AMBER, continue countdown.
	MCCC	a. Announce, "MARK - POWER INTERNAL GREEN". b. Start stopwatch for LO_2 COMMIT indicator to illuminate GREEN in approximately 60 seconds. c. Acknowledge EPPT announcement, "NONESSENTIAL POWER OFF".
	DMCCC	At MCCC announcement "MARK-POWER INTERNAL GREEN" or when COMMIT IN PROGRESS indicator illuminates GREEN return to launch control console and relieve BMAT after phase III pressures have stabilized.
	BMAT	a. Acknowledge MCCC announcement, "MARK - POWER INTERNAL GREEN". b. Observe RAPID TOPPING VALVE L-50 indicaator GREEN. If indicator fails to illuminate GREEN during a training launch, abort is required because of possible low head pressure at the turbopumps which may result due to improper chilldown.
	MFT	Monitor FRCP.
	EPPT	a. Acknowledge MCCC announcement, "MARK - POWER INTERNAL GREEN". b. If only one generator is operating, position FEEDER NUMBER 3 NON-ESSENTIAL BUS CONTROL SWITCH to TRIP and observe FEEDER NUMBER 3 NON-ESSENTIAL BUS CONTROL SWITCH GREEN indicator illuminates. c. Announce, "NONESSENTIAL POWER OFF" (if accomplished).
31	MCCC	PNEUMATICS INTERNAL indicator AMBER. Indicator will illuminate AMBER when missile power is internal and pneumatics has been selected for phase III pressures, and pneumatics is not internal. If indicator fails to illuminate AMBER and LO_2 tank pressure is not rising, ABORT is required. a. Observe PNEUMATICS INTERNAL indicator AMBER. b. Acknowledge BMAT announcement, "LO_2 PRESSURE RISING NORMALLY".

Table 3-16. Amplified Countdown Procedures (CONT)

STEP	CREW POS	REQUIREMENT
31 (CONT)	DMCCC	Stand by to relieve BMAT.
	BMAT	After observing RAPID TOPPING VALVE L-50 indicator illuminated GREEN: a. Observe LO$_2$ tank pressure rising. b. Announce, "LO$_2$ PRESSURE RISING NORMALLY". LO$_2$ tank pressure may increase to approximately 29 to 30 PSI during LO$_2$ COMMIT indicator illuminated AMBER. This indicates that LO$_2$ commit loading is in progress. Tank pressure will decrease to normal flight pressure after rapid topping sequence is complete.
	MFT	Monitor FRCP.
	EPPT	Monitor PRCP.
32		GUIDANCE COMMIT indicator AMBER. Indicator will illuminate AMBER when guidance is on memory, as verified by the countdown group, and is not inertial. If GUIDANCE COMMIT indicator fails to illuminate AMBER and LO$_2$ COMMIT indicator fails to illuminate GREEN, abort is required.
	MCCC	Observe GUIDANCE COMMIT indicator AMBER.
	DMCCC	Stand by to relieve BMAT.
	BMAT	Announce. "LO$_2$ PRESSURE 25 PSI".
	MFT	Monitor FRCP.
	EPPT	Monitor PRCP.
33		LO$_2$ COMMIT indicator AMBER. Indicator will illuminate AMBER if missile power is internal and the 60-second missile lift delay timer is not picked up.
	MCCC	a. Observe LO$_2$ COMMIT indicator AMBER. b. Acknowledge BMAT announcement "LO$_2$ PRESSURE 25 PSI".

Table 3-16. Amplified Countdown Procedures (CONT)

STEP	CREW POS	REQUIREMENT
33 (CONT)	DMCCC	Stand by to relieve BMAT.
	BMAT	Monitor phase III pressures.
	MFT	Monitor FRCP and TV.
	EPPT	Monitor PRCP.
34	MCCC	PNEUMATICS INTERNAL indicator GREEN. Indicator will illuminate GREEN if pneumatics are internal after phase III pressure are attained plus 5 seconds. If indicator fails to illuminate GREEN, abort is required. a. Announce, "PNEUMATICS INTERNAL GREEN". b. Acknowledge BMAT announcement, "TANK PRESSURES STABILIZED AT PHASE III".
	DMCCC	Stand by to relieve BMAT.
	BMAT	a. Observe missile tank pressures stabilized at flight pressures (LO$_2$ tank pressure 23.0 to 29.0 PSI, differential pressure greater than 5 PSI, and fuel tank pressure 59.5 to 65.6 PSI). b. Acknowledge MCCC announcement "PNEUMATICS INTERNAL GREEN". c. Announce, "TANK PRESSURES STABILIZED AT PHASE III".
	MFT	Monitor FRCP and TV.
	EPPT	Monitor PRCP.
35		LO$_2$ COMMIT indicator GREEN. Indicator will illuminate GREEN after expiration of the 60-second missile lift commit delay timer and commit internal. The LO$_2$ commit signal also starts a 165-second missile lift not up and locked timer, and initiates a pulsating reset signal to the propellant utilization computer assembly which continues until abort is initiated or until missile lift off has been accomplished. If indicator fails to illuminate GREEN, abort is required.

Table 3-16. Amplified Countdown Procedures (CONT)

STEP	CREW POS	REQUIREMENT
35 (CONT)		**CAUTION** During launcher platform up-run, if an emergency occurs which requires launcher platform to be manually stopped and immediate down-run is not desired, position RESET PROGRAMMER key switch to ON and depress ABORT pushbutton. When down-run is desired, position RESET PROGRAMMER key switch to OFF.
	MCCC	Observe LO₂ COMMIT indicator GREEN.
	DMCCC	Relieve BMAT at launch control console and monitor flight pressures.
	BMAT	Monitor commit sequence.
	MFT	Prepare to monitor missile lifting sequence.
	EPPT	Monitor PRCP.
36		**NOTE** If more than 165 seconds have elapsed since LO₂ COMMIT indicator illuminated GREEN and MISSILE LIFT UP & LOCKED indicator has not illuminated GREEN, ABORT indicator will illuminate AMBER. MISSILE UP & LOCKED indicator AMBER. Indicator will illuminate AMBER when launcher platform is not up and locked and site is soft. If indicator does not illuminate AMBER, monitor missile lifting system on TV. If silo overhead doors are opening or open and normal missile lift is apparent, wait for MISSILE LIFT UP & LOCKED indicator GREEN and continue countdown.
	MCCC	a. Announce, "MISSILE LIFT UP & LOCKED AMBER". b. Acknowledge MFT announcement, "CSMOL NORMAL".
	DMCCC	Monitor flight pressures. **NOTE** After PNEUMATICS INTERNAL indicator has illuminated GREEN, the pressurization RAISE and LOWER pushbuttons are ineffective unless ABORT pushbutton is depressed and ABORT EXTERNAL indicator has illuminated AMBER.

Table 3-16. Amplified Countdown Procedures (CONT)

STEP	CREW POS	REQUIREMENT
36 (CONT)	BMAT	Monitor commit sequence.
	MFT	a. Acknowledge MCCC announcement, "MISSILE LIFT UP & LOCKED AMBER". b. Observe the following on CSMOL: (1) HYDRAULIC 40 HP PUMP PRESSURE indicator GREEN. (2) CRIB VERTICAL LOCK indicator GREEN approximately 6 to 9 seconds after MISSILE LIFT UP & LOCKED indicator AMBER. (3) CRIB HORIZONTAL LOCK indicator GREEN approximately 9 to 15 seconds after MISSILE LIFT UP & LOCKED indicator AMBER (4) SILO DOORS OPEN indicator GREEN approximately 20 to 45 seconds after MISSILE LIFT UP & LOCKED indicator AMBER. (5) LAUNCHER PLATFORM CREEP DISABLED indicator extinguished simultaneously with SILO DOORS OPEN indicator GREEN. c. Announce, "CSMOL NORMAL".
	EPPT	Monitor PRCP.
37		PROGRAMMER ARMED indicator AMBER. Indicator will illuminate AMBER with programmer armed signal present 70 seconds after missile lift commit start. PROGRAMMER ARMED indicator GREEN. Indicator will illuminate GREEN when flight programmer is armed.
	MCCC	Observe PROGRAMMER ARMED indicator AMBER to GREEN.
	DMCCC	Monitor flight pressures.
	BMAT	Monitor commit sequence.
	MFT	Monitor TV and CSMOL.
	EPPT	Monitor PRCP.

Table 3-16.   Amplified Countdown Procedures (CONT)

STEP	CREW POS	REQUIREMENT
38		MISSILE LIFT UP & LOCKED indicator GREEN. Indicator will illuminate GREEN when missile lift is up and locked. The missile lift up and locked signal completes the commit lockup summary which starts the 15-second abort timer. If indicator does not illuminate GREEN, monitor missile lifting system on TV for missile lift motion and wait for ABORT indicator to illuminate AMBER or MISSILE LIFT FAIL indicator to illuminate RED.
		CAUTION
		If ABORT indicator illuminates AMBER, initiate abort sequence. If abort sequence is not initiated within 15 seconds after MISSILE LIFT UP & LOCKED indicator has illuminated GREEN, depress EMERGENCY pushbutton. Use nomogram (LO$_2$ or LN$_2$ as applicable) contained in section VI to determine boiloff time. Manual control of LO$_2$ tank pressure shall be established at the HCU. Remove missile lifting system from automatic sequence control by positioning RESET PROGRAMMER key to ON.
	MCCC	Announce, "MISSILE LIFT UP & LOCKED GREEN".
	DMCCC	**NOTE** After MISSILE LIFT UP & LOCKED indicator has illuminated GREEN, the EMERGENCY pushbutton is enabled after 15 seconds or by depressing ABORT pushbutton after ABORT indicator illuminates RED.
	BMAT	Monitor commit sequence.
	MFT	a. Acknowledge MCCC announcement, "MISSILE LIFT UP & LOCKED GREEN". b. Observe the following indications on CSMOL: (1) LAUNCHER PLATFORM UP COMPLETED RUN AND LOCKED indicator GREEN. (2) HYDRAULIC 40 HP PUMP ON indicator WHITE.
	EPPT	Monitor PRCP.

Table 3-16. Amplified Countdown Procedures (CONT)

STEP	CREW POS	REQUIREMENT
39		GUIDANCE COMMIT indicator GREEN. Indicator illuminates GREEN when guidance goes on inertial, which should occur within 1 second after MISSILE LIFT UP & LOCKED indicator illuminates GREEN.
	MCCC	Observe GUIDANCE COMMIT indicator GREEN.
	DMCCC	Monitor flight pressures.
	BMAT	Monitor commit sequence.
	MFT	Monitor TV and FRCP.
	EPPT	Monitor PRCP.
40.		ENGINE START indicator AMBER to GREEN. Indicator illuminates GREEN if the following conditions exist: a. Guidance on inertial. b. Engines not cut off. c. Abort not started d. Commit lockup.
	MCCC	Observe ENGINE START indicator GREEN.
	DMCCC	Monitor flight pressures.
	BMAT	Monitor commit sequence.
	MFT	Monitor TV and FRCP.
	EPPT	Monitor PRCP.

T.O. 21M-HGM16F-1

Table 3-16. Amplified Countdown Procedures (CONT)

STEP	CREW POS	REQUIREMENT
41		MISSILE AWAY indicator GREEN. Indicator illuminates GREEN when missile rises 1 inch.
	MCCC	a. Announce, "MISSILE AWAY" or "ABORT RED".
		b. Acknowledge EPPT announcement, "NONESSENTIAL POWER ON".
		c. If nontactiacl launch was performed (VAFB), refer to postlaunch securing procedures contained in T.O. 21M-HGM16F-3CL-1.
		d. If a tactical launch was performed, refer to postlaunch abort procedures (tactical) contained in table 3-18.
		e. If launch was not performed, refer to abort procedures contained in table 3-17.
		NOTE
		Launch or abort report should be made to command post as soon as practical.
	DMCCC	a. Stop stopwatch.
		b. Log Zulu Time of missile away.
	BMAT	Stand by
	MFT	Stand by
	EPPT	a. Acknowledge MCCC announcement, "MISSILE AWAY" or "ABORT RED".
		b. Verify nonessential power on by observing FEEDER NUMBER 3 NONESSENTIAL BUS CONTROL SWITCH RED indicator is illuminated. If red indicator is not illuminated, position FEEDER NUMBER 3 NONESSENTIAL BUS CONTROL SWITCH to CLOSE and observe red indicator illuminates.
		c. Announce, "NONESSENTIAL POWER ON".

Table 3-17. Amplified Abort Procedures

STEP	CREW POS	REQUIREMENT
		NOTE Normal indications shall be announced as specified in this table. Abnormal or emergency indications shall be announced in all instances, and actions necessary to return systems to a safe configuration following referecned tables shall be initiated. **NOTE** If ABORT pushbutton is depressed prior to MISSILE LIFT UP & LOCKED indicator illuminating AMBER, SITE HARD and ABORT EXTERNAL indicators will illuminate GREEN immediately. The DMCCC will start timing the LO$_2$ drain sequence when LO$_2$ DRAIN COMPLETE indicator illuminates AMBER. **NOTE** If launcher platform is within 33 inches of the up-and-locked position, pressurization system must be in the automatic mode and LO$_2$ tank pressure greater than 8 PSI before launcher platfrom lowering sequence will begin. **NOTE** If launcher platform is observed to stop during lowering sequence prior to DOWN COMPLETED RUN AND LOCKED indicator on CSMOL illuminating GREEN, see table 4-16.
1	MCCC	When abort is to be initiated: a. Announce, "ABORT START". b. Depress ABORT pushbutton. c. Acknowledge EPPT announcement, "NONESSENTIAL POWER ON".
	DMCCC	Monitor abort.
	BMAT	Monitor abort.
	MFT	Monitor abort.
	EPPT	a. Verify nonessential power on by observing FEEDER NUMBER 3 NONESSENTIAL BUS CONTROL SWITCH RED indicator is illuminated. If

Table 3-17. Amplified Abort Procedures (CONT)

STEP	CREW POS	REQUIREMENT
1 (CONT)	EPPT (CONT)	red indicator is not illuminated, position FEEDER NUMBER 3 NONESSENTIAL BUS CONTROL SWITCH to CLOSE and observe red indicator illuminates. b. Announce, "NONESSENTIAL POWER ON".
2		**WARNING** If abort sequence fails to start or LP fails to lower, if boiloff valve is closed, immediately depress EMERGENCY pushbutton. After LO_2 tank pressure decreases to phase II, depress AUTOMATIC pushbutton. Enable boiloff valve periodically to relieve pressure when LO_2 tank pressure increases to 12 PSI. Continue with this procedure until appropriate emergency table directs otherwise. ABORT COMPLETE indicator AMBER. Indicator illuminates AMBER when countdown has started, ABORT pushbutton has been depressed, abort is not complete, and when one or more of the following conditions exist: Guidance not inertial, engines not started, or not commit lockup, or engine cut off. If indicator fails to illuminate AMBER and standby status patch indicators fail to extinguish, abort sequence has failed to start. The MCCC will direct A-1 to position REMOTE LOCAL switch on LO_2 TANKING (PANEL 4) to LOCAL. After LO_2 tanking panels are in local, MCCC will refer to the point in countdown when abort was initiated and proceed to table 4-10 and enter the table at the applicable point: a. Countdown start to PNEUMATICS IN PHASE II indicator GREEN, enter at action 4. b. PNEUMATICS IN PHASE II indicator GREEN to COMMIT START, enter at action 3. c. From commit start to MISSILE LIFT UP & LOCKED indicator AMBER, enter at action 2, step 3. d. From MISSILE LIFT UP & LOCKED indicator AMBER to GREEN, enter at action 1.
	MCCC	Observe ABORT COMPLETE indicator AMBER.
	DMCCC	a. Log start-abort time. b. If MISSILE LIFT UP & LOCKED indicator was not illuminated AMBER prior to abort, start timing LO_2 drain sequence when LO_2 DRAIN COMPLETE indicator illuminates AMBER.

Table 3-17. Amplified Abort Procedures (CONT)

STEP	CREW POS	REQUIREMENT
2 (CONT)	BMAT	a. (Deleted) b. Monitor abort sequence and observe television for start of launcher platform down motion.
	MFT	 **CAUTION** If HYDRAULIC 40 HP PUMP PRESSURE indicator on CSMOL extinguishes during the launcher platform down run sequence, position RESET PROGRAMMER key to ON and depress HYDRAULIC 40 HP PUMP OFF pushbutton. See table 4-16 for emergency procedures. Observe the following indications on CSMOL: a. HYDRAULIC 40 HP PUMP PRESSURE indicator GREEN. b. LAUNCHER PLATFORM UP COMPLETED RUN LOCKED indicator extinguished. Indicator should extinguish approximately 1 to 5 seconds after ABORT COMPLETE indicator AMBER, signifying launcher platform down sequence has started.
	EPPT	Monitor PRCP.
3		**NOTE** If MISSILE LIFT UP & LOCKED indicator has illuminated GREEN and boiloff valve cannot readily be observed, the EMERGENCY pushbutton shall be depressed 35 seconds after SITE HARD indicator AMBER, or, if boiloff valve is monitored on television, the EMERGENCY pushbutton shall be depressed when launcher platform lowers approximately 7-½ feet out of the uplocks. **NOTE** If MISSILE LIFT UP & LOCKED indicator was illuminated AMBER but not GREEN, place pressurization system is EMERGENCY if or when launcher platform is below 33 inches of the up-and-locked position and the boiloff valve is above the silo cap. Return pressurization system to AUTOMATIC when the boiloff valve reaches the silo cap. If boiloff valve fails to open, see table 4-17.

Table 3-17. Amplified Abort Procedures (CONT)

STEP	CREW POS	REQUIREMENT
3 (CONT)		SITE HARD indicator AMBER. Indicator illuminates AMBER when a MLS sequence 2 or 3 has started, abort has started, not site hard, and launcher platform has started to lower. If the indicator fails to illuminate AMBER, observe launcher platform for down motion. If no motion is observed, see table 4-16.
	MCCC	a. Observe SITE HARD indicator AMBER.
		b. Start stopwatch to time opening and closing of boiloff valve.
	DMCCC	Monitor missile tank pressures and control the Boiloff valve.
	BMAT	Monitor abort sequence.
	MFT	Observe launcher platform lowering .
	EPPT	Monitor PRCP.
4		**⚡ CAUTION**
		A minimum of 20 PSI differential pressure between LO$_2$ and fuel tank pressure gauges must be maintained with a full LO$_2$ load. While the launcher platform is in the lowering sequence, LO$_2$ tank pressure may be controlled only by opening and closing the boiloff valve. During this period of time, the LO$_2$ tank derives pressurization from LO$_2$ boiloff. No manual control of the fuel tank is possible at this time. fuel tank pressure decays below 53 PSI, the helium control charging unit will supply regulated helium pressure to maintain fuel tank pressure at approximately 53 PSI. Failure to maintain proper differential pressure may result in bulkhead reversal and possible loss of missile.
		NOTE
		If the boiloff valve cannot be readily observed, the AUTOMATIC pushbutton shall be depressed 5 minutes 35 seconds after SITE HARD indicator illuminates AMBER, or if monitored on television, when the boiloff has lowered to ground level.
		NOTE
		If boiloff valve fails to close when AUTOMATIC pushbutton is depressed, allow down sequence to continue.

Table 3-17. Amplified Abort Procedures (CONT)

STEP	CREW POS	REQUIREMENT
4 (CONT)		ABORT EXTERNAL indicator AMBER. Indicator will illuminate AMBER when abort has been started, not abort external, and launcher platform down and locked. If indicator fails to illuminate AMBER, see table 4-16.
	MCCC	Observe ABORT EXTERNAL indicator AMBER.
	DMCCC	Monitor missile tank pressure gauges for phase II.
	BMAT	Monitor LO$_2$ TANKING (PANEL 1 and PANEL 2).
	MFT	Observe LAUNCHER PLATFORM DOWN COMPLETED RUN AND LOCKED indicator GREEN.
	EPPT	Monitor PRCP.
5		PNEUMATICS IN PHASE 1 indicator AMBER. Indicator will illuminate AMBER when launcher platform is down and locked and pneumatics not in phase I.
	MCCC	Observe PNEUMATICS IN PHASE 1 indicator AMBER.
	DMCCC	Observe LO$_2$ tank pressure decreases to approximately 4 PSI and fuel tank pressure rapidly increases to approximately 63.9 PSI (phase II).
	BMAT	Monitor LO$_2$ TANKING (PANEL 1 and PANEL 2).
	MFT	Monitor CSMOL and FRCP.
	EPPT	Monitor PRCP.
6		ABORT EXTERNAL indicator GREEN. Indicator will illuminate GREEN when launcher platform is down and locked, abort has been started, pneumatics in phase II selected, missile power not internal, and missile not away. If indicator fails to illuminate GREEN, observe LO$_2$ DRAIN COMPLETE indicator to illuminate AMBER.
	MCCC	Announce, "ABORT EXTERNAL GREEN".

Table 3-17. Amplified Abort Procedures (CONT)

STEP	CREW POS	REQUIREMENT
6 (CONT)	MCCC (CONT)	Acknowledge DMCCC announcement, "PRESSURES STABILIZED IN PHASE II".
	DMCCC	Monitor missile tank pressures for phase II pressures. Announce, "PRESSURES STABILIZED IN PHASE II" after MCCC announces ABORT EXTERNAL GREEN.
	BMAT	Observe LO$_2$ TANKING (PANEL 1 and PANEL 2) for the following indications: a. STORAGE TANK VENT VALVE N-5 indicator extinguished or GREEN. NOTE The following indications will not occur until LO$_2$ storage tank pressure decreases to less than 25 PSI: b. LO$_2$ STG TNK PRESSURE indicator GREEN. c. DRAIN VALVE L-16 indicator GREEN. d. AIRBORNE FILL & DRAIN VALVE indicator GREEN.
	MFT	Monitor CSMOL and FRCP.
	EPPT	Monitor PRCP.
7		**CAUTION** If LO$_2$ DRAIN COMPLETE indicator illuminates GREEN less than 40 minutes after being AMBER, depress EMERGENCY pushbutton, maintain fuel tank pressures greater than 55 PSI, and perform manual abort (table 4-10, action 3). Failure to comply will allow the PSC to go to phase I pressures which may cause bulkhead reversal and possible loss of missile. The LO$_2$ drain sequence must be closely monitored. NOTE A rise in missile fuel tank pressure, as indicated on FUEL TANK pressure gauge in pressurization patch, should occur approximately 25 to 29 minutes after LO$_2$ DRAIN COMPLETE indicator AMBER. This pressure rise indicates than LO$_2$ or LN$_2$ has drained sufficiently to uncover the bulkhead. The DMCCC should monitor the rise in pressure and notify the MCCC.

Table 3-17. Amplified Abort Procedures (CONT)

STEP	CREW POS	REQUIREMENT
7 (CONT)		LO$_2$ DRAIN COMPLETE indicator AMBER. Indicator will illuminate AMBER when drain valve L-16 is not closed, launcher platform is down and locked, abort external signal is present, LO$_2$ tank pressure is not less than 1.65 PSI, and missile is not away. If indicator fails to illuminate AMBER, depress EMERGENCY pushbutton, maintain phase II pressures, and see table 4-10, action 3.
	MCCC	Announce, "MARK - LO$_2$ DRAIN COMPLETE AMBER". Acknowledge BMAT announcement, "LO$_2$ TANKING PANELS NORMAL".
	DMCCC	a. Start timing LO$_2$ drain (LO$_2$ drain time 45 minutes). b. Monitor pressure gauges for phase II pressures and notify MCCC when fuel tank pressure rises, indicating that LO$_2$ or LN$_2$ has drained off the bulkhead.
	BMAT	a. Monitor drain sequence on LO$_2$ TANKING (PANEL 1 and PANEL 2). b. Announce, "LO$_2$ TANKING PANELS NORMAL".
	MFT	Observe the following indications as they occur on CSMOL: a. SILO DOORS OPEN indicator extinguished. b. SILO DOORS CLOSE indicator GREEN. c. CRIB HORIZONTAL LOCK indicator extinguished. d. CRIB HORIZONTAL UNLOCK indicator GREEN. e. CRIB VERTICAL LOCK indicator extinguished. f. All indicators extinguished.
	EPPT	Monitor PRCP.
8		HELIUM VENT COMPLETE indicator AMBER. Indicator will illuminate AMBER when inflight helium bottle pressure is above 50 PSI, bottles are venting, and abort external signal is present. If indicator fails to illuminate AMBER, the EMERGENCY pushbutton shall be depressed and phase II pressure maintained. After LO$_2$ DRAIN COMPLETE indicator illuminates GREEN, a manual abort shall be performed (table 4-10, action 4). HELIUM VENT COMPLETE indicator will remain AMBER approximately 17 minutes before illuminating GREEN.
	MCCC	Observe HELIUM VENT COMPLETE indicator AMBER.
	DMCCC	Monitor missile tank pressures.

Table 3-17. Amplified Abort Procedures (CONT)

STEP	CREW POS	REQUIREMENT
8 (CONT)	BMAT	Monitor LO$_2$ TANKING (PANEL 1 and PANEL 2).
	MFT	Monitor CSMOL and FRCP.
	EPPT	Monitor PRCP.
9	MCCC	SITE HARD indicator GREEN. Indicator will illuminate GREEN with launcher platform down and locked and site hard signals present. If indicator fails to illuminate GREEN, observe HYDRAULIC 40 HP PUMP PRESSURE indicator on CSMOL. If HYDRAULIC 40 HP PUMP PRESSURE indicator is extinguished RESET PROGRAMMER key switch on CSMOL shall be positioned to ON, 40 HP pump shall be stopped, and malfunction trouble shooting performed after LO$_2$ drain COMPLETE indicator is illuminated GREEN. If HYDRAULIC 40 HP PUMP PRESSURE indicator is GREEN, manually close silo overhead doors (table 4-10, action 2, step 2 only). a. Announce, "SITE HARD GREEN". b. Announce, "M-1, OPEN LCC BLAST CLOSURES." c. Acknowledge MFT announcement, "LCC BLAST CLOSURES OPEN". d. Announce, if both generators are operating, "L-1, SHUT DOWN ONE GENERATOR". e. Acknowledge EPPT announcement, "GENERATOR SHUT DOWN".
	DMCCC	Monitor missile tank pressures.

Changed 21 October 1964

Table 3-17. Amplified Abort Procedures (CONT)

STEP	CREW POS	REQUIREMENT
9 (CONT)	BMAT	Monitor drain sequence on LO$_2$ TANKING (PANEL 1 and PANEL 2).
	MFT	a. Acknowledge MCCC announcement, "M-1, OPEN LCC B L A S T CLOSURES". b. Depress LCC BLAST CLOSURE MANUAL OPERATION OPEN pushbutton on FRCP. Observe LCC AIR INTAKE OPEN. LCC AIR EXHAUST OPEN, and LCC STAIRWELL AIR EXHAUST OPEN (NA OSTF-2) indicators are illiuminated GREEN. c. Announce, "LCC BLAST CLOSURES OPEN".
	EPPT	a. Acknowledge MCCC announcement, "L-1, SHUT DOWN ONE GENERATOR". b. Shut down one generator at PRCP, if both generators are operating, in accordance with table 3-11, item 2. c. Announce, "GENERATOR SHUTDOWN".
10		HELIUM VENT COMPLETE indicator GREEN. Indicator illuminates GREEN when pressure in inflight helium spheres has been less than 50 PSI for more than 10 minutes. If indicator does not illuminate GREEN, the EMERGENCY pushbutton shall be depressed and phase II pressures shall be maintained. After LO$_2$ DRAIN COMPLETE indicator illuminates GREEN, a manual abort shall be performed (table 4-10, action 4).
	MCCC	Observe HELIUM VENT COMPLETE indicator GREEN.
	DMCCC	Monitor missile tank pressures.
	BMAT	Monitor abort sequence.
	MFT	Monitor FRCP.
	EPPT	Monitor PRCP.

T.O. 21M-HGM16F-1

Table 3-17. Amplified Abort Procedures (CONT)

STEP	CREW POS	REQUIREMENT
11		LO₂ DRAIN COMPLETE indicator GREEN. Indicator illuminates GREEN when valve L-16 is closed and the 45-minute timer has expired. If indicator does not illuminate GREEN 50 minutes after illuminating AMBER position REMOTE LOCAL switch on LO₂ TANKING (PANEL 1) to LOCAL. A manual abort shall then be performed (table 4-10, action 4). If indicator illuminates GREEN prior to 40 minutes of LO₂ drain, The DMCCC shall immediately depress EMERGENCY pushbutton and maintain phase II pressures. A manual abort shall then be performed (table 4-10, action 3).
	MCCC	Announce, "LO₂ DRAIN COMPLETE GREEN".
	DMCCC	a. Acknowledge MCCC announcement, "LO₂ DRAIN COMPLETE GREEN". b. Observe fuel tank pressure is decreasing to phase I pressure. Fuel tank pressure will decrease at a steady rate to approximately 12 PSI.
	BMAT	Observe the following indication on LO₂ TANKING (PANEL 2). b. AIRBORNE FILL & DRAIN VALVE indicator AMBER. a. DRAIN VALVE L-16 indicator AMBER.
	MFT	Monitor FRCP.
	EPPT	Monitor PRCP.
12		HYDRAULIC SYSTEM OFF indicator AMBER. Indicator illuminates AMBER with LO₂ drain complete signal present and hydraulic pressure greater than 1750 PSI.
	MCCC	Observe HYDRAULIC SYSTEM OFF indicator AMBER.
	DMCCC	Monitor missile tank pressures for phase I pressures.
	BMAT	Monitor abort sequence.
	MFT	Monitor FRCP.
	EPPT	Monitor PRCP.

Table 3-17.　Amplified Abort Procedures (CONT)

STEP	CREW POS	REQUIREMENT
13		HYDRAULIC SYSTEM OFF indicator GREEN. Indicator illuminates GREEN when LO$_2$ drain complete signal is present, oil not evacuated, and hydraulic pressure less than 1750 PSI. If indicator fails to illuminate GREEN, position HPU circuit breaker on essential motor control center on silo level 2 to OFF.
	MCCC	Observe HYDRAULIC SYSTEM OFF indicator GREEN.
	DMCCC	Monitor missile tank pressures for phase I pressures.
	BMAT	Monitor abort sequence.
	MFT	Monitor FRCP.
	EPPT	Monitor PRCP.
14		PNEUMATICS IN PHASE 1 indicator GREEN. Indicator illuminates GREEN when LO$_2$ drain is complete and pneumatics is in phase I. If indicator fails to illuminate GREEN, depress EMERGENCY pushbutton, adjust pressures to phase I, and return system to automatic. If system fails to return to phase I, see table 4-10, action 4, step 3.
	MCCC	Observe PNEUMATICS IN PHASE 1 indicator GREEN. Acknowledge DMCCC announcement, "TANK PRESSURES STABILIZED IN PHASE I".
	DMCCC	a.　Monitor missile tank pressures for phase I pressures. b.　Announce, "TANK PRESSURE STABILIZED IN PHASE I".
	BMAT	Monitor abort sequence.
	MFT	Monitor FRCP.
	EPPT	Monitor PRCP.

Table 3-17. Amplified Abort Procedures (CONT)

STEP	CREW POS	REQUIREMENT
15		ABORT COMPLETE indicator GREEN. Indicator illuminates GREEN when start abort signal is present, not start count-down, and not ready for countdown.
	MCCC	a. Announce "ABORT COMPLETE GREEN".
		b. Direct that communications be restored by positioning the communications disconnect panel key to RESTORE.
		c. Direct silo safety inspection.
	DMCCC	Monitor missile tank pressures.
	BMAT	Stand by.
	MFT	Stand by.
	EPPT	Stand by.
16	MCCC	NOTE Ambient helium bottle will retain pressure for approximately 2 hours to provide purge pressure to booster engine turbopump seals before auto-matically venting. Gaseous oxygen vent fan and engine thrust section heaters will continue to operate during this period. CAUTION Missile battery must be removed within 10 hours after activation. Failure to comply will result in overheating of the battery and possible fire.

Table 3-17. Amplified Abort Procedures (CONT)

STEP	CREW POS	REQUIREMENT
16 (CONT)	MCCC (CONT)	**NOTE** If missile guidance system was in aligned status when abort was initiated, the system must be removed from aligned status and re-aligned before another countdown can be initiated. Refer to table 3-9. (Sight tube must be in position.) Fluid and gas level must be verified to ensure sufficient supply. **NOTE** If missile guidance system was on memory during countdown due to a nuclear blast, the alignment group calibration must be verified prior to initiating another countdown. **CAUTION** If a training launch is not resumed within 72 hours, recharge the inert fluid injection modules and purge the booster thrust chambers. If launch is resumed within 72 hours, modules need only be vented (open vent valve until pressure is relieved then close vent valve and inlet valve).
	DMCCC	Monitor missile tank pressures in phase 1.
	BMAT	Stand by for instructions from MCCC.
	MFT	Stand by for instructions from MCCC.
	EPPT	Monitor PRCP. **NOTE** If FEEDER NUMBER 3 NON-ESSENTIAL BUS CONTROL SWITCH was tripped during countdown or abort, certain equipment shall be verified as operating or reset. See complex electrical system reset procedures table 4-7.

Table 3-18.   Amplified Postlaunch Abort Procedures (Tactical)

STEP	CREW POS	REQUIREMENT
1	MCCC	**CAUTION** Unpredicatable conditions, such as nuclear blast, fire, or damage to equipment may be present at this time. At discretion of MCCC, survey the silo and launcher platform prior to lowering launcher platform. a. Announce, "M-1, INITIATE MISSILE ENCLOSURE FOG SYSTEM". b. Announce, "ABORT START". c. Depress ABORT pushbutton. d. Deleted e. Observe the following indications: (1) ABORT COMPLETE indicator AMBER. (2) SITE HARD indicator AMBER.
	DMCCC	Stand by.
	BMAT	Stand by.
	MFT	a. Acknowledge MCCC announcement, "INITIATE MISSILE ENCLOSURE FOG SYSTEM". b. Depress MISSILE ENCLOSURE FOG ON pushbutton on FRCP. c. Monitor CSMOL.
	EPPT	a. Monitor PRCP. b. Deleted c. Deleted

Table 3-18. Amplified Postlaunch Abort Procedures (Tactical) (CONT)

STEP	CREW POS	REQUIREMENT
2	MCCC	a. Announce, "ABORT EXTERNAL AMBER". b. Observe, SITE HARD indicator GREEN. c. Announce, "M-1, OPEN LCC BLAST CLOSURES". d. If both generators are operating, announce, "L-1, SHUT DOWN ONE GENERATOR. **NOTE** At discretion of MCCC, silo inspection, applicable steps in section 7 of T.O. 21M-HGM16F-1CL-6, and postlaunch procedures in T.O. 21M-HGM16F-3CL-1 may be performed.
	DMCCC	Stand by.
	BMAT	Stand by.
	MFT	Stand by.
	EPPT	If two generators are operating, shut down one generator in accordance with table 3-11, item 2.

SECTION IV

EMERGENCY AND TROUBLE ANALYSIS PROCEDURES

TABLE OF CONTENTS

LIST OF TABLES

4-1. SCOPE.

4-2. This section contains the procedures to be used by the missile combat crew in the event a malfunction, abnormal indication, or emergency condition should occure during standby, countdown, or abort. These procedures are divided into two categories: procedures which difine steps to ensure safety of personnel and to place equipment in a safe configuiration when an emergency or malfunction occurs, and procedures for correcting or bypassing a malfunction to permit continuation of a tactical countdown. The first portion of this section contains a general discussion of safety and emergency considerations to aid as guidelines in the application of the emergency procedures which follow.

4-3. Missile crew members shall be thoroughly familiar with the contents and procedures in this section before performing missile crew alert duty. A crew member noting an abnormal or emergency situation is responsible for immediately reporting to the MCCC all relative information. Information reported should include details of the situation, actions being taken, recommendations and technical data reference if appropriate.

4-4. During a nontactical countdown, abort is mandatory when a malfunction occurs which prevents the sequence from continuing, unless specifically excepted in the emergency procedures. For malfunctions which do not prevent the sequence from continuing, countdown may be continued unless an abort is specified. Tactical countdown and abort sequence failures may require sending personnel into the silo to complete the sequence. Every safety precaution must be taken for hazardous conditions.

4-5. Disaster control procedures, as locally directed, will be limited to actions required to implement other directives and shall not duplicate actions outlined in this manual. Statements such as "Notify proper authorities" are used to aid in developing there supplemental procedures.

4-6. Abnormal or emergency conditions not covered by procedures in this manaul may occur. Sound judgment must be exercised in such cases. The correct handling of any emergency situation is best achieved by a thorough knowledge of equipment and procedures.

4-7. SILO ENTRY.

4-8. All personnel involved in these procedures must be aware at all times that these are emergency procedures and therefore are performed under circumstances of potentially extreme hazards.

4-9. Hazardous conditions may exist in the silo when fuel and liquid oxygen are aboard the missile or if the silo ventilation system has malfunctioned. Sending personnel to the silo during hazardous conditions is left to the discretion of the MCCC.

4-10. Prior to dispatching personnel to the silo the MCCC shall brief and ensure understanding of the existing emergency condition, safety precautions, communication procedures, emergency equipment and tools needed, and required crew coordination.

4-11. In addition to normal safety equipment requirements the following, although not all-inclusive, will assist in determing that crew members have the required safety and emergency items when entering the silo to accomplish emergency procedures contained in this section:

   a. Breathing equipment, and portable analyzers as indicated, shall be used when there is an indication or suspicion of the following:

     (1) High or low oxygen content (oxygen analyzer).

     (2) Diesel fuel vapors (explosimeter).

     (3) RP-1 fuel vapors (explosimeter).

     (4) Other toxic vapors.

     (5) Smoke.

   b. Leaks or spils involving cryogens require the following equipment:

     (1) Breathing equipment.

     (2) Protable oxygen analyzer.

     (3) Protective LO_2 safety coveralls, gloves, boots, and head piece.

   c. Emergency procedures directed towards bypassing a tactical countdown malfunction or abort malfunction may require the use of the following equipment:

     (1) PSM-6 multimeter.

     (2) Jumper wires.

     (3) Screwdriver.

     (4) Flashlight.

   d. Procedures for 400 CYCLE POWER indicator RED require a screwdriver.

   e. Missile Lift failure to lower procedures require a crowbar and a roll of tape.

4-12. Normally personnel shall not be allowed in the silo unaccompanied. However, due to the limited number of personnel assigned to a missile combat crew, it may not be possible to meet this criterion at all times. The MCCC will use discretion and consider the best interests of safety before sending a crew member into the silo alone. If possible, communications shall be maintained at all times between crew members in the silo and the launch control center (LCC).

4-13. EMERGENCY COMMUNICATION PROCEDURES.

4-14. Personnel entering the silo must establish and maintain contact with the LCC whenever possible. They must report arrival at a silo location and departure from the location, giving intended destination.

4-15. COMMUNICATION LOST BETWEEN SILO AND LAUNCH CONTROL CENTER DURING STANDBY. If communication is lost between silo and the LCC, observe COMMUNICATIONS DISABLE indicator on communications disconnect panel in launch control room.

　a. If indicator is extinguished, use public address system and direct personnel in the silo to contact the LCC using another maintenance net or an explosionproof jack. If contact cannot be made in this manner, use public address system and instruct personnel in the silo to return to the LCC. Dispatch available crew members to investigate the cause of communication failure.

　b. If COMMUNICATION DISABLE indicator is illuminated, position communications disconnect panel key in the COMM OVERRIDE position and re-establish communication on assigned maintenance net. If communication cannot be re-established, follow procedures in paragraph 4-16.

　c. If communication with the LCC on the assigned maintenance net is lost, personnel in the silo will immediately attempt to re-establish communication using an alternate net, explosionproof jacks, and(or) spare headsets. If communication cannot be re-established

and task being performed is not critical, return to LCC.

　d. During countdown, communication will automatically be disabled, except explosionproof jacks and panels. Do not override communication unless necessary during emergency conditions and then, only after consideration of possible explosion due to existing conditions.

4-16. Maintenance net 1 shall be reserved for emergency use or assistance and shall have answering priority over all other nets at the LCOC. Net 1 shall not be used for normal routine contact with the LCC. If signal on net 1 is observed at the LCOC and contact cannot be established, it will be assumed that someone in the silo needs emergency assistance and personnel shall be dispatched with a spare headset to determine the source of difficulty.

4-17. ALERT TONE FAILURE. If the 1000-cycle tone does not sound when the ALERT pushbutton is depressed, one of the following alternate methods shall be used to signal personnel to evacuate the silo and proceed to their countdown positions in the LCC:

　a. Public address system.

　b. Fire alarm warning system to sound a general alarm.

　c. Silo light switch in the silo entrance vestibule. (OFF and on three times at 1-second intervals.)

　d. WARNING HORN pushbutton on FRCP to signal the silo cap.

　e. Dispatch of persons to silo or silo cap to notify others.

4-18. EMERGENCY INTERSITE COMMUNICATIONS.

4-19. In the event of complete loss of AC power, intersite communication is no longer possible by normal methods due to loss of the required carrier signal. Emergency communication can be established with job control and other sites by use of the orderwire line located in the LCC communications room. During AC power loss this system is powered by the backup battery bank in the communications room.

4-20. SILO EVACUATION PROCEDURES

4-21. There are five primary reasons why it would become necessary to evacuate personnel from the silo under emergency conditions. These conditions are fire, high or low oxygen concentration, high nitrogen concentration, RP-1 fuel vapor concentration, and diesel fuel vapor concentration. Crew members shall immediately notify the LCC of any emergency condition in the silo. It is the responsibility of the MCCC to initiate and direct silo evacuation.

WARNING

Personnel shall not be allowed to enter or remain in an area where GOX concentration is above 35 percent except to rescue personnel or to perform essential emergency actions to safe the system. If high GOX concentration is indicated, frequent readings shall be taken with a portable oxygen analyzer to determine concentration level.

4-22. EMERGENCY EVACUATION OF SILO CAP OR SILO. Emergency evacuation of the silo cap or silo shall be accomplished at the discretion of the MCCC and shall be conducted as follows:

a. The MCCC shall initiate evacuation signals.

b. Deleted

c. When personnel on silo cap are alerted to evacuate, non-crew members shall proceed to the fall back area; crew members shall report to LCC level 2.

d. Whenever possible, personnel in the silo should return all equipment to a safe configuration before evacuating.

e. All personnel on silo levels 1 through 6 shall evacuate by the stairway and tunnel to LCC level 2.

f. All personnel on silo levels 7 and 8 shall immediately put on emergency breathing apparatus and use facility elevator to leave the silo. To facilitate evacuation procedures when working on these levels,

the elevator shall be left at the lowest level occupied when not otherwise in use.

g. The elevator operator shall visually check elevator entrance at each level enroute to level 2 and will stop at each entrance to the missile enclosure area (MEA) to determine if any personnel are in need of assistance. Personnel in the MEA should make themselves plainly visible.

h. All personnel shall report immediately to LCC level 2 for accountability. Non-crew members shall then proceed to level 1 of LCC.

i. The MFT shall ensure that everyone has evacuated the silo.

j. The MFT shall ensure that blast doors are closed as soon as evacuation of the silo has been completed.

4-23. SILO AND SILO CAP EVACUATION SIGNALS. One or all of the following methods, in order listed, (sub-paragraph a. through d.) shall be used to evacuate personnel from the silo during emergency situations. The silo cap evacuation signal (sub paragraph e.) shall be used when silo cap evacuation is necessary.

a. Public address system announcement directing that the silo be evacuated. Whenever possible, state the reason for evacuation.

b. Fire alarm warning system activated from fire alarm panel by depressing any operating fire zone TEST pushbutton.

c. Silo lights turned OFF and ON three times, at 1-second intervals, with switch in the silo entrance vestibule.

d. The 1000-cycle alert tone activated with the ALERT pushbutton on launch control console.

e. Activation of the PERSONNEL WARNING HORN and the PERSONNEL WARNING LIGHT with controls located on FRCP.

4-24. All evacuation signals should originate from the LCC. If contact cannot be established with the LCC and an emergency condition requiring evacuation is detected by personnel, one of the following methods of initiating evacuation signals may be used:

a. The public address system microphone located on silo level 2 near the tunnel entrance.

b. The 1000-cycle tone ALERT pushbuttons located near the explosionproof communication panel on silo level 7 and on the silo cap.

4-25. LAUNCH CONTROL CENTER EVACUATION PROCEDURES.

4-26. Emergency evacuation of the LCC may be necessary in the event of an imminent missile explosion or after an in-silo explosion which has resulted in contamination of the LCC. When the decision is made to evacuate, the following procedures will be adhered to:

a. All personnel shall don self-contained breathing apparatus.

b. MCCC shall notify command post of decision to evacuate, if practicable.

c. Appropriate steps shall be taken to safeguard classified materials.

d. If communications are available with the fall back area, personnel in the LCC shall determine when evacuation may be safely accomplished and of safest evacuation routes from the LCC entry.

e. Personnel will normally evacuate by the stairway to the aboveground entrance. If AC power has been lost, the outer entrapment door NO. 1 key must be used to open the entrapment door. In case evacuation cannot be made by the stairway, the escape hatch on level 1 of the LCC must be used.

f. Personnel shall proceed to the fallback area away from the silo. A ladder or other device may be required to clear the perimeter fence.

4-27. FIRES.

4-28. FIRES IN THE SILO. The entire complex can be monitored for fires through alarm and detector networks. There are a total of 38 fire detectors installed at essential points in the LCC and silo. Seven detectors are within the MEA. The detectors are the combined rate-of-rise and fixed temperature units, actuated by either a temperature increase exceeding 15°F per minute or by a temperature exceeding the fixed temperature limit. These detectors automatically initiate a fire alarm warning on the annunciator unit in the LCC and sound alarm bells and illuminate red lights if a fire is detected. In addition, four manually initiated fire alarm boxes are located on silo levels 2, 4, 6 and 8.

4-29. If a fire is discovered which has not automatically activated the fire detection system, immediately notify the LCC of the nature and location of the fire and proceed to combat the fire. At the discretion of the MCCC, assistance will be dispatched to aid in combating the fire. If contact with LCC cannot be established, the alarm may be activated at any one of the manual fire alarm boxes located on silo levels 2, 4, 6 and 8 adjacent to the elevator.

4-30. The three general classes of fires that may be encountered within the complex and the method to be used to combat the fire are as follows:

a. Class A fires. These are fires involving wood, paper, cloth or other cellulose materials. They will be combated with CO_2 or water.

b. Class B fires. These are fires involving petroleum hydrocarbons such as hydraulic fluid, RP-1 fuel and diesel fuel. Class B fires will be combated by isolating the fire from the fuel source if possible and extinguishing with CO_2. Water shall not be be used on class B fires.

c. Class C fires. These are fires involving electrical equipment such as motors, wiring, power panels, and transformers. Class C fires will be combated by isolating the fire from the electrical source if possible and extinguishing with CO_2. Water shall not be used on class C fires.

4-31. PYROPHORIC MATERIALS. Pyrophoric rocket engine ignitors must be handled with extreme care as they contain highly explosive material. These materials, triethylaluminum (TEA) and triethyloborane (TEB), are extremely destructive to living tissue and, on contact with the skin, produce a combined effect of dehydration and thermal burn. If Pyrophoric materials come in contact with the skin, flush the area with large amounts of water and administer medical treatment immediately. In the event a fire occurs, isolate the fire from its source if possible and control its spreading. CO_2 extinguishers may be used to extinguish small fires, however, water spray should be used for large fires. Do not use carbon tetrachloride (CCL_4) extinguishers as a violent reaction will result and poison phosgene gas may be formed. After extinguishing the fire, survey area closely for several hours in case of re-ignition. The following shall be observed when working with pyrophoric materials:

a. Wear protective clothing at all times when handling expended or unexpended ignitors.

b. Wear eye protection when handling missile rocket engine solid propellant gas generators and initiators.

4-32. MISSILE RE-ENTRY VEHICLE ACCIDENT OR FIRE. If the re-entry vehicle is involved in an accident or a fire occurs near or envelops the re-entry vehicle, danger from possible high explosive detonation and possible radioactive contamination exists. Any accident or fire involving the re-entry vehicle requires immediate notification of proper authorities and implementation of disaster control plans.

4-33. HAZARDS.

4-34. LIQUID OXYGEN HAZARDS. Considerations for liquid oxygen hazards include the following:

a. Immediate freezing of any part of human body it contacts.

b. Extreme fire hazard of any combustible material. Ignition is possible from any electrical static charge, impact, or heat source.

c. Explosion hazards when in contact with any hydrocarbon. Explosive gels may form.

d. Saturation of clothing or other porous materials with oxygen increasing their flammability.

e. Retention of high oxygen concentrations in porous materials, human hair, etc., for unknown period of time after atmosphere returns to normal. If clothing becomes saturated with gaseous oxygen use emergency water shower.

4-35. LIQUID NITROGEN HAZARDS. Considerations for liquid nitrogen hazards include the following:

a. Immediate freezing of any part of the human body it contacts.

b. Extreme cold and constant exaporation to the gaseous state resulting in a high concentration of nitrogen in the atmosphere. Oxygen deficiency, particularly at lower levels, may result in suffociation.

4-36. RP-1 AND DIESEL FUEL HAZARDS. Considerations for RP-1 and diesel fuel hazards include the following:

a. Toxic vapors can be fatal if in large concentrations.

b. Possibility of explosive vapors.

c. Possibility of explosive gels in conjunction with a LO_2 leak.

d. The need for adequate ventilation in areas when RP-1 and diesel fuel is stored or handled.

4-37. ELECTRICAL HAZARDS. Considerations for electrical hazards include the following:

a. Electrical shock, electrocution of personnel, or damage to equipment. Extreme caution must be exercised when performing operations in vicinity of voltages.

b. Adjustments or repairs in electrical cabinets shall not be made with power on, unless required for completion of the task.

c. Personnel shall not wear metal rings or watches when working on electrical equipment.

d. First consideration in rescue of personnel exposed to electrical shock is to turn off power source before rendering aid.

4-38. NOISE HAZARDS. Consideration for noise hazards include the following:

a. Permanent damage to hearing can result from exposure to excessive noise levels.

b. Protective ear devices will be worn during all venting operations involving high pressure gases.

c. Prolonged exposure to moderate noise levels may cause minor damage. Personnel shall wear protective ear devices when total accumulative time in the operating diesel generator room exceeds 10 minutes within a 24 hour period.

4-39. **HIGH PRESSURE HAZARDS.** Considerations for high pressure gas and liquids include the following:

a. Flexible high-pressure line shall be sandbagged every six feet and the ends of the line adequately secured to prevent line whipping in the event of line breakage or accidental disconnection.

b. Safety glasses and hard hats shall be worn at all times when performing adjustments or maintenance on pressurized equipment.

c. Fittings or components shall not be tightened or loosened while under high pressure.

d. Hands shall not be used in testing for high pressure leaks.

4-40. **EMERGENCY BREATHING APPARATUS.**

4-41. Emergency breathing apparatus containing 5 to 7 minutes supply of air are located throughout the silo. Breathing apparatus containing 30 minutes supply of air are located in the LCC. Breathing apparatus shall be donned whenever personnel suspect contamination of the admosphere within the silo or LCC. Breathing apparatus will be checked daily. Recharging may be accomplished using the air supply bottles (K bottles). Crew members must be proficient in the use of breathing apparatus to enable donning without reference to written instructions Procedures to don the 30-minute type breathing apparatus are in the lid of the storage container. To don the 5 to 7-minute emergency breathing apparatus, proceed as follows:

a. Grasp strap labeled PULL located at top of pack and pull down to expose breathing apparatus.

b. Verify supply valve is open (rotate counterclockwise).

c. Place neckband over head and around neck.

d. Place mask over nose and mouth and breathe normally.

e. Attach unit to chest with harness provided.

4-42. **FIRST AID.**

4-43. When practicable, apply first aid at location. If location is hazardous, it may be necessary to move the injured person to a safe area prior to administrating first aid. Personal judgment will determine how much first aid will be administered before moving injured persons. Observe the following first aid practices:

a. Do not aggravate injuries by hasty action.

b. For wounds: Stop bleeding, protect wound, and treat for shock.

c. For fractures: Immobilize fracture. Do not attempt to set or realign broken bones, and treat for shock.

d. For fire, pyrophoric, and cryogenic burns: Protect burned area, do not attempt to remove clothing or other material stuck to burned area. Deluge pyrophor and cryogen contacts with water. Treat for shock.

e. For unconsciousness: If breathing, treat for shock. If not breathing, apply artificial respiration.

f. For shock: Lower head, elevate feet, and cover with warm materials.

g. Notify medical department immediately.

4-44. **EMERGENCY PROCEDURES TABLES.**

4-45. Emergency procedures contained in the following paragraphs and tables are for detailed study and reference. Missile combat crew members must become thoroughly familiar with these procedures in order to understand and implement the step-by-step actions contained in the associated abbreviated emergency checklists. The numbered steps in the tables are identical to the steps in the corresponding sections of the emergency checklist. An explanation of the numbered step in the tables is given immediately preceding each step. Steps in procedures common to all emergencies do not normally appear in the abbreviated checklists or the tables. Steps, such as directing the MCCC to brief personnel prior to starting emergency actions, announcing indications of malfunctions, directing personnel to a location, notifying outside agencies, are omitted. These steps are left to be accomplished as normal procedures to effect quick, safe, and coordinated effort by the missile combat crew. Abbreviated checklist sections condensed from these paragraphs and tables are contained in T.O. 21M-HGM16F-1CL-6.

4-46. The abbreviated checklist derived from the tables are designated as sections and are numbered in the same sequence as their corresponding tables.

4-47. Table 4-1 through 4-4 (sections 1 through 4 of emergency checklists) direct emergency actions during operating periods listed below:

 a. Table 4-1 (emergency checklist section 1)— Standby.

 b. Table 4-2 (emergency checklist section 2)— Start countdown to MISSILE LIFT UP & LOCKED indicator AMBER.

 c. Table 4-3 (emergency checklist section 3)— MISSILE LIFT UP & LOCKED indicator AMBER to MISSILE LIFT UP & LOCKED indicator GREEN.

 d. Table 4-4 (emergency checklist section 4)— Abort sequence.

4-48. Each of these four tables and checklist sections are composed of items directing crew reactions to non-sequential emergencies or abnormal situations, as reflected by one or more of the following: Facilities remote control panel; power remote control panel; fire alarm panel; and the malfunction patch, standby status patch, and pressure patch of the launch control console. Although indications of a given emergency are generally the same irrespective of when it occurs, the crew reaction may vary according to the specific time phase.

4-49. The emergency checklist should be left opened to the section (1, 2, 3, or 4) corresponding to the current operation period. (See paragraph 4-47, steps a through d.) A standard item number is used in each of the four tables and checklist sections for a given emergency. For example: PRESSURE MODE RED is item NO. 1 in each of the tables and checklist sections. Each item contains immediate steps necessary to ensure a safe equipment configuration. The items contained in emergency checklist sections 1, 2, 3 and 4 and table 4-1 through 4-4 are resigned to direct crew reactions through all subsequent phases of countdown, commit, and abort when specific actions are necessary.

Therefore, if a malfunction occurs during countdown and the item in section 2 of the emergency checklist contains the crew actions through commit and abort, the emergency checklist will be kept open to that section throughout the commit and abort sequence. However, if another malfunction of a different nature should occur, the crew must refer to the appropriate checklist section corresponding with the time phase to determine reaction to that particular malfunction. To avoid duplication, operations common to several emergency procedures, such as emergency stretch or manual abort, are presented only once and as separately numbered tables and checklist sections. A table or checklist section item may therefore reference another table or checklist section which contains a needed sequence, common to other procedures. The remaining tables and checklists sections contain crew operating procedures used to correct or bypass the effects of an abnormal sequential light indication occurring occuring in the countdown or abort sequence.

4-50. Certain critical emergencies require immediate response actions on the part of the missile crew in order to avoid damage to equipment or the development of a hazardous condition. Such immediate response actions are distinguished as bold print items in the tables and emergency checklist and contain lettered or numbered steps which shall be committed to memory. Although word-for-word memorization is not mandatory, crew members must be able to accomplish all necessary actions quickly and accurately. All bold print items shall be performed from memory without reference to checklists and in the order in which the steps appear in the checklist. Upon completion of the memorized items, reference shall be made to the applicable checklist section which shall then be completed in its entirety. At this time, memory items which have already been completed shall be verified.

4-51. The normal countdown and abort checklists, T.O. 21M-HGM16F-1CL-1 through T.O. 21M-HGM16F-1CL-5, list immediate action or emergency checklist section references in the lift-hand column for sequential malfunctions which may appear on the countdown, commit, or abort patch of the launch control console, LO$_2$ tanking panels, or FRCP. Upon noting abnormal indications on the launch control console, the crew commander shall direct the crew to the appropriate action or section of the emergency checklist. These actions will place the equipment in a safe configuration or if during a tactical countdown, correct or bypass the malfunction to enable a launch.

4-52. The emergency checklists are the demand-response type. The demand-response steps are also contained in the tables. All crew member checklists are identical. Each step in the checklist and tables is preceded by a call sign identifier, indicating the crew member responsible for directing or accomplishing the step. The following call sign identifiers are used:

(C) — Commander.

(D) — Deputy commander.

(A-1) — BMAT.

(M-1) — MFT.

(L-1) — EPPT.

4-53. The commander shall direct appropriate actions and monitor crew accomplishment of these actions. Bold print items are memory items and shall be accomplished as applicable without direction or reference to the checklist.

4-54. EMERGENCY MISSILE TANK PRESSURIZATION.

*4-55. During standby, when PRESSURE MODE indicator is illuminated GREEN and an unscheduled rapid pressure change or any pressure outside of the allowable pressures is observed, depress EMERGENCY pushbutton immediately. Manually correct missile tank pressures by depressing the appropriate RAISE or LOWER pushbutton.

*4-56. If PRESSURE MODE indicator illuminates RED, the pressure control system has switched to emergency and immediate action shall be taken. The PRESSURE MODE indicator illuminates RED under one or more of the following conditions:

a. Prior to switching to internal pressurization (before PNEUMATICS INTERNAL indicator on launch control console illuminates GREEN):

(1) Pneumatics standby on and missile fuel tank pressure less than 8 PSI.

(2) Missile fuel tank pressure less than 53 PSI after having been greater than 53 PSI for 20 seconds.

(3) Missile fuel tank pressure greater than 67 PSI.

(4) Missile liquid oxygen tank pressure less than 1.65 PSI.

(5) Missile liquid oxygen tank pressure greater than 30.5 PSI.

(6) Pneumatics standby on and differential pressure less than 2.3 PSI.

(7) EMERGENCY pushbutton depressed.

(8) Instrument air pressure less than 50 PSI (INSTRUMENT AIR 50 PSI indicator on PNEUMATICS (PANEL 1) of control-monitor group 1 of 4 illuminated RED).

(9) Pressure system control normal supply pressure less than 1450 PSI (HELIUM SUPPLY 1450 PSI indicator on PNEUMATICS PANEL 1 of control-monitor group 1 of 4 illuminated RED during countdown or gaseous nitrogen supply below 1450 PSI during standby).

b. After switching to internal pressurization (after PNEUMATICS INTERNAL indicator on launch control consol illuminates GREEN) and prior to MISSILE LIFT UP & LOCKED indicator on launch control console illuminates GREEN:

(1) Missile fuel tank pressure greater than 67 PSI.

(2) Missile fuel tank pressure less than 53 PSI.

(3) Missile liquid oxygen tank pressure greater than 30.5 PSI.

(4) Missile liquid oxygen tank pressure less than 23 PSI.

(5) EMERGENCY pushbutton depressed.

c. After EMERGENCY pushbutton depressed and system is returned to automatic mode:

(1) Missile liquid oxygen tank pressure greater than 30.5 PSI.

(2) Missile fuel tank pressure greater than 67 PSI.

(3) Differential pressure less than 2.3 PSI.

(4) EMERGENCY pushbutton depressed.

d. After ABORT EXTERNAL indicator illuminates AMBER:

(1) Missile fuel tank pressure less than 8 PSI.

(2) Missile liquid oxygen tank pressure less than 1.65 PSI.

(3) Instrument air pressure less than 50 PSI.

(4) Missile fuel tank pressure less than 53 PSI 7 seconds after ABORT EXTERNAL indicator illuminates AMBER.

(5) Helium supply less than 1450 PSI 7 seconds after ABORT EXTERNAL indicator illuminates AMBER.

(6) Differential pressure less than 2.3 PSI.

(7) Missile fuel tank pressure greater than 67.0 PSI.

(8) Missile liquid oxygen tank pressure greater than 30.5 PSI.

(9) EMERGENCY pushbutton depressed.

4-57. Under normal conditions, missile fuel and oxidizer tank pressures are maintained automatically and no operator action is necessary. However, if the PRESSURE MODE indicator on the launch control console illuminates RED or if the automatic control system fails, manual actions are required. Therefore, tank pressure gauges and the PRESSURE MODE indicator shall be monitored at all times. The emergency circuit is automatically operated by pressure switches in the helium control charging unit which senses missile oxidizer and fuel tank pressures and by an air borne pressure switch which senses differential pressure. The pressure indicators on the launch control console have no automatic function in relation to activation of the emergency circuits and are obtained from airborne pressure transducers. The TANK DIFFERENTIAL PRESSURE gauge requires a missile-on-stand signal and 115-volt, 400-cycle power to operate a control-monitor group amplifier. Pressurization of the missile tanks is manually controlled by depressing the FUEL or LO$_2$ RAISE or LOWER pushbuttons. If the PRESSURE MODE indicator is illuminated RED, the EMERGENCY pushbutton need not be depressed to enable the RAISE or LOWER pushbuttons. If the

PRESSURE MODE indicator is not illuminated RED, the EMERGENCY pushbutton must be depressed to enable the RAISE or LOWER pushbuttons. Depressing the AUTOMATIC pushbutton after the emergency condition is corrected, returns the system to automatic control. After PNEUMATICS INTERNAL indicator has illuminated GREEN, the RAISE and LOWER pushbuttons are ineffective unless the ABORT pushbutton is depressed and ABORT EXTERNAL indicator has illuminated AMBER. After MISSILE LIFT UP and LOCKED indicator illuminates GREEN, the EMERGENCY pushbutton is enabled when MISSILE LIFT UP & LOCKED indicator illuminates GREEN plus 15 seconds, or by depressing the ABORT pushbutton if the ABORT indicator is illuminated RED. Item 1 of tables 4-1, 4-2, 4-3, and 4-4 provide procedures to be followed in the event PRESSURE MODE indicator illuminates RED.

NOTE

If EMERGENCY pushbutton is depresed during launcher platform up sequence, the sequence will stop and the boiloff valve will be enabled. During the launcher platform down sequence, if within 33 inches of the up and locked position, depressing the EMERGENCY pushbutton will stop the sequence. If the launcher platform is below 33 inches of the up and locked position, the down sequence will continue into the down locks, and boiloff valve will be enabled.

CAUTION

If TANK DIFFERENTIAL PRESSURE gauge should drop to zero due to a 400-cycle power failure, LO$_2$ TANK and FUEL TANK pressure gauges must be monitored closely to ensure that differential pressure does not fall below 2 PSI. If LO$_2$ tank loading has started or is completed, the head of liquid oxygen (or LN$_2$) will be exerting a maximum of 18 PSI pressure on the intermediate bulkhead. Therefore fuel tank pressure shall be maintained at least 20 PSI greater than liquid oxygen tank pressure to prevent bulkhead reversal.

4-58. BULKHEAD REVERSAL.

4-59. If the TANK DIFFERENTIAL PRESSURE gauge on the launch control console indicates zero, the PRESSURE MODE indicator is illuminated RED, and the LO$_2$ TANK PRESSURE is equal or higher than FUEL TANK PRESSURE, suspect bulkhead reversal. Follow procedures for PRESSURE MODE indicator RED and attempt to manually maintain proper missile tank pressures. If differential pressure cannot be maintained suspect bulkhead rupture and attempt to maintain 3.5 PSI in both tanks.

4-60. DIESEL GENERATOR MALFUNCTION.

4-61. Loss of a diesel generator may be anticipated if one of the following indications are observed:

a. AMMETER AC on PRCP indicates rapidly falling current.

b. DIESEL GEN D-60 (or D-61) OVERSPEED LO LUBE OIL HI-TEMP indicator on FRCP illuminates RED.

4-62. Item 2A of tables 4-1 through 4-4 provides procedures to be followed in the event of a diesel generator malfunction with the alternate generator in standby. Item 2B of tables 4-1 through 4-4 provides procedures to be followed in the event of a diesel generator malfunction with the alternate generator not in standby. Item 2C of tables 4-1 through 4-4 provides procedures to be followed in the event of a diesel generator malfunction when generators are operating in parallel.

4-63. LOSS OF AC POWER.

4-64. Loss of AC power is indicated by the following:

a. Loss of normal AC lighting.

b. Alarm sounds on the launch control console, fire detection panel, blast detection cabinet, and communications panel (in communications room).

c. 400 CYCLE POWER indicator RED.

d. ENGINES and GROUND POWER indicator RED.

e. TANK DIFFERENTIAL PRESSURE gauge indicating zero PSI.

f. GUIDANCE FAIL indicator RED.

g. FLIGHT CONTROL & R/V indicator RED.

h. POD AIR CONDITIONING indicator AMBER.

i. FACILITIES & MISSILE LIFT indicator RED.

j. NUCLEAR BLAST indicator AMBER.

k. MISSILE LIFT FAIL indicator RED.

l. All FRCP indicators extinguished.

NOTE

If the diesel has not shutdown, D-60 and/or D-61 ON indicator may be illuminated.

m. MAIN BREAKER CONTROL SWITCH on PRCP tripped.

n. AMMETER AC and(or) VOLTAGE meter on PRCP zero.

o. HYD PNEU & LN$_2$/HE indicator RED.

4-65. Item 3 of tables 4-1 through 4-4 contains procedures to be followed in the event of AC power loss.

4-66. LOSS OF UMBILICALS.

4-67. Loss of all umbilicals is positively indicated by all three missile tank pressure gauges indicating zero, R/V SAFE indicator RED, AUTOPILOT FAIL indicator RED, and LO$_2$ & FUEL indicator RED. Other console indications may be present depending on when umbilicals were ejected. Loss of 600U4 only will result in LO$_2$ & FUEL and AUTOPILOT FAIL indicators RED. Loss of 600U3, 600U2, and 600U1 as a group is indicated by the three missile tank pressure gauges indicating zero and R/V SAFE indicator RED. Emergency loss of umbilicals procedures for practical purposes are divided into three groups:

a. All umbilicals ejected.

b. 600U4 only ejected. (Thrust section umbilical.)

c. 600U1, 600U2, and 600U3 as a group ejected. (B-2 pod umbilicals.)

4-68. Item 4 of tables 4-1 through 4-4 contains procedures to be followed in the event of loss of umbilicals.

4-69. FIRE IN THE SILO.

4-70. Item 5 of tables 4-1 through 4-4 contains procedures to be followed in the event of a fire indication on the fire alarm panel.

4-71. FACILITY REMOTE CONTROL PANEL EMERGENCY AND MALFUNCTION INDICATIONS.

4-72. Items 6A through 6H of tables 4-1 through 4-4 provide procedures to be followed in the event the following indications appear on the FRCP.

a. SILO CONTROL CABINET HI-TEMP indicator illuminates RED. Item 6A of tables 4-1 thrugh 4-4.

b. MAIN EXHAUST FAN NOT OPERATING indicator illuminates RED. Item 6B of tables 4-1 through 4-4.

c. STORAGE AREA OXYGEN 25% or 19% indicator illuminates RED. Item 6C of tables 4-1 through 4-4.

NOTE

If STORAGE AREA OXYGEN 25%, GASEOUS OXYGEN VENT FAN ON, WATER CHILLER UNITS MALFUNCTION, SILO AIR INTAKE CLOSURES CLOSED, and SILO AIR EXHAUST CLOSURE CLOSED indicators illuminate simultaneously, suspect 120-VAC control voltage automatic transfer switch has switched to the standby control transformer due to a malfunction in the operating transformer. If the above indications occur, perform procedures for STORAGE AREA OXYGEN 25% indicator RED, and if the automatic transfer switch has switched to the standby control transformer, the complex electrical system reset procedures shall be performed to ensure proper operation of the silo electrical systems.

d. AIR WASHER DUST COLLECTING UNITS NOT OPERATING indicator illuminates RED. Item 6D of tables 4-1 through 4-4.

e. DIESEL VAPOR HIGH LEVEL indicator illuminates RED. Item 6E of tables 4-1 through 4-4.

f. LCC AIR RECEIVER/INSTRUMENT AIR RECEIVER LOW PRESSURE indicator illuminates RED. Item 6F of tables 4-1 through 4-4.

g. SILO AIR INTAKE CLOSED and SILO AIR EXHAUST CLOSED indicators illuminates RED. Item 6G of tables 4-1 through 4-4.

h. MISSILE POD AIR CONDITIONER MALFUNCTION indicator illuminates RED. Item 6H of tables 4-1 through 4-4.

i. EMERGENCY WATER PUMP P-32 ON indicator illuminates RED. Item 6I of tables 4-1 through 4-4.

4-73. ENGINES AND GROUND POWER.

4-74. Item 7 of tables 4-1 through 4-4 provides procedures to be followed in the event the ENGINES AND GROUND POWER indicator on the launch control console illuminates RED.

4-75. MALFUNCTION PATCH INDICATIONS ON LAUNCH CONTROL CONSOLE.

4-76. Items 8 through 11 of tables 4-1 through 4-4 provides procedures to be followed in the event the following indications appear on the malfunction patch of the launch control console:

a. 400 CYCLE POWER indicator illuminates RED, item 8 of tables 4-1 through 4-4.

b. 28 VDC POWER indicator illuminates AMBER or RED, item 9 of tables 4-1 through 4-4.

c. GUIDANCE FAIL indicator illuminates RED. Item 10 of tables 4-1 through 4-4.

d. MISSILE LIFT FAIL indicator illuminates RED. Item 11 of tables 4-1 through 4-4.

4-77. EMERGENCY MISSILE STRETCH.

4-78. Table 4-5 provides procedures to be followed in the event it becomes necessary to place the missile in stretch due to an emergency.

4-79. RESTORING AC POWER AFTER POWER LOSS (NA OSTF-2).

4-80. Table 4-6A provides procedures to be followed in the event it becomes necessary to restore AC power at operational sites.

4-81. RESTORING AC POWER AFTER POWER LOSS (OSTF-2).

4-82. Table 4-6B provides procedures to be followed in the event it becomes necessary to restore AC power at OSTF-2.

4-83. COMPLEX ELECTRICAL SYSTEM RESET.

4-84. Table 4-7 provides procedures to be followed anytime it becomes necessary to reset the complex real property installed equipment (RPIE) facilities and aerospace ground equipment (AGE) after a disruption of normal AC power.

4-85. EMERGENCY DIESEL GENERATOR PARALLELING AND SHUTDOWN.

4-86. Table 4-8 provides procedures to be followed in the event it becomes necessary to start and parallel the alternate diesel and (or) to shutdown the faulty diesel.

4-87. EMERGENCY USE OF HELIUM CONTROL CHARGING UNIT (HCU) FOR MANUAL CONTROL OF MISSILE TANK PRESSURES.

*4-88. Under emergency conditions, manual pressurization of the missile fuel and oxidizer tanks may be accomplished by using the HCU, regardless of launcher platform position, when the missile is properly mated to the launcher platform.

4-88A. The valves and gauges, and precautions in their use, associated with controlling missile tank pressures with the HCU are as follows:

CAUTION

If 28 VDC is lost and helium pressure to the HCU is being supplied by valve 149, manual bypass valve 148 must be opened immediately to maintain supply pressure to the HCU.

a. Valve 149 in the pressure system control (PSC) is a solenoid operated valve that is energized to the open position during countdown from POWER INTERNAL indicator illuminated GREEN until ABORT EXTERNAL indicator has illuminated AMBER. Valve 148 in the PSC is a manually operated valve that bypasses valve 149. One of these valves must be opened to supply helium pressure to the HCU. The valve that is opened will depend on the point in time when emergency use of the HCU is required .

b. Manual valve 337 in the HCU is normally open so that when helium pressure is supplied to the HCU the pressure will be allowed to reach HCU regulator 320 which will automatically supply approximately 53 PSI regulated pressure to the missile fuel tank. The only time that valve 337 should be closed is when phase I pressures are desired in the missile fuel tank and the airborne shrouded helium spheres are not charged.

c. The missile fuel tank pressure is manually controlled by using HCU manual valves 339 and 340 and pressure gauge 338. Valve 339 is used to raise pressure and valve 340 is used to lower pressure. Gauge 338 indicates the missile fuel tank pressure. Do not attempt to lower fuel tank pressure below 53 PSI if HCU valve 337 is open or if the airborne shrouded helium spheres are charged.

d. The missile oxidizer tank pressure is manually controlled by HCU manual valves 342 and 343 and pressure gauge 341. Valve 342 is used to raise pressure and valve 343 is used to lower pressure. Gauge 341 indicates the missile oxidizer tank pressure. Do not attempt to pressurize the missile oxidizer tank above 3.7 PSI if the boiloff valve is enabled.

CAUTION

Manual control of missile tank pressures must be maintained from the HCU at completion of LO_2 or LN_2 boiloff if launcher platform is not down and locked. Refer to section VI for boiloff time. Emergency tables and checklist do not normally indicate or direct when manual control of the missile tank pressures will be maintained from the HCU. The MCCC is responsible for ensuring that control of missile tank pressures is established at the HCU if required.

CAUTION

HCU vent valve 343 shall not be used to vent missile oxidizer tank pressure if LO_2 or LN_2 is in the missile oxidizer tank.

4-88B. During countdown, when the launcher platform is not down and locked, emergency manual control of missile oxidizer tank pressure with the HCU at completion of LO_2 or LN_2 boiloff is accomplished by operating HCU valves 342 and 343. Do not attempt to pressurize missile oxidizer tank above 3.7 PSI if the boiloff valve is enabled. Normally, manual pressurization of the fuel tank is not required as the HCU will automatically maintain approximately 53 PSI in the fuel tank.

4-88C. Prior to moving the launcher platform, during emergency manual control of missile tank pressures with the HCU, open valve 342 three full turns and clear personnel from the launcher platform. During launcher platform movement, control missile oxidizer tank pressure with the boiloff valve. Normally, manual pressurization of the fuel tank is not required as the HCU will automatically maintain approximately 53 PSI in the fuel tank. After the launcher platform is down and locked and the silo is safe, close HCU valve 342.

4-88D. During standby, regardless of whether the launcher platform is down and locked or in a maintenance configuration out of the down locks, the HCU may be used to maintain missile tank pressures during an emergency. To use the HCU under the above conditions, first close HCU valve 337 if phase I pressures are desired, then open PSC valve 148. Missile tank pressures shall be maintained by opening or closing HCU valves 339, 340, 342, and 343 as applicable.

> ## CAUTION
>
> When emergency manual pressurization from the HCU is terminated, close PSC Valve 148 and vent HSU line by opening PSC Valve 150. When venting ceases, close valve 150 and open HCU valve 337.

4-89. MANUAL ABORT.

4-90. Table 4-10 provides procedures to be followed in the event a manual abort becomes necessary.

4-91. TACTICAL COUNTDOWN MALFUNCTION PROCEDURES.

4-92. Tables 4-11 through 4-15 provides procedures to be followed in the event the following conditions occur during a tactical countdown:

a. LN₂ LOAD indicator does not illuminate AMBER, or does not illuminate GREEN after illuminating AMBER. Table 4-11.

b. HYDRAULIC PRESSURE indicator does not illuminate GREEN. Table 4-12.

c. LO₂ TANKING PANEL malfunctions during LO₂ CHILLDOWN. Table 4-13.

(1) Item 1 STORAGE TANK VENT VALVE N-5 indicator not AMBER.

(2) Item 2 TOPPING TANK VENT VALVE N-4 indicator not AMBER.

(3) TOPPING TANK PRESS. VALVE N-50 indicator not GREEN.

(4) TOPPING CHILL VALVE L-60 indicator not GREEN.

d. HELIUM LOAD indicator does not illuminate AMBER, or does not illuminate GREEN after illuminating AMBER. Table 4-14.

e. HYD-PNEU LN₂-HE ready indicator does not illuminate GREEN. Table 4-15.

4-93. LAUNCHER PLATFORM FAILS TO LOWER.

4-94. Table 4-16 provides procedures to be followed in the event launcher platform fails to lower as indicated on the TV monitor and(or) SITE HARD indicator not illuminating AMBER after ABORT pushbutton is depressed.

4-95. MISSILE ENCLOSURE PURGE UNIT.

4-96. A missile enclosure purge unit is provided to expel dangerous concentrations of fumes, gaseous oxygen, or gaseous nitrogen from the missile enclosure area and to prevent impingement of the tension equalizer with LO₂ or LN₂ boiloff. Actual impingment of the tension equalizer occurs only when the boiloff valve is open and approximately at the same level.

4-97. To use the missile enclosure purge unit, the purge unit must first be set up in accorance with T.O. 21M-HGM16F-3CL-1, the purge unit mobile generator set must be operating, and the silo overhead doors must be open. The purge unit is extended to the purge position and retracted to the standby position by depressing pushbuttons on a remote control unit, located in the launch control center grade entry stairwell. To extend the purge unit, depress the FOR pushbutton. To retract the purge unit, depress REV pushbutton.

CAUTION

Prior to closing the silo overhead doors or moving the launcher platform, the purge unit must be retracted. Emergency tables and checklist do not normally indicate or direct when the purge unit will be used or retracted. The MCCC is responsible for directing the use of the purge unit and ensuring it is retracted prior to silo door closure or launcher platform movement.

4-98. IMPINGEMENT OF TENSION EQUALIZER.

4-99. Gaseous oxygen (or gaseous nitrogen) impingement on the triangle plate and wire rope hanger on the north end of the tension equalizer assembly may occur when the boiloff valve is open and the launcher platform is less than 20 feet out of the down locks with turning vanes installed, or between 20 and 40 feet out of the down locks without the turning vanes installed. Impingement is considered to be a direct flow of gaseous oxygen (or gaseous nitrogen) from the boiloff valve onto the tension equalizer triangle plate and wire rope hanger. During an emergency which prevents launcher platform movement and with the boiloff valve in such a position that impingement on the tension equalizer may occur, action should be taken to prevent impingement as quickly as possible. Impingement may be prevented by placing the missile enclosure purge unit in operation, or if the purge unit is not available, by spraying the impinged area of the tension equalizer with water. If water is used, it should be sprayed from the south side of the silo overhead door opening in quadrant III, directly and only on the impinged area of the tension equalizer triangle plate and wire rope hanger (north end).

WARNING

Extreme care shall be taken to prevent water from spraying or splashing on the missile or boiloff valve. Water shall be used only when a missile enclosure purge unit is not available.

4-100. Water shall be sprayed only while the tension equalizer is being impinged upon (boiloff valve open). Intermittent impingement may be necessary to control missile oxidizer tank pressure prior to initiating missile enclosure purge unit operation or water spray.

4-101. LIQUID AND GAS SPILLAGE.

4-102. Emergency action required when liquid and gas spillage occurs as reflected as FRCP indication are contained in appropriate tables in this section. Only those units with operational diesel fuel vapor detector activated shall react as required by procedures contained in the appropriate sections.

4-102a. Emergency action required when a spillage of RP-1 fuel, hydraulic fluid or diesel fuel is visually observed are contained in appropriate tables in this section.

4-103. MANUAL STOP OF LAUNCHER PLATFORM DURING AN AUTOMATIC SEQUENCE.

4-104. Procedures contained in tables of this section require the initiation of abort if an emergency or abnormal condition occurs during countdown which requires a return to standby. During the launcher platform up-run sequence, condition may be observed which would require stopping the up-run movement without immediately lowering the launcher platform. To stop the up-run movement and prevent immediate lowering, position RESET PROGRAMMER key switch on CSMOL to ON then, immediately depress ABORT pushbutton. If automatic launcher platform lowering is to be initiated after a manual stop, ensure that ABORT COMPLETE indicator is illuminated AMBER prior to positioning RESET PROGRAMMER key switch to OFF. If it becomes necessary to manually stop an automatic down-run sequence position RESET PROGRAMMER key switch to ON. To continue the automatic down-run sequence after a manual stop, position RESET PROGRAMMER key switch to OFF.

4-105. BOILOFF VALVE FAILURE TO OPEN DURING ABORT.

4-106. Table 4-17 provides procedures to be followed in the event the boiloff valve fails to open when the EMERGENCY pushbutton is depressed when initiation of abort is delayed or during abort when the launcher platform is 7-1/2 feet below the uplocks.

4-107. LO2 GROWTH WITH BOILOFF VALVE CLOSED.

4-108. Prior to the commit sequence LO_2 in the missile is at its boiling temperature. During the commit sequence LO_2 tank pressure is increased to phase III pressure and boiling of LO_2 ceases because LO_2 boiling temperature increases with increased pressure. The increase in temperature results in an increase in volume. As the volume increases (LO_2 growth) there is a corresponding rise of LO_2 level in the missile. The rate at which the LO_2 level rises depends upon prevailing atmospheric conditions.

4-109. If the boiloff valve remains closed, LO_2 will eventually flow down the pressurization duct to the airborne relief valve. LO_2 flow will cause this valve to fail to an open position and result in little or no control of LO_2 tank pressure, causing a very serious condition to exist.

4-110. If the boiloff valve remains closed for an extended period of time at phase III pressure and is

then opened, LO_2 will be dumped through the boiloff valve. The amount of LO_2 dumped will depend on the length of time the boiloff valve was closed and on prevailing atmospheric conditions. LO_2 is dumped due to the following reason:

a. When the boiloff valve is closed the temperature of the LO_2 is approaching its boiling point at phase III pressure.

b. When the boiloff valve is opened, LO_2 tank pressure decreases very rapidly.

c. When the pressure decreases to phase II, the LO_2 is above its boiling temperature, resulting in violent boiling of LO_2. The end result is LO_2 being dumped through the boiloff valve into the silo.

Table 4-1. Emergencies During Standby

ITEM 1. PRESSURE MODE INDICATOR RED

This item contains procedures to place the missile in a safe configuration if PRESSURE MODE indicator illuminates RED during standby.

```
CAUTION
```

Maintain a minimum of 2 PSI differential pressure. To prevent fuel prevalves from opening, do not exceed 20 PSI in the fuel tank.

If PRESSURE MODE indicator illuminates RED, the pressurization monitor shall immediately determine if pressures are normal or abnormal by observing the pressure gauges. If pressures are normal and in phase 1 (LO_2 tank pressure 3.4 to 4.2 PSI, fuel tank pressure 11.9 to 13.0 PSI), do not adjust. If pressures are abnormal, adjust by depressing the applicable RAISE or LOWER pushbutton.

1 (D) TANK PRESSURES . . . . **ADJUSTED**

If pressures were adjusted to normal, or if pressures were normal when pressurization switched to emergency, the deputy shall depress AUTOMATIC pushbutton to return the system to automatic mode.

2. (D) AUTOMATIC PUSHBUTTON . . . . **DEPRESSED**

If pressure system remains in automatic mode, remaining steps in this table need not be accomplished. If system remains or returns to emergency mode after depressing AUTOMATIC pushbutton, deputy shall manually maintain phase 1 pressures. No further attempt should be made to return the system to automatic until missile is placed in stretch and troubleshooting is accomplished.

3. (D) PHASE 1 PRESSURES . . . . **MANUALLY MAINTAINED**

The MCCC shall direct crew the place missile in emergency stretch, using procedures contained in table 4-5.

4. (C) CREW, PLACE MISSILE IN STRETCH . . . . **DIRECTED**

Table 4-1.   Emergencies During Standby (CONT)

ITEM 2A.   GENERATOR MALFUNCTION (ALTERNATE IN STANDBY)

This item contains procedures to place the alternate generator on the line and to shut down the primary generator if generator malfunction is indicated during standby. Steps in this procedure will normally be performed by the L-1 at direction of the MCCC.

If a generator malfunction is indicated, at the direction of the MCCC, L-1 shall place alternate generator on the line and shut down the faulty generator using procedures contained in table 4-8. If L-1 is not present in the LCC, it may be necessary for other crew members to take corrective action.

1. (C) L-1, PLACE ALTERNATE GENERATOR ON THE LINE AND SHUT DOWN FAULTY GENERATOR .... **DIRECTED**

After AC power has been transferred to the alternate generator, at MCCC direction, L-1 assisted by

M-1 and A-1 shall troubleshoot the faulty system using applicable technical manuals.

2.   (C) L-1, TROUBLESHOOT .... DIRECTED

ITEM 2B.   GENERATOR MALFUNCTION (ALTERNATE NOT IN STANDBY)

This item contains procedures to shut down the faulty generator if generator malfunction is indicated during standby and the alternate generator is not available. Steps in this procedures will normally be performed by L-1 at the direction of the MCCC.

```
★  ┌─────────────┐
   ┊  CAUTION    ┊
   └─────────────┘
```

If the malfunction is due to generator low lube oil pressure or high temperature, delay shutdown of the generator and extend work platform NO. 1. This will allow the application of missile stretch if required during the period of AC power loss.

If a generator malfunction is indicated, at direction of the MCCC, L-1 shall shut down the faulty generator at the PRCP as follows:

(a) Position FEEDER NUMBER 3 NONESSENTIAL BUS CONTROL SWITCH to TRIP and observe FEEDER NUMBER 3 NONESSENTIAL BUS CONTROL SWITCH RED indicator extinguished and GREEN indicator illuminates.

(b) Position GENERATOR MAIN BREAKER CONTROL SWITCH to TRIP and observe GENERATOR MAIN BREAKER CONTROL SWITCH RED indicator extinguishes and GREEN indicator illuminates.

(c) Position ENGINE START-STOP SWITCH TO STOP and momentarily hold. Release switch and observe ENGINE START-STOP SWITCH RED indicator extinguishes and GREEN indicator illuminates.

1. (C) L-1, SHUT DOWN FAULTY GENERATOR .... **DIRECTED**

After faulty generator has been shut down, at MCCC direction, L-1 assisted by M-1 and A-1 shall trouble-

shoot faulty system and restore AC power using applicable technical manuals.

2.   (C) L-1, TROUBLESHOOT AND RESTORE AC POWER .... DIRECTED

After AC power is restored, at MCCC direction, L-1 shall restore nonessential power by positioning FEEDER NUMBER 3 NON-ESSENTIAL BUS CONTROL SWITCH to CLOSE and observing

FEEDER NUMBER 3 NON-ESSENTIAL BUS CONTROL SWITCH GREEN indicator extinguishes and RED indicator illuminates.

Table 4-1. Emergencies During Standby (CONT)

ITEM 2B. GENERATOR MALFUNCTION (ALTERNATE NOT IN STANDBY)   (CONT)

3. (C) L-1, NONESSENTIAL POWER ON .... DIRECTED

After AC power is restored, at MCCC direction, crew will reset complex electrical system using procedures contained in table 4-7:

4. (C) CREW, RESET ELECTRICAL SYSTEM .... DIRECTED

ITEM 2C. GENERATOR MALFUNCTION (GENERATORS IN PARALLEL)

This item contains procedures to remove a faulty generator from the line if a generator malfunction is indicated during standby while operating generators in parallel. Steps in this procedures will normally be performed by the L-1 at the direction of the MCCC.

If a generator malfunction is indicated, at MCCC direction, L-1 shall shut down faulty generator at PRCP:

(a) Position GENERATOR MAIN BREAKER CONTROL SWITCH to TRIP. Observe GENERATOR MAIN BREAKER CONTROL SWITCH RED inindicator extinguishes and GREEN indicator illuminates.

(b) Position ENGINE START-STOP SWITCH to STOP and momentarily hold. Release switch and observe ENGINE START-STOP SWITCH RED indicator extinguishes and GREEN indicator illuminates.

1. (C) L-1, SHUT DOWN FAULTY GENERATOR .... *DIRECTED*

After faulty generator has been shut down, at MCCC direction, L-1, assisted by M-1 and A-1, shall

troubleshoot the faulty system using applicable technical manuals.

2. (C) L-1, TROUBLESHOOT .... DIRECTED

ITEM 3. LOSS OF AC POWER

This item contains procedures for restoring the complex to a safe configuration after an AC power loss has occurred during standby. Steps in this procedure will normally be performed by L-1 at the

direction of the MCCC.
If AC power is lost, at direction of MCCC, L-1 shall attempt to restore AC power using procedures contained in table 4-6A or 4-6B as applicable.

1. (C) L-1 RESTORE AC POWER .... DIRECTED

After AC power is restored at direction of MCCC, L-1 shall restore nonessential power by positioning FEEDER NUMBER 3 NON-ESSENTIAL BUS CONTROL SWITCH to CLOSE and observing

FEEDER NUMBER 3 NON-ESSENTIAL BUS CONTROL SWITCH G R E E N indicator extinguishes and RED indicator illuminates.

2. (C) L-1, NONESSENTIAL POWER ON .... DIRECTED

After AC power is restored, at MCCC direction, crew shall reset complex electrical system using procedures contained in table 4-7.

3. (C) CREW, RESET ELECTRICAL SYSTEM .... DIRECTED

Table 4-1.   Emergencies During Standby (CONT)

ITEM 4.   LOSS OF UMBILICALS

WARNING

Maintenance support is required to safety ordnance items (if installed) and to reconnect umbilicals. Exception may be made to initiate a tactical launch if the missile crew can determine that no stray voltage exists at the umbilicals.

This item contains procedures to place the missile in a safe configuration if umbilical loss occurs during standby. If missile tank pressure gauges are indicating normal and R/V SAFE indicator is extinguished, only umbilical 600U4 has ejected and steps 1 and 2 need not be performed.

If all missile tank pressure gauges on launch control console indicate zero, MCCC shall direct deputy or M-1 to monitor and maintain missile tank pressures. If TV is not available to monitor PSMR missile tank pressure gauges, M-1 shall proceed to the HCU, launcher platform level 3, and close valve 337 to prevent fuel tank pressure from rising to 53.0 PSI when pressure is supplied to the HCU. M-1 shall then proceed to the PSC, level 8, and open valve 148 to supply pressure to the HCU. M-1 shall then return to the HCU and maintain phase I pressures (LO_2 tank 3.4 to 4.2 PSI, fuel tank pressure 11.9 to 13.0 PSI) using valves 339, 340, 342, and 343. If TV monitor is available to monitor PSMR missile tank pressures gauges, deputy shall maintain phase I pressures using the TV monitor and applicable RAISE or LOWER pushbutton on launch control console.

1.   (C) DEPUTY, OR M-1 MONITOR AND MAINTAIN TANK PRESSURES .... DIRECTED

The loss of umbilical 600U3 causes the logic units to lose the sensing capability of the missile tank pressure transducers. The resultant sensing of zero PSI differential pressure causes the pressurization system to switch to emergency mode. The MCCC shall direct A-1 and L-1, to place missile in emergency stretch using procedures contained in table 4-5.

2.   (C) A-1, L-1, PLACE MISSILE IN STRETCH .... DIRECTED

ITEM 5.   FIRES IN SILO

The following procedures contain immediate actions required if a fire is detected in the silo during standby. Methods of detecting a fire in the silo are the fire alarm panel, television monitor, and on-the-spot observation. Upon detection of a fire, it shall be the responsibility of the MCCC to use all available personnel to combat the fire by the most effective means possible. Paragraph 4-30 specifies the type of extinguisher to be used and identifies the classes of fires.

If a fire is detected, the individual who detects the fire shall immediately notify the MCCC as to the location and extent of the fire. Other details, such as actual or potential damage, should be included as time permits.

1.   (C) FIRE LOCATION .... IDENTIFED

The individual detecting a fire shall activate the fire alarm system from level 2, 4, 6, or 8 when expeditous to do so, or the MCCC shall activate the fire alarm at the fire alarm panel in the LCC.

2.   (C) FIRE ALARM .... ACTIVATED

The MCCC shall direct available personnel to attempt to contain and extinguish the fire by using available fire fighting equipment.

3.   (C) CREW, COMBAT THE FIRE .... DIRECTED

If the fire is located in the missile enclosure area, the MCCC shall direct M-1 or any other available crew member to activate the missile enclosure fog system at the FRCP. Fog system is activated by depressing MISSILE ENCLOSURE FOG ON pushbutton.

Table 4-1. Emergencies During Standby (CONT)

ITEM 5. FIRES IN SILO (CONT)

4. (C) M-1, ACTIVATE MISSILE ENCLOSURE FOG SYSTEM .... DIRECTED

The MCC shall ensure that all personnel not directly involved in combating the fire are evacuated from the silo.

5. (C) SILO EVACUATED OF NONESSENTIAL PERSONNEL .... VERIFIED

The MCCC shall ensure that local disaster control and potential hazard procedures are implemented at the earliest possible time but not to interfere with combating the fire. Personnel in the LCC not involved in fire fighting may be used to implement the disaster control procedures.

6. (C) DISASTER CONTROL AND POTENTIAL HAZARD PROCEDURES .... IMPLEMENTED

If the fire cannot be controlled the MCCC will, at his discretion, direct M-1 to manually open the silo overhead doors. Opening the silo overhead doors will relieve silo overpressure in the event of an explosion. It will also allow silo blast door opening for silo entry and exit if required and prevent possible damage to the LCC. If the fire is small and quickly brought under control, the silo overhead doors should not be opened. When directed by the the MCCC, M-1 shall open silo overhead doors from CSMOL as follows:

(a) Position RESET PROGRAMMER key switch to ON.

(b) Depress HYDRAULIC 40 HP PUMP ON pushbutton and observe HYDRAULIC 40

HP PUMP PRESSURE indicator illuminates GREEN.

(c) Depress CRIB VERTICAL LOCK pushbutton and observe CRIB VERTICAL LOCK indicator illuminates GREEN.

(d) Depress CRIB HORIZONTAL LOCK pushbutton and observe CRIB HORIZONTAL LOCK indicator illuminates GREEN.

(e) Depress and hold SILO DOORS OPEN pushbutton for 30 seconds and observe SILO DOORS CLOSE indicator extinguishes.

In approximately 5 minutes, the SILO DOORS OPEN indicator will illuminate GREEN.

7. (C) M-1, OPEN SILO DOORS .... DIRECTED

ITEM 6A. SILO CONTROL CABINET HI-TEMP INDICATOR RED

These procedures shall be used to prevent overheating of the control-monitor and countdown groups when a malfunction in the control cabinet air conditioning system is indicated on the FRCP. The MCCC shall direct M-1 to investigate and troubleshoot the cause of the malfunction indication. A tactical launch may be initiated with this malfunction indication.

1. (C) M-1, TROUBLESHOOT .... DIRECTED

ITEM 6B. MAIN EXHAUST FAN NOT OPERATING INDICATOR RED

These procedures shall be used to prevent a hazardous ventilation condition from occuring in the silo when the FRCP indicates that the main exhaust fan is not operating. The MCCC shall direct M-1 to investigate and troublshoot the cause of the malfunction indication. A tactical launch may be initiated with this malfunction indication.

1. (C) M-1, TROUBLESHOOT .... DIRECTED

Table 4-1. Emergencies During Standby (CONT)

ITEM 6C: STORAGE AREA 25% OR 19% INDICATOR RED

These procedures shall be used to return the silo to a safe configuration when a gaseous or liquid oxygen/nitrogen leak is indicated on the FRCP during standby. The MCCC shall direct evacuation of the silo and direct M-1 to investigate and troubleshoot the cause of the malfunction indication. Countdown shall not be initiated with high oxygen concentration indications. A tactical countdown may be initiated with low oxygen concentration indications.

The MCCC shall direct evacuation of the silo by Public Address announcement from the launch control console, by activating the fire alarm at the fire alarm panel, or by turning the silo lights off and on 3 times at 1 second intervals by using the silo lighting pushbutton in the silo entrance vestibule.

1. (C) EVACUATE SILO . . . . **DIRECTED**

The MCCC shall direct M-1 to investigate and troubleshoot the cause of the FRCP malfunction indication. Personnel investigating the malfunction shall use breathing apparatus and protective clothing when approaching any liquid oxygen/nitrogen or gaseous oxygen/nitrogen leak. The MCCC shall direct M-1 to check the oxygen detector TROUBLE indicator, and if illuminated troubleshoot the oxygen detector cabinet. If the TROUBLE indicator is extinguished, M-1 shall check the AREA ALARM indicator and the percentage meter to determine location and amount of oxygen concentration. The area generating the alarm should be checked first. When a leak is discovered, it should be isolated, if possible, referring to appropriate technical manuals. If the leak or spillage cannot be corrected, the MCCC shall notify appropriate agencies for assistance.

2. (C) M-1, TROUBLESHOOT . . . . DIRECTED

ITEM 6D. AIR WASHER DUST COLLECTING UNITS NOT OPERATING INDICATOR RED

These procedures shall be used to return the silo ventilation system to a normal configuration when a malfunction of the air washer duct collecting units is indicated on the FRCP. The MCCC shall direct M-1 to investigate and troubleshoot the cause of the malfunction indication. A tactical launch may be initiated with this malfunction indication.

1. (C) M-1, TROUBLESHOOT . . . . DIRECTED

ITEM 6E. DIESEL VAPOR HIGH LEVEL INDICATOR RED

These procedures shall be used to return the silo to a safe configuration when a high concentration of diesel vapor is indicated on the FRCP. The MCCC shall direct evacuation of the silo and direct M-1 to investigate and troubleshoot the cause of the malfunction indication. Countdown shall not be initiated due to danger of an explosion and fire.

The MCCC shall direct silo evacuation by making a public address announcement from the launch control console, by activating the fire alarm at the fire alarm panel, or by turning the silo lights of and on 3 times at 1 second intervals by using the silo lighting pushbutton in silo entrance vestibule.

1. (C) EVACUATE SILO . . . . **DIRECTED**

Table 4-1. Emergencies During Standby (CONT)

ITEM 6E. DIESEL VAPOR HIGH LEVEL INDICATOR RED (CONT)

The MCCC shall direct M-1 to investigate and troubleshoot the cause of the malfunction indication. Personnel investigating the malfunction shall use breathing apparatus when a contaminated atmosphere is encountered. The crew shall inspect the operating generator and all piping on that level. Then the alternate generator and associated piping shall be inspected. If a leak is discovered it should be isolated by turning appropriate valves, referring to appropriate technical manuals. The situation may require that the standby generator be placed on the line or that both diesels be shut down. Appropriate agencies will be notified if a substantial spillage is discovered. If no spillage is observed on level 5 or 6, proceed to level 7 and troubleshoot diesel vapor detector cabinet.

2. (C) M-1, TROUBLESHOOT .... DIRECTED

ITEM 6F. LCC AIR RECEIVER AND INSTRUMENT AIR RECEIVER LOW PRESSURE INDICATORS RED

These procedures shall be used to return the instrument air system to a safe configuration when the FRCP indicates a malfunction in the silo or LCC instrument air systems. When both indicators illuminate RED in close proximity, a serious leak is indicated. Countdown shall not be initiated until malfunction is corrected. The MCCC shall direct M-1 to troubleshoot the malfunction indication.

1. (C) M-1, TROUBLESHOOT .... DIRECTED

ITEM 6G. SILO AIR INTAKE CLOSURES CLOSED/SILO AIR EXHAUST CLOSURES CLOSED INDICATORS RED

These procedures shall be used to return the silo to a safe configuration if silo blast closures are closed electrically as indicated on the FRCP.

The MCCC shall direct evacuation of the silo by making a public address system announcement from the launch control console, by activating the fire alarm at the fire alarm panel, or by turning the silo lights off and on 3 times at 1 second intervals by using the silo lighting pushbutton in silo entrance vestibule.

1. (C) EVACUATE SILO .... **DIRECTED**

If the blast detection system has detected an optic signal but not a radio signal, the blast closures should open in 90 seconds and the blast detection cabinet will have to be reset.

If the blast detection system has detected both optic signal and a radio signal, the blast closures will close and the blast detection cabinet must be reset before the blast closures will open.

If no nuclear blast has been detected and a malfunction has occurred in the blast detection cabinet:

(a) Position OUTPUT RELAY switch to DISCONNECT.
(b) Depress RCVR 1 MANUAL TEST pushbutton.
(c) Depress ALARM RESET pushbutton.
(d) Position OUTPUT RELAY switch to CONNECT.
(e) Depress DETECTION MODE RESET pushbutton.
(f) Verify OPTIC MODE indicator is illuminated.
(g) Verify channel indicator is cycling.

2. (C) M-1, RESET BLAST DETECTION CABINET .... DIRECTED

Table 4-1.   Emergencies During Standby (CONT)

ITEM 6G.   SILO AIR INTAKE CLOSURES CLOSED/SILO AIR EXHAUST CLOSURES
CLOSED INDICATORS RED (CONT)

If blast detection cabinet fails to reset or if the blast closures fail to open, at direction of MCCC, M-1 shall position OUTPUT RELAY switch on blast detection cabinet to DISCONNECT.

3.   (C) M-1, OUTPUT RELAY SWITCH TO DISCONNECT . . . . DIRECTED

If blast closures were closed by an actual nuclear blast, emergency water pump P-32 may be operating. If diesel generators are operating in parallel, to prevent a possible overheat condition and possible AC power loss, one generator shall be shut down. At MCCC direction to shut down one diesel generator, L-1 shall:

   (a)   Position GENERATOR MAIN BREAKER CONTROL SWITCH to TRIP. Observe

       GENERATOR MAIN BREAKER CONTROL SWITCH RED indicator extinguishes and GREEN indicator illuminates.

   (b)   Position ENGINE START-STOP SWITCH to STOP and momentarily hold. Release switch and observe ENGINE START-STOP SWITCH RED indicator extinguishes and GREEN indicator illuminates.

Omit step 4 if generators are not operating in parallel.

4.   (C) L-1, SHUT DOWN ONE GENERATOR . . . . DIRECTED

5.   (Deleted)

If the blast closures are closed and blast detection cabinet is normal, a failure of relay control circuit or a tripped circuit breaker should be suspected. At direction of the MCCC to troubleshoot blast closure system, M-1 shall:

   (a)   Proceed to essential motor control panel C,

level 2 and verify control circuit breakers CB-1 and CB-3 are ON. Reset CB-1 and CB-3 if tripped. If circuit breakers had to be reset and blast closures open, omit step (b) through (e) and troubleshoot probable cause in accordance with applicable technical manual.

Table 4-1. Emergencies During Standby (CONT)

WARNING

Control circuit voltage is 120
VAC. Use extreme caution
while installing jumper wires.

If circuit breakers were not tripped or will
not reset, proceed to facilities interface
cabinet NO. 1 on level 3 to bypass the control
circuit of the blast closures as follows: (b)
Install jumper wire on terminal board E-1
between wire numbers 9013 and 9015.
(c) Install jumper wire on terminal board
E-1 between wires 9015 and 9017. (d) Install
jumper wire on terminal board E-1, between
wires 9017 and 9019. (e) Install jumper wire
from wire C-5 to wire 9013. Verify with
MCCC that blast closures are open. If blast

closures do not open, troubleshoot blast
closures system in accordance with applicable
technical manual. If blast closures remain
closed it will become necessary to shut down
the diesel generator to prevent excessive
carbon buildup in the exhaust system and contamination of air in the silo. Prior to shut
down of the diesel generator, extend work
platform NO. 1 to allow missile stretch to be
applied if an emergency occurs that may
require it.

6. (C) M-1, TROUBLESHOOT BLAST CLOSURES . . . . DIRECTED

The MCCC shall direct M-1 to troubleshoot the cause of emergency.

7. (C) M-1, TROUBLESHOOT . . . . DIRECTED

ITEM 6H. MISSILE POD AIR CONDITIONER MALFUNCTION INDICATOR RED

These procedures shall be used to return
the missile pod air conditioner to normal
configuration when a malfunction of the pod
air conditioner is indicated on the FRCP. The

MCCC shall direct M-1 to investigate and
troubleshoot the causes of the malfunction
indication using the applicable technical
manuals.

1. (C) M-1, TROUBLESHOOT . . . . DIRECTED

ITEM 6I. EMERGENCY WATER PUMP P-32 ON INDICATOR RED

This item contains procedures for shutting
down one diesel generator, if operating in
parallel, to prevent overheating. If emergency water pump P-32 is running, a
failure of condenser water flow from the
water cooling tower is indicated. The
capacity of pump P-32 is not sufficient to
cool both diesel engines. Steps in this
procedure will normally be performed by
L-1 at the direction of the MCCC. If
EMERGENCY WATER PUMP P-32 ON
indicator on FRCP illuminates, at direction of MCCC to shut down one generator, L-1 shall:

(a) Position GENERATOR MAIN BREAKER CON-
TROL SWITCH to TRIP. Observe GENERATOR
MAIN BREAKER CONTROL SWITCH RED indicator extinguishes and GREEN indicator
illuminates.

(b) Position ENGINE START-STOP SWITCH to
STOP and momentarily hold. Release switch
and observe ENGINE START-STOP SWITCH
RED indicator extinguishes and GREEN indicator illuminates.

A tactical countdown may be initiated.

1. (C) L-1, SHUT DOWN ONE GENERATOR . . . . DIRECTED

The MCCC shall direct M-1 and L-1 to troubleshoot the condenser water system and power
distribution system using applicable technical manuals.

2. (C) M-1, L-1, TROUBLESHOOT . . . . DIRECTED

Table 4-1. Emergencies During Standby (CONT)

ITEM 7. ENGINES AND GROUND POWER INDICATOR RED (28 VDC POWER AND 400 CYCLE POWER INDICATORS EXTINGUISHED)

This condition during standby indicates that the solid propellant gas generator (SPGG) temperature is out of tolerance or that the 28-VDC transformer-rectifier has failed. The MCCC shall direct A-1 and M-1 to proceed to level 3 to observe the SPGG indicators on the AC power distribution panel. If all SPGG indicators are GREEN, an assumption will be made that the power supply distribution set rectifier circuit has failed. This is considered to be an emergency condition and manual control of the PSC must be obtained in case 28 VDC is completely lost (step 2.) If any SPGG indicator is extinguished or RED the crew need only troubleshoot the SPGG malfunction (step 3).

1. (C) A-1, M-1, OBSERVE SPGG INDICATORS . . . . DIRECTED

If all SPGG indicators are GREEN, MCCC shall direct A-1 and M-1 to obtain manual control of the missile tank pressures at the PSC on level 8. Manual control of pressures is necessary to ensure proper missile tank pressures should 28 VDC be intermittently or gradually lost as the emergency batteries are depleted. Manual control at the PSC will prevent any failure in the automatic system from affecting missile tank pressurization. (a) M-1 shall obtain manual control of missile tank pressurization at the PSC by closing valves 105 and 106. (b) After valves 105 and 106 are closed, A-1 shall ensure that the intermittent or gradual loss of 28 VDC does not affect missile tank pressurization by positioning SYSTEM POWER switch on PNEUMATICS (PANEL 1) to OFF. (c) M-1 shall then maintain missile tank pressures at the PSC. Fuel tank pressure shall be maintained at 11.9 to 13.0 PSI using valve 123 to raise pressure or valve 125 to lower pressure. LO_2 tank pressure shall be maintained at 3.4 to 4.2 PSI using valve 124 to raise pressure or valve 126 to lower pressure.

2. (C) A-1, M-1, OBTAIN MANUAL CONTROL OF PSC . . . . DIRECTED

The MCCC shall without delay, direct L-1 to start and configure the pod air conditioner for local operation to prevent prolonged loss of pod cooling to the missile guidance system. L-1 shall start and configure the pod air conditioner for local operation by: (a) Positioning the UNIT STOP/START switch (S-1) to ON; (b) Positioning the REMOTE CONTROL switch (S-3) to OFF.

2A. (C) L-1, POSITION POD AIR CONDITIONER FOR LOCAL OPERATION . . . . DIRECTED

The MCCC shall direct A-1 and L-1 to troubleshoot existing malfunction, using appropriate technical manuals.

3. (C) A-1, L-1, TROUBLESHOOT . . . . DIRECTED

ITEM 8. 400 CYCLE POWER INDICATOR RED

This condition does not indicate an emergency situation during standby. The MCCC shall direct the crew to troubleshoot in accordance with Section V.

1. (C) CREW TROUBLESHOOT . . . . DIRECTED

ITEM 9. 28 VDC POWER INDICATOR AMBER OR RED

The 28 VDC POWER AMBER or RED indication is considered to be an emergency because either indication could lead to an eventual complete loss of 28 VDC. The AMBER indication means that cooling air is not being supplied to the unit or that the backup batteries have depleted past 40 ampere-hours due to a partial or complete rectifier failure. The RED indication means that 28-VDC power is out of tolerance. Manual control of the pressurization system shall be obtained at the pressure system control (PSC). Manual control is necessary should 28 VDC be intermittently or gradually lost as the emergency batteries are depleted. Manual control at the PSC will prevent any failure

---

ITEM 9.   28 VDC POWER INDICATOR AMBER OR RED (CONT)

in the automatic system from affecting missile tank pressurization. At MCCC direction, A-1 and M-1 shall obtain manual control of pressurization system as follows: (a) M-1, close valves 105 and 106 on PSC. (b) A-1, position SYSTEM POWER switch on PNEUMATIC (PANEL 1) to OFF. (c) M-1, maintain missile tank pressures at PSC. Fuel tank pressure shall be maintained at 11.9 to 13.0 PSI using valve 123 to raise pressure or valve 125 to lower pressure. LO_2 tank pressure shall be maintained at 3.4 to 4.2 PSI using valve 124 to raise pressure or valve 126 to lower pressure. After abort is complete, troubleshoot the malfunctioning system.

1. (C) A-1, M-1, OBTAIN MANUAL CONTROL OF PSC . . . . DIRECTED

The MCCC shall without delay, direct L-1 to start and configure the pod air conditioner for local operation to prevent prolonged loss of pod cooling to the missile guidance system. L-1 shall start and configure the pod air conditioner for local operation by: (a) Positioning the UNIT STOP/START switch (S-1) to ON; (b) Positioning the REMOTE CONTROL switch (S-3) to OFF.

1A. (C) L-1, POSITION POD AIR CONDITIONER FOR LOCAL OPERATION . . . . DIRECTED

After manual control is obtained the MCCC shall direct A-1 and L-1 to troubleshoot Power Supply Distribution Set on level 3.

2. (C) A-1, L-1, TROUBLESHOOT . . . . DIRECTED

ITEM 10.   GUIDANCE FAIL INDICATOR RED

This condition does not indicate an emergency during standby. The MCCC shall direct the crew to troubleshoot in accordance with section 5. If guidance has been on memory for more than 7 minutes and GUIDANCE FAIL indicator illuminates RED, troubleshooting should be delayed for 2 minutes. If no malfunction has occured, GUIDANCE FAIL indicator should extinguish in 2 minutes and no troubleshooting is necessary.

1. (C) CREW TROUBLESHOOT . . . . DIRECTED

ITEM 11.   MISSILE LIFT FAIL INDICATOR RED

This condition does not indicate an emergency during standby. The MCCC shall direct the crew to troubleshoot in accordance with section V.

1. (C) CREW TROUBLESHOOT . . . . DIRECTED

ITEM 12.   LIQUID SPILLAGE (VISUALLY OBSERVED)

The following procedure shall be used to return the silo to a safe configuration after spillage of RP-1 fuel, hydraulic fluid or diesel fuel has been visually observed. Reaction to a liquid spillage requires use of sound judgment after evaluation of all relative information. If the liquid spillage is of substantial quantity or nature to create a hazard to either personnel or equipment the MCCC shall evacuate the silo. Consideration shall be given to stopping, restricting or isolation of spillage source by personnel in process of silo evacuation.

1. (C) EVACUATE SILO . . . . DIRECTED

When the hazard created by liquid spillage may be reduced by activation of the RP-1 purge fan EF-1 or the fire fog system the MCCC at his discretion shall direct M-1 to activate one or both depending on the result of evaluation of all relative facts.

T. O. 21M-HGM16F-1

Table 4-1. Emergencies During Standby (CONT)

ITEM 12.   LIQUID SPILLAGE VISUALLY OBSERVED (CONT)

2. (C) M-1 ACTIVATE RP-1 PURGE FAN . . . . DIRECTED

The MCCC shall direct activation of the fire fog system as described above. Activation of the
fire fog system will cause extensive maintenance and clean up to be performed.

3. (C) M-1 DEPRESS FOG ON PUSHBUTTON . . . . DIRECTED

Immediately after evacuation of the silo or activation of the RP-1 purge fan and (or) fire fog system,
or in the event a hazard does not exist and the utilization of silo evacuation, RP-1 purge fan and
fire fog system was not accomplished the MCCC shall direct M-1 to investigate silo conditions and
cause of spillage. The crew shall investigate all sources of spillage and immediately attempt to
isolate, stop or restrict spillage by referring to appropriate technical manuals. During and after
correction of spillage source action shall be taken to prevent further hazards from occurring by
cleaning spillage by use of most appropriate means available, fire hose, drain, rags, bucket
brigade.

4. (C) M-1 TROUBLESHOOT . . . . DIRECTED

Table 4-2. Emergencies from Start Countdown to MISSILE LIFT UP & LOCKED Indicator Amber

ITEM 1.   PRESSURE MODE INDICATOR RED

This item contains procedures to follow if PRESSURE MODE indicator illuminates RED from start countdown to MISSILE LIFT UP & LOCKED indicator AMBER.

> **CAUTION**
>
> Maintain a minimum of 2 PSI differential pressure, as observed on TANK DIFFERENTIAL PRESSURE gauge. Failure to comply may result in bulkhead reversal.

> **CAUTION**
>
> If LO$_2$ loading has been started fuel tank pressure must be maintained at least 20 PSI greater than LO$_2$ tank pressure. Failure to comply may result in bulkhead reversal.

If PRESSURE MODE indicator illuminates RED, deputy shall immediately determine if pressures normal or abnormal by observing pressure gauges. If pressures are abnormal, deputy shall immediately attempt to adjust pressures to normal by depressing the applicable RAISE or LOWER pushbutton.

1. (D) TANK PRESSURES . . . . **ADJUSTED**

If pressures were adjusted to normal or if no adjustment was required due to normal pressures, deputy shall return system to automatic mode by depressing AUTOMATIC pushbutton.

2. (D) AUTOMATIC PUSHBUTTON . . . . **DEPRESSED**

If pressure system remains in automatic mode, continue countdown. Do not perform remaining steps in this table. If system remains or returns to emergency mode after depressing AUTOMATIC pushbutton abort countdown (Step 3). If pressure mode RED was caused by fuel tank pressure less than 53 PSI after LO$_2$ LINE FILLED indicator illuminated GREEN, the LO$_2$ LOAD sequence will not continue, successful countdown can not be completed.

3. (C) ABORT . . . . **INITIATED**

The abort sequence (if started) shall be monitored by all crew members. The deputy must maintain proper missile tank pressure manually throughout the abort sequence.

4. (D) PHASE II PRESSURES . . . . **MANUALLY MAINTAINED**

The missile shall be placed in emergency stretch as soon as LO$_2$ DRAIN COMPLETE indicator is GREEN. If abort was initiated due to pressurization system returning to emergency mode, the ABORT COMPLETE indicator will not illuminate GREEN because PNEUMATICS IN PHASE I and HELIUM VENT COMPLETE indicators will not illuminate GREEN. PRESSURE MODE RED indication will prevent the PSC from selecting a phase I valve configuration and will prevent the helium vent valve from opening and helium from venting. The MCCC shall direct crew members to place missile in emergency stretch (table 4-5) upon initial silo entry. The MCCC should observe P. S. 323, 324, 329, 330, and DIFFERENTIAL PRESS LOW indicators on the PLCP for an AMBER indication. These indicators should be extinguished and P. S. 332 indicator should be AMBER at this time. An AMBER indication is the most probable cause of the pressurization malfunction. All abnormal indications must be remembered for use when required in table 4-5. After malfunction is corrected and the pressurization system returned to automatic the abort sequence will continue.

5. (C) CREW PLACE MISSILE IN STRETCH. . . . . **DIRECTED**

Table 4-2.   Emergencies from Start Countdown to MISSILE LIFT UP & LOCKED Indicator Amber (CONT)

ITEM 2A.   GENERATOR MALFUNCTION (ALTERNATE IN STANDBY)

This item contains procedures to place the alternate generator on the line and shutdown the faulty generator if a diesel generator malfunction occurs during countdown prior to the missile lift sequence and the alternate generator is in standby. Steps in this procedure shall be performed by L-1 at the PRCP in an attempt to prevent AC power loss. A tactical countdown shall be continued with one generator.

1.   (C) COUNTDOWN (TACTICAL) . . . . **CONTINUED**

L-1 shall, at MCCC direction, place the alternate generator on the line and shutdown the faulty gen- erator, using procedures contained in table 4-8.

**2.   (C) L-1, PLACE ALTERNATE GENERATOR ON THE
       LINE AND SHUTDOWN FAULTY GENERATOR** . . . . **DIRECTED**

ITEM 2B.   GENERATOR MALFUNCTION (ALTERNATE NOT IN STANDBY)

This item contains procedures to place the complex in the safest possible configuration after diesel generator malfunction occurs during countdown prior to the missile lifting sequence with the alternate generator not in standby. Steps in this procedure shall be performed by crew members as indicated. The MCCC shall continue a tactical countdown (A non- tactical countdown would not be started with the alternate generator not in standby) until the missile is launched or an abort is required. The launcher platform shall not be lowered until L-1 has cor- rected the malfunction of the faulty generator and restored AC power in order to prevent a possible loss of AC power during a launcher platform lower- ing sequence.

1.   (C) COUNTDOWN (TACTICAL) . . . . **CONTINUED**

If the missile is not launched the MCC shall direct deputy to depress EMERGENCY pushbutton after the ABORT indicator illuminates AMBER or RED. This will relieve the pressure in the LO$_2$ tank.

2.   (C) DEPUTY, DEPRESS EMERGENCY PUSHBUTTON . . . . DIRECTED

The MCCC shall direct M-1 to position RESET PROGRAMMER key switch to ON after ABORT indicator illuminates AMBER or RED to prevent the launcher platform from immediately lowering after AC power is restored.

3.   (C) M-1, RESET PROGRAMMER KEY ON . . . . DIRECTED

The MCCC shall direct L-1 to shut down the faulty generator at PRCP: (a) Position GENERATOR MAIN BREAKER CONTROL SWITCH to TRIP. Observe GENERATOR MAIN BREAKER CON- TROL SWITCH RED indicator extinguishes and GREEN indicator illuminates. (b) Position EN- GINE START-STOP SWITCH to STOP and mo- mentarily hold. Release switch and observe ENGINE START-STOP SWITCH RED indicator extinguishes and GREEN indicator illuminates.

4.   (C) L-1, SHUT DOWN FAULTY GENERATOR . . . . DIRECTED

The MCCC shall depress the ABORT pushbutton to enable the abort circuit and ensure that missile power changes from internal to external.

Table 4-2.   Emergencies from Start Countdown to MISSILE LIFT UP & LOCKED Indicator Amber (CONT)

ITEM 2B.   GENERATOR MALFUNCTION (ALTERNATE NOT IN STANDBY) (CONT)

5.   (C) ABORT PUSHBUTTON . . . . DEPRESSED

The MCCC shall direct L-1 to troubleshoot the faulty generator. After troubleshooting is completed, L-1 shall correct the malfunction and restore AC power.

6.   (C) L-1, TROUBLESHOOT AND RESTORE AC POWER . . . . DIRECTED

After restoration of AC power, MCCC shall initiate an abort by directing M-1 to position RESET PRO-GRAMMER key switch to OFF.

7.   (C) M-1, RESET PROGRAMMER KEY OFF . . . . DIRECTED

After abort is complete, MCCC shall direct the crew to reset complex electrical system to restore the silo to a standby configuration using procedures contained in table 4-7.

8.   (C) CREW RESET ELECTRICAL SYSTEM . . . . DIRECTED

ITEM 2C.   GENERATOR MALFUNCTION (GENERATORS IN PARALLEL)

This item contains procedures to remove a faulty generator from the line if a diesel generator malfunction occurs during countdown prior to missile lifting sequence with the generators in parallel. Steps in this procedure shall be performed by crew members as indicated. If a generator malfunction is indicated, at MCCC direction, L-1 shall shut down the faulty generator at the PRCP:

(a) *Remove nonessential power by positioning FEEDER NUMBER 3 NON-ESSENTIAL BUS CONTROL SWITCH to TRIP, if malfunction* occurred after POWER INTERNAL illuminated GREEN.

(b) *Position GENERATOR MAIN BREAKER CONTROL SWITCH to TRIP. Observe GENERATOR MAIN BREAKER CONTROL SWITCH RED indicator extinguishes and GREEN indicator illuminates.*

(c) *Position ENGINE START-STOP SWITCH to STOP and momentarily hold. Release switch and observe ENGINE START-STOP SWITCH RED indicator extinguishes and GREEN indicator illuminates.*

1.   *(C) L-1, SHUTDOWN FAULTY GENERATOR* . . . . DIRECTED

A tactical launch countdown shall continue on one generator. A nontactical countdown shall be aborted.

2.   *(C) ABORT (NONTACTICAL COUNTDOWN)* . . . . INITIATED

Table 4-2.   Emergencies from Start Countdown to MISSILE LIFT UP & LOCKED Indicator Amber (CONT)

ITEM 3.   LOSS OF AC POWER

This item contains procedures for restoring the complex to a safe configuration after an AC power loss during countdown prior to missile lifting sequence. Steps in this procedure will be performed by crew members as indicated.

If AC power is lost MCCC shall immediately initiate abort sequence by depressing ABORT pushbutton. This will stop the countdown sequence and start the LO$_2$ drain sequence.

1. (C) ABORT . . . . INITIATED

At MCCC direction, L-1 shall attempt to restore AC power using procedures contained in table 4-6A or 4-6B as applicable.

2. (C) L-1, RESTORE AC POWER . . . . DIRECTED

After AC power is restored, at MCCC direction L-1 shall restore nonessential power: (a) Position FEEDER NUMBER 3 NON-ESSENTIAL BUS CONTROL SWITCH to C L O S E and observe

FEEDER NUMBER 3 NON-ESSENTIAL BUS CONTROL SWITCH GREEN indicator extinguishes and RED indicator illuminates. This will restore silo ventilation.

3. (C) L-1, NONESSENTIAL POWER ON . . . . DIRECTED

During abort sequence, at MCCC direction, M-1 shall reset LCC electrical systems to restore LCC

ventilation using procedures contained in table 4-7.

4. (C) M-1, RESET LCC ELECTRICAL SYSTEM . . . . DIRECTED

After abort sequence is complete, at MCCC direction, the crew shall complete the reset of the com-

plex electrical system, using procedures contained in table 4-7.

5. (C) CREW, RESET ELECTRICAL SYSTEM . . . . DIRECTED

ITEM 4.   LOSS OF UMBILICALS

ACTION 1

When umbilical loss is identified, the MCCC shall immediately initiate abort and direct deputy to de-

press EMERGENCY pushbutton in an attempt to open the boiloff valve.

1. (C) ABORT . . . . INITIATED

2. (C) DEPUTY, DEPRESS EMERGENCY PUSHBUTTON . . . . DIRECTED

There are three primary points of consideration beginning with loss of umbilicals prior to FINE LO$_2$ LOAD indicator GREEN. At this time the airborne fill and drain valve and boiloff valve are open, LO$_2$ drain may be accomplished when abort is initiated, refer to action 2.

If loss of the umbilicals occurs after FINE LO$_2$ LOAD indicator GREEN, but prior to POWER INTERNAL indicator GREEN, the airborne fill-and-drain valve is closed but the boiloff valve is open. If the LO$_2$ & FUEL indicator on the Launch Control Console is RED, umbilical

Table 4-2.   Emergencies from Start Countdown to MISSILE LIFT UP & LOCKED Indicator Amber (CONT)

ITEM 4.   LOSS OF UMBILICALS (CONT)

ACTION 1 (CONT)

600U4 has ejected. Refer to Action 3. If the LO$_2$ & FUEL indicator is GREEN, umbilical 600U4 did not eject. Refer to Action 2. If the loss of umbilicals occurred after POWER INTERNAL indicator illuminates GREEN, both the airborne fill and drain valve and the boiloff valve are closed. If the LO$_2$ & FUEL

indicator on the Launch control console is illuminated GREEN umbilical 600U4 did not eject. Refer to Action 2. If the LO$_2$ & FUEL indicator is illuminated RED umbilical 600U4 has ejected. Refer to Action 4.

ACTION 2

If umbilical 600U3 has ejected there will be no pressure indications on the gauges on the launch control console. Phase II fuel tank pressures can be maintained by observing PNEUMATICS IN PHASE II indicator in the countdown patch. If the indicator reverts from GREEN to AMBER (fuel tank pressure less than 53 PSI), the deputy shall depress FUEL RAISE pushbutton and hold until PNEU-

MATICS IN PHASE II indicator again illuminates GREEN. The deputy shall then release FUEL RAISE pushbutton immediately to avoid overpressurizing the missile. If TV is available to observe gauges on top of PSMR, tank pressures may be monitored on the gauges instead of using PNEUMATICS IN PHASE II indicator GREEN.

1.   (C) DEPUTY, MAINTAIN PNEUMATICS IN PHASE II INDICATOR GREEN   DIRECTED

At MCCC direction to manually drain LO$_2$ A-1 at LO$_2$ TANKING (PANEL 1 and PANEL 2) shall proceed as follows:

   (a) Position REMOTE LOCAL switch to LOCAL.

   (b) Position L-16 valve switch to OPEN and observe DRAIN VALVE L-16 indicator is illuminated GREEN. If indicator does not illuminate GREEN position L-1 valve switch to OPEN and observe FINE LOAD VALVE L-1 indicator is illuminated GREEN. Valve L-16 will be closed if valve L-1 is used to drain LO$_2$.

   (c) Position A/B F & D valve switch to OPEN. This will allow LO$_2$ to drain from the missile. If airborne fill and drain valve cannot be opened, LO$_2$ must be allowed to boiloff. If it becomes necessary to stop LO$_2$ draining, A-1 shall position A/B F & D valve switch to CLOSE. Position L-16 (or L-1 if used) valve switch if airborne fill-and-drain valve cannot be controlled.

   (d) Observe AIRBORNE FILL & DRAIN VALVE indicator illuminates GREEN. Start timing drain sequence when indicator illuminates GREEN.

2.   (C) A-1, MANUALLY DRAIN LO$_2$ . . . .   DIRECTED

After LO$_2$ has drained for 30 minutes, the system will be ready for securing. At MCCC direction secure the LO$_2$ tanking panels, A-1 shall: (a) Position L-16 (or L-1 if used) valve switch to CLOSE. (b) Observe DRAIN VALVE L-16 (or FINE

LOAD VALVE L-1) indicator is illuminated AMBER. (c) Position A/B F & D valve switch to CLOSE. (d) Observe AIRBORNE FILL & DRAIN VALVE indicator is illuminated AMBER.

3.   (C) A-1, SECURE LO$_2$ TANKING PANELS   . . . .   DIRECTED

The loss of umbilical 600U3 prevents the differential pressure signal from being sent to the logic units from the airborne differential pressure switch. The

pressurization system will switch to emergency mode. At MCCC direction, crew shall apply missile stretch in accordance with table 4-5.

Table 4-2.   Emergencies from Start Countdown to MISSILE LIFT UP & LOCKED Indicator Amber (CONT)

ITEM 4.   LOSS OF UMBILICALS (CONT)

ACTION 2 (CONT)

4.   (C) CREW, PLACE MISSILE IN STRETCH .... DIRECTED

After LO$_2$ drain is complete, MCCC shall direct deputy or M-1 to monitor and maintain missile tank pressures. If TV is not available to monitor PSMR missile tank pressure gauges, M-1 shall proceed to PSC, level 8, to open valve 148. This will ensure that pressure is available at the HCU. M-1 shall then proceed to the HCU, launcher platform level

3, and maintain missile tank pressures (LO$_2$ tank 3.4 to 8.0 PSI, fuel tank 51.0 to 63.9 PSI) using valves 339, 340, 342, and 343. If TV is available to monitor PSMR missile tank pressure gauges, deputy shall maintain phase II missile tank pressures using TV monitor and the applicable RAISE or LOWER pushbutton on launch control console.

5.   (C) DEPUTY, OR M-1 MONITOR AND MAINTAIN TANK PRESSURES .. .. DIRECTED

WARNING

After LO$_2$ is drained, maintenance support is required to safety ordnance items (if installed) and reconnect umbilicals. Exception may be made during a tactical launch if missile crew can determine that no stray voltage exists at the umbilicals.

ACTION 3

If pressure system is in emergency mode and pressure gauges on launch control console indicate zero,

missile fuel tank pressure shall be monitored and controlled during boiloff by maintaining PNEUMATICS IN PHASE II indicator illuminated GREEN. The deputy shall depress FUEL RAISE pushbutton when PNEUMATICS IN PHASE II indicator illuminates AMBER and release the pushbutton when indicator illuminates GREEN. After boiloff, monitor missile tank pressures at PSC and control pressures at the launch control console. If TV monitor is available to observe gauges on top of PSMR, tank pressures may be monitored on the gauges instead of using PNEUMATICS IN PHASE II indicator GREEN.

1.   (C) DEPUTY, MAINTAIN PNEUMATICS IN PHASE II INDICATOR GREEN .. DIRECTED

LO$_2$ shall be allowed to boil off as there is no way of draining the missile unless umbilical 600U4 is reconnected.

2.   (C) ALLOW LO$_2$ TO BOILOFF .... ACCOMPLISHED

WARNING

After LO$_2$ has completely boiled off, maintenance support is required to safety ordnance items (if installed) and reconnect umbilicals.

ACTION 4

The silo overhead doors shall be opened immediately to minimize blast effect if loss of missile occurs. At MCCC direction to open the silo overhead doors, M-1 at CSMOL shall: (a) Position PROGRAMMER RESET key switch to ON (b) Start 40 HP pump by depressing HYDRAULIC 40 HP PUMP

1.   (C) M-1, OPEN SILO DOORS ....

If 600U3 did not eject pressure indications will be present on launch control and the boiloff valve

ON pushbutton. (c) After HYDRAULIC 40 HP PUMP PRESSURE indicator is illuminated GREEN, depress CRIB VERTICAL LOCK pushbutton and observe CRIB VERTICAL LOCK indicator illuminates GREEN.

(d) Depress CRIB HORIZONTAL LOCK pushbutton and observe CRIB HORIZONTAL LOCK indicator illuminates GREEN. (e) Depress SILO DOORS OPEN pushbutton and hold for 30 seconds and observe SILO DOORS CLOSE indicator is extinguished. Approximately 5 minutes are required for the doors to open. Door opening sequence shall be accomplished simultaneously with other steps in this action.

DIRECTED

can be opened. Observe TV to determine if boiloff is present.

Table 4-2. Emergencies from Start Countdown to MISSILE LIFT UP & LOCKED Indicator Amber (CONT)

ITEM 4. LOSS OF UMBILICALS (CONT)

ACTION 4 (CONT)

2. (C) LO₂ BOILOFF .... OBSERVED

With the boiloff valve and air borne fill-and-drain valve closed, the LO₂ tank pressure can only be relieved through the airborne relief valve. Loss of missile can be expected and personnel shall remain in a safe area. At MCCC direction to drive launcher platform up, M-1 at the CSMOL shall depress LAUNCHER PLATFORM UP RUN pushbutton.

3. (C) M-1, DRIVE LAUNCHER PLATFORM UP .... DIRECTED

> ### WARNING
>
> Remain in LCC. Expect loss of missile.

ITEM 5. FIRES IN SILO

This item contains procedures to place the system in the safest possible configuration in the event of a fire. If a fire occurs after the start of countdown the MCCC shall continue a tactical or training launch. A PLX shall be aborted.

1. *(C) ABORT (PLX)* .... *INITIATED*

The MCCC shall attempt to determine the location and intensity of the fire using TV monitors and fire alarm panel.

2. (C) FIRE LOCATION .... IDENTIFIED

If the fire is located in the missile enclosure area, MCCC shall direct M-1 to activate the missile enclosure area fog system by depressing MISSILE ENCLOSURE FOG ON pushbutton on FRCP.

3. (C) M-1, ACTIVATE MISSILE ENCLOSURE AREA FOG SYSTEM .... DIRECTED

After missile has been launched or after abort was initiated during a PLX, at discretion of the MCCC, the crew shall attempt to contain and extinguish the fire using available fire fighting equipment.

4. (C) CREW, COMBAT THE FIRE .... DIRECTED

The MCCC shall ensure that local disaster control and potential hazard procedures are implemented at the earliest possible time that will not conflict with the launch sequence. Personnel in the LCC not involved in the launch or fire fighting procedures may be used to implement the disaster control and potential hazard procedures.

5. (C) DISASTER CONTROL AND POTENTIAL HAZARD PROCEDURES ...... IMPLEMENTED

Table 4-2. Emergencies from Start Countdown to MISSILE LIFT UP & LOCKED Indicator Amber
(CONT)

ITEM 6A. SILO CONTROL CABINET HI-TEMP INDICATOR RED

This item contains procedures to follow when a malfunction in the control air conditioning system is indicated on the FRCP. A high temperature in the control cabinets could cause an overheating of the pressure gauge amplifiers which would send erroneous indications to the launch control console pressure gauges. A nontactical countdown shall be aborted to prevent damage to these amplifiers. Tactical countdown shall be continued. After abort is complete, troubleshoot the malfunctioning system.

1. (C) ABORT (NONTACTICAL COUNTDOWN) . . . . INITIATED

ITEM 6B. MAIN EXHAUST FAN NOT OPERATING INDICATOR RED

This item contains procedures to follow to prevent a hazardous ventilation occuring in the silo when the FRCP indicates that the main exhaust fan is not operating. If the main exhaust fan fails during a countdown, an accumu-lation of fumes in the silo could create an explosion hazard. The MCCC shall continue a tactical countdown. A nontactical countdown shall be aborted. After abort is complete, troubleshoot malfunction system.

1. (C) ABORT (NONTACTICAL COUNTDOWN) . . . . INITIATED

ITEM 6C. STORAGE AREA 25% or 19% INDICATOR RED

This item contains procedures to return the silo to a safe configuration when a gaseous or liquid oxygen/liquid nitrogen leak is indicated on the FRCP. If 25% or 19% oxygen indication occurs, an immediate attempt shall be made using the TV monitors to determine if leakage or spillage is evident, by scanning LO_2 transfer lines/valves and by visual observation of logic and pressure indications. If leakage or spillage is evident the MCCC shall initiate abort. If it can be determined that no leakage or spillage is evident, the countdown may continue under the following conditions:

a. During a tactical countdown.

b. During a nontactical countdown when LN_2 is being used in place of LO_2, proceeding to ACTION 1 may be delayed at MCCC discretion until after MISSILE LIFT UP & LOCKED indicator illuminates GREEN and ABORT indicator illuminates RED.

c. At 576SMS only. During a nontactical countdown using LO_2 if 19% storage area oxygen is indicated on the FRCP and storage area oxygen is not less than 15% as indicated on the remote GOX detectors.

NOTE

At 576D and E sites the remote GOX detectors will alarm and lock up on station detecting 17% oxygen. The remote detector shall be reset after approximately 30 seconds by depressing the RESET pushbutton. The ALARM SILENCE pushbutton may be depressed at any time.

ACTION 1

1. (C) ABORT . . . . INITIATED

When ABORT EXTERNAL indicator illuminates AMBER, immediately stop LO_2 drain. At MCCC direction to stop LO_2 drain, A-1 shall:

(a) Position L-16 valve switch to OPEN to ensure continuous line drain and line pressure relief from airborne fill-and-drain valve to LO_2 storage tank.

(b) Position REMOTE-LOCAL SWITCH to LOCAL. This prevents the LO_2 automatic sequencing from continuing and provides local control of the LO_2 valves.

(c) Verify drain valve L-16 and vent valve N-4 are open by observing DRAIN VALVE L-16 and TOPPING TANK VENT VALVE N-4 indicators are GREEN. Verify vent valve N-5 is open by observing STORAGE TANK VENT VALVE N-5 indicator is GREEN or EXTINGUISHED. Vent valves N-4 and N-5 are automatically opened at start of abort.

(d) Verify all other valves on LO_2 Tanking Panels are CLOSED by observing all other valve indicators are AMBER.

2. (C) A-1 STOP LO_2 DRAIN . . . . DIRECTED

Table 4-2. Emergencies from Start Countdown to MISSILE LIFT UP & LOCKED Indicator Amber (CONT)

ITEM 6C. STORAGE AREA 25% or 19% INDICATOR RED (CONT)

ACTION 1 (CONT)

At MCCC direction for emergency pressurization, deputy shall:

(a) Depress EMERGENCY pushbutton.

This places the missile pressurization in emergency mode and prevents pressures from returning to phase 1 in the event of a LO_2 drain timer malfunction, or

if it is desired to maintain phase II missile pressures in excess of the 45-minute drain time.

(b) Maintain phase II missile tank pressure with the pressurization system in emergency mode.

Pressures can be regulated and maintained by depressing the approporate RAISE or LOWER pushbutton on the launch control console.

3. (C) DEPUTY, PLACE PRESSURIZATION IN EMERGENCY .... DIRECTED

At MCCC direction, A-1 and M-1 shall investigate cause of high oxygen alarm. Logical sources of high oxygen concentration at this time are at the LO_2 storage and LO_2 topping tanks and piping up to the

liquid oxygen prefabs on silo level 7. At launch control center, situation requires close observation of launch control console, fire alarm panel, and facility remote control panel.

4. (C) A-1, M-1, INVESTIGATE CAUSE OF HIGH (OR LOW) OXYGEN ALARM ... DIRECTED

A-1 and M-1 shall check oxygen detector TROUBLE indicator, level 7. If TROUBLE indicator is extinguished, check, AREA ALARM indicator and percentage meter to determine location and amount of

oxygen concentration. Locate leakage and isolate if possible, referring to appropriate technical manuals to ensure valve positioning will not damage or cause additional malfunctions.

5. (C) CAUSE OF OXYGEN ALARM .... DETERMINED

At direction of MCCC to reset the oxygen detector, A-1 and M-1 shall: (a) Depress SYSTEM RESET and HORN SILENCE pushbuttons on oxygen detector cabinet. This will silence audible alarm and restart the station sampling unit. Verify the station sampling is cycling by observing the station indi-

cators. (b) Reset oxygen purge system by depressing OXYGEN PURGE RESET (OR CR-44 RELAY RESET) pushbutton on Facilities Terminal Cabinet NO. 2, level 2. This will reset the gaseous oxygen blast closure and vent fan control circuits.

6. (C) A-1, M-1, RESET OXYGEN DETECTOR SYSTEM .... DIRECTED

If, after investigation of the high oxygen alarm, it is determined that it would be unsafe to continue LO_2 drain, procedures contained in action 2 shall

be followed. If determined safe to continue LO_2 drain, procedures contained in action 3 shall be followed.

ACTION 2

If determined that it would be unsafe to continue LO_2 drain for reasons such as drain line leakage, MCCC shall allow LO_2 to boiloff.

1. (C) ALLOW LO_2 TO BOILOFF .... ACCOMPLISHED

Table 4-2.   Emergencies from Start Countdown to MISSILE LIFT UP & LOCKED Indicator Amber (CONT)

ITEM 6C.   STORAGE AREA 25% or 19% INDICATOR RED (CONT)

ACTION 3

If determined that it would be safe to continue LO$_2$ drain, or after malfunction is corrected at MCCC direction to start LO$_2$ drain, A-1 shall: (a) Cycle LO$_2$ TANKING (PANEL 1) SYSTEM POWER switch OFF for 1 second then ON. This deenergizes the LO$_2$ tanking panels, removes all previous tank-

ing panel commands and restarts the LO$_2$ drain timers (if abort external signal is present): (b) Position REMOTE-LOCAL switch on LO$_2$ TANK-ING (PANEL 1) to REMOTE. This action will reposition airborne fill-and-drain valve to open and automatic drain sequence will start.

1.   (C) A-1, START LO$_2$ DRAIN   .... DIRECTED

MCCC shall direct the deputy to return missile tank pressures to automatic mode by depressing AUTO-MATIC pushbutton on the launch control console.

This will allow the tank pressures to revert to phase 1 at completion of 45-minute drain sequence.

2.   (C) DEPUTY, AUTOMATIC PUSHBUTTON DEPRESSED   .... DIRECTED

ITEM 6D.   AIR WASHER DUST COLLECTING UNITS NOT OPERATING INDICATOR RED

This item contains procedures to follow when a malfunction of the air washer dust collecting units is indicated on the FRCP. A tactical countdown shall be continued. A nontactical countdown shall be

aborted. After abort is completed, troubleshoot the malfunctioning system. An improper operating ventilation system could create an explosion hazard.

1.   *(C) ABORT (NONTACTICAL COUNTDOWN)*   .... *INITIATED*

ITEM 6E.   DIESEL VAPOR HIGH LEVEL INDICATOR RED

This item contains procedures to follow when a high concentration of diesel fuel vapor is indicated on the FRCP. There is danger of an explosion and

fire if an excessive amount of diesel fuel leaks into the silo. A tactical countdown shall be continued. A nontactical countdown shall be aborted.

1.   *(C) ABORT (NONTACTICAL COUNTDOWN)*   .... *INITIATED*

After abort has been initiated or missile has been launched, at MCCC direction to visually check silo levels 5 and 6, A-1 and M-1 shall investigate the cause of diesel fuel vapor high level alarm. If a spil-

lage is discovered, actions necessary to isolate the leak and prevent a fire will be taken. If no leak is discovered, M-1 shall troubleshoot the diesel fuel vapor detector cabinet after abort complete.

2.   (C) A-1, M-1, VISUALLY CHECK SILO LEVELS 5 AND 6   .... DIRECTED

Table 4-2. Emergencies from Start Countdown to MISSILE LIFT UP & LOCKED Indicator Amber (CONT)

ITEM 6F. LCC AIR RECEIVER AND INSTRUMENT AIR RECEIVER LOW PRESSURE INDICATORS RED

This item contains procedures to follow when the FRCP indicates a malfunction in the silo and LCC instrument air systems. If the LCC AIR RECEIVER and the INSTRUMENT AIR RECEIVER LOW PRESSURE indicators illuminate separately, countdown may be continued. If both indicators illuminate in close proximity, a severe leak in the system is indicated and a nontactical countdown shall be aborted. After abort complete troubleshoot the malfunction indication.

1. (C) ABORT (NONTACTICAL COUNTDOWN) . . . . **INITIATED**

ITEM 6G. SILO AIR INTAKE CLOSURES CLOSED/SILO AIR EXHAUST CLOSURES CLOSED INDICATORS RED

These procedures shall be used to return the silo to a safe configuration if the silo blast closures are closed electrically as indicated on the FRCP. During a nontactical countdown MCCC shall initiate abort. Continue a tactical countdown.

1. (C) ABORT (NONTACTICAL COUNTDOWN) . . . . **INITIATED**

If the blast detection system has detected an optic signal but not a radio signal the blast closures should open in 90 seconds and the blast detection cabinet will have to be reset.

If the blast detection system has detected both an optic signal and a radio signal, the blast closures will close and the blast detection cabinet must be reset before the blast closures will open.

If no nuclear blast has been detected and a malfunction has occurred in the blast detection system the blast detection cabinet must be reset.

When directed by the MCCC, M-1 shall reset blast detection cabinet: (a) Position OUTPUT RELAY switch to DISCONNECT. (b) Depress RCVR 1 MANUAL TEST pushbutton. (c) Depress ALARM RESET pushbutton. (d) Position OUTPUT RELAY switch to CONNECT. (e) Depress DETECTION MODE RESET pushbutton. (f) Verify OPTIC MODE indicator is illuminated. (g) Verify channel indicator is cycling.

2. (C) M-1, RESET BLAST DETECTION CABINET . . . . DIRECTED

If blast detection cabinet fails to reset or if the blast closures fail to open, at direction of MCCC, M-1 shall position OUTPUT RELAY switch on blast detection cabinet to DISCONNECT.

3. (C) M-1, OUTPUT RELAY SWITCH TO DISCONNECT . . . . DIRECTED

If blast closures were closed by an actual nuclear blast, emergency water pump P-32 may be operating. If diesel generators are operating in parallel, to prevent a possible overheat condition and possible AC power loss, one generator shall be shut down after SITE HARD indicator GREEN. At MCCC direction to shut down one diesel generator, L-1 shall: (a) Position GENERATOR MAIN BREAKER CONTROL SWITCH to TRIP. Observe GENERATOR MAIN BREAKER CONTROL SWITCH RED indicator extinguishes and GREEN indicator illuminates. (b) Position ENGINE START-STOP SWITCH to STOP and momentarily hold. Release switch and observe ENGINE START-STOP SWITCH RED indicator extinguishes and GREEN indicator illuminates.

Table 4-2.   Emergencies from Start Countdown to MISSILE LIFT UP & LOCKED Indicator Amber (CONT)

ITEM 6G.   SILO AIR INTAKE CLOSURE CLOSED/SILO AIR EXHAUST CLOSURES
CLOSED INDICATORS RED (CONT)

Omit step 4 if diesel generators are not operating in parallel. Do not perform step 4 until SITE HARD indicator is illuminated GREEN.

4.   (C) L-1, SHUT DOWN ONE GENERATOR ....   DIRECTED

5.   (Deleted)

Do not perform step 6 until LO$_2$ has drained, as indicated by LO$_2$ DRAIN COMPLETE indicator illuminated GREEN Or by a slight rise in missile fuel tank pressure, which should occur approximately 25 minutes (missile LO$_2$ tank at 100%) after start of LO$_2$ drain.

If the blast closures are closed and blast detection cabinet is normal, a failure of relay control circuit or a tripped circuit breaker should be suspected.

At direction of the MCCC to troubleshoot blast closure system, M-1 shall: (a) Proceed to essential motor control panel C, silo level 2 and verify control circuit breakers CB-1 and CB-3 are ON. Reset CB-1 and CB-3 if tripped. If circuit breakers had to be reset and blast closures open, omit step (b) through (e) and troubleshoot probable cause in accordance with applicable technical manual.

WARNING

Control circuit voltage is 120 VAC. Use extreme caution while installing jumper wires.

Table 4-2.   Emergencies from Start Countdown to MISSILE LIFT UP & LOCKED Indicator Amber (CONT)

ITEM 6G.   SILO AIR INTAKE CLOSURES CLOSED/SILO AIR EXHAUST CLOSURES CLOSED INDICATOR RED (CONT)

If circuit breakers were not tripped or will not reset, proceed to facilities interface cabinet NO. 1 on level 3 to bypass the control circuit of the blast closures as follows: (b) Install jumper wire on terminal board E-1 between wire numbers 9013 and 9015. (c) Install jumper wire on terminal board E-1 between wires 9015 and 9017. (d) Install jumper wire on terminal board E-1, between wire 9017 and 9019. (e) Install jumper wire from C-5 to wire 9013.

Verify with MCCC that blast closures are open. If blast closures do not open, troubleshoot blast closures system in accordance with applicable technical manual. If blast closures remain closed it may become necessary to shut down the diesel generator to prevent excessive carbon buildup in the exhaust system and contamination of air in the silo. Prior to shut down of the diesel generator, extend work platform NO. 1 to allow missile stretch to be applied if an emergency occurs that may require it.

6.   (C) M-1, TROUBLESHOOT CLOSURES   DIRECTED

7.   (C) M-1, TROUBLESHOOT   DIRECTED

ITEM 6H.   MISSILE POD AIR CONDITIONER MALFUNCTION INDICATOR RED

During a nontactical countdown, if either indicator illuminates RED prior to MISSILE LIFT UP & LOCKED indicator illuminated AMBER, abort sequence shall be initiated. Initiating the abort sequence reduces the length of time that the guidance

system is subjected to an overheat environment and will possibly prevent major damage to the airborne guidance components. No action is required during a tactical countdown; allow the sequence to continue.

1.   *(C) ABORT (NONTACTICAL COUNTDOWN)*   *INITIATED*

ITEM 6I.   EMERGENCY WATER PUMP P-32 ON INDICATOR RED

This is a normal indication for single generator operation, if nonessential power is removed after POWER INTERNAL indicator illuminated GREEN, and no action is required. If generators are operating in parallel, MCCC shall direct L-1 to shut down one generator at the PRCP since emergency water pump P-32 does not have sufficient capacity to cool both diesel engines. At MCCC direction to shut down one generator, L-1 shall:

   (a)   Remove nonessential power, if after POWER INTERNAL indicator is illuminated GREEN, by positioning FEEDER NUMBER 3 NON-ESSENTIAL BUS CONTROL SWITCH to TRIP

and observing FEEDER NUMBER 3 NON-ESSENTIAL BUS CONTROL SWITCH RED indicator extinguishes and GREEN indicator illuminates.

  (b)   Position GENERATOR MAIN BREAKER CONTROL SWITCH to TRIP. Observe GENERATOR MAIN BREAKER CONTROL SWITCH RED indicator extinguishes and GREEN indicator illuminates.

  (c)   Position ENGINE START-STOP SWITCH to STOP and momentarily hold. Release switch and observe ENGINE START-STOP SWITCH RED indicator extinguishes and GREEN indicator illuminates.

1.   *(C) L-1, SHUT DOWN ONE GENERATOR*   *DIRECTED*

A tactical countdown shall be continued on one generator. A nontactical countdown shall be aborted.

2.   *(C) ABORT (NONTACTICAL COUNTDOWN)*   *INITIATED*

Table 4-2. Emergencies from Start Countdown to MISSILE LIFT UP & LOCKED Indicator
Amber (CONT)

ITEM 7.   ENGINES AND GROUND POWER INDICATOR RED (28 VDC POWER AND 400 CYCLE POWER INDICATORS EXTINGUISHED)

This single indication indicates that the solid propellant gas generators (SPGG) temperature is out of tolerance, or that the 28-VDC transformer-rectifier has failed. Particular attention should be given to the fact that the 28 VDC POWER and 400 CYCLE POWER indicators remain extinguished. Erratic temperatures or overheating of the SPGG could affect the condition of the pyrotechnic material within the SPGG. Abort must be initiated to prevent possible loss of missile when the SPGG are ignited at engines start. If the transformer-rectifier has failed, abort shall be initiated immediately to prevent further depletion of emergency batteries at missile inverter start.

1. (C) ABORT . . . . INITIATED

After abort is complete, MCCC shall direct A-1 and M-1 to proceed to level 3 to observe SPGG indicators on AC power distribution panel. If all SPGG indicators are GREEN, an assumption will be made that the power supply-distribution set rectifier circuit has failed.

This is considered to be an emergency condition and manual control of the PSC must be obtained in case 28 VDC is completely lost (step 2). If any of the SPGG indicators are extinguished or RED, the crew need only troubleshoot the SPGG malfunction (step 2).

2. (C) A-1, M-1, OBSERVE SPGG INDICATORS . . . . DIRECTED

If all SPGG indicators are GREEN, MCCC shall direct A-1 and M-1 to obtain manual control of the missile tank pressure at the PSC on level 8. Manual control of pressure is necessary to ensure proper missile tank pressures should 28 VDC be intermittently or gradually lost as the emergency batteries are depleted. Manual control at the PSC will prevent any failure in the automatic system from affecting missile tank pressurization. (a) M-1 shall obtain manual control of missile tank pressurization at the PSC by closing valves 105 and 106. (b) After valves 105 and 106 are closed, A-1 will ensure that the intermittent or gradual loss of 28 VDC does not affect missile tank pressurization by positioning SYSTEM POWER switch on PNEUMATICS (PANEL 1) to OFF. (c) M-1 shall then maintain missile tank pressures at the PSC. Fuel tank pressure shall be maintained at 11.9 to 13.0 PSI using valve 123 to raise pressure or valve 125 to lower pressure. LO_2 tank pressure shall be maintained at 3.4 to 4.2 PSI using valve 124 to raise pressure or valve 126 to lower pressure.

3. (C) A-1, M-1, OBTAIN MANUAL CONTROL OF PSC . . . . DIRECTED

The MCCC shall without delay, direct L-1 to start and configure the pod air conditioner for local operation to prevent prolonged loss of pod cooling to the missile guidance system. L-1 shall start and configure the pod air conditioner for local operation by: (a) Positioning the UNIT STOP/START switch (S-1) to ON; (b) Positioning the REMOTE CONTROL switch (S-3) to OFF.

4. (C) L-1, POSITION POD AIR CONDITIONER FOR LOCAL OPERATION . . . . . DIRECTED

ITEM 8.   400 CYCLE POWER INDICATOR RED

This item contains procedures to follow when 400-cycle power frequency or voltage are out of tolerance, as indicated by the 400 CYCLE POWER indicator illuminating RED. A nontactical countdown shall be aborted immediately. After abort is complete, troubleshoot the malfunctioning system.

1. (C) ABORT (NONTACTICAL COUNTDOWN) . . . . INITIATED

During a tactical countdown, at MCCC direction M-1 shall observe GROUND FREQUENCY and GROUND AC VOLTAGE indicators on MISSILE GROUND POWER (PANEL 1), level 3, to determine which condition needs to be corrected. If GROUND AC VOLTAGE indicator is illuminated, M-1 shall position 400 CPS VOLTAGE SELECTOR SWITCH on MISSILE GROUND POWER (PANEL 2) to MOTOR GENERATOR OUTPUT and observe AC VOLTMETER indicates 116.7

Table 4-2. Emergencies from Start Countdown to MISSILE LIFT UP & LOCKED Indicator Amber
(CONT)

ITEM 8.   400 CYCLE POWER INDICATOR RED (CONT)

($\pm$ 1.4) volts. If GROUND FREQUENCY indicator is illuminated, M-1 shall observe FREQUENCY METER on 400-cycle motor generator panel indicates 400 ($\pm$ 5) CPS. Report conditions to MCCC. If both frequency and voltage are in tolerance, suspect a sensor failure, do not attempt to make any adjustment, continue countdown.

2. (C) M-1, REPORT STATUS OF GROUND FREQUENCY AND GROUND
     AC VOLTAGE INDICATORS ON MISSILE GROUND POWER (PANEL 1)......DIRECTED

If voltage is out of tolerance, at MCCC direction, M-1 shall adjust voltage output of 400-cycle motor generator by rotating ADJUST VOLTAGE rheostat on 400-cycle motor generator control panel. If frequency is out of tolerance, at MCCC direction, L-1 shall slowly raise or lower frequency by adjusting both GENERATOR GOVERNOR MOTOR CONTROL SWITCHES on PRCP.

3. M-1/L-1, ADJUST VOLTAGE/FREQUENCY ....DIRECTED

ITEM 9.   28 VDC POWER INDICATOR AMBER OR RED

This item contains the procedures to follow when 28-VDC power failure is indicated by the 28 VDC POWER indicator illuminating AMBER or RED. If during a tactical launch after missile inverter has been started and it is anticipated that launch can be completed prior to inverter shutdown, abort shall be delayed until missile has been launched. Immediate abort shall be initiated in all other cases.

1. (C) ABORT ....INITIATED

The 28 VDC POWER AMBER or RED indication is considered to be an emergency because either indication could lead to an eventual complete loss of 28 VDC. The AMBER indication means that cooling air is not being supplied to the unit or that the backup batteries have depleted past 40 ampere-hour due to a partial or complete rectifier failure. The RED indication means that 28-VDC power is out of tolerance. After LO_2 drain is complete, manual control of the pressurization system shall be obtained at the pressure system control (PSC). Manual control is necessary should 28 VDC be intermittently or gradually lost as the emergency batteries are depleted. Manual control at the PSC will prevent any failure in the automatic system from affecting missile tank pressurization. At MCCC direction, A-1 and M-1 shall obtain manual control of pressurization system as follows: (a) M-1, close valves 105 and 106 on PSC. (b) A-1, position SYSTEM POWER switch on PNEUMATICS (PANEL 1) to OFF. (c) M-1, maintain missile tank pressures at PSC. Fuel tank pressures shall be maintained at 11.9 to 13.0 PSI using valve 123 to raise pressure or valve 125 to lower pressure. LO_2 tank pressure shall be maintained at 3.4 to 4.2 PSI using valve 124 to raise pressure or valve 126 to lower pressure. After abort is complete, troubleshoot the malfunctioning system.

2. (C) A-1, M-1, OBTAIN MANUAL CONTROL OF PSC ....DIRECTED

The MCCC shall without delay, direct L-1 to start and configure the pod air conditioner for local operation to prevent prolonged loss of pod cooling to the missile guidance system. L-1 shall start and configure the pod air conditioner for local operation by: (a) Positioning the UNIT STOP/START switch (S-1) to ON; (b) Positioning the REMOTE CONTROL switch (S-3) to OFF.

3. (C) L-1, POSITION POD AIR CONDITIONER FOR LOCAL
     OPERATION ....DIRECTED

Table 4-2.   Emergencies from Start Countdown to MISSILE LIFT UP & LOCKED Indicator Amber (CONT)

ITEM 10. GUIDANCE FAIL INDICATOR RED

This item provides a means of recycling the alignment countdown group in an attempt to obtain a successful countdown if guidance failure is indicated by GUIDANCE FAIL indicator illuminating RED. Allow a minimum of 6 minutes from start of countdown before proceeding to action 1.

ACTION 1

At launch control console, MCCC shall recycle guidance system: (a) Depress alternate TARGET SELECT pushbutton. (b) Observe GUIDANCE FAIL indicator extinguish. (c) Observe alternate TARGET indicator illuminates GREEN. (d) Depress launch TAGET SELECT pushbutton. (e) Observe launch TARGET indicator GREEN. The countdown may be continued if GUIDANCE FAIL indicator remains extinguished. If GUIDANCE FAIL indicator illuminates again, the MCCC shall continue with procedures in action 2 if during a tactical launch. If during a nontactical countdown, abort is required. After abort is complete, MCCC shall direct crew to troubleshoot system.

1. (C) TARGET RECYCLE .... INITIATED

ACTION 2

During a tactical launch, if GUIDANCE FAIL indicator illuminates after targets were cycled at MCCC direction to check the guidance countdown group: (a) A-1 shall proceed to level 3, check and report status of fuses in alignment countdown group. If fuses are not blown, an abort is required. (b) If any fuses are abnormal, A-1 shall replace blown fuses.

1. (C) A-1, CHECK GUIDANCE COUNTDOWN GROUP FUSES ... . DIRECTED

After fuses have been replaced, MCCC shall again recycle guidance: (a) Depress alternate TARGET SELECT pushbutton. (b) Observe GUIDANCE FAIL indicator extinguishes. (c) Observe alternate TARGET indicator illuminates GREEN. (d) Depress launch TARGET SELECT pushbutton. (e) Observe launch TARGET indicator GREEN. If GUIDANCE FAIL indicator illuminates again, abort is required. After abort is complete, MCCC shall direct crew to troubleshoot system.

2. (C) TARGET RECYCLE .... INITIATED

ITEM 11. MISSILE LIFT FAIL INDICATOR RED

This item contains procedures to provide a means of correcting a missile lift malfunction and continuing a tactiacl countdown. This procedure will remedy only quickly corrected malfunctions such as low return line pressure, GN$_2$ storage bottles recharge, and tripped circuit breakers. If malfunction occurred during a nontactical countdown, abort is required. After abort is complete, troubleshoot the malfunctioning system.

1. (C) ABORT (NONTACTICAL COUNTDOWN) .... *INITIATED*

If the malfunction occurred during a tactical launch after commit start, allow the countdown to continue. If malfunction occurred during a tactical launch prior to commit start, at MCCC direction to troubleshoot missile lifting system, A-1 and M-1 shall: (a) Position RESET PROGRAMMER key switch on CSMOL to ON. (b) Proceed to level 2 and observe NOT RECHARGED indicato ron local control hy-

Table 4-2. Emergencies from Start Countdown to MISSILE LIFT UP & LOCKED Indicator Amber
(CONT)

ITEM 11.   MISSILE LIFT FAIL INDICATOR RED (CONT)

draulic panel to determine if the accumulator GN$_2$ storage bottles require recharge. (c) If NOT RECHARGED indicator is extinguished, M-1 shall change 28-VDC supply source by depressing the applicable TRANSFER, pushbutton located on missile lifting motor control center on level 1. (d) Reset all circuit breakers located on the missile lifting motor control center, level 1 and on missile lifting electrical control system. This step may correct the malfunction by resetting a tripped circuit breaker. (e) Correlate missile lifting system logic by simultaneously depressing LAUNCHER STATUS and TEST START pushbuttons located on logic cabinet A4A2. This action could correct a logic malfunction and return system to standby. (f) Depress HYDRAULIC 40 HP PUMP ON pushbutton on local control hydraulic

panel. Observe pump starts and that operating hydraulic pressure is reached. This could correct low return line pressure caused by failure of a pressure switch, check valve, or standby pump. (g) After the 40 HP pump has operated for 30 seconds, M-1 shall stop pump by depressing HYDRAULIC 40 HP PUMP OFF pushbutton located on local control hydraulic panel. If return line pressure was the cause of the malfunction, MISSILE LIFT FAIL indicator should extinguish. (h) After the malfunction has or has not been corrected M-1 shall position the RESET PROGRAMMER key switch to the OFF position.

2. (C) M-1, A-1, TROUBLESHOOT MISSILE LIFTING SYSTEM . . . . DIRECTED

If M-1 reports that NOT RECHARGED indicator is illuminated, MCCC shall direct a GN$_2$ recharge. A GN$_2$ recharge shall be initiated only after LN$_2$ LOAD indicator is illuminated GREEN. This is necessary because the valve configuration required by this procedure is dependent upon the liquid nitrogen-helium functions in progress after LN$_2$ LOAD indicator is illuminated GREEN. If GN$_2$ recharge is directed, A-1 shall: (a) Position valve switches 7, 13, 14, 26, 37, 215, 52, and 54 on the LN$_2$-HELIUM (PANEL 1) on silo level 3 to OPEN. (b) Position all other valves switches on LN$_2$-HELIUM (PANEL 1) to CLOSED. (c) Position

LN$_2$-HELIUM (PANEL 1) LOCAL REMOTE switch to LOCAL and standby until GN$_2$ recharge is complete. (d) M-1 shall depress RECHARGE pushbutton on local control hydraulic panel. (e) When M-1 reports that NOT RECHARGED indicator is extinguished, A-1 shall return LN$_2$-HELIUM (PANEL 1) LOCAL REMOTE switch to REMOTE. (f) After the malfunction has or has not been corrected M-1 shall position the RESET PROGRAMMER key switch to the OFF position.

3. (C) A-1, M-1, INITIATE GN$_2$ RECHARGE . . . . DIRECTED

ITEM 12.   LIQUID SPILLAGE (VISUALLY OBSERVED)

This procedure outlines procedures to be accomplished when a spillage of RP-1 fuel, hydraulic fluid or diesel fuel is visually observed during a non-tactical countdown. When a liquid spillage occurs, the MCCC shall depress the ABORT

pushbutton. The abort presence is initiated to immediately start the abort sequence function and attempt to prevent compounding of a hazardous situation involving the sequence of explosive vapors and high GOX concentration.

1. (C) ABORT (NONTACTICAL CD) . . . . INITIATED

The MCCC shall direct M-1 to activate the RP-1 purge fan on the FRCP. Activating the

RP-1 purge fan will exhaust vapors from liquid spillages out of the missile enclosure.

2. (C) M-1 ACTIVATE RP-1 PURGE FAN . . . . DIRECTED

The MCCC at his discretion shall direct M-1 to activate the fire fog system after evaluation of all facts relative to spillage observed. Acti-

vation of the fire fog system will result in extensive water fogging within the missile enclosure and should limit hazards caused by liquid spillage.

Table 4-2. Emergencies from Start Countdown to MISSILE LIFT UP & LOCKED Indicator Amber
(CONT)

ITEM 12.   LIQUID SPILLAGE (VISUALLY OBSERVED) (CONT)

3. (C) M-1 DEPRESS FOG ON PUSHBUTTON . . . . DIRECTED

After the abort sequence is complete the MCCC shall direct M-1 to investigate silo condition and cause of spillage. The crew shall investigate all sources of spillages and immediately attempt to isolate, stop or restrict spillage by referring to appropriate technical manuals.

During and after correction of spillage source, action shall be taken to prevent further hazards from occuring by cleaning spillage by use of most appropriate means available, fire hose, drains, rags, bucket brigade.

4. (C) M-1 TROUBLESHOOT . . . . DIRECTED

Table 4-3. Emergencies from MISSILE LIFT UP & LOCKED Indicator Amber through
MISSILE LIFT UP & LOCKED Indicator Green

ITEM 1. PRESSURE MODE INDICATOR RED

This item contains procedures to follow if PRESSURE MODE indicator illuminates RED during the missile lift sequence. The launcher platform up sequence will stop, the boiloff valve will be enabled, and LO_2 tank pressure should decrease to 3.4 to 8.0 PSI. If PRESSURE MODE indicator illuminates RED,

MCCC shall initiate abort. Missile tank pressures shall be monitored and controlled as in a normal abort while launcher platform is lowering. After ABORT EXTERNAL indicator illuminates AMBER, adjust tank pressures to phase II if necessary, then return pressurization system to automatic mode by depressing AUTOMATIC pushbutton.

1. (C) ABORT . . . . INITIATED

ITEM 2A. GENERATOR MALFUNCTION (ALTERNATE IN STANDBY)

This item contains procedures to place the alternate generator on the line and shut down the faulty generator if a diesel generator malfunction occurs during the missile lift sequence

and the alternate generator is in standby. Steps in this procedure shall be performed by L-1 at the PRCP in attempt to prevent AC power loss. A tactical countdown shall be continued with one generator.

1. (C) COUNTDOWN . . . . CONTINUED

At MCCC direction, L-1 shall place the alternate generator on the line and shutdown the

faulty generator using procedures contained in table 4-8.

2. (C) L-1 PLACE ALTERNATE GENERATOR ON THE LINE AND SHUTDOWN FAULTY
GENERATOR   . . . . . . DIRECTED

ITEM 2B. GENERATOR MALFUNCTION (ALTERNATE NOT IN STANDBY)

This item contains procedures to place the complex in the safest possible configuration after a diesel generator malfunction occurs during the missile lifting sequence with the alternate generator not in standby. Steps in this procedure shall be performed by crew members as indicated. The MCCC shall continue a tactical countdown (a nontactical count-

down would not be started with the alternate generator not in standby) until the missile is launched or an abort is required. The launcher platform shall not be lowered until L-1 has corrected the malfunction of the faulty generator, and restored AC power in order to prevent a possible loss of AC power during the launcher platform lowering sequence.

1. (C) COUNTDOWN (TACTICAL)   CONTINUED

If missile is not launched, MCCC shall direct deputy to depress EMERGENCY pushbutton

after ABORT indicator illuminates AMBER or RED. This will relieve the pressure in the LO_2 tank.

2. (C) DEPUTY, DEPRESS EMERGENCY PUSHBUTTON . . . . DIRECTED

The MCCC shall direct M-1 to position RESET PROGRAMMER key switch on CSMOL to ON after ABORT indicator illuminates AMBER

or RED to prevent the launcher platform from immediately lowering after AC power is restored.

Table 4-3. Emergencies from MISSILE LIFT UP & LOCKED Indicator
Amber through MISSILE LIFT UP & LOCKED Indicator Green (CONT)

ITEM 2B. GENERATOR MALFUNCTION (ALTERNATE NOT IN STANDBY) (CONT)

3. (C) M-1, RESET PROGRAMMER KEY ON .... DIRECTED

The MCCC shall direct L-1 to shutdown the faulty generator at PRCP: (a) Position GENERATOR MAIN BREAKER CONTROL SWITCH to TRIP. Observe GENERATOR MAIN BREAKER CONTROL SWITCH RED indicator extinguishes and GREEN indicator illuminates. (b) Position ENGINE START-STOP SWITCH to STOP and momentarily hold. Release switch and observe ENGINE START-STOP SWITCH RED indicator extinguishes and GREEN indicator illuminates.

4. (C) L-1, SHUTDOWN FAULTY GENERATOR .... DIRECTED

The MCCC shall depress ABORT pushbutton to enable the abort circuit and ensure that missile power changes from internal to external.

5. (C) ABORT PUSHBUTTON .... DEPRESSED

The MCCC shall direct L-1 to troubleshoot the faulty generator. After troubleshooting is completed, L-1 shall correct the malfunction and restore AC power.

6. (C) L-1, TROUBLESHOOT AND RESTORE AC POWER .... DIRECTED

After restoration of AC power, MCCC shall initiate an abort by directing M-1 to position RESET PROGRAMMER key to OFF.

7. (C) M-1, RESET PROGRAMMER KEY OFF .... DIRECTED

After abort is complete, MCCC shall direct the crew to reset complex electrical system to restore the silo to a standby configuration using procedures contained in table 4-7.

8. (C) CREW, RESET ELECTRICAL SYSTEM .... DIRECTED

ITEM 2C. GENERATOR MALFUNCTION (GENERATORS IN PARALLEL)

This item contains procedures to remove a faulty generator from the line if a diesel generator malfunction occurs during missile lifting sequence and the diesels are in parallel. Steps in this procedure shall be performed by crew members as indicated. If a generator malfunction is indicated, countdown shall be allowed to continue (both nontactical and tactical).

1. (C) COUNTDOWN .... CONTINUED

> ┌──────────────────┐
> │   CAUTION   │
> └──────────────────┘
>
> If rapid falling current is indicated on PRCP, perform step 2. If no rapid falling current is indicated, delay step 2 until missile away, or ABORT indicator is illuminated AMBER or RED.

If indications on PRCP are not normal (for example, rapid falling current), MCCC shall immediately direct L-1 to shut down the faulty generator. If the malfunction indications are indicated only on the FRCP (DIESEL GEN D-60 or D-61 OVERSPEED, LOW LUBE OIL PRESS, HI-TEMP indicator illuminated) and PRCP indications are normal, the faulty

Table 4-3. Emergencies from MISSILE LIFT UP & LOCKED Indicator
Amber through MISSILE LIFT UP & LOCKED Indicator Green (CONT)

ITEM 2C. GENERATOR MALFUNCTION (GENERATORS IN PARALLEL) (CONT)

generator shall not be removed from the line until missile is launched or after ABORT indicator is illuminated AMBER or RED. At direction of MCCC to shut down faulty generator, L-1 at PRCP shall:

(a) *Remove nonessential power by positioning FEEDER NUMBER 3 NON-ESSENTIAL BUS CONTROL SWITCH to TRIP if rapid falling current is indicated.*

(b) *Position GENERATOR MAIN BREAKER CONTROL SWITCH to TRIP. Observe GENERATOR MAIN BREAKER CONTROL SWITCH RED indicator extinguishes and GREEN indicator illuminates.*

(c) *Position ENGINE START-STOP SWITCH to STOP and momentarily hold. Release switch and observe ENGINE START-STOP SWITCH RED indicator extinguishes and GREEN indicator illuminates.*

2. (C) L-1, SHUT DOWN FAULTY GENERATOR . . . . **DIRECTED**

ITEM 3. LOSS OF AC POWER

This item contains procedures for attempting to restore the complex to a safe configuration after an AC power loss occurred during the missile lifting sequence. Steps in this procedure shall be performed by crew members as indicated.

ACTION 1

If AC power is lost, MCCC shall immediately initiate abort sequence by depressing ABORT pushbutton. This will stop the countdown sequence and ensure that missile power changes from internal back to external.

1. (C) ABORT . . . . INITIATED

At MCCC direction, M-1 shall position RESET PROGRAMMER key switch on CSMOL to ON. This will prevent the launcher platform from lowering immediately after AC power is restored causing a possible overload and loss of AC power again.

2. (C) M-1, RESET PROGRAMMER KEY ON . . . . DIRECTED

At MCCC direction, L-1 shall attempt to restore AC power using procedures contained in table 4-6A or 4-6B as applicable. If AC power is restored at the PRCP, procedures in Action 2 shall be performed. If AC power cannot be immediately restored at the PRCP, procedures in action 3 will be performed.

3. (C) L-1, RESTORE AC POWER . . . . DIRECTED

ACTION 2

If AC power was restored after performing action 1, at MCCC direction, L-1 shall apply nonessential power by positioning FEEDER NUMBER 3 NON-ESSENTIAL BUS CONTROL SWITCH on PRCP to CLOSE.

1. (C) L-1, NONESSENTIAL POWER ON . . . . DIRECTED

Table 4-3. Emergencies from MISSILE LIFT UP & LOCKED Indicator
Amber through MISSILE LIFT UP & LOCKED Indicator Green (CONT)

ITEM 3.   LOSS OF AC POWER (CONT)

ACTION 2 (CONT)

At MCCC direction, M-1 shall position RESET
PROGRAMMER key switch on CSMOL to OFF.

This will allow abort sequence to continue.

2.   (C) M-1, RESET PROGRAMMER KEY OFF .... DIRECTED

During abort sequence, at MCCC direction, M-1
shall reset LCC electrical system to restore LCC

ventilation using procedures contained in table 4-7.

3.   (C) M-1, RESET LCC ELECTRICAL SYSTEM .... DIRECTED

After the abort sequence is complete, at MCCC
direction, crew shall reset the complex electrical

system using procedures contained in table 4-7.

4.   (C) CREW, RESET ELECTRICAL SYSTEM .... DIRECTED

ACTION 3

If AC power cannot be restored at the PRCP and
if boiloff will not impinge on tension equalizer, deputy shall depress EMERGENCY pushbutton to
open the boilof valve. If boiloff will impinge on
tension equalizer, the tension equalizer must be
sprayed with water or the missile enclosure silo
purge unit (MEPU) extended (placed in operation)
prior to opening the boiloff valve. However, the
LO_2 tank pressure must be maintained at least 20
PSI less than the fuel tank pressure regardless of
the boiloff valve location. As the pressure difference
approaches 20 PSI, deputy shall depress EMER-
GENCY pushbutton and allow sufficient time for
the LO_2 tank pressure to decrease before depressing
AUTOMATIC pushbutton. This shall be repeated.
After water spray or the MEPU is extended, the
boiloff valve shall be opened. Water spray is not
required if boiloff does not impinge on the tension
equalizer.

WARNING

Regardless of LP position, if boiloff valve
is closed, immediately depress E M E R-
GENCY pushbutton. After LO_2 tank pressure decreases to phase II, depress AUTO-
MATIC pushbutton. Enable boiloff valve
periodically to relieve pressure when LO_2
tank pressure increases to 12 PSI.

1.   (C) A-1, SPRAY TENSION EQUALIZER OR EXTEND MEPU .... DIRECTED

If boiloff will not impinge on tension equalizer or
after the tension equalizer is being sprayed or the

MEPU is extended, MCCC shall direct deputy to
depress EMERGENCY pushbutton.

2.   (C) DEPUTY, DEPRESS EMERGENCY PUSHBUTTON .... DIRECTED

3.   (Deleted)

Table 4-3.   Emergencies from MISSILE LIFT UP & LOCKED Indicator
Amber through MISSILE LIFT UP & LOCKED Indicator Green (CONT)

ITEM 3.   LOSS OF AC POWER (CONT)

ACTION 3 (CONT)

At direction of MCCC, L-1 assisted by all available crew members, shall troubleshoot the power generation system and restore AC power.

4.   (C) L-1, TROUBLESHOOT AND RESTORE AC POWER . . . . DIRECTED

5.   (Deleted)

After AC power is restored, at MCCC direction, M-1 shall reset LCC electrical systems to restore LCC ventilation using procedures in table 4-7.

6.   (C) M-1, RESET LCC ELECTRICAL SYSTEM . . . . DIRECTED

If AC power is restored, at MCCC direction, M-1 shall position RESET PROGRAMMER key switch on CSMOL to OFF. This will allow abort sequence to continue.

7.   (C) M-1, RESET PROGRAMMER KEY OFF . . . . DIRECTED

After abort is complete, at MCCC direction, crew shall reset complex electrical system using procedures contained in table 4-7.

8.   (C) CREW, RESET ELECTRICAL SYSTEM . . . . DIRECTED

ITEM 4.   LOSS OF UMBILICALS

At this time the missile lifting system is in sequence 1, pneumatics are internal, and flight programmer may be armed. The first thing to consider in a tactical or training launch is the flight programmer position as an umbilical loss could cause the programmer to run, which in turn will cause booster staging and retarding-rocket firing. If umbilical loss occurs during a nontactical countdown, the primary concern is the LO_2 airborne fill-and-drain valve position as indicated on LO_2 TANKING (PANEL 2). Because of the physical location of umbilical 600U4, if the AIRBORNE FILL & DRAIN VALVE indicator is GREEN, it is most probable that a complete electrical umbilical ejection has occurred.

WARNING

During a tactical or training launch if umbilical loss occurs after PROGRAMMER ARMED indicator has been GREEN, the flight programmer in the missile may be running and the retarding rockets may fire within 5 minutes after loss of umbilicals. Loss of the missile may occur. Do not initiate abort. Allow countdown to continue and leave the missile up and locked. All personnel shall remain in the LCC.

Table 4-3. Emergencies from MISSILE LIFT UP & LOCKED Indicator
Amber through MISSILE LIFT UP & LOCKED Indicator Green (CONT)

ITEM 4.   LOSS OF UMBILICALS (CONT)

WARNING

If AIRBORNE FILL & DRAIN VALVE indicator is GREEN, continue countdown and do not initiate abort. All personnel shall remain in the LCC. Expect loss of the missile.

If M C C C observes that AIRBORNE FILL & DRAIN VALVE indicator remains AMBER, which indicates that umbilical 600U4 was not ejected and the airborne fill-and-drain valve can be controlled, he shall immediately initiate abort and direct A-1 to place LO_2 tanking panels to local control by positioning R E M O T E - L O C A L switch on LO_2 TANKING (PANEL 1) to LOCAL.

1.   (C) AIRBORNE FILL & DRAIN VALVE INDICATOR AMBER .... OBSERVED

2.   (C) ABORT .... INITIATED

3.   (C) A-1, LO_2 TANKING PANEL TO LOCAL .... DIRECTED

When launcher platform reaches the down locks and ABORT EXTERNAL indicator is AMBER, MCCC shall direct deputy to depress EMERGENCY pushbutton in a attempt to open the boiloff valve and also enable the RAISE and LOWER pushbuttons in case the pressurization remains in automatic mode when the launcher platform reaches the down locks.

4.   (C) DEPUTY, DEPRESS EMERGENCY PUSHBUTTON .... DIRECTED

If umbilical 600U3 has ejected there will be no pressure indications on the gauges on the launch control console. Phase II fuel tank pressures can be maintained by observing PNEUMATICS IN PHASE II indicator in the countdown patch. If the indicator reverts from GREEN to AMBER (fuel tank pressure less than 53 PSI), the deputy shall depress FUEL RAISE pushbutton and hold until PNEUMATICS IN PHASE II indicator again illuminates GREEN. The deputy shall then release FUEL RAISE pushbutton immediately to avoid overpressurizing the missile. If TV to monitor missile tank pressure gauges on PSMR is available, the gauges may be used instead of PNEUMATICS IN PHASE II indicator GREEN.

5.   (C) DEPUTY, MAINTAIN PNEUMATICS IN PHASE II INDICATOR GREEN .... DIRECTED

At MCCC direction to manually drain LO_2 A-1 at LO_2 TANKING (PANEL 1 and PANEL 2) shall proceed as follows:

(a) Position L-16 valve switch to OPEN and observe DRAIN VALVE L-16 indicator is illuminated GREEN. If indicator does not illuminate GREEN position L-1 valve switch to OPEN and observe FINE LOAD VALVE L-1 indicator is illuminated GREEN. Valve L-16 will be closed if valve L-1 is used to drain LO_2.

(b) Position A/B F&D valve switch to OPEN. This will allow LO_2 to drain from the missile. If airborne fill and drain valve cannot be opened, LO_2 must be allowed to boiloff. If it becomes necessary to stop LO_2 draining, A-1 shall position A/B F&D valve switch to CLOSED.

(c) Observe AIRBORNE FILL & DRAIN VALVE indicator illuminates GREEN. Start timing drain sequence when indicator illuminates GREEN.

6.   (C) A-1, MANUALLY DRAIN LO_2 .... DIRECTED

After LO_2 has drained for 30 minutes, the system will be ready for securing. At MCCC direction to secure the LO_2 tanking panels, A-1 shall: (a) Position L-16 (or L-1 if used) valve switch to CLOSE. (b) Observe DRAIN VALVE L-16 (or FINE LOAD VALVE L-1) indicator is illuminated AMBER. (c) Position A/B F&D valve switch to CLOSE. (d) Observe AIRBORNE FILL & DRAIN VALVE indicator is illuminated AMBER.

Table 4-3. Emergencies from MISSILE LIFT UP & LOCKED Indicator
Amber through MISSILE LIFT UP & LOCKED Indicator Green (CONT)

ITEM 4. LOSS OF UMBILICALS (CONT)

7. (C) A-1, SECURE LO$_2$ TANKING PANELS .... DIRECTED

The loss of umbilical 600U3 prevents the differential pressure signal from being sent to the logic units from the airborne differential pressure switch. The pres-surization system will switch to emergency mode. At MCCC direction, crew shall apply missile stretch in accordance with table 4-5.

8. (C) CREW, PLACE MISSILE IN STRETCH .... DIRECTED

After LO$_2$ drain is complete, MCCC shall direct deputy or M-1 to monitor and maintain missile tank pressures. If TV is not available to monitor PSMR missile tank pressure gauges, M-1 shall proceed to PSC, level 8, to open valve 148. This will ensure that pressure is available at the HCU. M-1 shall then proceed to the HCU, launcher platform level 3, and maintain missile tank pressures (LO$_2$ tank 3.4 to 8.0 PSI, fuel tank 51.0 to 63.9 PSI) using valves 339, 340, 342, and 343. If TV is available to monitor PSMR missile tank pressure gauges, deputy shall maintain phase II missile tank pressures using TV monitor and the applicable RAISE or LOWER pushbutton on launch control console.

9. (C) DEPUTY OR M-1, MONITOR AND MAINTAIN TANK PRESSURES .... DIRECTED

WARNING

After LO$_2$ is drained, maintenance support is required to safety ordnance items (if installed) and reconnect umbilicals. Exception may be made during tactical launch if missile crew can determine that no stray voltage exists at the umbilicals.

ITEM 5. FIRES IN SILO

This item contains procedures to place the system in the safest possible configuration in the event of a fire during the missile lift sequence. If a fire occurs, MCCC shall allow the sequence to continue as this will remove the missile from the danger area in the shortest possible time.

1. (C) COUNTDOWN .... CONTINUED

The MCCC shall attempt to determine the location and intensity of the fire using TV monitors and fire alarm panel.

2. (C) FIRE LOCATION .... IDENTIFIED

If the fire is located in the missile enclosure area, the MCCC will direct M-1 to activate the missile en-closure area fog system by depressing MISSILE ENCLOSURE FOG ON pushbutton on FRCP.

3. (C) M-1, ACTIVATE MISSILE ENCLOSURE AREA FOG SYSTEM .... DIRECTED

After MISSILE LIFT UP & LOCKED GREEN, do not initiate abort until fire has been extinguished and the silo has been inspected and assured to be safe. If the missile was not launched, missile LO$_2$ tank pressure shall be properly maintained by the deputy by depressing EMERGENCY pushbutton to enable the boiloff valve. The EMERGENCY pushbutton is enabled 15 seconds after MISSILE LIFT UP & LOCKED indicator GREEN. If the launcher platform fails to reach the up and locked

Table 4-3. Emergencies from MISSILE LIFT UP & LOCKED Indicator
Amber through MISSILE LIFT UP & LOCKED Indicator Green (CONT)

ITEM 5. FIRES IN SILO (CONT)

position, position the RESET PROGRAMMER key switch on CSMOL to ON and depress ABORT pushbutton. This will allow partial abort, but prevent the launcher platform from lowering until the silo is safe. Missile LO_2 tank pressures shall be controlled by the EMERGENCY pushbutton. The

MCCC shall determine, by any and all means available, if an attempt to combat the fire is feasible. At the direction of the MCCC, the crew shall attempt to contain and extinguish the fire using available fire fighting equipment.

4. (C) CREW, COMBAT THE FIRE .... DIRECTED

MCCC shall ensure that local disaster control and potential hazard procedures are implemented at the earliest possible time that will not conflict with the

launch sequence. Personnel in the LCC not involved in the launch or fire fighting procedures may be used to implement the disaster control procedures.

5. (C) DISASTER CONTROL AND POTENTIAL HAZARD PROCEDURES ...... IMPLEMENTED

ITEM 6A. SILO CONTROL CABINET HI TEMP INDICATOR RED

This item contains procedures to follow when a malfunction in the control cabinet air conditioning system is indicated on the FRCP. Allow the count-

down to continue as it will be completed within a short time. M-1 shall investigate the cause of the malfunction indication after abort complete.

1. (C) COUNTDOWN .... CONTINUED

ITEM 6B. MAIN EXHAUST FAN NOT OPERATING INDICATOR RED

This item contains procedures to follow to prevent a hazardous ventilation condition in the silo when the FRCP indicates that the main exhaust fan is

not operating. The MCCC shall continue the countdown and direct M-1 to troubleshoot the malfunction indication after abort complete.

1. (C) COUNTDOWN .... CONTINUED

ITEM 6C. STORAGE AREA OXYGEN 25% OR 19% INDICATOR RED

ACTION 1

This item contains procedures to return the silo to a safe configuration when a gaseous or liquid oxygen/nitrogen leak or spillage is indicated on the FRCP during the missile lift sequence. If 25% or 19% oxygen indication occurs, an immediate attempt shall be made using the TV monitor to deter-

mine if leakage from the launcher platform disconnect is apparent. If leakage from the launcher platform is not apparent, the countdown may continue to missile lift up and locked or missile away, at which time an immediate abort shall be initiated.

1. (C) COUNTDOWN .... CONTINUED

Table 4-3. Emergencies from MISSILE LIFT UP & LOCKED Indicator
Amber through MISSILE LIFT UP & LOCKED Indicator Green (CONT)

ITEM 6C. STORAGE AREA OXYGEN 25% OR 19% INDICATOR RED (CONT)

If leakage is apparent from the launcher platform LO_2 disconnect, an immediate abort shall be initiated.

2. *(C) ABORT* .... ***INITIATED***

When ABORT EXTERNAL indicator illuminates AMBER, immediately stop LO_2 drain. At MCCC direction to stop LO_2 drain, A-1 shall:

(a) Position DRAIN VALVE L-16 switch to OPEN.

This ensures continuous line drain and line pressure relief from airborne fill-and-drain valve to LO_2 storage tank when LOCAL REMOTE switch is positioned to LOCAL.

(b) Position LOCAL REMOTE switch on LO_2 TANKING (PANEL 1) to LOCAL.

This prevents the LO_2 automatic sequencing from continuing and provides local control of the LO_2 valves.

(c) Verify drain valve L-16 and vent valve N-4 are open by observing DRAIN VALVE L-16 and TOPPING TANK VENT VALVE N-4 indicators are GREEN. Verify vent valve N-5 is open by observing STORAGE TANK VENT VALVE N-5 indicator is GREEN or EXTINGUISHED.

Vent valves N-5 and N-4 are automatically opened at start of abort.

(d) Verify all other valves on LO_2 tanking panels are closed by observing all other valve indicators are illuminated AMBER.

When LOCAL REMOTE switch was positioned to LOCAL, airborne fill-and-drain valve closed and stopped LO_2 drain.

3. *(C) A-1, STOP LO_2 DRAIN* .... ***DIRECTED***

At MCCC direction, deputy shall: (a) *Depress EMERGENCY pushbutton* on the launch control console after ABORT EXTERNAL indicator illuminates AMBER. This places the missile pressurization in emergency mode and prevents pressures from returning to phase I in the event of a LO_2 drain timer malfunction, or if it is desired to main-

tain phase II missile pressures in excess of the 45-minute drain time. (b) *Maintain phase II missile tank pressures*. With the missile tank pressure in emergency mode, pressures can be regulated and maintained by depressing the appropriate LO_2 and FUEL RAISE or LOWER pushbuttons on the launch control console.

4. (C) DEPUTY, PLACE PRESSURIZATION IN EMERGENCY .... DIRECTED

At MCCC direction, A-1 and M-1 shall investigate cause of high (or low) oxygen alarm. Logical sources of high (or low) oxygen concentration at this time are at the LO_2 storage and LO_2 topping tanks, and

piping up to the liquid oxygen prefabs on silo level 7. At launch control center, situation requires close observation of launch control console, fire alarm panel, and facility remote control panel.

5. (C) A-1, M-1, INVESTIGATE CAUSE OF HIGH (OR LOW) OXYGEN ALARM .... DIRECTED

A-1 and M-1 shall check oxygen detector TROUBLE indicator, level 7, and if illuminated, troubleshoot the oxygen detector cabinet. If TROUBLE indicator is extinguished, check area alarm indicator and percentage meter to determine location and amount

of oxygen concentration. Locate leakage and isolate if possible, referring to appropriate technical manual, to ensure valve positioning shall not damage or cause additional malfunctions.

6. (C) CAUSE OF OXYGEN ALARM .... DETERMINED

At direction of the MCCC to reset oxygen detector, A-1 and M-1 shall: (a) Depress SYSTEM RESET and HORN SILENCE pushbuttons at oxygen de-

tector cabinet. This will silence audible alarm and restart the station sampling unit. Verify the station sampling is cycling by observing the station indi-

Table 4-3. Emergencies from MISSILE LIFT UP & LOCKED Indicator
Amber through MISSILE LIFT UP & LOCKED Indicator Green (CONT)

ITEM 6C. STORAGE AREA OXYGEN 25% OR 19% INDICATOR RED (CONT)

ACTION 1 (CONT)

cators. (b) Reset oxygen purge system by depressing OXYGEN PURGE RESET (or CR-44 RELAY RESET) pushbutton at the facilities terminal cabinet NO. 2, level 2. This will reset the gaseous oxygen blast closure and vent fan control circuits. Perform action 2 if system is unsafe for LO_2 drain. Perform action 3 if system is safe for LO_2 drain.

7.   (C) A-1, M-1, RESET OXYGEN DETECTOR SYSTEM .... DIRECTED

ACTION 2

If the MCCC determines system is unsafe for LO_2 drain, crew shall refer to nomograms in section VI and allow LO_2 to boiloff.

1.   (C) ALLOW LO_2 TO BOILOFF .... ACCOMPLISHED

ACTION 3

If the MCCC determines the system is safe for LO_2 drain, at his direction, A-1 shall restart LO_2 drain at LO_2 tanking panels: (a) Cycle LO_2 tanking power at LO_2 TANKING (PANEL 1) by positioning SYSTEM POWER switch OFF for one second then ON. This de-energizes the LO_2 tanking panels and removes all previous tanking panel commands and restarts the LO_2 drain timers (if abort external signal is present). (b) Return LO_2 tanking panels to remote by positioning LOCAL REMOTE switch on LO_2 TANKING (PANEL 1) to REMOTE. This action will reposition airborne fill-and-drain vlave to open and the automatic drain sequence will start.

1.   (C) A-1, START LO_2 DRAIN .... DIRECTED

After LO_2 DRAIN COMPLETE indicator illuminates AMBER, MCCC shall direct Deputy to return pressurization system to automatic mode by depressing A U T O M A T I C pushbutton on the launch control console. This will allow the tank pressures to revert to phase I at completion of 45-minute drain sequence.

2.   (C) DEPUTY, DEPRESS AUTOMATIC PUSHBUTTON .... DIRECTED

ITEM 6D. AIR WASHER DUST COLLECTING UNITS NOT OPERATING INDICATOR RED

This item contains procedures to follow when a malfunction of the air washer dust collecting units is indicated on the FRCP. The countdown shall be continued. After abort complete, M-1 will troubleshoot the malfunction indication.

1. (C) COUNTDOWN .... CONTINUED

Table 4-3. Emergencies from MISSILE LIFT UP & LOCKED Indicator
Amber through MISSILE LIFT UP & LOCKED Indicator Green (CONT)

ITEM 6E. DIESEL VAPOR HIGH LEVEL INDICATOR RED

This item contains procedures to follow when a high concentration of diesel fuel vapor is indicated on the FRCP. There is danger of an explosion and fire if an excessive amount of diesel fuel leaks into the silo. The countdown shall be continued so that the missile will be as far from the spillage as possible.

1. (C) COUNTDOWN .... **CONTINUED**

After MISSILE LIFT UP & LOCKED indicator illuminates GREEN, at MCCC direction, L-1 and M-1 shall visually check silo levels 5 and 6 for cause of diesel fuel vapor high level alarm. Abort shall not be initiated until the silo has been determined safe from potential fire or explosion. If the missile was not launched, proper missile LO$_2$ tank pressures shall be maintained by enabling the boiloff valve. If the launcher platform failed to reach the up and locked position. RESET PROGRAMMER key switch on CSMOL shall be positioned on ON, ABORT pushbutton depressed, and proper missile LO$_2$ tank pressure maintain. At MCCC direction, L-1 and M-1 shall inspect levels 5 and 6. If a spillage is discovered, action necessary to isolate the leak and prevent a fire potential shall be taken. If no leak is discovered, M-1 shall troubleshoot the diesel vapor detection cabinet after abort complete.

2. (C) L-1, M-1, VISUALLY CHECK SILO LEVELS 5 AND 6 .... **DIRECTED**

When the silo is determined safe, the MCCC shall initiate the abort sequence.

3. (C) ABORT .... **INITIATED**

ITEM 6F. LCC AIR RECEIVER AND INSTRUMENT AIR RECEIVER LOW PRESSURE INDICATORS RED

This item contains procedures to follow when the FRCP indicates a malfunction in the silo and LCC instrument air systems. If the LCC AIR RECEIVER and the INSTRUMENT AIR RECEIVER LOW PRESSURE indicators illuminate separately, count-down may be continued. If both indicators illuminate in close proximity, a severe leak in the system is indicated and a nontactical countdown shall be aborted. After abort complete, troubleshoot the malfunction indication.

1. (C) ABORT (NONTACTICAL COUNTDOWN) .... **INITIATED**

ITEM 6G. SILO AIR INTAKE CLOSURES CLOSED/SILO AIR EXHAUST CLOSURES CLOSED INDICATORS RED

These procedures shall be used to return the silo to a safe configuration if the silo blast closures are closed electrically as indicated on the FRCP. The MCCC shall allow the sequence to continue and immediately after ABORT indicator illuminates RED, depress ABORT pushbutton.

1. (C) COUNTDOWN .... **CONTINUED**

2. (C) ABORT .... **INITIATED**

If the blast detection system has detected an optic signal but not a radio signal the blast closures should open in 90 seconds and the blast detection cabinet will have to be reset.

Table 4-3. Emergencies from MISSILE LIFT UP & LOCKED Indicator
Amber through MISSILE LIFT UP & LOCKED Indicator Green (CONT)

ITEM 6G. SILO AIR INTAKE CLOSURE CLOSED/SILO AIR EXHAUST CLOSURES CLOSED INDICATORS RED (CONT)

If the blast detection system has detected both an optic signal and a radio signal, the blast closures will close and the blast detection cabinet must be reset before the blast closures will open.

If no nuclear blast has been detected and a malfunction has occurred in the blast detection system the blast detection cabinet must be reset.

When directed by the MCCC, M-1 shall reset blast detection cabinet: (a) Position OUTPUT RELAY switch to DISCONNECT. (b) Depress RCVR 1 MANUAL TEST pushbutton. (c) Depress ALARM RESET pushbutton. (d) Position OUTPUT RELAY switch to CONNECT. (e) Depress DETECTION MODE RESET pushbutton. (f) Verify OPTIC MODE indicator is illuminated. (g) Verify channel indicator is cycling.

3. (C) M-1, RESET BLAST DETECTION CABINET . . . . DIRECTED

If blast detection cabinet fails to reset or if the blast closures fail to open, at direction of

MCCC, M-1 shall position OUTPUT RELAY switch on blast detection cabinet to DISCONNECT.

4. (C) M-1, OUTPUT RELAY SWITCH TO DISCONNECT . . . . DIRECTED

If blast closures were closed by an actual nuclear blast, or by a malfunction in the blast detection or blast closures system, emergency water pump P-32 may be operating. If diesel generators are operating in parallel, to prevent a possible overheat condition and possible AC power loss, one generator shall be shut down after performing step 4 and blast closures are not open. A MCCC direction to shut down one diesel generator, L-1 shall: (a) Position GENERATOR MAIN BREAKER CONTROL SWITCH

to TRIP. Observe GENERATOR MAIN BREAKER CONTROL SWITCH RED indicator extinguishes and GREEN indicator illuminates. (b) Position ENGINE START-STOP SWITCH to STOP and momentarily hold. Release switch and observe ENGINE START-STOP SWITCH RED indicator extinguishes and GREEN indicator illuminates.

Omit step 5 if diesel generators are not operating in parallel.

5. (C) L-1, SHUT DOWN ONE GENERATOR . . . . DIRECTED

6. Deleted

Do not perform step 7 until LO$_2$ has drained, as indicated by LO$_2$ DRAIN COMPLETE indicator illuminated GREEN or by a slight rise in missile fuel tank pressure, which should occur approximately 25 minutes (missile LO$_2$ tank at 100%) after start of LO$_2$ drain.

If the blast closures are closed and blast detection cabinet is normal, a failure of relay control circuit or a tripped circuit breaker should be suspected.

At direction of the MCCC to troubleshoot blast closure system, M-1 shall: (a) Proceed to essential motor control panel C, silo level 2 and verify control circuit breakers CB-1 and CB-3 are ON. Reset CB-1 and CB-3 if tripped. If circuit breakers had to be reset and blast closures open, omit step (b) through (e) and troubleshoot probable cause in accordance with applicable technical manual.

T.O. 21M-HGM16F-1

Section IV   Table 4-3. Emergencies from MISSILE LIFT UP & LOCKED Indicator
Amber through MISSILE LIFT UP & LOCKED Indicator Green (CONT)

ITEM 6G. SILO AIR INTAKE CLOSURE CLOSED/SILO AIR EXHAUST CLOSURES
CLOSED INDICATORS RED (CONT)

WARNING

Control circuit voltage is 120 VAC. Use extreme caution while installing jumper wires.

If circuit breakers were not tripped or will not reset, proceed to facilities interface cabinet NO. 1 on level 3 to bypass the control circuit of the blast closures as follows: (b) Install jumper wire on terminal board E-1 between wire numbers 9013 and 9015. (c) Install jumper wire on terminal board E-1 between wires 9015 and 9017. (d) Install jumper wire on terminal board E-1, between wires 9017 and 9019. (e) Install jumper wire from C-6 to wire 9013. Verify with MCCC that blast closures are open. If blast closures do not open, troubleshoot blast closures system in accordance with applicable technical manual. If blast closures remain closed it may become necessary to shut down the diesel generator to prevent excessive carbon buildup in the exhaust system and contamination of air in the silo. Prior to shut down of the diesel generator, extend work platform NO. 1 to allow missile stretch to be applied if an emergency occurs that may require it.

7. (C) M-1, TROUBLESHOOT BLAST CLOSURES .... DIRECTED

The MCCC shall direct M-1 to troubleshoot the cause of emergency.

8. (C) M-1, TROUBLESHOOT   DIRECTED

ITEM 6H. MISSILE POD AIR CONDITIONER MALFUNCTION INDICATOR RED

This condition is not an emergency at this time because of the short time remaining prior to completion of the missile up-run sequence.

1. *(C) COUNTDOWN* .... *CONTINUED*

ITEM 6I. EMERGENCY WATER PUMP P-32 ON INDICATOR RED

This is a normal indication and no action is required if nonessential power has been removed for single generator operation. If generators are operating in parallel, allow both generators to remain on the line to ensure successful completion of the up-run sequence.

1. *(C) COUNTDOWN* .... *CONTINUED*

After missile is launched or ABORT indicator has illuminated AMBER or RED, MCCC shall direct L-1 to shut down one generator at the PRCP since emergency water pump P-32 does not have suficient capacity to cool both diesel engines. At MCCC direction to shut down one generator, L-1 shall:

(a) Position GENERATOR MAIN BREAKER CONTROL SWITCH to TRIP. Observe GENERATOR

MAIN BREAKER CONTROL SWITCH RED indicator extinguishes and GREEN indicator illuminates.

(b) Position ENGINE START-STOP SWITCH to STOP and momentarily hold. Release switch and observe ENGINE START-STOP SWITCH RED indicator extinguishes and GREEN indicator illuminates.

2. *(C) L-1, SHUT DOWN ONE GENERATOR* .... *DIRECTED*

Table 4-3. Emergencies from MISSILE LIFT UP & LOCKED Indicator
Amber through MISSILE LIFT UP & LOCKED Indicator Green (CONT)

ITEM 7. ENGINES AND GROUND POWER INDICATOR RED (28 VDC POWER AND 400 CYCLE POWER INDICATORS EXTINGUISHED)

This indication can occur only because of a logic malfunction and will not affect countdown.

1. (C) COUNTDOWN . . . . **CONTINUED**

ITEM 8. 400 CYCLE POWER INDICATOR RED

This condition does not affect a countdown at this time as this indicator monitors external frequency and voltage, neither of which are required to be in tolerance after power change over to internal. The MCCC shall allow the countdown to continue.

1. (C) COUNTDOWN . . . . **CONTINUED**

ITEM 9. 28 VDC POWER INDICATOR AMBER OR RED

This condition does not constitute an emergency at this time as missile power is internal and launch will not be affected by an external 28 VDC malfunction. The MCCC shall allow the countdown to continue.

1. (C) COUNTDOWN . . . . **CONTINUED**

After ABORT indicator illuminates RED, MCCC shall immediately initiate abort in order to lower the missile, drain LO$_2$, and place the pressurization system in a safe configuration. If launcher platform fails to reach the up and locked position, MCCC shall delay initiating abort sequence until MISSILE LIFT FAIL indicator illuminates RED and ABORT indicator illuminates AMBER. If missile was not launched, proper missile tank pressures must be maintained by enabling the boiloff valve.

2. (C) ABORT . . . . INITIATED

Table 4-3. Emergencies from MISSILE LIFT UP & LOCKED Indicator
Amber through MISSILE LIFT UP & LOCKED Indicator Green (CONT)

ITEM 9. 28 VDC POWER INDICATOR AMBER OR RED (CONT)

If missile was not launched, the 28VDC POWER AMBER or RED indication is considered to be an emergency because either indication could lead to an eventual complete loss of 28VDC. The AMBER indication means that cooling air is not being supplied to the unit or that the backup batteries have depleted past 40 ampere-hours due to a partial or complete rectifier failure.   The RED indication means that 28-VDC power is out of tolerance. After LO_2 drain is complete, manual control of the pressurization system shall be obtained at the pressure system control (PSC). Manual control is necessary should 28 VDC be intermittently or gradually lost as the emergency batteries are depleted. Manual control at the PSC will

prevent any failure in the automatic system from affecting missile tank pressurization. At MCCC direction, A-1 and M-1 shall obtain manual control of pressurization system as follows: (a) M-1, close valves 105 and 106 on PSC. (b) A-1, position SYSTEM POWER switch on PNEUMATICS (PANEL 1) to OFF. (c) M-1, maintain missile tank pressures at PSC. Fuel tank pressure shall be maintained at 11.9 to 13.0 PSI using valve 123 to raise pressure or valve 125 to lower pressure. LO_2 tank pressure shall be maintained at 3.4 to 4.2 PSI using valve 124 to raise pressure or valve 126 to lower pressure. After abort is complete, troubleshoot the malfunctioning system.

3. (C) A-1, M-1, OBTAIN MANUAL CONTROL OF PSC . . . . DIRECTED

The MCCC shall without delay, direct L-1 to start and configure the pod air conditioner for local operation to prevent prolonged loss of pod cooling to the missile guidance system. L-1 shall start and configure the pod air

conditioner for local operation by: (a) Positioning the UNIT STOP/START switch (S-1) to ON: (b) Positioning the REMOTE CONTROL switch (S-3) to OFF.

4. (C) L-1, POSITION POD AIR CONDITIONER FOR LOCAL OPERATION . . . . . DIRECTED

ITEM 10. GUIDANCE FAIL INDICATOR RED

This indication can occur only because of a logic malfunction and will not affect countdown.
1. (C) COUNTDOWN . . . . CONTINUED

ITEM 11. MISSILE LIFT FAIL INDICATOR RED

This item contains procedures to follow when MISSILE LIFT FAIL indicator illuminates RED during a sequence 1 operation of the missile lifting system.

> **CAUTION**
>
> If ABORT indication illuminates AMBER, abort shall be initiated to prevent cavitation of the engine turbopumps when the engines are started after a prolonged up-run sequence.

Observe silo doors and launcher platform on TV monitor. If motion is observed, the countdown will continue. If launcher platform motion stops or if no motion is observed, wait for ABORT indicator to illuminate AMBER then initiate abort.

1. (C) COUNTDOWN . . . . CONTINUED

Table 4-3. Emergencies from MISSILE LIFT UP & LOCKED Indicator
Amber through MISSILE LIFT UP & LOCKED Indicator Green (CONT)

ITEM 12. LIQUID SPILLAGE (VISUALLY OBSERVED)

This procedure outlines procedures to be accomplished when a spillage of RP-1 fuel, hydraulic fluid or diesel fuel is visually observed during a non-tactical countdown. The launcher platform should normally be rising or in the fully up and locked position. When a liquid spillage occurs, the MCCC shall depress the ABORT pushbutton. Normally, consideration shall be given to have a boil off valve above the silo cap prior to depressing the ABORT pushbutton.

The abort sequence is initiated to immediately start abort sequence function and attempt to prevent compounding of a hazardous situation involving the presence of explosive vapors and high GOX concentration.

1. (C) ABORT . . . . INITIATED

The MCCC shall direct M-1 to activate the RP-1 purge fan on the FRCP. Activating the RP-1 purge fan will exhaust vapors from liquid spillage out of the missile inclosure.

2. (C) ACTIVATE RP-1 PURGE FAN . . . . DIRECTED

The MCCC at his discretion shall direct M-1 to activate the fire fog system after the ABORT EXTERNAL indicator is AMBER after evaluation of all facts relative to spillage observed. Activation of the fire fog system will result in extensive water fogging within the missile inclosure and should limit hazards caused by liquid spillage. Activation of the fire fog system is delayed until after ABORT EXTERNAL indicator AMBER to prevent contamination and icing of disconnects.

3. (C) M-1 DEPRESS FOG ON PUSHBUTTON . . . . DIRECTED

After the abort sequence is complete the MCCC shall direct M-1 to investigate silo condition and cause of spillage. The crew shall investigate all sources of spillage and immediately attempt to isolate, stop or restrict spillage by referring to appropriate technical manuals. During and after correction of spillage source action shall be taken to prevent further hazards from occuring by cleaning spillage by use of most appropriate means available, fire hose, drains, rags, bucket brigade.

4. (C) M-1 TROUBLESHOOT . . . . DIRECTED

Table 4-4. Emergencies During Abort

ITEM 1. PRESSURE MODE INDICATOR RED

This item contains procedures to correct missile LO$_2$ and fuel tank pressures and to contain the emergency or malfunction until standby troubleshooting and maintenance can be accomplished.

> **CAUTION**
>
> Maintain a minimum of 2 PSI differential pressure, as observed on **TANK DIFFERENTIAL PRESSURE** gauge. Failure to comply may result in bulkhead reversal.

a. The missile is within 33 inches of the up-and-locked position and PRESSURE MODE indicator RED will stop the lowering sequence. When this condition exists, the boiloff valve is enabled and LO$_2$ tank pressure should decrease to 3.4 to 8.0 PSI. Fuel tank pressure will begin to gradually decay until the helium control charging unit (HCU) pressure controller maintains pressure at 53 PSI. The malfunction was caused by a missile tank pressure or emergency circuit malfunction. Pressures cannot be controlled by depressing the RAISE or LOWER pushbutton. If the deputy depresses the AUTOMATIC pushbutton, the boiloff valve will close and LO$_2$ tank pressure will gradually increase. After it increases to greater than 8 PSI, launcher platform lowering sequence should continue.

b. The missile boiloff valve is below the silo cap prior to ABORT EXTERNAL indicator AMBER. Possible impingement of the tension equalizer (LO$_2$ boiloff within the silo) and automatic phase II pressurization cannot be selected. When this condition exists, the lowering sequence will continue, the boiloff valve is enabled to open, and LO$_2$ tank pressure will decrease 3.4 to 8.0 PSI. Fuel tank pressure

If the pressurization system switches to emergency during abort, the deputy shall immediately determine if pressures are normal or abnormal by observing pressure gauges and correct pressures when necessary. If the malfunction occurs before ABORT EXTERNAL indicator illuminates AMBER, an attempt shall be made to return the system to automatic mode by depressing AUTOMATIC pushbutton. The above procedures are necessary because one of the following conditions exist:

will gradually decrease until the HCU pressure controller maintains pressure at 53 PSI. The PRESSURE MODE indicator RED condition was caused by a missile tank pressure or emergency circuit malfunction. LO$_2$ and fuel tank pressure cannot be controlled by depressing the RAISE or LOWER pushbutton. Depressing the AUTOMATIC pushbutton will close the boiloff valve and prevent boiloff within the silo and impingement of the tension equalizer. LO$_2$ tank pressures will gradually begin to increase. When the launcher platform is down and locked, pressures will automatically be controlled by the pressure system control. The pressurization system will normally be in emergency mode when 7-1/2 feet out of the uplocks until the boiloff valve reaches the silo cap.

If PRESSURE MODE indicator illuminates RED prior to the launcher platform lowering 33-1/2 inches from the up-and-locked position, or after pressurization system is returned to automatic when boiloff valve is at silo cap, perform procedures contained in action 1.

If PRESSURE MODE indicator illuminates RED after ABORT EXTERNAL indicator illuminates AMBER, perform procedures contained in action 2.

ACTION 1

The pressurization system must be in automatic mode and LO$_2$ tank pressure greater than 8 PSI during the first 33-1/2 inches of down movement or the launcher platform will not lower. After the boiloff valve reaches the silo cap the pressurization system should be

in automatic mode to prevent LO$_2$ boiloff within the silo. However, if automatic pressurization mode cannot be selected, allow abort sequence to continue in emergency mode. If PRESSURE MODE indicator is illuminated RED, deputy shall depress automatic pushbutton.

1. (D) **AUTOMATIC PUSHBUTTON** . . . . **DEPRESSED**

Table 4-4. Emergencies During Abort (CONT)

ITEM 1.   PRESSURE MODE INDICATOR RED (CONT)

ACTION 1 (CONT)

Allow abort sequence to continue.

2. (C) ABORT . . . . **CONTINUED**

ACTION 2

If PRESSURE MODE indicator illuminates RED after ABORT EXTERNAL indicator illuminates AMBER, deputy shall immediately attempt to adjust pressures (phase II pressures if prior to LO_2 DRAIN COMPLETE indicator GREEN, phase I pressure if after LO_2 DRAIN COMPLETE indicator GREEN) if abnormal.

1. (D) TANK PRESSURES . . . . **ADJUSTED**

If pressures were adjusted to normal, or if pressures were normal when pressurization system switched to emergency, deputy shall depress AUTOMATIC pushbutton to return the system to automatic mode.

2. (D) AUTOMATIC PUSHBUTTON . . . . **DEPRESSED**

If pressurization system remains in automatic mode, troubleshoot system after abort sequence is complete.

If pressurization system returns to emergency mode after AUTOMATIC pushbutton is depressed, deputy shall manually maintain proper missile tank pressures. All indicators for LO_2, fuel, and differential pressure switches on the PLCP that could activate the emergency circuit, shall be observed and abnormal indications noted. All abnormal indications noted shall be remembered and used when required in table 4-5.

3. (D) PROPER PRESSURES . . . . **MANUALLY MAINTAINED**

After LO_2 DRAIN COMPLETE indicator illuminates GREEN, at MCCC direction, crew shall place missile in stretch and troubleshoot the system in accordance with table 4-5.

4. (C) CREW, PLACE MISSILE IN STRETCH . . . . **DIRECTED**

ITEM 2A.   GENERATOR MALFUNCTION (ALTERNATE IN STANDBY)

This item contains procedures to place the alternate generator on the line and shut down the faulty generator if a diesel generator malfunction occurs during abort and the alternate generator is in standby. Steps in the procedure shall be performed by L-1 at the PRCP in an attempt to prevent AC power loss. The L-1 at MCCC direction shall place the alternate generator on the line and shut down the faulty generator using table 4-8.

1. (C) L-1, PLACE ALTERNATE GENERATOR ON THE LINE AND SHUTDOWN FAULTY GENERATOR . . . . **DIRECTED**

Table 4-4. Emergencies During Abort (CONT)

ITEM 2B. GENERATOR MALFUNCTION (ALTERNATE NOT IN STANDBY)

This item contains procedures to place the complex in a safe configuration if a diesel generator malfunction occurs during the abort sequence with the alternate generator not in standby. Steps in this procedures shall be performed by crew members as indicated. The MCCC shall allow the abort sequence to continue until SITE HARD indicator illuminates GREEN in order for the missile lowering sequence to complete and the silo doors to close prior to shutting down AC power. LO₂ drain will continue during the AC power loss since the airborne fill-and-drain valve and drain valve L-16 are open at this time.

1. (C) ABORT . . . . **CONTINUED**

After SITE HARD indicator illuminates GREEN, at MCCC direction to shutdown generator, L-1 shall: (a) Position FEEDER NUMBER 3 NON-ESSENTIAL BUS CONTROL SWITCH to TRIP. (b) Position GENERATOR MAIN BREAKER CONTROL SWITCH to TRIP. Observe GENERATOR MAIN BREAKER CONTROL SWITCH RED indicator extinguishes and GREEN indicator illuminates. (c) Position ENGINE START-STOP SWITCH to STOP and momentarily hold. Release switch and observe ENGINE START-STOP SWITCH RED indicator extinguishes and GREEN indicator illuminates. If malfunction is due to low lube oil pressure or high temperature, delay shutdown of generator until work platform NO. 1 is extended.

2. (C) L-1, SHUT DOWN FAULTY GENERATOR . . . . **DIRECTED**

At MCCC direction, L-1 shall troubleshoot and restore AC power after LO₂ drain is complete.

3. (C) L-1, TROUBLESHOOT AND RESTORE AC POWER . . . . **DIRECTED**

After abort is complete, at MCCC direction, crew shall reset complex eletrical system using procedures contained in table 4-7.

4. (C) CREW, RESET ELECTRICAL SYSTEM . . . . **DIRECTED**

ITEM 2C. GENERATOR MALFUNCTION (GENERATORS IN PARALLEL)

This item contains the procedures to remove a faulty generator from the line if a diesel generator malfunction occurs during the abort sequence and the diesel generators are in parallel. If a diesel generator malfunction occurs, at MCCC direction to shut down the faulty generator, L-1 shall:

(a) Position GENERATOR MAIN BREAKER CONTROL SWITCH to TRIP. Observe GENERATOR MAIN BREAKER CONTROL SWITCH RED indicator extinguishes and GREEN indicator illuminates.

(b) Position ENGINE START-STOP SWITCH to STOP and momentarily hold. Release switch and observe ENGINE START-STOP SWITCH RED indicator extinguishes and GREEN indicator illuminates.

1. (C) L-1, SHUT DOWN FAULTY GENERATOR . . . . **DIRECTED**

The abort sequence will be continued on one generator.

2. (C) ABORT . . . . **CONTINUED**

Table 4-4. Emergencies During Abort (CONT)

ITEM 3. LOSS OF AC POWER

This item contains procedures for restoring the complex to a safe configuration after an AC power loss during the abort sequence. Steps in this procedure shall be performed by crew members as indicated.

ACTION 1

If AC power loss occurs prior to ABORT EXTERNAL indicator illuminating AMBER, at MCCC direction, M-1 shall position RESET PROGRAM-

1. (C) M-1, RESET PROGRAMMER KEY ON .... DIRECTED

At MCCC direction, L-1 shall attempt to restore AC power using procedures contained in table 4-6A or 4-6B. If AC power is restored at the PRCP, per-

2. (C) L-1, RESTORE AC POWER ..... DIRECTED

ACTION 2

If AC power was restored after performing action 1, at MCCC direction, L-1 shall apply nonessential power by positioning FEEDER NUMBER 3 NON-

1. (C) L-1, NONESSENTIAL POWER ON ..... DIRECTED

At MCCC direction, M-1 shall position RESET PROGRAMMER key switch on CSMOL to OFF. This restores the automatic abort sequence.

2. (C) M-1, RESET PROGRAMMER KEY OFF..... DIRECTED

During abort sequence, at MCCC direction M-1 shall reset LCC electrical system to restore the LCC ventilation system using procedures contained in table 4-7.

3. (C) M-1, RESET LCC ELECTRICAL SYSTEM ..... DIRECTED

After abort sequence is complete, at MCCC direction, crew shall reset the complex electrical system using procedures contained in table 4-7.

4. (C) CREW, RESET ELECTRICAL SYSTEM ..... DIRECTED

ACTION 3

If AC power cannot be restored at the PRCP and if boiloff will not impinge on the tension equalizer, the deputy shall depress the EMERGENCY pushbutton to open the boiloff valve. If boiloff will

MER key switch on CSMOL to ON to prevent the launcher platform from lowering after AC power is restored

If AC power was lost after ABORT EXTERNAL indicator has illuminated GREEN, LO$_2$ drain will begin. The LO$_2$ drain sequence shall be monitored closely throughout performance of the remaining actions.

form procedures contained in action 2. If AC power cannot be restored immediately at the PRCP, perform procedures contained in action 3.

ESSENTIAL BUS CONTROL SWITCH on PRCP to CLOSE.

impinge on the tension equalizer, the tension equalizer must be sprayed with water or the missile enclosure purge unit (MEPU) must be extended. The fuel tank pressure must

Table 4-4. Emergencies During Abort (CONT)

ITEM 3. LOSS OF AC POWER (CONT)

ACTION 3 (CONT)

be maintained at least 20 PSI greater than LO_2 tank pressure regardless of the boiloff valve location. As the pressure difference approaches 20 PSI, deputy shall depress EMERGENCY pushbutton and allow sufficient time for the LO_2 tank pressure to decrease before depressing AUTOMATIC pushbutton. This shall be repeated as necessary. After water spray or the MEPU has been placed in operation, the boiloff valve shall be opened. Water spray is not required if boiloff does not impinge on the tension equalizer.

> **WARNING**
>
> Regardless of LP position, if boiloff valve is closed, immediately depress EMERGENCY pushbutton. After LO_2 tank pressure decreases to phase II, depress AUTOMATIC pushbutton. Enable boiloff valve periodically to relieve pressure, when LO_2 tank pressure increases to 12 PSI.

1. (C) A-1, SPRAY TENSION EQUALIZER OR EXTEND MEPU .... DIRECTED

If boiloff will not impinge on tension equalizer, or after the tension equalizer is being sprayed or the MEPU has been placed in operation, at MCCC direction, deputy shall depress EMERGENCY pushbutton to enable the boiloff valve. If AC power loss occurs after ABORT EXTERNAL indicator illuminating GREEN the LO_2 drain sequence will have started and pressurization system will be in phase II. The boiloff valve will be enabled. The EMERGENCY pushbutton need not be depressed.

2. (C) DEPUTY, DEPRESS EMERGENCY PUSHBUTTON .... DIRECTED

3. (Deleted)

After LO_2 has boiled off to a safe condition, at MCCC discretion, he shall direct L-1, assisted by available crew members, to troubleshoot the malfunction and restore AC power.

4. (C) L-1, TROUBLESHOOT AND RESTORE AC POWER .... DIRECTED

5. (Deleted)

After AC power is restored, at MCCC direction, M-1 shall reset LCC electrical systems to restore LCC ventilation using procedures in table 4-7.

6. (C) M-1, RESET LCC ELECTRICAL SYSTEM .... DIRECTED

Table 4-4. Emergencies During Abort (CONT)

ITEM 3.   LOSS OF AC POWER (CONT)

ACTION 3 (CONT)

If AC power is restored, at MCCC direction M-1 shall position RESET PROGRAMMER key switch on CSMOL to OFF. This will allow the abort sequence to continue.

7.   (C) M-1, RESET PROGRAMMER KEY OFF .... DIRECTED

After abort is complete, at MCCC direction, crew shall reset the complex electrical system using procedures contained in table 4-7.

8.   (C) CREW RESET ELECTRICAL SYSTEM .... DIRECTED

ITEM 4.   LOSS OF UMBILICALS

The actions required to attempt to return the missile to a safe configuration is determined by the AIRBORNE FILL & DRAIN VALVE indicator on LO$_2$ TANKING (PANEL 2) and the condition of the boiloff valve.

If umbilical loss occurred prior to LO$_2$ DRAIN COMPLETE indicator illuminated AMBER, a LO$_2$ drain may be accomplished with the boiloff valve open or closed, if AIRBORNE FILL & DRAIN VALVE indicator is AMBER. The steps in action 3 shall be accomplished to return the complex to a safe configuration. The AIRBORNE F I L L & DRAIN VALVE amber indication on the LO$_2$ tanking panel indicates that umbilical 600U4 has not been ejected and the fill-and-drain valve can be controlled. If the AIRBORNE FILL & DRAIN VALVE indicator is illuminated GREEN, the condition of the boiloff valve determines which action will be followed. If the boiloff valve is closed the steps in action 1 shall be accomplished. The conditions of action 1 will eventually result in the probable loss of the missile. If the boiloff valve is open, the steps in action 2 shall be accomplished.

If umbilical loss occurred after LO$_2$ DRAIN COMPLETE indicator illuminated AMBER, an automatic drain sequence will be in progress and will probably continue, regardless of which umbilicals were ejected. The steps in action 4 shall be accomplished if umbilical loss occurred during the automatic drain sequence.

ACTION 1

If umbilical loss occurred prior to LO$_2$ DRAIN COMPLETE indicator illuminated AMBER, with the boiloff valve closed and AIRBORNE FILL & DRAIN VALVE indicator illuminated GREEN, the MCCC shall direct M-1 to position RESET PROGRAMMER key switch on CSMOL to ON and depress the UP RUN pushbutton. The launcher platform may or may not rise depending on whether PRESSURE MODE indicator on the launch control console is RED or GREEN. If indicator is RED, M/L STOP indicator on CSMOL will normally be RED, which will prevent the launcher platform up movement. The missile in all probability will eventually destroy itself.

> **WARNING**
>
> Expect loss of the missile. all personnel shall remain in the LCC.

Table 4-4. Emergencies During Abort (CONT)

ITEM 4. LOSS OF UMBILICALS (CONT)

ACTION 1 (CONT)

1. (C) M-1, RESET PROGRAMMER KEY ON .... DIRECTED

2. (C) M-1, DEPRESS UP RUN PUSHBUTTON .... DIRECTED

ACTION 2

If umbilical loss occurred prior to LO_2 DRAIN COMPLETE indicator illuminated AMBER with the boiloff valve open and AIRBORNE FILL & DRAIN VALVE indicator illuminated GREEN, LO_2 shall be allowed to boiloff. The MCCC shall allow the down-run sequence to continue. When ABORT EXTERNAL indicator illuminates AMBER, MCCC shall direct deputy to depress EMERGENCY pushbutton to ensure enabling the RAISE and LOWER pushbuttons.

1. (C) ABORT .... CONTINUED

2. (C) DEPUTY, DEPRESS EMERGENCY PUSHBUTTON .... DIRECTED

Missile fuel tank pressure shall be monitored and controlled during boiloff by maintaining PNEUMATICS IN P H A S E II indicator illuminated GREEN. The deputy shall depress FUEL RAISE pushbutton when PNEUMATICS IN PHASE II indicator illuminates AMBER and Release the pushbutton when indicator illuminates GREEN. After boiloff, monitor and control missile tank pressures at HCU. If TV is available to monitor missile tank pressure gauges on PSMR, they may be used instead of PNEUMATICS IN PHASE II indicator GREEN.

3. (C) DEPUTY, MAINTAIN PNEUMATIC IN PHASE II INDICATOR GREEN ...... DIRECTED

LO_2 shall be allowed to boiloff as there is no way of draining the missile unless umbilical 600U4 is reconnected.

4. (C) ALLOW LO_2 TO BOILOFF .... ACCOMPLISHED

WARNING

After LO_2 has completely boiled off, maintenance support is required to safety ordnance items (if installed) and reconnect umbilicals.

ACTION 3

If umbilical loss occurred prior to LO_2 DRAIN COMPLETE indicator illuminated AMBER and it is determined that LO_2 airborne fill-and-drain valve can be opened, as indicated by AIRBORNE FILL & DRAIN VALVE indicator on LO_2 TANKING (PANEL 2) illuminated AMBER, abort should be allowed to continue and the MCCC shall direct A-1 to place LO_2 tanking panels in local control by positioning REMOTE L O C A L switch on LO_2 TANKING (PANEL 1) to LOCAL.

1. (C) ABORT .... CONTINUED

2. (C) A-1, LO_2 TANKING PANEL TO LOCAL .... DIRECTED

After launcher platform has reached the down and locked position and ABORT EXTERNAL indicator is illuminated AMBER, MCCC shall direct deputy to depress EMERGENCY pushbutton in an attempt to open the boiloff valve (if closed) and also to ensure that RAISE and LOWER pushbuttons are enabled.

3. (C) DEPUTY, DEPRESS EMERGENCY PUSHBUTTON .... DIRECTED

Table 4-4. Emergencies During Abort (CONT)

ITEM 4. LOSS OF UMBILICALS (CONT)

ACTION 3 (CONT)

If umbilical 600U3 has ejected there will be no pressure indications on the gauges on the launch control console. Phase II fuel tank pressures can be maintained by observing PNEUMATICS IN PHASE II indicator in the countdown patch. If the indicator reverts from GREEN to AMBER (fuel tank pressure less than 53 PSI), the deputy shall depress FUEL RAISE pushbutton and hold until PNEU-

MATICS IN PHASE II indicator again illuminates GREEN. The deputy shall then release FUEL RAISE pushbutton immediately to avoid overpressurizing the missile. If TV monitor is available to observe gauges on top of PSMR, tank pressures may be monitored on the gauges instead of using PNEUMATICS IN PHASE II indicator GREEN.

4. (C) DEPUTY, MAINTAIN PNEUMATICS IN PHASE II INDICATOR GREEN ..... DIRECTED

At MCCC direction to manually drain LO$_2$, A-1 at LO$_2$ TANKING (PANEL 1 and PANEL 2) shall proceed as follows:

(a) Position L-16 valve switch to OPEN and observe DRAIN VALVE L-16 indicator is illuminated GREEN. If indicator does not illuminate GREEN position L-1 valve switch to OPEN and observe FINE LOAD VALVE L-1 indicator is illuminated GREEN. Valve L-16 will be closed if valve L-1 is used to drain LO$_2$.

(b) Position A/B F&D valve switch to OPEN. This will allow LO$_2$ to drain from the missile. If airborne fill-and-drain valve cannot be opened, LO$_2$ must be allowed to boiloff. If it becomes necessary to stop LO$_2$ draining, A-1 shall position A/B F&D valve switch to CLOSE.

(c) Observe AIRBORNE FILL & DRAIN VALVE indicator illuminates GREEN. Start timing drain sequence when indicator illuminates GREEN.

5. (C) A-1, MANUALLY DRAIN LO$_2$ .... DIRECTED

After LO$_2$ has drained for 30 minutes, the system will be ready for securing. At MCCC direction to secure the LO$_2$ tanking panels, A-1 shall: (a) Position L-16 (or L-1 if used) valve switch to CLOSE. (b) Observe DRAIN VALVE L-16 (or FINE

LOAD VALVE L-1) indicator is illuminated AMBER. (c) Position A/B F&D valve switch to CLOSE. (d) Observe AIRBORNE FILL & DRAIN VALVE indicator is illuminated AMBER.

6. (C) A-1, SECURE LO$_2$ TANKING PANELS .... DIRECTED

The loss of umbilical 600U3 prevents the differential pressure signal from being sent to the logic units from the airborne differential pressure switch. The

pressurization system will remain in emergency mode. At MCCC direction, crew shall apply missile stretch in accordance with table 4-5.

7. (C) CREW, PLACE MISSILE IN STRETCH .... DIRECTED

After LO$_2$ drain is complete, MCCC shall direct deputy or M-1 to monitor and maintain missile tank pressures. If TV is not available to monitor PSMR missile tank pressure gauges, M-1 shall proceed to PSC, level 8, to open valve 148. This will ensure that pressure is available at the HCU. M-1 shall then proceed to the HCU, launcher platform level 3, and

maintain missile tank pressures (LO$_2$ tank 3.4 to 8.0 PSI, fuel tank 51.0 to 63.9 PSI) using valves 339, 340, 342, and 343: If TV is available to monitor PSMR missile tank pressure gauges, deputy shall maintain phase II missile tank pressures using TV monitor and the applicable RAISE or LOWER pushbutton on launch control console.

8. (C) DEPUTY OR M-1, MONITOR AND MAINTAIN TANK PRESSURES .... DIRECTED

Table 4-4. Emergencies During Abort (CONT)

ITEM 4.   LOSS OF UMBILICALS (CONT)

WARNING

After LO₂ is drained, maintenance support is required to safety ordnance items
(if installed) and reconnect umbilicals. Exception may be made during a tactical
launch if missile crew can determine that no stray voltage exists at the umbilicals.

ACTION 4

If umbilical loss occurred after LO₂ DRAIN COM-
PLETE indicator illuminated AMBER, drain sequence will continue. However, fuel tank pressure
will have to be controlled manually. The MCCC
shall direct deputy to depress EMERGENCY pushbutton to ensure RAISE and LOWER pushbuttons
are enabled.

1.   (C) DEPUTY, DEPRESS EMERGENCY PUSHBUTTON .... DIRECTED

If pressure system is in emergency mode and pressure gauges on launch control console indicate zero,
missile fuel tank pressure shall be monitored and
controlled during LO₂ drain by maintaining PNEU-
MATICS IN PHASE II indicator illuminated
GREEN. The deputy shall depress FUEL RAISE
pushbutton when PNEUMATICS IN PHASE II
indicator illuminates AMBER and release the pushbutton when indicator illuminates GREEN.

2.   (C) DEPUTY, MAINTAIN PNEUMATICS IN PHASE II INDICATOR GREEN ...... DIRECTED

The normal drain will continue until LO₂ DRAIN
COMPLETE indicator illuminates GREEN, at
which time crew shall proceed to step 4.

3.   (C) NORMAL LO₂ DRAIN .... CONTINUED

After LO₂ drain is complete, MCCC shall direct deputy or M-1 to monitor and maintain missile tank
pressures. If TV is not available to monitor PSMR
missile tank pressure gauges, M-1 shall proceed to
PSC, level 8, to open valve 148. This will ensure that
pressure is available at the HCU. M-1 shall then
proceed to the HCU, launcher platform level 3,
and maintain missile tank pressures (LO₂ tank 3.4
to 8.0 PSI, fuel tank 51.0 to 63.9 PSI) using valves
339, 340, 342, and 343. If TV is available to monitor PSMR missile tank pressure gauges, deputy shall
maintain phase II missile tank pressures using TV
monitor and the applicable RAISE or LOWER pushbutton on launch control console.

4.   (C) DEPUTY OR M-1, MONITOR AND MAINTAIN TANK PRESSURES .... DIRECTED

The loss of umbilical 600U3 prevents the differential
pressure signal from being sent to the logic units
from the airborne differential pressure switch. The
pressurization system will remain in emergency mode.
At MCCC direction, A-1 and L-1 shall apply missile
stretch in accordance with table 4-5.

5.   (C) A-1, L-1, PLACE MISSILE IN STRETCH .... DIRECTED

WARNING

After LO₂ is drained, maintenance support is required to safety ordnance items
(if installed) and reconnect umbilicals. Exception may be made during a tactical
launch if missile crew can determine that no stray voltage exists at the umbilicals.

Table 4-4.   Emergencies During Abort (CONT)

ITEM 5.   FIRES IN SILO

This item contains procedures to place the system in the safest possible configuration if a fire occurs during the abort sequence. If the fire occurs prior to ABORT EXTERNAL indicator illuminated AMBER, MCCC shall direct M-1 to stop the lowering sequence and to manually raise the launcher platform. (Refer to action 1.) If the fire occurs after ABORT EXTERNAL indicator is illuminated AMBER, refer to action 2.

Table 4-4. Emergencies During Abort (CONT)

ITEM 5. FIRES IN SILO (CONT)

ACTION 1

If a fire occurs prior to ABORT EXTERNAL indicator illuminating AMBER, MCCC shall attempt to determine the location and intensity of the fire using TV monitors and fire alarm panel.

1. (C) FIRE LOCATION .... IDENTIFIED

At MCCC direction, deputy shall depress AUTOMATIC pushbutton so that launcher platform can be raised.

2. (C) DEPUTY, DEPRESS AUTOMATIC PUSHBUTTON .... DIRECTED

At MCCC direction, M-1 shall stop and drive the launcher platform up: (a) Position RESET PRO-GRAMMER key switch on CSMOL to ON; (b) Depress UP RUN pushbutton.

3. (C) M-1, DRIVE LAUNCHER PLATFORM UP .... DIRECTED

If the fire is located in the missile enclosure area, MCCC will direct M-1 to activate missile enclosure area fog system by depressing MISSILE ENCLOSURE FOG ON pushbutton on FRCP.

4. (C) M-1, ACTIVATE MISSILE ENCLOSURE AREA FOG SYSTEM .... DIRECTED

After the launcher platform is up, MCCC shall direct crew to attempt to contain and extinguish the fire by using available fire fighting equipment.

5. (C) CREW, COMBAT THE FIRE .... DIRECTED

MCCC shall ensure that local disaster control and potential hazard procedures are implemented at the earliest possible time. Personnel in the LCC not involved in fire fighting procedures may be used to implement the disaster control and potential hazard procedures.

6. (C) DISASTER CONTROL AND POTENTIAL HAZARD PROCEDURES ..... IMPLEMENTED

MCCC shall direct crew to perform a manual abort in accordance with table 4-10, action 1, after the fire has been extinguished. The launcher platform will not be lowered until the silo has been determined safe.

7. (C) CREW, PERFORM MANUAL ABORT .... DIRECTED

ACTION 2

If the fire indication occurred after ABORT EXTERNAL indicator illuminated AMBER, MCCC shall attempt to determine the location and intensity of the fire using TV monitors and fire alarm panel.

1. (C) FIRE LOCATION .... IDENTIFIED

If fire is in missile enclosure area on lower silo (fire alarm panel zones 5 or 7) MCCC shall direct M-1 to activate missile enclosure area fog system by depressing MISSILE ENCLOSURE FOG ON pushbutton on FRCP.

2. (C) M-1, ACTIVATE MISSILE ENCLOSURE AREA FOG SYSTEM .... DIRECTED

Table 4-4.   Emergencies During Abort (CONT)

ITEM 5.   FIRES IN SILO (CONT)

ACTION 2 (CONT)

After the SITE HARD indicator has illuminated GREEN, at MCCC discretion, M-1 open the silo overhead doors at CSMOL: (a) Position RESET PROGRAMMER key switch to ON. (b) Depress HYDRAULIC 40 HP PUMP ON pushbutton. (c)

Depress CRIB VERTICAL LOCK pushbutton. (d) Depress CRIB HORIZONTAL LOCK pushbutton. (e) Depress and hold SILO DOORS OPEN pushbutton for 30 seconds. Observe SILO DOORS CLOSED indicator extinguishes.

3. (C) M-1, OPEN SILO DOORS .... DIRECTED

MCCC shall direct crew to contain and extinguish the fire using available fire fighting equipment.

4. (C) CREW, COMBAT THE FIRE .... DIRECTED

MCCC shall ensure local disaster control and potential hazard procedures are implemented at the earliest possible time the emergency permits. Personnel in the LCC not involved in fire fighting procedures may be used to implement the disaster control and potential hazard procedures.

5. (C) DISASTER CONTROL AND POTENTIAL HAZARD PROCEDURES ...... IMPLEMENTED

ITEM 6A.   SILO CONTROL CABINET HI TEMP INDICATOR RED

This item contains procedures to follow when a malfunction in the control cabinet air conditioning system is indicated on the FRCP. Continue the abort sequence and troubleshoot the cause of the malfunction indication after abort complete.

1. (C) ABORT .... **CONTINUED**

ITEM 6B.   MAIN EXHAUST FAN NOT OPERATING INDICATOR RED

This item contains procedures to follow to prevent a hazardous ventilation condition occurring in the silo when the FRCP inicates that the main exhaust fan is not operating. The MCCC shall continue the abort sequence. After abort complete he shall direct M-1 to troubleshoot the malfunction indication.

1. (C) ABORT .... **CONTINUED**

ITEM 6C.   STORAGE AREA OXYGEN 25% OR 19% INDICATOR RED

This item contains procedures to return the silo to a safe configuration when a gaseous or liquid oxygen/nitrogen leak or spillage is indicated on the FRCP during abort. The MCCC shall allow the abort sequence to continue.

When or if the ABORT EXTERNAL indicator illuminates AMBER, at MCCC direction, A-1 shall stop LO₂ drain so that the cause of high or low oxygen concentration can be determined before allowing automatic drain to continue.

Table 4-4. Emergencies During Abort (CONT)

ITEM 6C. STORAGE AREA OXYGEN 25% OR 19% INDICATOR RED (CONT)

ACTION 1

The abort sequence shall be allowed to continue until ABORT EXTERNAL indicator illuminates AMBER. Immediately after ABORT EXTERNAL indicator illuminate AMBER, at MCCC direction to stop LO_2 drain, A-1 shall:

(a) Position drain valve L-16 switch to OPEN.

This ensures continuous line drain and line pressure relief from airborne fill-and-drain valve to LO_2 storage tank when LOCAL REMOTE switch is positioned to LOCAL.

(b) Position LOCAL REMOTE switch on LO_2 TANKING (PANEL 1) to LOCAL.

This prevents the LO_2 automatic sequencing from continuing and provides local control of the LO_2 valves.

(c) Verify drain valve L-16 and vent valve N-4 are open by observing DRAIN VALVE L-16 and TOPPING TANK VENT VALVE N-4 indicators are GREEN. Verify vent valve N-5 is open by observing STORAGE TANK VENT VALVE N-5 indicator is GREEN or EXTINGUISHED.

Vent valves N-5 and N-4 are automatically opened at start of abort.

(d) Verify all other valves on LO_2 tanking panels are closed by observing all other valve indicators are illuminated AMBER.

When LOCAL REMOTE switch was positioned to LOCAL the airborne fill-and-drain valve closed and stopped LO_2 drain.

---

1. (C) ABORT .... **CONTINUED**

2. (C) A-1, STOP LO_2 DRAIN .... **DIRECTED**

At MCCC direction deputy shall: (a) *Depress EMERGENCY pushbutton* on the launch control console after ABORT EXTERNAL indicator illuminates AMBER. This places the missile pressurization in emergency mode and prevents pressures from returning to phase I in the event of a LO_2 drain timer malfunction, or if it is desired to maintain phase II missile pressures in excess of the 45-minute drain time. (b) *Maintain phase II missile tank pressures.* With the missile tank pressure in emergency mode, pressures can be regulated and maintained by depressing the appropriate LO_2 and FUEL RAISE or LOWER pushbuttons on the launch control console.

---

3. (C) DEPUTY, PLACE PRESSURIZATION IN EMERGENCY .... DIRECTED

At MCCC direction, A-1 and M-1 shall investigate cause of high (or low) oxygen alarm. Logical sources of high (or low) oxygen concentration at this time are at the LO_2 storage and LO_2 topping tanks, and piping up to the liquid oxygen prefabs on silo level 7. At launch control center, situation requires close observation of launch control console, fire alarm panel, and facility remote control panel.

---

4. (C) A-1, M-1, INVESTIGATE CAUSE OF HIGH (OR LOW) OXYGEN ALARM   DIRECTED

A-1 and M-1 shall check oxygen detector TROUBLE indicator, level 7, and if illuminated troubleshoot the oxygen detector cabinet. If TROUBLE indicator is extinguished, check area alarm indicator and percentage meter to determine location and amount of oxygen concentration. Locate leakage and isolate if possible, referring to appropriate technical manual to ensure valve positioning shall not damage or cause additional malfunctions.

---

5. (C) CAUSE OF OXYGEN ALARM .... DETERMINED

Table 4-4. Emergencies During Abort (CONT)

ITEM 6C. STORAGE AREA OXYGEN 25% OR 19% INDICATOR RED (CONT)

ACTION 1 (CONT)

At direction of the MCCC to reset oxygen detector, A-1 and M-1 shall: (a) Depress SYSTEM RESET and HORN SILENCE pushbuttons at oxygen detector cabinet. This will silence audible alarm and restart the station sampling unit. Verify the station sampling is cycling by observing the station indicators. (b) Reset oxygen purge system by depress-ing OXYGEN PURGE RESET (or CR-44 RELAY RESET) pushbutton at the facilities terminal cabinet NO. 2, level 2. This will reset the gaseous oxygen blast closure and vent fan control circuits. Perform action 2 if system is unsafe for LO$_2$ drain. Perform action 3 if system is safe for LO$_2$ drain.

6.   (C) A-1, M-1, RESET OXYGEN DETECTOR SYSTEM .... DIRECTED

ACTION 2

If the MCCC determines system is unsafe for LO$_2$ drain, crew shall refer to nomograms in section VI and allow LO$_2$ to boiloff.

1.   (C) ALLOW LO$_2$ TO BOILOFF .... ACCOMPLISHED

ACTION 3

If tthe MCCC determines the system is safe for LO$_2$ drain, at his direction, A-1 shall restart LO$_2$ drain at LO$_2$ tanking panels: (a) Cycle LO$_2$ tanking power at LO$_2$ TANKING (PANEL 1) by positioning SYSTEM POWER switch OFF for one second then ON. This de-energizes the LO$_2$ tanking panels and removes all previous tanking panel commands and restarts the LO$_2$ drain timers (if abort external signal is present). (b) Return LO$_2$ tanking panels to remote by positioning LOCAL REMOTE switch on LO$_2$ TANKING (PANEL 1) to REMOTE. This action will reposition airborne fill-and-drain valve to open and the automatic drain sequence will start.

1.   (C) A-1, START LO$_2$ DRAIN .... DIRECTED

After LO$_2$ DRAIN COMPLETE indicator illuminates AMBER, MCCC shall direct deputy to return pressurization system to automatic mode by depressing AUTOMATIC pushbutton on the launch control console. This will allow the tank pressures to revert to phase 1 at completion of 45-minute drain sequence.

2.   (C) DEPUTY, DEPRESS AUTOMATIC PUSHBUTTON .... DIRECTED

ITEM 6D. AIR WASHER DUST COLLECTING UNITS NOT OPERATING INDICATOR RED

This item contains procedures to follow when a malfunction of the air washer dust collecting units is indicated on the FRCP. The MCCC shall continue the abort sequence. After abort complete he shall direct M-1 to troubleshoot the malfunction indication.

1.   *(C) ABORT* .... **CONTINUED**

Table 4-4. Emergencies During Abort (CONT)

ITEM 6E. DIESEL VAPOR HIGH LEVEL INDICATOR RED

This item contains procedures to follow when a high concentration of diesel fuel vapor is indicated on the FRCP. There is danger of an explosion and fire if an excessive amount of diesel fuel leaks into the silo. The abort sequence shall be continued, but the MCCC shall direct A-1 and M-1 to inspect levels 5 and 6.

1. *(C) ABORT* .... **CONTINUED**

The MCCC shall direct M-1 and A-1 to inspect levels 5 and 6. If a spillage is discovered, action necessary to isolate the leak and prevent a fire potential shall be taken. If no leak is discovered, M-1 shall troubleshoot the diesel fuel vapor detection cabinet after abort complete.

2. (C) A-1, M-1, VISUALLY CHECK SILO LEVELS 5 AND 6 .... DIRECTED

ITEM 6F. LCC AIR RECEIVER AND INSTRUMENT AIR RECEIVER LOW PRESSURE INDICATORS RED

This item contains procedures to follow when the FRCP indicates a malfunction in the silo and LCC instrument air systems. If malfunction occurs, the abort sequence shall be continued. After abort complete M-1 shall troubleshoot the malfunction indications.

1. *(C) ABORT* .... **CONTINUED**

ITEM 6G. SILO AIR INTAKE CLOSURES CLOSED/SILO AIR EXHAUST CLOSURES CLOSED INDICATORS RED

These procedures shall be used to return the silo to a safe configuration if the silo blast closures are closed electrically as indicated on the FRCP. The MCCC shall allow abort sequence to continue.

1. *(C) ABORT* .... **CONTINUED**

If the blast detection system has detected an optic signal but not a radio signal the blast closures should open in 90 seconds and the blast detection cabinet will have to be reset.

If the blast detection system has detected both an optic signal and a radio signal, the blast closures will close and the blast detection cabinet must be reset before the blast closures will open.
If no nuclear blast has been detected and a malfunction has occurred in the blast detection system the blast detection cabinet must be reset.

When directed by the MCCC, M-1 shall reset blast detection cabinet: (a) Position OUTPUT RELAY switch to DISCONNECT. (b) Depress RCVR 1 MANUAL TEST pushbutton. (c) Depress ALARM RESET pushbutton. (d) Position OUTPUT RELAY switch to CONNECT. (e) Depress DETECTION MODE RESET pushbutton. (f) Verify O P T I C MODE indicator is illuminated. (g) Verify channel indicator is cycling.

2. (C) M-1, RESET BLAST DETECTION CABINET .... DIRECTED

If blast detection cabinet fails to reset or if the blast closures fail to open, at direction of MCCC, M-1 shall position OUTPUT RELAY switch on blast detection cabinet to DISCONNECT.

3. (C) M-1, OUTPUT RELAY SWITCH TO DISCONNECT .... DIRECTED

Table 4-4. Emergencies During Abort (CONT)

ITEM 6G.   SILO AIR INTAKE CLOSURES CLOSED/SILO AIR EXHAUST
CLOSURES CLOSED INDICATOR RED (CONT)

If blast closures were closed by an actual
nuclear blast, or by a malfunction in the
blast detection or blast closures system,
emergency water pump P-32 may be operating. If diesel generators are operating
in parallel, to prevent a possible overheat
condition and possible AC power loss, one
generator shall be shut down after SITE
HARD indicator GREEN. At MCCC direction
to shut down one diesel generator, L-1 shall:
(a) Position GENERATOR MAIN BREAKER
CONTROL SWITCH to TRIP. Observe

GENERATOR MAIN BREAKER CONTROL
SWITCH RED indicator extinguishes and
GREEN indicator illuminates. (b) Position
ENGINE START-STOP SWITCH to STOP and
momentarily hold. Release switch and observe ENGINE START-STOP SWITCH RED
indicator extinguishes and GREEN indicator
illuminates.

Omit step 4 if diesel generators are not operating in parallel. Do not perform step 4
until SITE HARD indicator is illuminated
GREEN.

4.   (C) L-1, SHUT DOWN ONE GENERATOR . . . . DIRECTED

5.   Deleted

Do not perform step 6 until LO$_2$ has drained,
as indicated by LO$_2$ DRAIN COMPLETE
indicator illuminated GREEN or by a slight
rise in missile fuel tank pressure, which
should occur approximately 25 minutes (missile
LO$_2$ tank at 100%) after start of LO$_2$ drain.

If the blast closures are closed and blast
detection cabinet is normal, a failure of
relay control circuit or a tripped circuit
breaker should be suspected.

At direction of the MCCC to troubleshoot blast
closure system, M-1 shall: (a) Proceed to
essential motor control panel C, silo level 2
and verify control circuit breakers CB-1 and
CB-3 are ON. Reset CB-1 and CB-3 if
tripped. If circuit breakers had to be reset
and blast closures open, omit step (b)
through (e) and troubleshoot probable cause in
accordance with applicable technical manual.

If circuit breakers were not tripped or will
not reset, proceed to facilities interface
cabinet NO. 1 on level 3 to bypass control
circuit of the blast closures as follows: (b)
Install jumper wire on terminal board E-1
between wire numbers 9013 and 9015. (c)
Install jumper wire on terminal board E-1,
between wire 9015 and 9017. (d) Install
jumper wire on terminal board E-1, between
wires 9017 and 9018. (e) Install jumper wire
from C-5 to wire 9013. Verify with MCCC
that blast closures are open. If blast closures
do not open, troubleshoot blast closures system
in accordance with applicable technical manual.
If blast closures remain closed it may become
necessary to shut down the diesel generator
to prevent excessive carbon buildup in the
exhaust system and contamination of air in
the silo. Prior to shut down of the diesel
generator, extend work platform NO. 1 to
allow missile stretch to be applied if an
emergency occurs that may require it.

WARNING

Control circuit voltage is 120 VAC.
Use extreme caution while installing
jumper wires.

6.   (C) M-1, TROUBLESHOOT BLAST CLOSURES . . . . DIRECTED
The MCCC shall direct M-1 to troubleshoot the cause of emergency.

7.   (C) M-1, TROUBLESHOOT . . . . DIRECTED

Table 4-4.   Emergencies During Abort (CONT)

ITEM 6H. MISSILE POD AIR CONDITIONER MALFUNCTION INDICATOR RED

This condition is not an emergency during the abort sequence. No corrective action shall be accomplished at this time. The MCCC shall allow the abort sequence to continue.

1. **(C) ABORT** .... **CONTINUED**

ITEM 6-I.   EMERGENCY WATER PUMP P-32 ON INDICATOR RED

This item contains procedures for shutting down one diesel generator, if operating in parallel, to prevent overheating. If generators are operating in parallel, MCCC shall direct L-1 to shut down one generator at the PRCP since emergency water pump P-32 does not have sufficient capacity to cool both diesel engines. At MCCC direction to shut down one generator, L-1 shall:

(a) Position GENERATOR MAIN BREAKER CONTROL SWITCH to TRIP. Observe GENERATOR MAIN BREAKER CONTROL SWITCH RED indicator extinguishes and GREEN indicator illuminates.

(b) Position ENGINE START-STOP SWITCH to STOP and momentarily hold. Release switch and observe ENGINE START-STOP SWITCH RED indicator extinguishes and GREEN indicator illuminates.

1. **(C) L-1, SHUT DOWN ONE GENERATOR** .... **DIRECTED**

The MCCC shall continue abort sequence. M-1 and L-1 shall troubleshoot the condenser water system and power distribution system after LO$_2$ drain is complete using applicable technical manuals.

2. **(C) ABORT** .... **CONTINUED**

Table 4-4. Emergencies During Abort (CONT)

ITEM 7. ENGINES AND GROUND POWER INDICATOR RED

This indication can occur only because of a logic malfunction and will not affect abort sequence.

1. (C) ABORT　　　　　　　　　　　　　　　　　　　　　　　　　　　CONTINUED

ITEM 8. 400 CYCLE POWER INDICATOR RED

This condition does not indicate an emergency situation during the abort sequence. The MCCC shall allow the abort sequence to continue and direct troubleshooting of malfunction after abort is complete. The TANK DIFFERENTIAL PRESSURE gauge will indicate zero if 400-cycle power is lost.

1. (C) ABORT . . . . CONTINUED

ITEM 9. 28 VDC POWER INDICATOR AMBER OR RED

This malfunction indicates that the 28 VDC batteries may be depleted in a short time. The MCCC shall closely monitor and continue the abort sequence. After abort is complete, MCCC shall direct troubleshooting in accordance with section V.

1. (C) ABORT . . . . CONTINUED

The 28 VDC POWER AMBER or RED indication is considered to be an emergency because either indication could lead to an eventual complete loss of 28 VDC. The AMBER indication means that cooling air is not being supplied to the unit or that the backup batteries have depleted past 40 ampere-hours due to a partial or complete rectifier failure. The RED indication means that 28-VDC power is out of tolerance. After LO$_2$ drain is complete, manual control of the pressurization system shall be obtained at the pressure system control (PSC). Manual control is necessary should 28 VDC be intermittently or gradually lost as the emergency batteries are depleted. Manual control at the PSC will prevent any failure in the automatic system from affecting missile tank pressurization. At MCCC direction, A-1 and M-1 shall obtain manual control of pressurization system as follows: (a) M-1, close valves 105 and 106 on PSC. (b) A-1, position SYSTEM POWER switch on PNEUMATICS (PANEL 1) to OFF. (c) M-1, maintain missile tank pressures at PSC. Fuel tank pressure shall be maintained at 11.9 to 13.0 PSI, using valve 123 to raise pressure or valve 125 to lower pressure. LO$_2$ tank pressure shall be maintained at 3.4 to 4.2 PSI, using valve 124 to raise pressure or valve 126 to lower pressure. After abort is complete troubleshoot the malfunctioning system.

2. (C) A-1, M-1, OBTAIN MANUAL CONTROL OF PSC . . . . DIRECTED

The MCCC shall without delay, direct L-1 to start and configure the pod air conditioner for local operation to prevent prolonged loss of pod cooling to the missile guidance system. L-1 shall start and configure the pod air conditioner for local operation by: (a) Positioning the UNIT STOP/START switch (S-1) to ON; (b) Positioning the REMOTE CONTROL switch (S-3) to OFF.

3. (C) L-1, POSITION POD AIR CONDITIONER FOR LOCAL OPERATION . . . . . DIRECTED

ITEM 10. GUIDANCE FAIL INDICATOR RED

This condition is not an emergency at this time. The MCCC shall allow the abort sequence to continue. After the abort is complete, the MCCC shall direct the crew to troubleshoot.

1. (C) ABORT . . . . CONTINUED

Table 4-4. Emergencies During Abort (CONT)

ITEM 11.   MISSILE LIFT FAIL INDICATOR RED

This item contains procedures to follow when MISSILE LIFT FAIL indicator illuminates RED during the abort sequence. The launcher platform shall be observed on TV monitor. If no down movement is apparent it must be assumed that the missile lifting system has failed and procedures contained in table 4-16 shall be followed.

1. (C) MISSILE DOWN MOVEMENT .... OBSERVED

ITEM 12.   LIQUID SPILLAGE (VISUALLY OBSERVED)

This procedure outlines procedures to be accomplished when a spillage of RP-1 fuel, hydraulic fluid or diesel fuel is visually observed during non-tactical countdown. The launcher platform should normally be lowering or in the down and locked position. When a liquid spillage occurs the MCCC shall direct M-1 to activate the RP-1 purge fan on the FRCP. Activating the RP-1 purge fan will exhaust vapors from liquid spillage out of the missile inclosure.

1. (C) M-1 ACTIVATE RP-1 PURGE FAN .... DIRECTED

The MCCC at his discretion shall direct M-1 to activate the fire fog system after the ABORT EXTERNAL indicator is AMBER after evaluation of all facts relative to spillage observed. Activation of the fire fog system will result in extensive water fogging within the missile inclosure and should limit hazards caused by liquid spillage. Activation of the fire fog system is delayed until after ABORT EXTERNAL indicator AMBER to prevent contamination and icing of disconnect.

2. (C) M-1 DEPRESS FOG ON PUSHBUTTON .... DIRECTED

After the abort sequence is completed, the MCCC shall direct M-1 to investigate silo conditions and cause of spillage. The crew shall investigate all sources of spillage and immediately attempt to isolate, stop or restrict spillage by referring to appropriate technical manuals. During and after correction of spillage source, action shall be taken to prevent further hazards from occuring by cleaning spillage by use of most appropriate means available, fire hose, drains, rags, bucket brigade.

3. (C) M-1 TROUBLESHOOT .... DIRECTED

Changed 21 October 1964

Table 4-5. Emergency Missile Stretch

This procedure establishes a sequence of actions required to place missile in emergency stretch. Normally, A-1, M-1 and L-1 shall accomplish these actions; however, the task shall not be delayed if all crew members are not immediately available. When the pressurization system switches to emergency or stretch is required to prevent missile collapse, the MCCC shall direct available crew members to proceed to level 2 and place missile in emergency stretch. Emergency stretch procedures shall be accomplished by memory, without the aid of a checklist. Communications are desirable but not mandatory. After missile stretch is accomplished, other actions in this checklist shall be performed to determine if missile pressurization system can be returned to automatic mode if reason for placing missile in stretch was PRESSURE MODE indicator illuminated RED.

ACTION 1

When emergency stretch application is necessary, MCCC or DMCCC shall : (a) Position RESET PROGRAMMER key switch on CSMOL to ON. (b) Depress HYDRAULIC 40 HP PUMP ON pushbutton on CSMOL. The RESET PROGRAMMER key switch being on will prevent inadvertent activation of other sequences involving the missile lifting system when hydraulic pressure is applied. When crew members reach level 2, they shall: (c) Establish communication with the launch control center on either a maintenance net or the public address system to advise the MCCC that stretch application is in process. (d) Lower work platform NO. 1 by positioning work platform key switch to EXTEND. When the indicator illuminates GREEN, the platform is down and ready for access. (e) Stop 40 HP pump by depressing HYDRAULIC 40 HP pump OFF pushbutton at the hydraulic local control panel after work platform is down. (f) Proceed to work platform and extend side leaf. (g) Raise and lock folding rails over stretch mechanism. (h) Rotate (up) and lock stretch mechanism. (i) Position lift pin support housing and rotate into the missile. Pins are properly engaged when there is 9/16-inch pin surface show-

ing in Quadrants 1 and 4, and 1/4-inch pin surface showing in Quadrants 2 and 3. (j) Remove locking pin and open stretch pump grating. (k) Close release valve. (l) Open stretch pump hand valve. Actuator caps are then positioned under the cavities of the pin support housing while stretchpump is being operated. If sufficient personnel are not available to position actuator caps and operate pump, actuators may be manually positioned into the cavities and held in position by locking the collars, then operate the pump. Monitor exposed pin surface carefully during pump operation. If stretch pin enters into missile socket up to the fillet radius before stretch pressure is obtained, pins must be withdrawn slightly to prevent damage to missile. (m) Operate stretch pump until 2000 PSI is indicated on hydraulic pump gauge. (n) Close stretch pump hand valve to maintain hydraulic fluid in actuators. (o) Rotate stretch locking collars to the full-up position. (p) Open stretch pump hand valve. (q) Open release valve to relieve hydraulic pressure. (r) Report missile in stretch to LCC. If missile was placed in stretch due to pressurization switching to emergency, continue with action 2.

1.   (C) PLACE MISSILE IN STRETCH .... DIRECTED

Table 4-5.   Emergencies Missile Stretch (CONT)

ACTION 2

After missile has been placed in stretch, at MCCC direction to observe PNEUMATICS (PANEL 1) A-1 shall proceed to PNEUMATICS (PANEL 1) on level 3 and: (a) Report status of EMERGENCY indicator. (b) Report whether any other indicators are illuminated RED. Observing PLCP pressure switch indicators for abnormal indications may assist in troubleshooting the malfunction.

1.   (C) A-1, OBSERVE PNEUMATICS (PANEL 1) .... DIRECTED

If only EMERGENCY indicator is RED and PLCP pressure switch indicators are normal, at MCCC direction to observe tank pressure, A-1 shall proceed to PSC, level 8, and verify fuel and LO$_2$ tank pressures are within applicable phase limits. If other indicators on PNEUMATICS (PANEL 1) are illuminated RED or PLCP pressure switch indicators are indicating abnormal conditions, troubleshoot the malfunction.

2.   (C) A-1, OBSERVE TANK PRESSURES AT PSC .... DIRECTED

If applicable phase pressures are abnormal, troubleshoot malfunction. If pressures on PSC are normal the MCCC shall direct A-1 to check PSMR for normal gauge readings. A-1 shall observe and report the following gauge readings: (a) INST AIR SUPPLY PRESSURE (must be greater than 50 PSI); (b) AIRBORNE HELIUM SUPPLY NO. 1 (must be greater than 1450 PSI), (c) AIRBORNE AIRBORNE HELIUM SUPPLY NO. 2 (must be greater than 1450 PSI), (d) NITROGEN SUPPLY PRESSURE (must be greater than 1450 PSI). If any of the gauge indications are abnormal, troubleshoot the particular system which indicates abnormal pressure.

3.   (C) A-1, OBSERVE PRESSURE GAUGES AT PSMR .... DIRECTED

If no abnormal indications are noted at the PLCP, PSC and PSMR, MCCC shall direct deputy to return pressurization system to automatic mode. If the system remains in automatic mode, remove missile from stretch and return to normal alert configuration. If the system returns to emergency mode, troubleshoot faulty pneumatic system. If pressurization system switched to emergency during countdown or abort sequence and airborne helium NO. 1 and 2 pressures less than 1450 PSI were the cause, a manual abort shall be initiated, entering table 4-10 at action 4, in an attempt to transfer missile tank pressurization to the nitrogen supply. If malfunction was caused by supply pressures other than above, the faulty system must be corrected, prior to returning to automatic mode. After malfunction is corrected and the pressurization system is returned to automatic mode, abort sequence should continue.

4.   (C) DEPUTY, DEPRESS AUTOMATIC PUSHBUTTON .... DIRECTED

Table 4-6A.   Restoring AC Power After AC Power Loss (NA OSTF-2)

The following procedures provide the means for L-1 to restore AC power at the PRCP under the following conditions:

a.   When operating on one generator, GENERATOR MAIN BREAKER CONTROL SWITCH trips and the generator is operating. Perform item 1.

b.   When operating on one generator, generator automatically shuts down and alternate generator is in standby status. Perform item 2.

c.   When operating with generators in parallel, both GENERATOR MAIN BREAKER CONTROL SWITCHES trip and one or both generators are operating. Perform item 1.

When a diesel generator is automatically shut down by its safety devices it cannot be restarted until MASTER CONTROL SWITCH, located on engine control panel (silo level 5 or 6), has been positioned to OFF then returned to ON. Steps in these procedures shall normally be accomplished from memory by L-1 when directed by the MCCC to restore AC power.

```
CAUTION
```

If power cannot be restored at the PRCP, operating generators must be shut down immediately.

```
CAUTION
```

Do not operate generators in parallel (operate only one diesel) after AC power loss until complex electrical system has been reset.

ITEM 1.   GENERATOR MAIN BREAKER CONTROL SWITCHES
TRIPPED, GENERATOR(S) OPERATING

At MCCC direction to restore AC power, L-1 shall perform the following steps at PRCP if generator is operating, as determined by observing GENERATOR D-60 (or D-61) ON indicator on FRCP is illuminated GREEN:

(a) Position FEEDER NUMBER 3 NON-ESSENTIAL BUS CONTROL SWITCH to TRIP. Observe FEEDER NUMBER 3 NON-ESSENTIAL BUS CONTROL SWITCH RED indicator extinguishes and GREEN indicator illuminates. (b) Insert synchronizing switch handle in SYNCHRONIZING SWITCH of primary generator and position switch to ON. (c) Observe INCOMING

VOLTAGE meter indicates 460 (+2, -8) volts. (d) Observe INCOMING FREQUENCY meter indicates 60 CPS. (e) Position GENERATOR MAIN BREAKER CONTROL SWITCH to CLOSE. Observe GENERATOR MAIN BREAKER CONTROL SWITCH GREEN indicator extinguishes and RED indicator illuminates. (f) Position SYNCHRONIZING SWITCH to OFF.

NOTE

If generators were operating in parallel and power was not restored, shut down faulty generator and repeat action 1, steps b through f, on the operating generator.

1.   (C) L-1, RESTORE AC POWER . . . . **DIRECTED**

ITEM 2.   GENERATOR SHUT DOWN AND ALTERNATE GENERATOR IN STANDBY

At MCCC direction to restore AC power, L-1 shall perform the following steps at the PRCP if primary generator has shut down, as determined by observing RUNNING VOLTAGE and RUNNING FREQUENCY meters for zero indications or GENERATOR D-60 (or D-61) ON indicator on FRCP is extinguished: (a) Position FEEDER NUMBER 3 NON-ESSENTIAL BUS CONTROL SWITCH to TRIP. Observe

FEEDER NUMBER 3 NON-ESSENTIAL BUS CONTROL SWITCH RED indicator extinguishes and GREEN indicator illuminates. (b) Position GENERATOR MAIN BREAKER CONTROL SWITCH of faulty generator to TRIP. Observe GENERATOR MAIN BREAKER CONTROL SWITCH RED indicator extinguishes and GREEN indicator illuminates. (c) Position faulty generator ENGINE START-STOP SWITCH

Table 4-6A. Restoring AC Power After AC Power Loss (NA OSTF-2) (CONT)

ITEM 2. GENERATOR SHUT DOWN AND ALTERNATE GENERATOR IN STANDBY (CONT)

---

to STOP and momentarily hold. Observe ENGINE START-STOP SWITCH RED indicator extinguishes and GREEN indicator illuminates. (d) Insert synchronizing switch handle in alternate generator SYNCHRONIZING SWITCH and position switch to ON. (e) Position alternate generator ENGINE START STOP SWITCH to START and momentarily hold. Observe ENGINE START-STOP SWITCH GREEN indicator extinguishes and RED indicator illuminates.

(f) Observe INCOMING VOLTAGE meter indicates 460 (+2, -8) volts. (g) Observe INCOMING FREQUENCY meter indicates 60 CPS. (h) Position alternate GENERATOR MAIN BREAKER CONTROL SWITCH to CLOSE. Observe GENERATOR MAIN BREAKER CONTROL SWITCH GREEN indicator extinguishes and RED indicator illuminates. (i) Position SYNCHRONIZING SWITCH to OFF.

1. (C) L-1, RESTORE AC POWER . . . . **DIRECTED**

Table 4-6B. Restoring AC Power After AC Power Loss (OSTF-2 only)

The following procedures provide the means for L-1 to restore AC power from the PRCP at OSTF-2 under the following conditions:

a. When operating with diesel generator on the line, GENERATOR MAIN BREAKER CONTROL SWITCH trips, and the generator is still operating. Perform item 1.

b. When operating with diesel generator on the line, diesel generator automatically shuts down, and POWER CO. power is available. Perform item 2.

c. When operating with P O W E R CO power, POWER CO LINE MAIN BREAKER CONTROL

SWITCH trips, and POWER CO power is still available. Perform item 3.

d. When operating with P O W E R CO power, Power CO power is lost and diesel generator is in standby. Perform item 4.

When the diesel generator is automatically shut down by its safety devices it cannot be restarted until MASTER CONTROL, switch located on engine control panel (silo level 5), has been positioned to OFF then returned to ON. Steps in these procedures shall normally be accomplished from memory by L-1 when directed by the MCCC to restore AC power.

ITEM 1. GENERATOR MAIN BREAKER CONTROL SWITCH TRIPPED,
GENERATOR STILL OPERATING

At MCCC direction to restore AC power, L-1 shall perform the following steps at the PRCP if generator is still operating, as determined by observing GENERATOR FREQUENCY AC meter indicates 60 CPS and GENERATOR VOLTMETER AC indicates 460 (+2, -8) volts: (a) Position FEEDER NO. 3 NON-ESSENTIAL BUS CONTROL SWITCH to TRIP. Observe FEEDER NO. 3 NON-ESSENTIAL BUS CONTROL

SWITCH RED indicator extinguishes and GREEN indicator illuminates. (b) Observe GENERATOR VOLTMETER AC meter indicates 460 (+2, -8) volts. (c) Observe GENERATOR FREQUENCY AC meter indicates 60 CPS. (d) Position GENERATOR MAIN BREAKER CONTROL SWITCH to CLOSE. Observe GENERATOR MAIN BREAKER CONTROL SWITCH GREEN indicator extinguishes and RED indicator illuminates.

1. (C) L-1, RESTORE AC POWER . . . . **DIRECTED**

Table 4-6B.   Restoring AC Power After AC power Loss (OSTF-2 only) (CONT)

ITEM 2.   GENERATOR SHUT DOWN, POWER CO. POWER AVAILABLE

At MCCC direction to restore AC power, L-1 shall perform the following steps at the PRCP if POWER CO power is available, as determined by observing POWER CO. LINE FREQUENCY AC meter indicates 60 CPS and POWER CO. LINE VOLTMETER AC meter indicates 460 (+2, -8) volts: (a) Position FEEDER NO. 3 NON-ESSENTIAL BUS CONTROL SWITCH to TRIP. Observe FEEDER NO. 3 NON-ESSENTIAL BUS CONTROL SWITCH RED indicator extinguishes and GREEN indicator illuminates. (b) Position GENER-ATOR MAIN BREAKER CONTROL SWITCH to TRIP. Observe GENERATOR MAIN BREAKER CONTROL SWITCH RED indicator extinguishes and GREEN indicator illuminates. (c) Position POWER CO. LINE MAIN BREAKER CONTROL SWITCH to CLOSE. Observe POWER CO. LINE MAIN BREAKER CONTROL SWITCH GREEN indicator extinguishes and RED indicator illuminates.

1. (C) L-1, RESTORE AC POWER . . . .   **DIRECTED**

ITEM 3.   POWER CO POWER LOST BUT STILL AVAILABLE

At MCCC direction to restore AC power, L-1 shall perform the following steps at the PRCP if POWER CO power is still available, as determined by observing POWER CO. LINE FREQUENCY AC meter indicates 60 CPS and POWER CO. LINE VOLTMETER AC meter indicates 460 (+2, -8) volts: (a) Position FEEDER NO. 3 NON-ESSENTIAL BUS CONTROL SWITCH to TRIP. Observe FEEDER NO. 3 NON-ES-SENTIAL BUS CONTROL SWITCH RED indicator ex-tinguishes and GREEN indicator illuminates. (b) Position POWER CO. LINE MAIN BREAKER CON-TROL SWITCH to CLOSE. Observe POWER CO. LINE MAIN BREAKER CONTROL SWITCH GREEN indi-cator extinguishes and RED indicator illuminates.

1. (C) L-1, RESTORE AC POWER . . . .   **DIRECTED**

ITEM 4.   POWER CO POWER LOST AND NOT AVAILABLE, GENERATOR IN STANDBY

At MCCC direction to restore AC power, L-1 shall perform the following steps at the PRCP if POWER CO. power is not available, as determined by observing POWER CO. LINE FREQUENCY AC meter and POWER CO. LINE VOLTMETER AC meter indi-cate zero: (a) Position FEEDER NO. 3 NON-ESSEN-TIAL BUS CONTROL SWITCH to TRIP. Observe FEEDER NO. 3 NON-ESSENTIAL BUS CONTROL SWITCH RED indicator extinguishes and GREEN indicator illuminates. (b) Position POWER CO. LINE MAIN BREAKER CONTROL SWITCH to TRIP. Observe POWER CO. LINE MAIN BREAKER CONTROL SWITCH RED indicator extinguishes and GREEN indicator illuminates. (c) Position ENGINE START-STOP SWITCH to START and momentarily hold. Observe ENGINE START-STOP SWITCH GREEN indicator ex-tinguishes and RED indicator illuminates. (d) Ob-serve GENERATOR VOLTMETER AC meter indicates 460 (+2, -8) volts. (e) Observe GENERATOR FRE-QUENCY AC meter indicates 60 CPS. (f) Position GENERATOR MAIN BREAKER CONTROL SWITCH to CLOSE. Observe GENERATOR MAIN BREAKER CON-TROL SWITCH GREEN indicator extinguishes and RED indicator illuminates.

1. (C) L-1, RESTORE AC POWER . . . .   **DIRECTED**

Table 4-7. Complex Electrical System Reset

The following procedures contain the required actions to reset complex electrical system and shall be accomplished after AC power is lost and then restored. When directed by the MCCC, crew members (normally L-1 and M-1) shall complete the task of resetting complex electrical systems without further directions. The steps identified by a double asterisk (**) must be accomplished or verified any time feeder NO. 3 nonessential power bus has been de-energized and then re-energized.

In launch control room, silence fire alarm panel by depressing RESET pushbutton on fire alarm panel.

1. FIRE ALARM RESET PUSHBUTTON .... DEPRESSED

In launch control room, reactivate blast detection cabinet as follows: (a) Position OUTPUT RELAY switch to DISCONNECT. (b) Depress RCVR 1 MANUAL TEST pushbutton. (c) Depress ALARM RESET pushbutton. (d) Position OUTPUT RELAY switch to CONNECT. (e) Depress DETECTION MODE RESET pushbutton. (f) Verify OPTIC MODE indicator is illuminated.

2. BLAST DETECTION CABINET .... RESET

In communications room, restore communications with outside agencies (if not automatically restored) by depressing FUSE AND ALARM RESET pushbuttons. There are two carrier fuses and alarm panels (12 at alternate command post), each with four FUSE AND ALARM RESET pushbuttons.

3. FUSE AND ALARM RESET PUSHBUTTONS .... DEPRESSED

In communications room, restart fan coil unit by depressing START pushbutton.

4. FAN COIL UNIT START PUSHBUTTON .... DEPRESSED

At LCC level 1, depress supply fan and exhaust fan START pushbuttons to reactivate LCC ventilation system.

5. SUPPLY FAN START PUSHBUTTON .... DEPRESSED

6. EXHAUST FAN START PUSHBUTTON .... DEPRESSED

At vestibule of silo entrance, depress silo lighting START pushbutton to reactivate silo lighting.

**7. SILO LIGHTING START PUSHBUTTON .... DEPRESSED

At silo level 2, facilities terminal cabinet (FTC) NO. 2, depress OXYGEN PURGE RESET (or CR-44 RELAY RESET) pushbutton to reset oxygen purge system fans (EF-40 and EF-41). Observe OXYGEN PURGE RESET RUN (or CR-44 RELAY) indicator illuminates. If indicator fails to illuminate, reset gaseous oxygen detector, then reset oxygen purge system fans.

8. OXYGEN PURGE RESET (OR CR-44 RELAY RESET) .... DEPRESSED

At FTC NO. 2, depress LAUNCH PLATFORM FAN COIL UNIT FC-40 START pushbutton to restore MEA air supply. Observe RUN indicator is illuminated.

**9. LAUNCH PLATFORM FAN COIL UNIT FC-40 START PUSHBUTTON .... DEPRESSED

Table 4-7. Complex Electrical System Reset (CONT)

At FTC NO. 2 depress EMERGENCY WATER PUMP P-32 STOP pushbutton to stop loss of utility water being pumped through the condenser water system to the sump. Observe E M E R G E N C Y

WATER PUMP P-32 RUN indicator is extinguished and CONDENSER WATER PUMP P-30 (or P-31) RUN indicator is illuminated.

**10.   EMERGENCY WATER PUMP P-32 STOP PUSHBUTTON .... DEPRESSED

**11.   CONDENSER WATER PUMP P-30 (OR P-31) RUN INDICATOR .... ILLUMINATED

At FTC NO. 2, depress C H I L L E D W A T E R PUMP P-50 (or P-51) START pushbutton to re-

activate chilled water system. Observe RUN indicator is illuminated .

**12.   CHILLED WATER PUMP P-50 (OR P-51) START PUSHBUTTON .... DEPRESSED

At FTC NO. 2, depress CONTROL CABINET FAN COIL UNIT FC-10 START pushbutton to

restore cooling air to the logic unit cabinet. Observe RUN indicator is illuminated.

**13.   CONTROL CABINET FAN COIL UNIT FC-10 START PUSHBUTTON .... DIRECTED

At FTC NO. 2 depress RESET SETTLING TANK pushbutton reset motor operated bypass valve to

permit air washer dust collector water to bypass the settling tank.

**14.   RESET SETTLING TANK PUSHBUTTON .... DEPRESSED

On silo level 3, start 400-cycle motor generator and restore 400-cycle power by performing the following on motor generator control panel: (a) Depress MOTOR STARTER START pushbutton. (b) Ob-

serve OUTPUT AC VOLT meter indicates 116 volts nominal (c) Depress OUTPUT CONTACTOR ON pushbutton.

15.   400-CYCLE MOTOR GENERATOR .... STARTED

On silo level 3, verify power supply distribution set is operating properly by observing: (a) POWER SUPPLY ON indicator illuminated GREEN. (b) STANDBY BUS ON indicator illuminated GREEN. (c) B A T T E R Y DISCHARGE indicator extinguished. (d) DC VOLT-METER indicates 29.5 volts nominal. (e) DC AMMETER indicates 600

amperes, maximum. The 28-VDC batteries will be discharged an amount dependent on the time AC power was not available. If B A T T E R Y D I S-CHARGE indicator is illuminated RED, batteries must be recharged in accordance with T.O. 21M-HGM16F-2-6.

16.   POWER SUPPLY DISTRIBUTION SET OPERATION .... VERIFIED

MGS P O W E R switch on MISSILE GROUND POWER (PANEL 1) shall be positioned to OFF for approximately 1 second, then ON. This will de-energize the MGS checkout complete signal and en-able guidance warmup to automatically start. If a

pod air conditioner malfunction exists, as indicated on the FRCP, MGS POWER switch shall be posi-tioned to OFF until pod air conditioner malfunction is corrected.

16A.   MGS POWER SWITCH .... OFF-1 SECOND-ON

When pod air conditioner is operating within band, position MGS POWER switch on M I S S I L E POWER (PANEL 1) to ON. If G U I D A N C E POWER indicator is illuminated GREEN and MGS

WARMUP indicator on guidance countdown group is extinguished, initiate warmup in accordance with T.O. 21M-HGM16F-2-4.

17.   GUIDANCE WARMUP .... INITIATED

Table 4-7. Complex Electrical System Reset (CONT)

On silo levels 5 and 6 depress ENGINE ALARM STOP pushbutton located on engine control panel of generator D-60 and (or D61) to silence alarm.

18. ENGINE ALARM STOP PUSHBUTTON .... DEPRESSED

On silo level 6 reset 48 Volt battery charger circuit as follows:

(a) Position AUXILIARY FAILURE RELAY RESET switch to OFF then ON.

18A. 48 VOLT BATTERY CHARGER   RESET

On silo level 7, reactivate diesel fuel vapor and gaseous oxygen detector cabinets as follows: (a) Depress HORN SILENCE pushbutton on diesel fuel vapor detector cabinet. (b) Depress SYSTEM RESET pushbutton on diesel fuel vapor detector cabinet. (c) Depress HORN SILENCE pushbutton on gaseous oxygen detector cabinet. (d) Depress SYSTEM RESET pushbutton on gaseous oxygen detector cabinet.

19. DIESEL FUEL VAPOR AND GASEOUS OXYGEN DETECTOR CABINETS .... RESET

On silo level 8 restart LO_2 storage, LO_2 topping, and LN_2 storage tank vacuum pumps as follows: (a) Depress LO_2 storage tank vacuum pump START pushbutton. (b) Depress LO_2 topping tank vacuum pump START pushbutton. (c) Depress LN_2 storage tank vacuum pump START pushbutton.

**20. STORAGE TANK VACUUM PUMPS .... STARTED

```
CAUTION
```

RESET PROGRAMMER key switch on CSMOL shall be in ON position before correlating missile lifting systtem.

Correlate missile lifting system as follows to ensure proper correlation: (a) Position RESET PROGRAMMER key switch on CSMOL to ON. (b) Observe HYDRAULIC 40 HP PUMP PRESSURE indicator on CSMOL is extinguished (c) At missile lifting system logic units on silo level 1 position D I R E C T O R Y switch on chassis A4A2 to 6. (d) On chassis A4A2, simultaneously depress LAUNCHER STATUS and TEST START pushbutton. (e) Observe LAUNCHER STATUS indicator is illuminated GREEN. (f) Position DIRECTORY switch to 1. (g) Position RESET PROGRAMMER key switch on CSMOL to OFF.

21. MISSILE LIFTING SYSTEM .... CORRELATED

Thirty minutes after AC power is restored, depress LCC BLAST CLOSURES OPEN pushbutton on FRCP. Observe LCC AIR INTAKE OPEN and LCC AIR EXHAUST OPEN indicators illuminate GREEN.

22. LCC BLAST CLOSURES OPEN PUSHBUTTON .... DEPRESSED

When missile guidance system (MGS) warmup is complete, as indicated by MGS R E A D Y F O R CHECKOUT indicator on guidance countdown group illuminated, An MGS checkout shall be performed in accordance with T.O. 21M-HGM16F-2-4 to place system in a standby status.

23. MGS CHECKOUT .... INITIATED

Table 4-8.   Emergency Diesel Generator Paralleling and Shutdown (NA OSTF-2)

The following procedures shall be used to start and parallel a standby generator and to shut down the operating generator when it has malfunctioned. These steps will normally be performed from memory, by L-1 at the PRCP. In the absence of L-1, any crew member may perform these procedures. At MCCC direction to place alternate generator on the line and to shut down faulty generator, L-1 shall perform the following steps:

a.   Position alternate generator AMMETER SWITCH to 1, 2, or 3. This allows the current load to be indicated on the ammeter when alternate generator is placed on the line.

b.   Insert synchronizing switch handle in SYN-CHRONIZING SWITCH of alternate genera-tor and position switch handle to ON. If the SYNCHRONIZING SWITCH is not ON. IN-COMING VOLTAGE meter, INCOMING FRE-QUENCY meter, and SYNCHROSCOPE will not operate, and GENERATOR MAIN BREAKER CONTROL SWITCH cannot be closed.

c.   Verify ENGINE START-STOP SWITCH GREEN indicator for alternate generator is illumi-ated.

d.   Position ENGINE START-STOP SWITCH of alternate generator to START and momen-tarily hold. ENGINE START-STOP SWITCH indicators are disabled until ENGINE START-STOP SWITCH is released

e.   After releasing ENGINE START-STOP SWITCH, verify ENGINE START-STOP SWITCH RED indicator illuminates, indicating that diesel engine has started.

f.   Monitor INCOMING VOLTAGE meter until valtage has stablized at 460 (+2, -8) volts.

g.   Operate alternate generator GOVERNOR

MOTOR CONTROL SWITCH to RAISE or LO-WER so that INCOMING FREQUENCY meter indicates 60 CPS.

h.   Operate alternate generator GOVERNOR MOTOR CONTROL SWITCH so that SYN-CHROSCOPE hand rotates slowly clockwise. When SYNCHROSCOPE hand indicates 11:55 (clock position), position GENERATOR MAIN BREAKER CONTROL SWITCH of alternate generator to CLOSE.

i.   Observe GENERATOR MAIN BREAKER CON-TROL SWITCH RED indicator is illuminated and GREEN indicator is extinguished.

j.   Observe AMMETER AC indication for in-coming generator increases and AMMETER AC indication of running generator de-creases. Position alternate generator SYN-CHRONIZING SWITCH to OFF.

k.   Observe RUNNING VOLTAGE meter indicates 460 (+2, -8) volts.

l.   Observe RUNNING FREQUENCY meter indi-cates 60 CPS.

m   Remove faulty generator from the line by positioning faulty generator GENERATOR MAIN BREAKER CONTROL SWITCH to TRIP.

n.   Observe GENERATOR MAIN BREAKER CON-TROL SWITCH GREEN indicator is illuminated and RED indicator is extinguished.

o.   Shut down faulty diesel by positioning EN-GINE START-STOP SWITCH to STOP and mo-mentarily hold. ENGINE START-STOP SWITCH indicators are disabled until ENGINE START-STOP SWITCH is released.

p.   After ENGINE START-STOP SWITCH is re-leased, verify ENGINE START-STOP SWITCH GREEN indicator illuminates.

1.   (C) L-1, PLACE ALTERNATE GENERATOR ON THE LINE SHUT DOWN FAULTY GENERATOR . . . . DIRECTED

Table 4-10.   Manual Abort

The following procedures provide the means for manually aborting the weapon system and returning to the safest condition possible prior to troubleshooting. The procedures are organized and oriented so that they may be entered either at an action or at a step within an action. Entry to the table shall depend on the exact place in the countdown, commit or abort sequence a malfunction or emergency occurred. If entry is directed at a certain action and steps within this action have previously been accomplished, either automatically or manually, they shall be omitted and only those steps required to continue the manual abort flow shall be accomplished.

All crew members are used to perform the actions in this table. While direction is given as to individual crew actions, the MCCC can and should assign the direct accomplishment of certain steps to other than the indicated crew member if conditions and safety so dictate.

The table is in a sequential flow beginning with the missile up and locked and continuing through the final step which is manually venting ambient helium. What has transpired prior to manually aborting shall dictate the remaining steps necessary to safety and successfully complete the manual abort.

ACTION 1

The MCCC directs M-1 to postion RESET PROGRAMMER key switch on CSMOL to ON. This enables the launcher platform to be lowered manually from the CSMOL and will prevent automatic sequencing of the missile lifting system.

1.   (C) M-1, RESET PROGRAMMER KEY SWITCH ON .... DIRECTED

Cycling countdown system power switch will remove the start countdown and start abort signals to prevent any further automatic sequencing. Normally the relay logic activated by the start of an automatic abort sequence will ensure that the necessary subsystem logic is returned to either an abort or standby condition. Once the countdown system power is cycled, automatic abort sequencing is blocked, since the necessary logic signals have dropped out.

When the countdown system power is turned off, the relay holding the missile pressurization system in emergency mode is de-energized and would normally allow the system to return to automatic mode when power is reapplied. To prevent this, the EMERGENCY pushbutton shall be depressed and held while countdown power is being cycled.

To confirm that the power changeover switch has returned to the external position, the CHANGE OVER SW EXTERNAL indicator on MISSILE GROUND POWER (PANEL 1) will be observed. If changeover has not occurred, power may have remained internal and missile battery depletion may be expected. The time for depletion depends on the power load. At MCCC direction to shut down the HPU and cycle countdown system power, the deputy at the launch control console and A-1 at the logic unit on level 3 shall perform step 2 in close coordination as follows: (a) A-1 shall depress and hold HYD PUMP STOP pushbutton. (b) Deputy shall depress and hold EMERGENCY pushbutton. (c) A-1 shall cycle COUNTDOWN (PANEL 1) SYSTEM POWER switch to OFF for 1 second then ON. (d) After PRESSURE MODE indicator RED, deputy shall release EMERGENCY pushbutton. (e) A-1 shall release HYD PUMP STOP pushbutton. (f) A-1 shall observe CHANGE OVER SW EXTERNAL indicator and report indication prior to returning to LCC. Indications on the launch control console

Table 4-10. Manual Abort (CONT)

ACTION 1 (CONT)

will reflect the above actions as they are performed. While countdown system power is off, indicators on the launch control console will extinguish. If missile power has been internal and changes to external, POWER INTERNAL indicator will extinguish when countdown system power is reapplied.

When the HPU is stopped and hydraulic pressure decreases to below 1750 PSI, HYDRAULIC PRESSURE indicator will change from GREEN to extinguished.

2. (C) Deputy, A-1, SHUT DOWN HPU AND CYCLE COUNTDOWN SYSTEM POWER .. DIRECTED

Before the launcher platform (LP) can be lowered, from within 33 inches of the up and locked position, the pressurization system shall be returned to automatic mode. This closes the boiloff valve and allows LO$_2$ tank pressure to increase above 8.0 PSI. When LO$_2$ tank pressure is above 8.0 PSI, logic summaries allow LP down motion.

At MCCC direction to manually lower the LP, M-1 shall accomplish step 3 in accordance with the following at CSMOL: (a) Depress HYDRAULIC 40 HP PUMP ON pushbutton and observe HYDRAU-

LIC 40 HP PUMP PRESSURE indicator is illuminated GREEN (pressure within limits). (b) Depress DOWN RUN pushbutton. (c) Observe down motion of LP. If LP was in the full up-and-locked position, CREEP DISABLED indicator will change from RED to extinguished. Down motion shall be observed and confirmed by TV cameras or other means. The launcher platform should be monitored throughout the drive down sequence.

During the lowering sequence, the deputy shall control boiloff valve as in a normal abort sequence.

3. (C) M-1, MANUALLY LOWER LAUNCHER PLATFORM .... DIRECTED

Upon completion of the manual drive down of the LP, DOWN COMPLETED RUN AND LOCKED indicator on CSMOL will illuminate GREEN. This shall be observed by the MCCC.

4. (C) DOWN COMPLETED RUN AND LOCKED INDICATOR GREEN .... OBSERVED

When all steps in action 1 have been accomplished, the crew shall continue manual abort in action 2 when directed by the MCCC.

ACTION 2

When LP has reached the down and locked position, MCCC shall direct deputy to depress EMERGENCY pushbutton. This will enable the boiloff valve, The RAISE or LOWER pushbuttons may be ineffective until step 4 is completed .Therefore steps 2 through 4 should be completed as rapidly as possible.

> **WARNING**
>
> To prevent a possible bulkhead reversal, the pressurization system must remain in emergency until LO$_2$ is drained and it is specifically directed to return to automatic.

1. (C) DEPUTY, DEPRESS EMERGENCY PUSHBUTTON .... DIRECTED

The MCCC shall direct M-1 to close silo doors if missile enclosure purge unit (MEPU) is retracted. Closing silo doors when MEPU is not retracted will cause damage to MEPU and missile. This table may be continued, if necessary, without closing the silo doors. At MCCC direction to close silo doors, M-1

shall: (a) Verify that the RESET PROGRAMMER KEY switch is in the ON position. (b) Depress silo DOORS CLOSE pushbutton and hold for 30 seconds. The silo doors will not start closing until a 30 second warning delay period is completed. When SILO DOORS OPEN indicator extinguishes or observing on T. V. silo doors closing, SILO DOORS CLOSE pushbutton will be released. The closing cycle will continue until doors

Table 4-10.   Manual Abort (CONT)

ACTION 2 (CONT)

are fully closed. (c) Observe SILO DOORS CLOSE indicator illuminates GREEN, indicating silo doors are closed. (d) Depress CRIB HORIZONTAL UN-LOCK pushbutton. (e) Observe CRIB HORIZONTAL UNLOCK indicator illuminates GREEN. (f) Depress CRIB VERTICAL UNLOCK pushbutton. (g) Observe CRIB VERTICAL UNLOCK indicator illuminates GREEN. (h) Depress HYDRAULIC 40HP PUMP OFF pushbutton (Hydraulic pressure is no longer needed). All CSMOL indicators will extinguish as pressure decays below 3000 psi.

> **CAUTION**
>
> Do not close silo doors if MEPU is extended.

2.   (C) M-1, CLOSE SILO DOORS .... DIRECTED

MCCC shall direct deputy and A-1 to obtain control of missile tank pressures and maintain phase II pressures. At MCCC direction, A-1 shall proceed to level 3, PNEUMATICS (PANEL 1) and standby: (a) When A-1 is in position on silo level 3, deputy shall depress and hold EMERGENCY pushbutton. (b) A-1 shall then cycle PNEUMATICS (PANEL 1) SYSTEM POWER switch OFF for 1 second then ON. (c) When PRESSURE MODE indicator illuminates RED after cycling pneumatics system power, deputy shall relsease EMERGENCY pushbutton. (d) The deputy shall manually adjust missile tank pressures to phase II (LO$_2$ tank pressure 3.4 to 8.0 PSI, fuel tank pressure 62.5 to 63.9 PSI). Fuel tank pressure may require frequent adjustment to compensate for normal pressure decay.

3.   (C) DEPUTY, A-1, CYCLE PNEUMATICS
     SYSTEM POWER AND ADJUST TO PHASE II PRESSURES .... DIRECTED

If HPU has been shutdown and countdown system power was previously cycled during the manual abort procedures, the next step will be omitted. If not, the amplified reasons and procedures for accomplishing this step are contained in action 1, step 2, of this table. The MCCC shall direct deputy and A-1 to shutdown HPU and cycle countdown system power.

4.   (C) DEPUTY, A-1, SHUT DOWN HPU AND
     CYCLE COUNTDOWN SYSTEM POWER .... DIRECTED

The MCCC shall direct M-1 to open LCC blast closures from the FRCP if they have been previously closed. This restores outside air circulation to the launch control center.

5.   (C) M-1, OPEN LCC BLAST CLOSURES .... DIRECTED

If both diesel generators have been operating, at MCCC direction to shutdown one diesel generator, L-1 shall: (a) Trip GENERATOR MAIN BREAKER CONTROL SWITCH on PRCP and observe GENERATOR MAIN BREAKER CONTROL SWITCH RED indicator extinguished and GREEN indicator illuminated. (b) Position ENGINE START-STOP SWITCH to STOP and momentarily hold. Release switch and observe ENGINE START-STOP SWITCH RED indicator extinguished and GREEN indicator illuminated.

6.   (C) L-1, SHUTDOWN ONE GENERATOR .... DIRECTED

When all steps in action 2 have been accomplished, the crew will continue manual abort in action 3 when directed by the MCCC.

Table 4-10. Manual Abort (CONT)

ACTION 3

At MCCC direction, A-1 shall manually drain LO₂ at LO₂ TANKING (PANEL 1 and PANEL 2) as follows: (a) Position REMOTE LOCAL switch to LOCAL. (b) Observe LOCAL POWER indicator is illuminated AMBER. (The valves will now position to the configuration of the switches on the LO₂ tanking panel and the main LO₂ storage tank and topping tank will vent transfer pressure.)

> **CAUTION**
>
> If LINE VENT VALVE N-80 indicator fails to illuminate AMBER, the N-80 vent valve plug must be installed prior to continuing manual LO₂ drain.

> **CAUTION**
>
> Monitor LO₂ DISCONNECT MATED SWITCH "A" and "B" indicators during LO₂ drain. If indicators illuminate RED, stop LO₂ drain until main LO₂ disconnect is verified to be properly mated for drain.

(c) Observe LINE VENT VALVE N-80 indicator illuminates AMBER. (d) After approximately 2 minutes, observe LO₂ STG TNK PRESSURE indicator is illuminated GREEN, indicating that pressure in the main LO₂ storage tank is less than 25 PSI. (LO₂ can now be drained.) If LO₂ STG TNK PRESSURE indicator does not illuminate GREEN after 2 minutes, observe STORAGE TANK VENT VALVE N-5 indicator. If valve N-5 indicator is AMBER, manually open valve N-5 by disconnecting the instrument air line between N-5 controller and valve bonnet at the controller. (e) A-1 shall position L-16 valve switch OPEN and observe DRAIN VALVE L-16 indicator illuminates GREEN.

If indicator does not illuminate GREEN, A-1 shall position L-1 valve switch to OPEN and observe FINE LOAD VALVE L-1 indicator illuminates GREEN. Valve L-16 shall be closed if valve L-1 is used to drain LO₂. (f) Position A/B F&D valve switch to OPEN to allow LO₂ to drain from missile. (g) Observe AIRBORNE FILL & DRAIN VALVE indicator illuminates GREEN. If airborne fill and drain valve cannot be opened, LO₂ must be allowed to boiloff. If it becomes necessary to stop LO₂ draining, A-1 shall position A/B F&D valve switch CLOSE. A-1 shall start timing drain when indicator illuminates GREEN.

1. (C) A-1, MANUALLY DRAIN LO₂ .... DIRECTED

After SITE HARD GREEN illuminates green if both generators are operating, at MCCC direction to shutdown one diesel generator, L-1 shall: (a) Trip GENERATOR MAIN BREAKER CONTROL SWITCH on PRCP and observe GENERATOR MAIN BREAKER CONTROL SWITCH RED indicator extinguished and GREEN indicator illuminated. (b) Position ENGINE START-STOP SWITCH to STOP and momentarily hold. Release switch and observe ENGINE START-STOP SWITCH RED indicator extinguished and GREEN indicator illuminated.

2. (C) L-1, SHUTDOWN ONE GENERATOR .... DIRECTED

When 50 minutes have elapsed after opening the airborne fill and drain valve, at MCCC direction to secure from LO₂ drain, A-1 at LO₂ tanking panels shall: (a) Position valve, switch L-16 (or L-1) to CLOSE. (b) Observe that DRAIN VALVE L-16 (or FINE LOAD VALVES L-1) indicator illuminates AMBER. (c) Position A/B F&D switch to CLOSE. (d) Observe that AIRBORNE FILL & DRAIN VALVE indicator illuminates AMBER. The manual LO₂ drain sequence is now complete.

3. (C) A-1, SECURE FROM LO₂ DRAIN .... DIRECTED

When all steps in action 3 have been accomplished, the crew will continue manual abort in action 4 when directed by the MCCC.

ACTION 4

If HPU has been shutdown and countdown system power was previously cycled during the manual abort procedure, the next step shall be omitted. If not, the amplified reasons and procedures for accomplishing this step are contained in action 1, step 2, of this table. The MCCC shall direct deputy and

A-1 to shutdown HPU and cycle countdown system power. Prior to accomplishing this step, if the LO₂ drain complete indicator is illuminated GREEN, cycle LO₂ tanking (PANEL 1) power switch to OFF for 1 second then ON.

1. (C) DEPUTY, A-1, SHUTDOWN HPU AND CYCLE COUNTDOWN SYSTEM POWER .... DIRECTED

Helium which has been stored in the airborne shrouded spheres must now be manually vented. The helium is used to maintain missile fuel tank pressure and will not be vented automatically when missile pressurization is in emergency mode. If conditions were such that an automatic helium vent cycle was completed, as indicated by HELIUM VENT COMPLETE indicator on launch control console G R E E N, manual helium vent may be omitted. At M C C C direction to manually vent helium, A-1 shall proceed to LN₂-HELIUM (PANEL 1) on level 3 and M-1 to HCU on launcher platform, level 3. At LN₂-HELIUM (PANEL 1), A-1 shall: (a) Position valve 14, 201 and 50 switches to OPEN. Opening valve 14 will allow helium to be supplied for emergency missile pressurization from inflight helium cylinder NO. 2 through the PSMR and PSC. Opening valve 201 will vent the LN₂ storage tank. Opening valve 50 will allow GN₂ to be supplied for normal missile pressurization through the PSMR and PSC when the missile pressurization system is returned to automatic mode. (d) Position all other valve switches on LN₂-HELIUM (PANEL 1) to CLOSED. This will prevent any undesired valve position changes from oc-

curring during manual helium venting. (c) When all valve switches are properly positioned A-1 shall position REMOTE-LOCAL SWITCH to LOCAL. (d) A-1 shall observe LOCAL POWER indicator illuminates AMBER. When A-1 has accomplished the above actions, M-1 shall manually vent helium at the HCU by steps (e) through (i).

WARNING

Venting Helium at the HCU creates a high noise level. Wear ear protectors to prevent ear injury.

(e) Mannually open valve 302 (helium dump valve). (f) Observe MISSILE HELIUM STORAGE pressure gauge 301. (g) Ten minutes after pressure gauge 301 reads less than 50 PSI manually close valve 302. (h) After valve 302 is closed, manually open valve 313 (manual vent valve) to allow venting of helium from the control storage bottle on the HCU. (i) After helium has vented valve 313 shall be closed. Manual helium venting of the shrouded spheres is now complete.

2. (C) A-1, M-1, MANUALLY VENT HELIUM .... DIRECTED

At MCCC direction to cycle pressurization to phase I, deputy at launch control console and A-1 at PNEUMATICS (PANEL 1), level 3 shall be in close coordination: (a) A-1 depress and hold STANDBY pushbutton on PNEUMATICS (PANEL 1) and (b) Cycle PNEUMATICS (PANEL 1) SYSTEM POWER switch to OFF for 1 second then ON. When pneumatics system power is cycled, pressurization system will switch to automatic and pressures will automatically adjust to phase I when S T A N D B Y STARTED indicator illuminates

G R E E N. (c) A-1 shall verify S T A N D B Y STARTED indicator is illuminated GREEN. (d) After STANDBY STARTED indicator is illuminated GREEN, A-1 shall release STANDBY pushbutton. Since fuel pressure was greater than 53 PSI when the pneumatics system power was cycled, a signal to start a 20-second time d e l a y p i c k u p r e l a y is locked in. When this relay times out, it will cause the pneumatics system to switch to emergency mode since fuel pressure was greater than 53 PSI but did not remain at 53 PSI after the 20-second

ACTION 4 (CONT)

time delay expired. This is a normal indication be-cause phase I controllers were selected. (e) After PRESSURE MODE indicator illuminates RED, deputy shall depress AUTOMATIC pushbutton. (f) Deputy shall verify that pressurization system has automatically switched to phase I pressures.

3. (C) DEPUTY, A-1 CYCLE PRESSURIZATION SYSTEM TO PHASE I .... DIRECTED

If helium was vented automatically, it is evident that the LN₂-helium panel received the abort external signal and de-energized the commit internal hold-in circuit. If a manual abort was started at any point requiring a manual helium vent, abort external signal was not received. Therefore, LN₂-HELIUM (PANEL 1) SYSTEM POWER switch shall be cycled to OFF for 1 second then ON.

4. (C) A-1, CYCLE LN₂-HELIUM SYSTEM POWER SWITCH .... DIRECTED

If HPU has not previously been stopped or failed to stop when HYD PUMP STOP pushbutton was de-pressed, the MCCC shall direct A-1 to proceed to essential motor control center, level 2, A-1 will position HYDRAULIC PUMPING UNIT circuit breaker to OFF. Hydraulic pressure will now decay below 1750 PSI.

5. (C) A-1, SHUTDOWN HPU .... DIRECTED

The missile and associated subsystems are now in the safest condition possible and malfunction trouble-shooting procedures may begin. The malfunction or emergency that necessitated the manual abort will be corrected using organizational maintenance tech-nical data applicable to the malfunctioning system. Maintenance assistance should be requested as re-quired.

6. (C) CREW TROUBLESHOOT .... DIRECTED

Two hours after phase I pressures have been ob-tained, the ambient helium sphere will be vented. Omit this step if shrouded helium spheres were vented automatically and ABORT COMPLETE indicator illuminated GREEN. Helium is maintained in the ambient sphere under both normal and manual abort conditions to provide a purge for the booster engine turbopumps to separate LO₂ and lubricant oil for a period of two hours. At MCCC direction to vent ambient helium sphere A-1 shall proceed to LN₂-HELIUM (PANEL 1) on level 3 and M-1 to HCU on launcher platform, level 3: (a) At LN₂-HELIUM (PANEL 1) A-1 shall position valve switch 26 to OPEN. Opening valve 26 will allow helium from the ambient sphere to be vented at the HCU. When A-1 has accomplished the above action, M-1 will manually vent helium at the HCU by performing steps (b) through (d).

WARNING

Venting helium at the HCU creates a high noise level. Wear ear protectors to prevent ear injury.

(b) Manually open valve 302 (helium dump valve) and valve 313 (manual vent valve). (c) Observe MISSILE HELIUM STORAGE PRESSURE gauge 301. (d) Ten minutes after pressure gauge 301 reads less than 50 PSI, manually close valves 302 and 313. When valves 302 and 313 are closed A-1 shall perform steps (e) through (h). (e) Posi-tion valve switches 26 and 14 CLOSED. (f) Posi-tion valve switches 15 and 54 to OPEN. Valve switch positions are now in standby configuration. (g) Position REMOTE-LOCAL SWITCH to RE-MOTE. (h) Observe LN₂-HELIUM (PANEL 1) S Y S T E M IN STANDBY indicator illuminates GREEN.

Table 4-10. Manual Abort (CONT)

ACTION 4 (CONT)

7. (C) A-1, M-1, VENT AMBIENT HELIUM SPHERE .... DIRECTED

> **WARNING**
>
> Hydraulic recharge is not initiated at the end of manual door closing sequence. Therefore, if silo doors were manually closed, loud venting will occur when GN_2 recharge cycle is initiated at hydraulic local control panel.

Table 4-11. LN₂ LOAD Indicator Not Amber or Not Green After Being Amber

These procedures provide a means of circumventing a liquid nitrogen loading malfunction to permit continuation of a tactical countdown. If LN₂ LOAD indicator fails to illuminate A M B E R, wait for HELIUM LOAD indicator to illuminate AMBER then proceed to action 1. If both LN₂ LOAD and HELIUM LOAD indicators fail to illuminate AMBER 2 minutes after countdown start, abort is required. If LN₂ LOAD indicator fails to illuminate GREEN after illuminating AMBER, proceed to action 2.

Liquid nitrogen rapid load valve 214 must be open to complete the summary required for LN₂ LOAD indicator AMBER. If valve 214 has opened, the microswitch that indicates valve 214 is open has failed and the microswitch shall be jumpered to obtain the sequence logic summary. Transfer pressure of 75 PSI is an additional signal required for the LN₂ complete summary. If pressure switch 96 which sends this signal has malfunctioned, it must be bypassed.

ACTION 1

The MCCC shall direct A-1 to proceed to LN₂-HELIUM (PANEL 1), level 3, to cycle system power.

1. (C) A-1, PROCEED TO LN₂-HELIUM (PANEL 1) .... DIRECTED

The MCCC shall direct M-1 to proceed to LN₂ prefab, level 7, to observe position of valves 213 and 214.

2. (C) M-1, PROCEED TO LN₂ PREFAB .... DIRECTED

When A-1 and M-1 are in position, MCCC shall direct A-1 to position SYSTEM POWER switch on LN₂-HELIUM (PANEL 1) to OFF. This will enable resetting of the timers that close valves 213 and 214.

3. (C) A-1, LN₂-HELIUM SYSTEM POWER OFF .... DIRECTED

The MCCC shall direct A-1 to position SYSTEM POWER switch on LN₂-HELIUM (PANEL 1) to ON. This will restart the timers that close valve 213 and 214. These valves should open after system power is on.

Table 4-11. LN$_2$ LOAD Indicator Not Amber or Not Green After Being Amber
(CONT)

ACTION 1 (CONT)

4. (C) A-1, LN$_2$-HELIUM SYSTEM POWER ON . . . . DIRECTED

After A-1 position SYSTEM POWER switch to
ON, MCCC shall direct M-1 to observe and
report position of valves 213 and 214. Both
valves and valve 215 should be open.

5. (C) M-1, REPORT VALVE 213 AND 214 POSITIONS . . . . DIRECTED

If both valves 213 and 214 are closed abort is
required due to a logic malfunction which cannot be corrected or bypassed. If either or
both valves are open, the rapid load valve 214
open microswitch shall be jumpered. This will
eliminate valve microswitch troubleshooting.
The loading of LN$_2$ through either valve 213 or
214 is sufficient to permit continuation of a
tactical countdown. The MCCC shall direct
A-1 to position LN$_2$-HELIUM (PANEL 1)
SYSTEM POWER switch to OFF. This will
close valves 213, 215, and 214, enable the
valve timers to be restarted, and take power
from the terminal boards while jumper is installed.

6. (C) A-1, LN$_2$-HELIUM SYSTEM POWER OFF. . . . . DIRECTED

The MCCC shall direct M-1 to install jumper
between pins 2 and 3 of TB54. This jumper will
circumvent valve 214 microswitch failure and
send the required open logic summary.

7. (C) M-1, JUMPER PINS 2 AND 3 OF TB54 . . . . DIRECTED

The MCCC shall direct A-1 to position SYSTEM
POWER switch on LN$_2$HELIUM (PANEL 1) to
ON. The LN$_2$ LOAD indicator should illuminate
AMBER, then GREEN in 3 minutes. If the
indicator fails to illuminate AMBER, abort is
required. M-1 shall remove jumper wire and
return to the LCC with A-1 before MCCC
initiates abort sequence.

8. (C) A-1, LN$_2$-HELIUM SYSTEM POWER ON . . . . DIRECTED

ACTION 2

If LN$_2$ LOAD indicator failed to illuminate
GREEN after having been AMBER the probable
cause is a malfunction of pressure switch 96
which senses that LN$_2$ transfer pressure is less
than 75 PSI. Observe P.S. 96 LN$_2$ PRESS LOW
indicator on PLCP. If this indicator is extinguished, abort is required because pressure
switch 96 has sensed greater than 75 PSI
transfer pressure in the LN$_2$ storage tank.
This indicates that a logic malfunction that
cannot be corrected has occurred which will
prevent the countdown from continuing. If
P.S. 96 LN$_2$ PRESS LOW indicator on PLCP
is illuminated AMBER, this indicates that a
possible internal failure of P.S. 96 has occurred which may be corrected. MCCC shall
direct M-1 to proceed to LN$_2$ prefab, level 7,
to observe storage tank pressure on gauge 227.
During countdown gauge 227 should normally
indicate 100 PSI. If gauge 227 indicates less
than 70 PSI, abort is required since a pressure controller malfunction is causing in-
sufficient transfer pressure or vent valve 201
is not seated properly, causing a loss of
transfer pressure.

1. (C) M-1, REPORT GAUGE 227 INDICATION . . . . DIRECTED

If gauge 227 indicates greater than 70 PSI,
MCCC shall direct A-1 to disconnect wire
from pin 14 of A22J1 at rear of LN$_2$-HELIUM
(PANEL 1). This will interrupt the signal
(LN$_2$ transfer pressure less than 75 PSI) sent
from pressure switch 96.

2. (C) A-1, DISCONNECT WIRE FROM PIN 14 OF A22J1 . . . . DIRECTED

Table 4-11. LN₂ LOAD Indicator Not Amber or Not Green After Being Amber (CONT)

ACTION 2 (CONT)

When wire from pin 14 has been disconnected, MCCC shall direct A-1 to cycle SYSTEM POWER switch on LN₂HELIUM (PANEL 1) to OFF for 1 second, then ON. Cycling SYSTEM POWER switch is required to drop out the transfer pressure less than 75 PSI signal and to restart the 3-minute LN₂ load timer. LN₂ LOAD indicator should illuminate GREEN 3 minutes after cycling SYSTEM POWER switch. If indicator fails to illuminate GREEN, an abort is required. The wire to pin 14 of A22J1 should be reconnceted after abort is complete.

3. (C) A-1, CYCLE LN₂-HELIUM SYSTEM POWER OFF-1 SECOND-ON .... DIRECTED

Table 4-12. HYDRAULIC PRESSURE Indicator Not Green

The following procedures provide the means of circumventing a hydraulic pumping unit (HPU) malfunction to continue a tactical countdown. Possible causes of HYDRAULIC PRESSURE indicator not illuminated GREEN are:

a. Failure of HPU to start. In this event, an attempt will be made to start the HPU, first at the essential motor control center (EMCC), then locally at the HPU.

b. Hydraulic pressure not within the normal range of 1750 to 2250 PSI. If this condition exists, an attempt will be made to adjust pressures at the HPU, either by closing the bypass valve if open, or by adjusting the pressure compensator.

c. Failure of a pressure switch (49S, 49aS, 49B, or 49aB) to sense pressure. To correct this malfunction, an attempt will be made to jumper the hydraulic pressure signal terminals in the rear of HYDRAULIC (PANEL 1).

Any of the following actions may correct the malfunction. HYDRAULIC PRESSURE indicator on the launch control console should be observed at the completion of each action. If the indicator illuminates GREEN, no further actions shall be accomplished.

ACTION 1

The MCCC shall direct M-1 to EMCC, level 2, to reset HYDRAULIC PUMPING UNIT circuit breaker.

1. (C) M-1, RESET HYDRAULIC PUMPING UNIT CIRCUIT BREAKER .... DIRECTED

The MCCC shall direct M-1 to HPU on launcher platform level 3 to report the status of the HPU: (a) HPU running or not running. (b) HPU stage pressures. If M-1 reports that HPU is not running, perform procedures contained in action 2. If HPU is running and stage pressures are not between 1750 and 2250 PSI, perform procedures contained in action 3. If HPU is running and stage pressures are between 1750 and 2250 PSI, perform procedures contained in action 4.

2. (C) M-1, REPORT STATUS OF HPU .... DIRECTED

Table 4-12. HYDRAULIC PRESSURE Indicator Not Green (CONT)

ACTION 2

If HPU is not running after performing procedures in action 1, at MCCC direction to attempt to start HPU, M-1 shall: (a) Reset circuit breaker NO. 1 inside of HPU. (b) If HPU fails to start, position REMOTE LOCAL switch on HPU to LOCAL. (c) Depress hydraulic pump START pushbutton.

If HPU fails to start, abort is required. If HPU starts, REMOTE LOCAL switch shall be left in LOCAL position. (d) If HPU starts, depress OIL EVACUATE pushbuttons for both the first and second stages when pressure in both stages is greater than 1750 PSI.

1. (C) M-1, ATTEMPT TO START HPU .... DIRECTED

ACTION 3

If HPU is reported to be running and stage pressures are not between 1750 and 2250 PSI, as indicated on the high pressure gauges, At MCCC direction, M-1

shall observe and report the position of bypass valves 13B and 13S.

1. (C) M-1, REPORT POSITION OF VALVE 13 ON MALFUNCTIONING STAGE .... DIRECTED

If bypass valve 13 of the malfunctioning stage is reported closed, do not perform step 2, proceed to step 3. If bypass valve 13 is reported open, at MCCC direction, M-1 shall: (a) Position manual control lever of the open bypass valve to close. (b) If by-

pass valve electrically opens after manually closing, disconnect electrical connector at the valve. If the malfunctioning bypass valve does not close and remain closed, abort is required.

2. (C) M-1, CLOSE VALVE 13 ON MALFUNCTIONING STAGE .... DIRECTED

If pressures are not between 1750 and 2250 PSI and both bypass valves are closed, MCCC shall direct M-1 to adjust pump compensator of the mal-

functioning stage to bring the pressure within limits. If pressure cannot be adjusted, abort is required.

3. (C) M-1, ADJUST COMPENSATOR TO OBTAIN 2000 PSI .... DIRECTED

ACTION 4

If HPU is running and pressures are between 1750 and 2250 PSI, MCCC shall direct A-1 to jumper the hydraulic pressure signal terminals by installing a jumper wire between pins 13 and 23 of A38J1 at

rear of HYDRAULIC (PANEL 1). If HYDRAULIC PRESSURE indicator on launch control console does not illuminate GREEN, abort is required.

1. (C) A-1, JUMPER PINS 13 AND 23 OF A38J1 .... DIRECTED

Table 4-13.   LO$_2$ Tanking Panel Malfunction During LO$_2$ Chilldown

The following procedures provide the means of correcting an electrical malfunction of N-5, storage tank vent valve (item 1); N-4, topping tank vent valve (item 2); N-50, topping tank pressurization valve (item 3); and L-60, topping chilldown valve (item 4) to enable the completion of a tactical countdown.

Failure of valve N-5 will prevent start of main LO$_2$ loading. Failure of valves N-4 or N-50 will prevent the completion of LO$_2$ loading as the 15-second topping timer must have energized at least once to get LO$_2$ READY indicator GREEN. Failure of valve L-60 will prevent proper chilldown of top-

ping lines and engine turbopumps which could result in possible cavitation of the turbopumps when the engine start signal is sent.

WARNING

If any jumpers are installed in the performance of these procedures and an abort is initiated, immediately position LO$_2$ TANKING (PANEL 1) SYSTEM POWER switch to OFF. Remove jumpers and return LO$_2$ TANKING (PANEL 1) SYSTEM POWER switch to ON.

ITEM 1.   STORAGE TANK VENT VALVE N-5 INDICATOR NOT AMBER

ACTION 1

During countdown, after PNEUMATIC IN PHASE II indicator illuminates G R E E N, if STORAGE TANK VENT VALVE N-5 indicator fails to illuminate AMBER, A-1 shall immediately position LO$_2$ TANKING (PANEL 1) SYSTEM POWER switch to OFF and announce the malfunction and action

taken to the MCCC. Switching the power off the LO$_2$ tanking panel stops any LO$_2$ loading that may be in progress and sets the LO$_2$ loading system in proper configuration for troubleshooting and malfunction correction.

1.   (A-1) LO$_2$ TANKING POWER .... **OFF**

The MCCC shall direct M-1 to proceed to silo level 7 and standby at pressurization prefab to report the

position of valve N-5 when LO$_2$ system power is turned on.

2.   (C) M-1, PROCEED TO PRESSURIZATION PREFAB .... DIRECTED

When M-1 is in position at pressurization prefab, the MCCC shall direct A-1 to position LO$_2$ TANK-

ING (PANEL 1) SYSTEM POWER switch to ON.

3.   (C) A-1, LO$_2$ TANKING POWER ON .... DIRECTED

The MCCC shall direct M-1 to report position of valve N-5. If valve N-5 is reported closed perform

steps contained in action 2. If valve N-5 is reported not closed, perform steps contained in action 3.

4.   (C) M-1, REPORT POSITION OF N-5 .... DIRECTED

ACTION 2

If valve N-5 is closed the MCCC shall direct A-1 to position LO$_2$ TANKING (PANEL 1) SYSTEM POWER switch to OFF. This is necessary to reset

the line chilldown timer so chilldown will be accomplished after malfunction is corrected.

1.   (C) A-1, LO$_2$ TANKING POWER OFF .... DIRECTED

Table 4-13. LO$_2$ Tanking Panel Malfunction During LO$_2$ Chilldown (CONT)

ITEM 1. STORAGE TANK VENT VALVE N-5 INDICATOR NOT AMBER (CONT)

ACTION 2 (CONT)

The MCCC shall direct M-1 to bypass valve N-5 closed microswitch by connecting a jumper between pins 11 and 12 of TB42 on pressurization prefab.

This will provide the required valve N-5 closed signal to the LO$_2$ tanking subsystem.

2. (C) M-1, JUMPER 11 AND 12 OF TB42 .... DIRECTED

The MCCC shall direct A-1 to position LO TANKING (PANEL 1) SYSTEM POWER switch to ON. This will restart the LO$_2$ chilldown and loading sequence. A-1 shall report the status of STORAGE

TANK VALVE N-5 indicator. If indicator is NOT AMBER, an abort is required. If indicator illuminates AMBER, proceed to step 5.

3. (C) A-1, LO$_2$ TANKING POWER ON .... DIRECTED

4. (C) A-1, REPORT N-5 INDICATOR .... DIRECTED

The MCCC shall direct A-1 to position LO$_2$ TANKING (PANEL 1) SYSTEM POWER switch to OFF and M-1 to return to the LCC. When M-1 returns to the LCC the MCCC shall direct A-1 to position LO$_2$

TANKING (PANEL 1) SYSTEM POWER switch to ON. Normal LO$_2$ loading should continue at this time.

5. (C) A-1, LO$_2$ TANKING POWER OFF .... DIRECTED

6. (C) M-1, RETURN TO LCC .... DIRECTED

7. (C) A-1, LO$_2$ TANKING POWER ON .... DIRECTED

ACTION 3

If valve N-5 is not closed, the MCCC shall direct M-1 to check and report the presence or absence of voltage between pins 3 and 4 of TB41 on pres-

surization prefab. If 20 to 28 VDC is measured, an abort is required since the closing solenoid has malfunctioned.

1. (C) M-1, REPORT VOLTAGE BETWEEN 3 AND 4 OF TB41 .... DIRECTED

If no voltage is present between pins 3 and 4 of TB41, the MCCC shall direct A-1 to position LO$_2$ TANKING (PANEL 1) SYSTEM POWER switch

to OFF. This is necessary to reset the line chilldown timer to allow chilldown after malfunction has been corrected.

2. (C) A-1, LO$_2$ TANKING POWER OFF .... DIRECTED

The MCCC shall direct M-1 to energize valve N-5 solenoid by connecting a jumper between pin 4

of TB41 and pin 8 of TB42 on pressurization prefab.

3. (C) M-1, JUMPER 4 OF TB41 AND 8 OF TB42 .... DIRECTED

Table 4-13. LO₂ Tanking Panel Malfunction During LO₂ Chilldown (CONT)

ITEM 1. STORAGE TANK VENT VALVE N-5 INDICATOR NOT AMBER (CONT)

ACTION 3 (CONT)

The MCCC shall direct A-1 to position LO₂ TANKING (PANEL 1) SYSTEM POWER switch to ON. This will restart the line chilldown and LO₂ loading sequence. A-1 shall report the status of STORAGE TANK VENT VALVE N-5 indicator. If indicator is NOT AMBER, abort is required. If indicator illuminates AMBER, proceed to step 6.

4. (C) A-1, LO₂ TANKING POWER ON .... DIRECTED

5. (C) A-1, REPORT N-5 INDICATOR .... DIRECTED

The MCCC shall direct A-1 to position LO₂ TANKING (PANEL 1) SYSTEM POWER switch to OFF and M-1 to return to the LCC. When M-1 returns to the LCC the MCCC shall direct A-1 to position LO₂ TANKING (PANEL 1) SYSTEM POWER switch to ON. Normal LO₂ loading should continue at this time.

6. (C) A-1, LO₂ TANKING POWER OFF .... DIRECTED

7. (C) M-1, RETURN TO LCC .... DIRECTED

8. (C) A-1, LO₂ TANKING POWER ON .... DIRECTED

ITEM 2. TOPPING TANK VENT VALVE N-4 INDICATOR NOT AMBER

ACTION 1

During countdown, after P N E U M A T I C S IN PHASE II indicator illuminates GREEN, if TOPPING TANK VENT VALVE N-4 indicator fails to illuminate AMBER, A-1 shall immediately position LO₂ TANKING (PANEL 1) S Y S T E M POWER switch to OFF and announce the malfunction and action taken to the MCCC. Switching the power of the LO₂ tanking panel stops any LO₂ loading that may be in progress and sets the LO₂ loading system in proper configuration for troubleshooting and malfunction correction.

1. (A-1) LO₂ TANKING POWER .... **OFF**

The MCC shall direct M-1 to preceed to silo level 7 and stand by at pressurization prefab to report the position of valve N-4 when LO₂ system power is turned on.

2. (C) M-1, PROCEED TO PRESSURIZATION PREFAB .... DIRECTED

When M-1 is in position at pressurization prefab, the M C C C shall direct A-1 to position LO₂ TANKING (PANEL 1) SYSTEM POWER switch to ON.

3. (C) A-1, LO₂ TANKING POWER ON .... DIRECTED

The MCCC shall direct M-1 to report position of valve N-4. If valve N-4 is reported closed, perform steps contained in action 2. If valve N-4 is reported open, perform steps contained in action 3.

4. (C) M-1, REPORT POSITION OF N-4 .... DIRECTED

Table 4-13. LO$_2$ Tanking Panel Malfunction During LO$_2$ Chilldown (CONT)

ITEM 2.   TOPPING TANK VENT VALVE N-4 INDICATOR NOT AMBER (CONT)

ACTION 2

If valve N-4 is closed, the MCCC shall direct A-1 to position LO$_2$ TANKING (PANEL 1) SYSTEM POWER switch to OFF. This is necessary to reset the line chilldown timer so chilldown will be accomplished after malfunction is corrected.

1.   (C) A-1, LO$_2$ TANKING POWER OFF .... DIRECTED

The MCCC shall direct M-1 to bypass valve N-4 closed microswitch by connecting a jumper between pins 5 and 6 of TB42 on pressurization prefab. This will provide the required valve N-4 closed signal to the LO$_2$ tanking subsystem.

2.   (C) M-1, JUMPER 5 AND 6 OF TB42 .... DIRECTED

The MCCC shall direct A-1 to position LO$_2$ TANKING (PANEL 1) SYSTEM POWER switch to ON. This will restart the LO$_2$ chilldown and loading sequence. A-1 shall report the status of TOPPING TANK VENT VALVE N-4 indicator. If indicator is NOT AMBER an abort is required. If indicator illuminates AMBER, proceed to step 5.

3.   (C) A-1, LO$_2$ TANKING POWER ON .... DIRECTED

4.   (C) A-1, REPORT N-4 INDICATOR .... DIRECTED

The MCCC shall direct A-1 to position LO$_2$ TANKING (PANEL 1) SYSTEM POWER switch to OFF and M-1 to return to the LCC. When M-1 returns to the LCC, the MCCC shall direct A-1 to position LO$_2$ TANKING (PANEL 1) SYSTEM POWER switch to ON. Normal LO$_2$ loading should continue at this time.

5.   (C) A-1, LO$_2$ TANKING POWER OFF .... DIRECTED

6.   (C) M-1, RETURN TO LCC .... DIRECTED

7.   (C) A-1, LO$_2$ TANKING POWER ON .... DIRECTED

ACTION 3

If valve N-4 is open, the MCCC shall direct M-1 to check and report the presence or absence of voltage between pins 1 and 2 of TB41 on pressurization prefab. If 20 to 28 VDC is measured, an abort is required since the closing solenoid has malfunctioned.

1.   (C) M-1, REPORT VOLTAGE BETWEEN 1 AND 2 OF TB41 .... DIRECTED

If no voltage is present between pins 1 and 2 of TB41, the MCCC shall direct A-1 to position LO$_2$ TANKING (PANEL 1) SYSTEM POWER switch to OFF. This is necessary to reset the line chilldown timer to allow chilldown after malfunction has been corrected.

2.   (C) A-1, LO$_2$ TANKING POWER OFF .... DIRECTED

Table 4-13.   LO₂ Tanking Panel Malfunction During LO₂ Chilldown (CONT)

ACTION 3 (CONT)

The MCCC shall direct M-1 to energize valve N-4 solenoid by connecting a jumper between pin 2 of TB41 and pin 8 of TB42 on pressurization prefab.

3.   (C) M-1, JUMPER 2 OF TB41 AND 8 OF TB42 .... DIRECTED

The MCCC shall direct A-1 to position LO₂ TANK-ING (PANEL 1) SYSTEM POWER switch to ON. This will restart the line chilldown and LO₂ loading sequence. A-1 shall report the status of TOPPING TANK VENT VALVE N-4 indicator. If indicator is NOT AMBER, abort is required. If indicator illuminates AMBER, proceed to step 6.

4.   (C) A-1, LO₂ TANKING POWER ON .... DIRECTED

5.   (C) A-1, REPORT N-4 INDICATOR .... DIRECTED

The MCCC shall direct A-1 to position LO₂ TANK-ING (PANEL 1) SYSTEM POWER switch to OFF and M-1 to return to LCC. When M-1 returns to the LCC, the MCCC shall direct A-1 to position LO₂ TANKING (PANEL 1) SYSTEM POWER switch to ON. Normal LO₂ loading should continue at this time.

6.   (C) A-1, LO₂ TANKING POWER OFF .... DIRECTED

7.   (C) M-1, RETURN TO LCC .... DIRECTED

8.   (C) A-1, LO₂ TANKING POWER ON .... DIRECTED

ITEM 3.   TOPPING TANK PRES VALVE N-50 INDICATOR NOT GREEN

ACTION 1

During countdown, after PNEUMATICS IN PHASE II indicator illuminates G R E E N, if TOPPING TANK PRES VALVE N-50 indicator fails to illuminate GREEN, A-1 shall immediately position LO₂ TANKING (PANEL 1) SYSTEM POWER switch to OFF and announce the malfunction and action taken to the MCCC. Switching the power off the LO₂ tanking panel stops any LO₂ loading that may be in progress and sets the LO₂ loading that may be in progress and sets the LO₂ loading system in proper configuration for troubleshooting and malfunction correction.

1.   (A-1) LO₂ TANKING POWER .... OFF

The MCCC shall direct M-1 to proceed to silo level 7 and standby at pressurization prefab to report the position of valve N-50 when LO₂ system power is turned on.

2.   (C) M-1, PROCEED TO PRESSURIZATION PREFAB .... DIRECTED

When the M-1 is in position at the pressurization prefab the MCCC shall direct A-1 to position LO₂ TANKING (PANEL 1) SYSTEM POWER switch to ON.

3.   (C) A-1, LO₂ TANKING POWER ON .... DIRECTED

Table 4-13. LO₂ Tanking Panel Malfunction During LO₂ Chilldown (CONT)

ITEM 3. TOPPING TANK PRES VALVE N-50 INDICATOR NOT GREEN (CONT)

ACTION 1 (CONT)

The MCCC shall direct M-1 to report position of valve N-50. If valve N-50 is reported closed, perform steps contained in action 2. If valve N-50 is reported open, proceed to step 5.

4. (C) M-1, REPORT POSITION OF N-50 .... DIRECTED

The MCCC shall direct A-1 to position LO₂ TANKING (PANEL 1) SYSTEM POWER switch to OFF and M-1 to return to the LCC. When M-1 returns to the LCC the MCCC shall direct A-1 to position LO₂ TANKING (PANEL 1) SYSTEM POWER switch to ON. Normal LO₂ loading should continue at this time.

5. (C) A-1, LO₂ TANKING POWER OFF .... DIRECTED

6. (C) M-1, RETURN TO LCC .... DIRECTED

7. (C) A-1, LO₂ TANKING POWER ON .... DIRECTED

ACTION 2

If valve N-50 is closed, the MCCC shall direct M-1 to check and report the presence or absence of voltage beween pins 11 and 12 of TB41 on pressurization prefab. If 20 to 28 VDC is measured, an abort is required since the opening solenoid has malfunctioned.

1. (C) M-1, REPORT VOLTAGE BETWEEN 11 AN 12 OF TB41 .... DIRECTED

If no voltage is present between pins 11 and 12 of TB41, the MCCC shall direct A-1 to position LO₂ TANKING (PANEL 1) SYSTEM POWER switch to OFF. This is necessary to reset the line chilldown timer to allow chilldown after the malfunction has been corrected.

2. (C) A-1, LO₂ TANKING POWER OFF .... DIRECTED

The MCCC shall direct M-1 to energize valve N-50 solenoid by connecting a jumper between pin 11 of TB41 and pin 8 OF TB42 on pressurization prefab.

3. (C) M-1, JUMPER 11 OF TB41 AND 8 OF TB42 .... DIRECTED

The MCC shall direct A-1 to position LO₂ TANKING (PANEL 1) SYSTEM POWER switch to ON. This will restart the line chilldown and LO₂ loading sequence. M-1 shall report the position of TOPPING TANK PRES VALVE N-50. If valve N-50 is not open, abort is required. If valve N-50 is open, proceed to step 6.

4. (C) A-1, LO₂ TANKING POWER ON .... DIRECTED

5. (C) M-1, REPORT POSITION OF N-50 .... DIRECTED

Table 4-13. LO₂ Tanking Panel Malfunction During LO₂ Chilldown (CONT)

ITEM 3. TOPPING TANK PRES VALVE N-50 INDICATOR NOT GREEN (CONT)

ACTION 2 (CONT)

The MCCC shall direct A-1 to position LO₂ TANK-ING (PANEL 1) SYSTEM POWER switch to OFF and M-1 to return to the LCC. When M-1 returns to the LCC the MCCC shall direct A-1 to position LO₂ TANKING (PANEL 1) SYSTEM POWER switch to ON. Normal LO₂ loading should continue at this time.

6. (C) A-1, LO₂ TANKING POWER OFF .... DIRECTED

7. (C) M-1, RETURN TO LCC .... DIRECTED

8. (C) A-1, LO₂ TANKING POWER ON .... DIRECTED

ITEM 4. TOPPING CHILL VALVE L-60 INDICATOR NOT GREEN

ACTION 1

During countdown, after P N E U M A T I C S IN PHASE II indicator illuminates GREEN, if TOP-PING CHILL VALVE L-60 indicator fails to illu-minate GREEN A-1 shall immediately position LO₂ TANKING (PANEL 1) SYSTEM POWER switch to OFF and announce the malfunction and action taken to the MCCC. Switching the power off the LO₂ tanking panel stops any LO₂ loading that may be in progress and sets the LO₂ loading system in pro-per configuration for troubleshooting and malfunc-tion correction.

1. (A-1) LO₂ TANKING POWER .... OFF

The MCCC shall direct M-1 to proceed to silo level 7 and standby at topping prefab to report the posi-tion of valve L-60 when LO₂ system power is turned on.

2. (C) M-1, PROCEED TO TOPPING PREFAB .... DIRECTED

When the M-1 is in position at topping prefab the MCCC shall direct A-1 to position LO₂ TANK-ING (PANEL 1) SYSTEM POWER switch to ON.

3. (C) A-1, LO₂ TANKING POWER ON .... DIRECTED

The MCCC shall direct M-1 to report position of valve L-60. If valve L-60 is reported closed, abort is required. If valve L-60 is reported open, proceed to step 5.

4. (C) M-1, REPORT POSITION OF L-60 .... DIRECTED

The MCCC shall direct A-1 to position LO₂ TANK-ING (PANEL 1) SYSTEM POWER switch to OFF and M-1 to return to the LCC. When M-1 returns to the LCC the MCCC shall direct A-1 to position LO₂ TANKING (PANEL 1) SYSTEM POWER switch to ON. Normal LO₂ loading should continue at this time.

5. (C) A-1, LO₂ TANKING POWER OFF .... DIRECTED

6. (C) M-1, RETURN TO LCC .... DIRECTED

7. (C) A-1, LO₂ TANKING POWER ON .... DIRECTED

Table 4-14. HELIUM LOAD Indicator Not Amber or Not
Green After Being Amber

The following procedures provide the means of circumventing a helium load malfunction to continue a tactical countdown.

Gauge 301 on the helium control charging unit (HCU) shall be observed to determine whether pressure switch 321 or the automatic loading sequence has malfunctioned. If pressure switch 321 has malfunctioned, it will be jumpered at the rear of PNEUMATICS (PANEL 1) to complete the helium load logic summary. If automatic loading sequence has malfunctioned, a manual helium load will be performed at LN$_2$-HELIUM (PANEL 1), level 3.

ACTION 1

Observe P.S. 321 SPHERES FULL indicator on PLCP. If indicator is illuminated AMBER, abort is required. Pressure switch 321 has sensed greater than 2950 PSI and a logic malfunction has occurred which cannot be corrected and will prevent the countdown from continuing. If P.S. 321 SPHERES FULL indicator is extinguished, MCCC shall direct M-1 to proceed to HCU on level 3 of launcher platform to observe and report pressure indication on gauge 301.

1. (C) M-1, REPORT GAUGE 301 INDICATION .... DIRECTED

ACTION 2

If M-1 reports gauge 301 indicates greater than 2950 PSI, pressure switch 321 has malfunctioned and must be jumpered to complete the helium load summary. The MCCC shall direct A-1 to proceed to the rear of PNEUMATICS (PANEL 1), level 3, to connect a jumper wire between pin 48 of A8J1 and pin 1 of A8J3. HELIUM LOAD indicator should illuminate GREEN after jumper wire is installed. If HELIUM LOAD indicator fails to illuminate GREEN, abort is required. Prior to initiating abort, A-1 shall remove installed jumper wire and return to the LCC with M-1.

1. (C) A-1, JUMPER PIN 48 OF A8J1 AND PIN 1 OF A8J3 .... DIRECTED

ACTION 3

If gauge 301 on HCU indicates 0 PSI, automatic helium loading has failed and will have to be manually loaded. The LN$_2$ LOAD indicator on launch control console should be illuminated GREEN prior to positioning helium and LN$_2$ system valve switches to ensure that liquid nitrogen is available in the system for chilling helium during loading. The MCCC shall direct A-1 to proceed to LN$_2$-HELIUM (PANEL 1), level 3, to manually load helium as follows: (a) Position valve switches 7, 13, 14, 26, 37, 213, 215, 52, and 54 to Open. (b) Position all other valve switches to CLOSED. (c) Position LOCAL REMOTE switch to LOCAL.

1. (C) A-1, MANUALLY LOAD HELIUM .... DIRECTED

M-1 shall monitor gauge 301 on HCU to verify helium load has started, as indicated by a increase in pressure. HELIUM LOAD indicator on launch control console should illuminate GREEN in approximately 5 minutes. If HELIUM LOAD indicator fails to illuminate GREEN in approximately 5 minutes and pressure gauge 301 indicates pressure stabilized at less than 2950 PSI, abort is required. Prior to initiating abort, A-1 shall position LN$_2$-HELIUM (PANEL 1) LOCAL REMOTE switch to REMOTE and return with M-1 to the LCC.

2. (C) HELIUM LOAD INDICATOR .... GREEN

If manual loading is successful, it shall not be stopped until READY FOR COMMIT indicator illuminated GREEN.

3. (C) READY FOR COMMIT INDICATOR .... GREEN

Table 4-14. HELIUM LOAD Iindicator Not Amber or Not
Green After Being Amber (CONT)

ACTION 3 (CONT)

A-1 shall position LOCAL REMOTE switch on LN₂-HELIUM (PANEL 1) to REMOTE when READY FOR COMMIT indicator is illuminated GREEN and 6 minutes have leapsed since LOCAL REMOTE switch was placed in LOCAL.

4. (C) A-1, PLACE LN₂-HELIUM SWITCH TO REMOTE .... DIRECTED

Return personnel to LCC and continue countdown as required.

5. (C) COUNTDOWN .... CONTINUED

Table 4-15. HYD-PNEU & LN₂-HE READY Indicator Not Green

The following procedures provide the means of circumventing a pressure switch malfunction or low pressure condition in the helium control charging unit (HCU) to continue a tactical countdown.

If HYD-PNEU & LN₂-HE READY indicator fails to illuminate GREEN and all other indicators necessary for this summary condition are illuminated GREEN, the probable cause is a failure of pressure switch 328 in the HCU or a low pressure in inflight helium bottle NO. 2. Pressure switch 328 is jumpered at the PNEUMATICS (PANEL 3). This should complete the H Y D-PNEU & L N₂-H E READY indicator logic summary.

ACTION 1

Observe P.S. 328 HCU SOURCE OK indicator on PLCP. If indicator is illuminated AMBER, abort is required. Pressure switch 328 has sensed greater than 4500 PSI and a logic malfunction has occurred which cannot be corrected and will prevent the countdown from continuing. If P.S. 328 HCU SOURCE OK indicator is extinguished, MCCC shall direct A-1 to proceed to rear of PNEUMATICS (PANEL 3), level 3, to connect a jumper wire between pin 34 and pin 50 of A10J2. HYD-PNEU & LN₂-HE READY indicator should illuminate GREEN after the jumper wire is installed. If the indicator fails to illuminate GREEN, abort is required. Prior to initiating abort, A-1 shall remove installed jumper wire and return to the LCC.

1. (C) A-1, JUMPER PINS 34 AND 50 OF A10J2 .... DIRECTED

Table 4-16. Launcher Platform Fails to Lower

The following procedures provide the means of returning the launcher platform and missile to a safe configuration when the launcher platform fails to lower automatically after abort start.

A missile lifting sequence II failure can be identified by A B O R T COMPLETE indicator illuminated AMBER (abort sequence has started) and SITE HARD indicator extinguished (no launcher platform down motion) or ABORT EXTERNAL indicator extinguished (no launcher platform not down and locked). If manual lowering of the launcher platform is unsuccessful, throublshooting shall be accomplished. When lowering the launcher platform, opening and closing the boiloff valve shall be accomplished as in a normal abort sequence.

WARNING

Regardless of LP position, if boiloff valve is closed, immediately depress EMERGENCY pushbutton. After LO₂ tank pressure decreases to phase II, depress AUTOMATIC pushbutton. Enable boiloff valve periodically to relieve pressure when LO₂ tank pressure increases to 12 PSI.

Table 4-16. Launcher Platform Fails to Lower (CONT)

If the launcher platform is within 33 inches of the up and locked position, the pressurization system must be in automatic and LO₂ tank pressure greater than 8 PSI before the launcher platform can be lowered. P.S. 326 LO₂ OVER 8 PSI indicator on PLCP must be illuminated AMBER.

ACTION 1

The MCCC shall direct M-1 to position RESET PROGRAMMER key switch on CSMOL to ON. This prevents the automatic sequence from being activated during a manual operation, malfunction isolation operation, or when pneumatic test set (PTS) is connected.

1. (C) M-1, RESET PROGRAMMER KEY ON .... DIRECTED

> **CAUTION**
>
> If HYDRAULIC 40 HP PUMP PRESSURE indicator on CSMOL fails to illuminate GREEN, depress HYDRAULIC 40 HP PUMP OFF pushbutton. The 40 HP pump or a pressure switch has malfunctioned or a line has ruptured or is leaking. If boiloff will impinge on tension equalizer it must be sprayed with water or the missile enclosure purge unit (MEPU) must be extended. The hydraulic line or valve malfunction must be corrected prior to activating the system. Proceed to action 4 if the above condition has occurred and boiloff will impinge on the tension equalizer. Proceed to action 5 if the above condition has occurred and boil-off will not impinge on the tension equalizer.

The MCCC shall direct M-1 to manually lower launcher platform at CSMOL: (a) Depress HYDRAULIC 40 HP PUMP ON pushbutton (if pump is off). (b) Observe HYDRAULIC 40 HP PUMP PRESSURE indicator is illuminated GREEN. (c) Depress stop pushbutton. (d) Depress DOWN RUN pushbutton.

If launcher platform down motion is observed or launcher platform is verified as down and locked, perform action 2. Normally the launcher platform is verified as down and locked by observing DOWN COMPLETED RUN AND LOCKED indicator on CSMOL is illuminated GREEN, or by observing either LO₂ DISCONNECT MATED SWITCH "A" or LO₂ DISCONNECT MATED SWITCH "B" indicator on LO₂ TANKING (PANEL 1) is illuminated GREEN. If after observing CSMOL and LO₂ TANKING (PANEL 1) the launcher platform position cannot be positively determined, visually observe launcher platform to determine the down-and-locked status. To visually determine that the launcher platform is down and locked, MCCC shall direct A-1 and M-1 to proceed to launcher platform level 2 and observe that the main locks are extended and fully engaged to the downlock strikers. If no launcher platform motion is observed and launcher platform is not up and locked, proceed to action 3. If no launcher platform motion is observed and launcher platform is up and locked, proceed to action 5.

2. (C) M-1, MANUALLY LOWER LAUNCHER PLATFORM .... DIRECTED

ACTION 2

When launcher platform is verified as down and locked, MCCC shall direct M-1 to position RESET PROGRAMMER key switch on CSMOL to OFF. This will allow the automatic abort sequence to continue.

1. (C) M-1, RESET PROGRAMMER KEY OFF .... DIRECTED

All crew members will use T.O. 2-M-HGM16F-1CL-1 through T.O. 21M-HGM16F-1CL-5 to monitor the abort sequence. If the automatic abort sequence does not start or fails to continue, as indicated by LO₂ DRAIN COMPLETE indicator remaining extinguished, perform manual abort (table 4-10, action 2).

2. (C) CREW, MONITOR ABORT .... DIRECTED

Changed 15 April 1964

Table 4-16.   Launcher Platform Fails to Lower (CONT)

ACTION 3

If launcher platform cannot be lowered, MCCC shall direct M-1 to manually raise launcher platform at CSMOL: (a) Depress LAUNCHER PLATFORM STOP pushbutton to erase down run command. (b) Depress UP RUN pushbutton. If launcher platform cannot be raised and boiloff will impinge on tension equalizer, spray tension equalizer or extend MEPU (action 4). If launcher platform can be raised, or if launcher platform cannot be raised and boiloff will not impinge on tension equalizer, depress EMERGENCY pushbutton (action 5).

1.   (C) M-1, MANUALLY RAISE LAUNCHER PLATFORM .... DIRECTED

ACTION 4

> **WARNING**
>
> LO$_2$ tank pressure must be relived each time tank pressure increases to 12 PSI by depressing EMERGENCY pushbutton even if tension equalizer is impinged. Depress AUTOMATIC pushbutton immediately after LO$_2$ tank pressure decreases to phase II.

> **WARNING**
>
> If water is not available to spray tension equalizer, or MEPU is not available, do not proceed with step 1.

The MCCC shall direct A-1 to the silo cap to spray the impinged area of tension equalizer with water or to extend MEPU prior to opening boiloff valve. This is necessary to prevent the tension equalizer from being crystalized by gaseous oxygen.

1.   (C) A-1, SPRAY TENSION EQUALIZER OR EXTEND MEPU .... DIRECTED

A-1 shall notify MCCC when tension equalizer is being sprayed or the MEPU is extended. MCCC shall proceed to action 5.

2.   (C) REPORT THAT TENSION EQUALIZER
       IS BEING SPRAYED OR MEPU IS EXTENDED .... RECEIVED

ACTION 5

The MCCC shall direct deputy to depress EMERGENCY pushbutton on launch control console to open the boiloff valve and relieve LO$_2$ tank pressure.

Deputy shall observe PRESSURE MODE indicator illuminates RED.

1.   (C) DEPUTY, DEPRESS EMERGENCY PUSHBUTTON .... DIRECTED

Although the helium control charging unit (HCU) should pressurize the missile fuel tank for approximately 72 hours, MCCC should request PTS to stand by. Normally the PTS would not be connected at this time because the boiloff valve is open and missile fuel tank pressures are being maintained by the HCU. If troubleshooting procedures in step 4 are successful, the launcher platform can be lowered into the down and locked position and an automatic abort continued without connecting PTS.

2.   (Deleted)

Table 4-16.   Launcher Platform Fails to Lower (CONT)

ACTION 5 (CONT)

> **CAUTION**
>
> The LAUNCHER PLATFORM S T O P
> pushbutton CSMOL must be depressed prior
> to troubleshooting the missile lifting system
> (MLS) if a manual drive has been attemped.

3.   (C) M-1, DEPRESS LAUNCHER PLATFORM STOP PUSHBUTTON .... DIRECTED

The MCCC shall direct M-1, and A-1 if available, formed of the results of troubleshooting.
to troubleshoot MLS. M-1 shall kept MCCC in-

4.   (C) M-1, A-1, TROUBLESHOOT MLS .... DIRECTED

ACTION 6

M-1 shall perform only those steps necessary to correct the malfunction. An attempt to drive the launcher platform may be made, using procedures contained in action 7, whenever the malfunction has been, or is believed to be, corrected.

If the launcher platform is within 33 inches of the up and locked positioned, M/L STOP indicator on CSMOL must extinguish after AUTOMATIC pushbutton is depressed and LO$_2$ tank pressure is greater

than 8 PSI before the lowering sequence will begin. If M/L STOP indicator fails to extinguish after LO$_2$ tank pressure is greater than 8 PSI, pressure switch 326, LO$_2$ tank pressure greater than 8 PSI, has failed. This prevents a missile lift go signal. The missile lift go relay is energized by connecting a jumper wire between pins 17 and 30 of terminal board 8 in electrical missile lifting control system chassis A-2 on level 1. Omit step 1 if launcher platform is below 33 inches of the up and locked position.

1.   (C) M-1, A-1, VERIFY M/L STOP INDICATOR
     EXTINGUISHED OR RED .... DIRECTED

M-1 shall: (a) Observe 40 HP pump and determine that it is running. The ACCUM. PRESSURE SYSTEM gauge on local control hydraulic panel shall be observed for a minimum indication of 2700 PSI. (b) If launcher platform failed to lower due to loss of 40 HP pump pressure, ensure that no hydraulic leaks or ruptures exist prior to starting the 40 HP pump. If HYDRAULIC 40 HP PUMP PRESSURE indicator is extinguished and ACCUM. PRESSURE SYSTEM gauge indicates greater than 2700 PSI, a pressure switch failure should be suspected. M-1

shall proceed to electrical missile lifting control system cabinet A1A2 and install a jumper wire from terminal board 4, pin 24 to terminal board 8, pin 18. M-1 shall then verify that HYDRAULIC 40 HP PUMP PRESSURE indicator on local control hydraulic panel is illuminated. (c) If 40 HP pump will not start, reset 40 HP pump circuit breakers on MLS motor control center and attempt to restart the pump at local control hydraulic panel. If hydraulic pressure is below 2700 PSI, continue troubleshooting.

2.   (C) M-1, VERIFY HYDRAULIC 40 HP PUMP IS ON AND ACCUM.
     PRESSURE SYSTEM GAUGE INDICATES 2700 PSI .... DIRECTED

Table 4-16, Launcher Platform Fails to Lower (CONT)

ACTION 6 (CONT)

M-1 shall proceed to fault tape register on electrical missile lifting control system, level 1, and remove tape, then continue troubleshooting. If troubleshooting fails to correct the malfunction, examine tape printout and troubleshoot in accordance with T.O. 21M-HGM16F-2-23.

3. (C) M-1, REMOVE FAULT TAPE .... DIRECTED

M-1 shall: (a) Observe 28 VOLT DC SUPPLY IN USE indicators on MLS motor control center, level 1, and determine that one 28 VOLT DC SUPPLY IN USE indicator is illuminated and that 28 volts is being supplied to the MLS. (b) If both 28 VOLT DC SUPPLY IN USE indicators are extinguished, troubleshooting is required.

4. (C) M-1, VERIFY ONE 28 VDC SUPPLY IN USE INDICATOR ILLUMINATED .... DIRECTED

M-1 shall reset LAUNCHER PLATFORM DRIVE MOTOR circuit breaker on MLS motor control center, level 1. If circuit breaker will reset, proceed to step 6. If circuit breaker will not reset, troubleshooting is required.

5. (C) M-1, RESET LAUNCHER PLATFORM
    DRIVE MOTOR CIRCUIT BREAKER .... DIRECTED

M-1 shall reset all tripped circuit breakers on electrical missile lifting control system circuit breaker chassis A3A1.

6. (C), RESET TRIPPED CIRCUIT BREAKERS .... DIRECTED

M-1 shall observe REGULATED POWER SUPPLY POWER switch in MLS drive assembly cabinet NO. 1, level 1, and ensure that switch is ON.

7. (C) M-1, REGULATED POWER SUPPLY POWER SWITCH ON .... DIRECTED

M-1 shall observe POWER ON indicator in MLS drive assembly cabinet NO. 1, level 1, and ensure that indicator is illuminated. If POWER ON indicator is extinguished, replace the 2-ampere fuse.

8. (C) M-1, VERIFY POWER ON INDICATOR ILLUMINATED .... DIRECTED

M-1 shall proceed to launcher platform drive assembly cabinet NO. 2, level 1, and depress thermal overload relay OL-1, OL-2, and BMOL reset pushbuttons to enable the drive motor contactors.

8A. (C) M-1, RESET THERMAL OVERLOAD RELAYS OL-1 AND OL-2, AND BMOL .... DIRECTED

M-1 shall proceed to overspeed control box; level 1, and depress RESET pushbutton. This will reset the indicator.

9. (C) M-1, DEPRESS RESET PUSHBUTTON (OVERSPEED CONTROL BOX) .... DIRECTED

If OVERSPEED CONTROL OPERATE indicator on overspeed control box is extinguished, M-1 shall proceed to step 11. If OVERSPEED CONTROL OPERATE indicator remains illuminated, M-1 shall disconnect cables: (a) 907U16A27P02 (b) 907U16A27P03. (c) 907U16A27P04. (d) 907U16A27P05.

10. (C) M-1, DISCONNECT CABLES FROM OVERSPEED CONTROL BOX .... DIRECTED

T.O. 21M-HGM16F-1

Table 4-16. Launcher Platform Fails to Lower (CONT)

ACTION 6 (CONT)

M-1 shall proceed to MLS drive assembly, level 1, and determine if drive coupling is engaged and locked. The drive coupling is engaged and locked if the coupling cannot be moved in the opposite direction and the hydraulic cylinder is extended with approximately 1-11/16 inch of the drive hub next to the main speed decreaser exposed. If drive coupling is engaged and locked, M-1 shall proceed to step 13. If drive coupling is not engaged and locked, M-1 shall proceed to step 12.

11. (C) M-1, VERIFY DRIVE COUPLING POSITION .... DIRECTED

If drive coupling is not engaged and locked, M-1 shall (a) Position circuit breaker CB-16 on electrical missile lifting control system circuit breaker chassis A3A1 to off, de-energizing the coupling shift solenoid valve. (b) Slightly rotate the flexible coupling between the low speed motor and the auxiliary speed decrease to allow the coupling gear teeth to align with the teeth of the main speed decreaser shaft. Manual engagements may be attempted at this time, if the coupling is not locked in the disengaged position, by shifting the coupling toward the auxiliary speed decreaser until the coupling locks in the engaged position. (c) Position Circut breaker CB-16 to ON. (d) Restart 40 HP pump by depressing HYDRAULIC 40 HP PUMP ON pushbutton on local control hydraulic panel or CSMOL. If the drive coupling was not manually engaged during step (b), automatic engagement should take place when manual drive down of the launcher platform is attempted. (Action 7)

12. (C) M-1, ENGINE DRIVE COUPLING .... DIRECTED

M-1 shall observe drive coupling engaged limit switch and determine if limit switch is actuated. If limit switch is actuated, M-1 shall proceed to step 14. If limit switch is not actuated, M-1 shall insert spacer or tape the limit switch closed.

13. (C) M-1, VERIFY DRIVE COUPLING ENGAGE LIMIT SWITCH ACTUATED .. DIRECTED

M-1 shall proceed to MLS drive cabinet NO. 2 level 1, and disconnect wire 53-1 from 4TB. This will result in circumventing a failure of the slow motor blower fan limit switch, overload relay or circuit which would prevent the slow motor contactors from closing. After removing wire, attempt a manual drive of the launcher platform.

14. (C) M-1, DISCONNECT WIRE 53-1 FROM 4TB .... DIRECTED

ACTION 7

The MCCC shall direct M-1 to manually lower launcher platform at CSMOL: (a) Position RESET PROGRAMMER key switch to ON. (b) Depress HYDRAULIC 40 HP PUMP ON pushbutton if pump is off. (c) Verify HYDRAULIC 40 HP PUMP PRESSURE indicator is illuminated GREEN. (d) Depress DOWN RUN pushbutton. If launcher platform will not lower, M-1 shall depress LAUNCHER PLATFORM STOP pushbutton on CSMOL and MCCC shall direct M-1 and A-1 to troubleshoot MLS. If launcher platform lowers (DOWN COMPLETED RUN AND LOCKED indicator on CSMOL illuminates), MCCC shall proceed to action 8.

1. (C) M-1, MANUALLY LOWER LAUNCHER PLATFORM .... DIRECTED

Table 4-16.   Launcher Platform Fails to Lower (CONT)

ACTION 8

The MCCC shall direct M-1 to position **RESET PROGRAMMER** key switch on CSMOL to OFF to reinitiate the automatic abort sequence.

1.   (C) M-1, RESET PROGRAMMER KEY OFF.. DIRECTED

The MCCC shall direct crew members to monitor abort sequence.

2.   (C) CREW, MONITOR ABORT .... DIRECTED

Table 4-17.   Boiloff Valve Failure to Open During Abort

This table contains procedures to follow in the event that the boiloff valve fails to open when the EMERGENCY pushbutton is depressed. Failure of the boiloff valve to open when the EMERGENCY pushbutton is depressed is indicated by no decrease in LO_2 tank pressure or lack of boiloff as observed on TV monitor. If boiloff valve fails to open when EMERGENCY pushbutton is depressed, and abort has not been initiated, MCCC shall immediately depress ABORT pushbutton.

WARNING

If boiloff valve fails to open during abort, loss of missile may occur due to overflow of LO_2 into the pressurization duct and airborne relief valve.

Under certain conditions of temperature, humidity, and wind velocity with the boiloff valve failed closed, heat input to the missile LO_2 tank can cause LO_2 to overflow into the missile pressurization duct and cause the airborne LO_2 tank pressure relief valve to fail in the open position. If this occurs, GO_2 or LO_2 may be exhausted into the MEA through the failed airborne relief valve.

ACTION 1

1.   (C) ABORT .... *INITIATED*

The pressurization system must be in automatic mode during the first 33 inches of down travel of the launcher platform. If or when the launcher platform has reached 7½ feet of the uplocks, deputy shall cycle AUTOMATIC and EMERGENCY pushbuttons twice in an attempt to open the boiloff valve.

CAUTION

If the boiloff valve fails to open after performing step 2, immediately return the pressurization system to AUTOMATIC MODE. The pressurization system must be returned to AUTOMATIC MODE before fuel tank pressure decreases below 56 PSI.

2.   (D) CYCLE AUTOMATIC AND EMERGENCY PUSHBUTTONS .... *ACCOMPLISHED*

Table 4-17. Boiloff Valve Failure to Open During Abort (CONT)

ACTION 1 (CONT)

```
CAUTION
```

If the boiloff valve opens when the pressurization system is cycled while performing step 2, allow automatic abort to continue. Do not proceed to step 3.

If the attempt to open the boiloff valve fails, the automatic drive of the launcher platform shall be

stopped and manual lowering accomplished. This will stop the down-and-locked signal to the launch control logic, prevent the pressure system control from automatically controlling missile tank pressures, and prevent an automatic LO₂ drain.

At direction of MCCC, M-1 shall stop, then lower the launcher platform manually at CSMOL: (a) Position RESET PROGRAMMER key switch to ON. (b) Depress DOWN RUN pushbutton.

3.   (C) M-1 STOP LAUNCHER PLATFORM AND DRIVE DOWN MANUALLY .... DIRECTED

When the launcher platform is down and locked, as indicated by DOWN COMPLETED RUN AND LOCKED indicator on CSMOL illuminating GREEN or by visual observation, the pressurization system

shall be placed in emergency mode. At M C C C direction the Deputy shall depress the emergency pushbutton.

4.   (C) DEPUTY, DEPRESS EMERGENCY PUSHBUTTON .... DIRECTED

After the emergency pushbutton has been depressed immediately start a manual drain of LO₂ to enlarge the ullage space in the tank. Manual drain shall be continued for ten minutes. At MCCC direction to start manual LO₂ drain, A-1 shall: (a) Position L O C A L R E M O T E switch on LO₂ TANKING (PANEL 1) to LOCAL. (b) Position L-16 valve switch to OPEN. Observe DRAIN VALVE L-16

indicator illuminated GREEN. If indicator is not illuminated GREEN, position L-16 valve switch to CLOSE and position L-1 valve switch to OPEN and observe FINE LOAD VALVE L-1 indicator illuminated GREEN. (c) Position A/B F&D valve switch to OPEN. Observe AIRBORNE FILL & DRAIN VALVE indicator illuminated GREEN.

5.   (C) A-1, START LO₂ DRAIN .... DIRECTED

As soon as LO₂ drain has started the silo overhead doors shall be manually closed and the crib unlocked. This will allow stretch to be applied to the missile after ten minutes of LO₂ drain. At MCCC direction to close silo doors, M-1 shall: (a) Depress silo DOORS CLOSE pushbutton and hold for 30-seconds. The silo doors will not start closing until a 30-second warning delay period is completed. When SILO DOORS OPEN indicator extinguishes or observing on TV silo doors closing, SILO DOORS

CLOSE pushbutton will be released. The closing cycle will continue until doors are fully closed. (b) Observe SILO DOORS CLOSE indicator illuminates GREEN, indicating silo doors are closed. (c) Depress CRIB HORIZONTAL UNLOCK pushbutton. (d) Observe CRIB HORIZONTAL UNLOCK indicator illuminated GREEN. (e) Depress CRIB VERTICAL UNLOCK pushbutton. (f) Observe CRIB VERTICAL U N L O C K indicator is illuminated GREEN.

6.   (C) M-1 CLOSE SILO DOORS & UNLOCK CRIB .... DIRECTED

While the doors are closing personnel should prepare to put the missile in emergency stretch. When LO₂ has .drained for 10 minutes, drain valve L-16 (or L-1 if used) shall be closed to stop drain. At

MCCC direction, A-1 shall close valve L-16 (or L-1 if used by positioning L-16 (or L-1) valve switch on LO₂ TANKING (PANEL 2) to CLOSE.

7.   (C) A-1, CLOSE L-16 (OR L-1 IF USED) .... DIRECTED

Table 4-17. Boiloff Valve Failure to Open During Abort (CONT)

ACTION 1 (CONT)

WARNING

While placing missile in stretch open L-16 to lower LO₂ tank pressure as required to maintain LO₂ tank pressure at least 20 PSI less than fuel tank pressure. The fuel raise and lower pushbutton are not effective at this time.

At this time the crew shall immediately place the missile in emergency stretch to prevent collapse of missile in case the airborne LO₂ relief valve has ruptured. At the MCCC direction the crew shall place the missile in emergency stretch (see table 4-5).

8.   (C) CREW, PLACE MISSILE IN STRETCH .... DIRECTED

Immediately after missile is in stretch and the crew shall return to the LCC, manual LO₂ drain shall be initiated again by opening drain valve L-16 (or L-1 if used). Manual drain shall continue for 20

minutes. At MCCC direction, A-1 shall open valve L-16 (or L-1 if used) by positioning L-16 (or L-1) valve switch on LO₂ TANKING (PANEL 2) to OPEN.

9.   (C) A-1, OPEN L-16 (OR L-1 IF USED) .... DIRECTED

Start timing LO₂ drain sequence when L-16 is opened. Allow drain to continue for 20 minutes. At MCCC direction, A-1 shall position L-16 (or L-1

if used) valve switch on LO₂ TANKING (PANEL 2) to CLOSE.

10.   (C) A-1, CLOSE L-16 (OR L-1 IF USED) .... DIRECTED

When LO₂ drain has been stopped LO₂ tank pressure shall be lowered to approximately 4 PSI. At MCCC direction to lower LO₂ tank pressure to approximately 4 PSI, A-1 shall: (a) Position N-60

valve switch to OPEN. (b) When LO₂ tank pressure reaches approximately 4 PSI, position N-60 valve switch to CLOSE. (c) Position A/B F&D valve switch to CLOSE.

11.   (C) A-1, LOWER LO₂ TANK PRESSURE TO 4 PSI .... DIRECTED

After the LO₂ tank pressure has been lowered to approximately 4 PSI the MCCC shall direct M-1 to

position the reset programmer key switch to OFF. This action may allow the boiloff valve to open.

12.   (C) M-1, RESET PROGRAMMER KEY OFF .... DIRECTED

After the reset programmer key switch has been positioned to OFF the fuel raise and lower pushbuttons are enabled. At MCCC direction to raise fuel

tank pressure to 62 PSI the Deputy shall depress the fuel raise pushbutton until the fuel tank gauge indicates 62 PSI.

13.   (C) DEPUTY RAISE FUEL PRESSURE TO 62 PSI .... DIRECTED

After the fuel tank pressure has been adjusted to 62 PSI the pressurization system shall be returned to automatic mode. At MCCC direction the Deputy

shall depress the AUTOMATIC pushbutton. If pressurization system fails to return to automatic mode, helium must be vented manually.

14.   (C) DEPUTY, DEPRESS AUTOMATIC PUSHBUTTON .... DIRECTED

Table 4-17.   Boiloff Valve Failure to Open During Abort (CONT)

ACTION 1 (CONT)

At MCCC direction A-1 shall position the LO$_2$ TANKING (PANEL 1) REMOTE-LOCAL switch to RE-MOTE.

15.   (C) A-1, LO$_2$ REMOTE-LOCAL SWITCH REMOTE .... DIRECTED

The ABORT COMPLETE indicator should illumi-nate GREEN in approximately 45 minutes. When abort is complete, at MCCC direction the Deputy shall depress the EMERGENCY pushbutton.

16.   (C) DEPUTY, DEPRESS EMERGENCY PUSHBUTTON .... DIRECTED

To determine if the LO$_2$ relief valve has failed LO$_2$ tank pressure shall be closely monitored for several minutes. If a noticeable decrease is observed assume a ruptured relief valve If a ruptured relief valve is suspected do not attempt to maintain LO$_2$ tank pressure. Remain in emergency mode. If relief valve appears normal return pressurization system to automatic mode. In either instance maintenance as-sistance is required.

17.   (C) STATUS OF RELIEF VALVE .... DETERMINED

SECTION V

MALFUNCTION PROCEDURES

TABLE OF CONTENTS

LIST OF TABLES

5-1. SCOPE.

5-2. This section contains the analysis procedures to be used by the missile combat crew to isolate malfunctions which may be indicated on the launch control console or facilities remote control panel (FRCP) during standby and countdown. The procedures in Section III and emergency procedeures in Section IV will take precedence over this section for analysis or actions.

5-3. MALFUNCTION ANALYSIS PROCEDURES.

5-4. During a countdown, if an immediate abort is not required, the appropriate table in this section shall be referred to and analysis procedures in column 3 completed prior to initiating abort. The actions or troubleshooting reference in column 4 will not be used until after abort is complete.

5-5. During a PLX or training launch, analysis procedures shall only be completed to the point where silo entry is necessary prior to initiating abort. If abort is initiated prior to complete malfunction analysis as outlined in column 3, some or all of the indications listed may not be present after abort is complete.

5-6. MALFUNCTION PATCH INDICATIONS, LAUNCH CONTROL CONSOLE.

5-7. Table 5-1 presents the malfunction analysis procedures for abnormal indications which may appear in the malfunction patch of the launch control console. Item numbers are in column 1, the malfunction indication is in column 2, the malfunction analysis or probable cause for malfunction is in column 3, and the action or troubleshooting reference to correct malfunction is in column 4. No actual troubleshooting procedures are given as the references listed are for this purpose.

5-8. STATUS PATCH INDICATIONS, LAUNCH CONTROL CONSOLE.

5-9. Table 5-2 presents the malfunction analysis procedures for abnormal indications which may appear in the status patch with no other malfunction indications on the launch control console. The column designation and use are the same as outlined in paragraph 5-7.

5-10. COUNTDOWN PATCH INDICATIONS, LAUNCH CONTROL CONSOLE.

5-11. Table 5-3 presents the malfunction analysis procedures for abnormal indications which may appear during a countdown in the countdown patch on the launch control console. The column designation and use are the same as outlined in paragrph 5-7. Column 3 in many cases is left blank as internal sequencer failures could be the only cause, and because of the high degree of reliability of sealed relays, only a troubleshooting reference in column 4 is given.

5-12. FACILITIES REMOTE CONTROL PANEL INDICATIONS.

5-13. Table 5-4 presents the malfunction analysis procedures for abnormal indications which may appear on the FRCP during standby, countdown and abort.

5-14. Certain malfunction indications may appear that are considered normal. These indications and the time period when they are considered normal are listed in column 2 under the malfunction indication. The column designation and use are the same as outlined in paragraph 5-7.

5-15. Malfunctions which do not prohibit the initiation of a tactical countdown are indicated in column 2 by the statement: Tactical countdown may be initiated.

Table 5-1. Malfunction Analysis Procedures, Malfunction Patch Indicators

1	2	3	4
ITEM NO.	MALFUNCTION INDICATION	MALFUNCTION ANALYSIS OR PROBABLE CAUSE	ACTION OR TROUBLESHOOTING REFERENCE
1	400 CYCLE POWER indicator RED	400-CYCLE MOTOR GEN Frequency not 400($\pm$6) CPS	21M-HGM16F-2-6
		MISSILE GROUND POWER (PANEL 2)	
		Motor generator output not 116.7 ($\pm$1.4) VAC	21M-HGM16F-2-6
		MISSILE GROUND POWER (PANEL 1)	
		GROUND AC VOLTAGE indicator RED	21M-HGM16F-2-9
		GROUND FREQUENCY indicator RED	21M-HGM16F-2-9
2	28 VDC POWER indicator AMBER	POWER SUPPLY DISTR SET Cooling fan not operating	A tactical countdown may be initiated. T.O. 35C3-3-39-2
		AMPERE HOURS meter indicating greater than 40 ampere hours discharge	21M-HGM16F-2-6
		BATTERY DISCHARGED indicator FLASHING RED	21M-HGM16F-2-6
3	28 VDC POWER indicator RED	POWER SUPPLY DISTR SET	
		DC voltmeter not indicating 29.5 ($\pm$1.1) VDC	21M-HGM16F-2-6
		MISSILE GROUND POWER (PANEL 2)	
		Standby bus voltage not 29.5 ($\pm$1.5) VDC	21M-HGM16F-2-6
		MISSILE GROUND POWER (PANEL 1)	
		GROUND DC VOLTAGE indicator RED	21M-HGM16F-2-9

Table 5-1. Malfunction Analysis Procedures, Malfunction Patch Indicators (CONT)

1	2	3	4
ITEM NO.	MALFUNCTION INDICATION	MALFUNCTION ANALYSIS OR PROBABLE CAUSE	ACTION OR TROUBLESHOOTING REFERENCE
4	MISSILE INVERTER Indicator RED **NOTE** If indication appears prior to commit start, delay commit start until the malfunction indication extinguishes, then initiate commit.	**NOTE** The malfunction indication will appear during a countdown only. Malfunction analysis must be accomplished within 4 minutes after the malfunction indication appears, as analysis indications will extinguish when the inverter shuts down after the 4-minute period. MISSILE GROUND POWER (PANEL 2) Inverter output not 115.3 $(+1.4)$ VAC MISSILE GROUND POWER (PANEL 1) INTERNAL FREQUENCY indicator RED	 21M-HGM16F-2-6 21M-HGM16F-2-9
5	GUIDANCE FAIL indicator AMBER	During standby, guidance has been on memory longer than 30 minutes due to a nuclear blast During countdown, a marginal MGS malfunction has occurred	A tactical countdown may be initiated Continue countdown

6	GUIDANCE FAIL indicator RED	LAUNCH CONTROL CONSOLE
		
		   POD AIR CONDITIONING indicator AMBER
		
		COUNTDOWN GROUP
		
		   Fuse holder(s) illuminated
		
		   Malfunction indicator(s) illuminated RED
			Refer to T. O. 21M-HGM16F-2-4-1, Safety Precautions for proper procedures concerning IGS.
			
			21M-HGM16F-2-4

Table 5-1. Malfunction Analysis Procedures, Malfunction Patch Indicators (CONT)

1	2	3	4
ITEM NO.	MALFUNCTION INDICATION	MALFUNCTION ANALYSIS OR PROBABLE CAUSE	ACTION OR TROUBLESHOOTING REFERENCE
7	R/V SAFE indicator RED **NOTE** Prelaunch monitor panel power switch must be on to check indications on pre launch monitor panel.	PRELAUNCH MONITOR A&F SAFETY BAD and(or) WAR HEAD SAFETY BAD indicators RED RE-ENTRY VEHICLE (PANEL 1) R/V SAFE indicator RED	Immediately notify re-entry vehicle personnel 21M-HGM16F-2-9
8	AUTOPILOT FAIL indicator AMBER	AUTOPILOT (PANEL 1) GYRO FINE HEATERS indicator AMBER	Continue countdown 21M-HGM16F-2-5
9	AUTOPILOT FAIL indicator RED	AUTOPILOT (PANEL 1) **NOTE** Those indications marked, (C/D) appear only when the countdown bus is energized. Those indications marked (A/P TEST) appear after autopilot test has started and remain until the countdown bus is de- energized. F/P SAFE indicator RED RATE GYRO TEMP MONITOR indicator RED GYRO FINE HEATERS indicator RED F/P Zero indicator RED (C/D) ENGINE POSITION indicator RED (C/D) GYRO SPIN MOTORS indicator RED (A/P TEST) SIGNAL AMPLIFIERS indicator RED (A/P TEST)	 21M-HGM16F-2-5 21M-HGM16F-2-5 21M-HGM16F-2-5 21M-HGM16F-2-5 21M-HGM16F-2-5 21M-HGM16F-2-5 21M-HGM16F-2-5

Table 5-1. Malfunction Analysis Procedures, Malfunction Patch Indicators (CONT)

1	2	3	4
ITEM NO.	MALFUNCTION INDICATION	MALFUNCTION ANALYSIS OR PROBABLE CAUSE	ACTION OR TROUBLESHOOTING REFERENCE
10	POD AIR CONDITION-ING indicator AMBER	FRCP MISSILE POD AIR CONDITIONER MALFUNCTION indicator RED	Table 5-4
11	IN FLIGHT HE SUPPLY LOW indicator AMBER **NOTE** This indication will appear during countdown only.	PSMR AIRBORNE HELIUM SUPPLY NO. 1 gauge less than 3600 PSI and AIR-BORNE HELIUM SUPPLY NO. 2 gauge less than 4000 PSI Helium leak evident	Continue countdown 21M-HGM16F-2-8
12	IN FLIGHT HE SUPPLY LOW indicator RED	PSMR AIRBORNE HELIUM SUPPLY NO. 1 and NO. 2 gauges less than 1450 PSI	Check for leak in helium system using T.O. 21M-HGM16F-2-8
13	FUEL LEVEL indicator AMBER	During countdown, missile fuel tank high transducer B has failed or missile fuel level is too high. During standby, missile fuel tank high transducer B has failed during a local control check of missile fuel tank level or missile fuel level is too high.	Continue countdown 21M-HGM16F-2-12

Table 5-1. Malfunction Analysis Procedures, Malfunction Patch Indicators (CONT)

1	2	3	4
ITEM NO.	MALFUNCTION INDICATION	MALFUNCTION ANALYSIS OR PROBABLE CAUSE	ACTION OR TROUBLESHOOTING REFERENCE
14	LO_2 LEVEL SENSING indicator AMBER **NOTE** Malfunction indication can appear in countdown only and will automatically extinguish after 2 minutes of rapid LO_2 load.	A single transducer failure has occurred in the missile LO_2 tank at one or more levels	Continue countdown
15	LO_2 LEVEL SENSING indicator RED **NOTE** Malfunction indication can appear in countdown only and will automatically extinguish after 2 minutes of rapid LO_2 load.	A double ttransducer failure on one or more levels of the missile LO_2 tank has occurred.	Abort is required. Refer to T. O. 21M-HGM16F-2-12
16	NUCLEAR BLAST indicator AMBER	BLAST DETECTION CABINET Actual nuclear blast detected Blast detection cabinet malfunction FACILITY INTERFACE CABINET Relay TR-4 de-energized	**WARNING** Do not start commit sequence until after ground shock has passed. SACCEM 35R-1-252 SACCEM 35R-1-252

Table 5-1. Malfunction Analysis Procedures, Malfunction Patch Indicators (CONT)

1	2	3	4
ITEM NO.	MALFUNCTION INDICATION	MALFUNCTION ANALYSIS OR PROBABLE CAUSE	ACTION OR TROUBLESHOOTING REFERENCE
17	MISSILE LIFT FAIL indicator RED **NOTE** Analysis procedures for this malfunction are valid during standby and countdown until missile lift commit start.	HYD LOCAL CONTROL PANEL NOT RECHARGED indicator RED MISSILE LIFTING LOGIC UNIT-A4A2 LAUNCHER STATUS indicator RED	21M-HGM16F-2-23 21M-HGM16F-2-23
18	RESPONDER MODE indicator RED	Cabling or switches not completely in STANDBY or not completely in RESPONDER mode.	21M-HGM16F-2-9 **NOTE** Do not initiate counttown.
19	THRUST SECTION HEATERS indicator AMBER **NOTE** This indication will appear in countdown only.	FACILITIES TERMINAL CABINET NO. 2 PF-70 RUN indicator extinguished	Continue countdown. SACCEM 21-SM65F-2-20-*
	* Use the manual applicable to the base of operation		

Changed 15 April 1964

Table 5-2.　Malfunction Analysis Procedures, Status Patch Indicators

1	2	3	4
ITEM NO.	MALFUNCTION INDICATION	MALFUNCTION ANALYSIS OR PROBABLE CAUSE	ACTION OR TROUBLESHOOTING REFERENCE
1	ENGINES AND GROUND POWER indicator RED	ENGINE (PANEL 2)	
		SYSTEM IN STANDBY indicator extinguished	21M-HGM16F-2-9
		MISSILE GROUND POWER (PANEL 1)	
		SYSTEM IN STANDBY indicator extinguished	21M-HGM16F-2-6
2	FLIGHT CONTROL and R/V indicator RED	PRELAUNCH MONITOR (PANEL POWER ON)	
		115 VAC POWER ON indicator extinguished	11N-RV4F-16
		COUNTDOWN GROUP	
		MGS CHECKDOWN COMPLETE indicator extinguished	21M-HGM16F-2-4
3	HYD PNEU & LN$_2$-HE indicator RED	PLCP (Standby only)	
		VALVE 308 CLOSED indicator AMBER	21M-HGM16F-2-8
		VALVE 305 indicator NOT AMBER	21M-HGM16F-2-8
		VALVE 119 LO$_2$ PHASE 1 & 2 indicator NOT GREEN	21M-HGM16F-2-8
		VALVE 117 FUEL PHASE 1 indicator NOT GREEN	21M-HGM16F-2-8
		VALVE 108 and VALVE 107 indicator NOT GREEN	21M-HGM16F-2-8
		VALVE 50 OPEN indicator NOT GREEN	21M-HGM16F-2-8

Table 5-2.   Malfunction Analysis Procedures, Status Patch Indicators
(CONT)

1 ITEM NO.	2 MALFUNCTION INDICATION	3 MALFUNCTION ANALYSIS OR PROBABLE CAUSE	4 ACTION OR TROUBLESHOOTING REFERENCE
3 (CONT)	HYD PNEU & LN$_2$-HE indicator RED (CONT)	PLCP (CONT)	
		VALVE 52 OPEN indicator GREEN	21M-HGM16F-2-8
		VALVE 7 indicator NOT AMBER	21M-HGM16F-2-8
		VALVE 13 indicator NOT AMBER	21M-HGM16F-2-8
		VALVE 26 indicator NOT AMBER	21M-HGM16F-2-8
		HYDRAULIC (PANEL 1)	
		SYSTEM IN STANDBY indicator extinguished	21M-HGM16F-2-7
		PNEUMATICS (PANEL 1) (Standby only)	
		AUTOMATIC VALVES READY indicator extinguished	21M-HGM16F-2-8
		PNEUMATICS IN PHASE 1 indicator extinguished	21M-HGM16F-2-8
		STANDBY STARTED indicator extinguished	21M-HGM16F-2-8
		LN$_2$-HE (PANEL 1)	
		LN$_2$ VALVE STANDBY indicator extinguished	21M-HGM16F-2-8
		HE VALVE STANDBY indicator extinguished	21M-HGM16F-2-8
		SUPPLY PRESSURE indicator extinguished	21M-HGM16F-2-8
4	LO$_2$ & FUEL indicator RED	LO$_2$ TANKING (PANEL 1)	
		MAIN FILL VALVE L-7 indicator NOT AMBER	21M-HGM16F-2-12
		LO$_2$ TANKING (PANEL 2) (Standby only)	
		TOPPING TANK VENT VALVE N-4 indicator NOT GREEN	21M-HGM16F-2-12
		RAPID LOAD VALVE L-2 indicator NOT AMBER	21M-HGM16F-2-12

Table 5-2.   Malfunction Analysis Procedures, Status Patch Indicators
(CONT)

1	2	3	4
ITEM NO.	MALFUNCTION INDICATION	MALFUNCTION ANALYSIS OR PROBABLE CAUSE	ACTION OR TROUBLESHOOTING REFERENCE
4 (CONT)	LO$_2$ & FUEL indicator RED (CONT)	LO$_2$ TANKING (PANEL 1) (CONT) DRAIN VALVE L-16 indicator NOT AMBER FINE LOAD VALVE L-1 indicator NOT AMBER LINE DRAIN PRES VALVE N-60 indicator NOT AMBER	21M-HGM16F-2-12 21M-HGM16F-2-12 21M-HGM16F-2-12

Table 5-2.   Malfunction Analysis Procedures, Status Patch Iindicators (CONT)

1	2	3	4
ITEM NO.	MALFUNCTION INDICATION	MALFUNCTION ANALYSIS OR PROBABLE CAUSE	ACTION OR TROUBLESHOOTING REFERENCE
4 (CONT)		LO$_2$ TANKING (PANEL 2) (Standby only) (cont) PROPELLANT LEVEL (PANEL 2) SYSTEM IN STANDBY indicator extinguished FUEL TANKING (PANEL 2) AIRBORNE F AND D VALVE indicator NOT AMBER	21M-HGM16F-2-9 21M-HGM16F-2-12
5	FACILITIES & MISSILE LIFT indicator RED	FACILITIES (PANEL 1) SYSTEM POWER indicator extinguished	21M-HGM16F-2-9

Table 5-3.   Malfunction Analysis Procedures, Countdown Patch Indicators

1	2	3	4
ITEM NO.	MALFUNCTION INDICATION	MALFUNCTION ANALYSIS OR PROBABLE CAUSE	ACTION OR TROUBLESHOOTING REFERENCE
1	MISSILE POWER indicator NOT GREEN	MISSILE GROUND POWER (PANEL 1) DC MISSILE LOADS indicator RED AC MISSILE LOADS indicator RED	21M-HGM16F-2-6 21M-HGM16F-2-6
2	HEATERS ON indicator NOT GREEN	AC POWER DISTR PANEL ENGINE VALVE HEATERS indicator extinguished	21M-HGM16F-2-6

Table 5-3. Malfunction Analysis Procedures, Countdown Patch Indicators (CONT)

1	2	3	4
ITEM NO.	MALFUNCTION INDICATION	MALFUNCTION ANALYSIS OR PROBABLE CAUSE	ACTION OR TROUBLESHOOTING REFERENCE
3	MISSILE BAT ACTIV-ATED indicator NOT AMBER		21M-HGM16F-2-9
4	MISSILE BAT ACTIV-ATED indicator NOT GREEN (2 minutes after countdown start)	MISSILE GROUND POWER (PANEL 2) Internal DC less than 25.6 VDC MISSILE GROUND POWER (PANEL 1) INTERNAL DC VOLTAGE indicator RED	21M-HGM16F-2-6 21M-HGM16F-2-9
5	ENG & MISSILE POWER READY indicator NOT GREEN	ENGINE (PANEL 2) MISSILE POWER indicator extinguished	21M-HGM16F-2-9
6	AUTOPILOT ON indicator NOT AMBER	MISSILE GROUND POWER (PANEL 2) Ground Supply (Missile Loads) not greater than 113.0 VAC	21M-HGM16F-2-6
7	AUTOPILOT ON indicator NOT GREEN (4 minutes after count-down start)		21M-HGM16F-2-9
8	AUTOPILOT TEST indicator NOT AMBER		21M-HGM16F-2-9
9	AUTOPILOT TEST indicator NOT GREEN (90 seconds after AMBER)	LAUNCH CONTROL CONSOLE AUTOPILOT FAIL indicator RED	Table 5-1
10	R/V BATTERY TEMPERATURE indicator NOT GREEN		11N-RV4F-16 21M-HGM16F-2-9

Table 5-3. Malfunction Analysis Procedures, Countdown Patch Indicators (CONT)

1	2	3	4
ITEM NO.	MALFUNCTION INDICATION	MALFUNCTION ANALYSIS OR PROBABLE CAUSE	ACTION OR TROUBLESHOOTING REFERENCE
11	GUIDANCE READY indicator NOT AMBER		21M-HGM16F-2-4
12	GUIDANCE READY indicator NOT GREEN	LAUNCH CONTROL CONSOLE GUIDANCE FAIL indicator RED	Table 5-1
13	FLIGHT CONTROL & R/V READY indicator NOT GREEN	LAUNCH CONTROL CONSOLE The selected target indicator NOT GREEN GUIDANCE (PANEL 1) GUIDANCE TARGET SELECTED indicator NOT GREEN R/V TARGET SELECTED indicator NOT GREEN	21M-HGM16F-2-9 21M-HGM16F-2-4 21M-HGM16F-2-9 11N-RV4F-16 21M-HGM16F-2-9
14	PNEUMATICS IN PHASE II indicator NOT AMBER (5 seconds after countdown start)		**NOTE** If missile fuel tank pressure starts to increase 5 seconds after countdown starts, continue countdown.
15	PNEUMATICS IN PHASE II indicator NOT GREEN	LAUNCH CONTROL CONSOLE FUEL TANK pressure gauge less than 53 PSI LO_2 TANK pressure gauge greater than 17 PSI	21M-HGM16F-2-8 21M-HGM16F-2-8
16	HYDRAULIC PRESSURE indicator NOT AMBER		21M-HGM16F-2-9

Table 5-3. Malfunction Analysis Procedures, Countdown Patch Indicators (CONT)

1	2	3	4
ITEM NO.	MALFUNCTION INDICATION	MALFUNCTION ANALYSIS OR PROBABLE CAUSE	ACTION OR TROUBLESHOOTING REFERENCE
17	HYDRAULIC PRES-SURE indicator NOT GREEN		(Tactical) Section 1V (PLX) 21M-HGM16F-2-7
18	LN$_2$ LOAD indicator NOT AMBER		(Tactical) Section IV (PLX) 21M-HGM16F-2-8
19	LN$_2$ LOAD indicator NOT GREEN (3 minutes after AMBER)		(Tactical) Section IV (PLX) 21M-HGM16F-2-8
20	HELIUM LOAD indicator NOT AMBER (2 minutes after countdown start)		(Tactical) Section IV (PLX) 21M-HGM16F-2-9
21	HELIUM LOAD indicator NOT GREEN		(Tactical) Section IV (PLX) 21M-HGM16F-2-8
22	HYD-PNEU & LN$_2$-HE READY indicator NOT GREEN		(Tactical) Section IV (PLX) 21M-HGM16F-2-8 21M-HGM16F-2-7
23	LO$_2$ LINE FILLED indicator NOT AMBER		(Tactical) Section IV (PLX) 21M-HGM16F-2-12

Table 5-3. Malfunction Analysis Procedures, Countdown Patch indicators (CONT)

1	2	3	4
ITEM NO.	MALFUNCTION INDICATION	MALFUNCTION ANALYSIS OR PROBABLE CAUSE	ACTION OR TROUBLESHOOTING REFERENCE
24	LO_2 LINE FILLED indicator NOT GREEN		Section III 21M-HGM16F-2-9
25	RAPID LO_2 LOAD indicator NOT AMBER		21M-HGM16F-2-9
26	RAPID LO_2 LOAD indicator NOT GREEN		21M-HGM16F-2-9 21M-HGM16F-2-12
27	FINE LO_2 LOAD indicator NOT AMBER		21M-HGM16F-2-9
28	FINE LO_2 LOAD indicator NOT GREEN		21M-HGM16F-2-9
29	LO_2 READY indicator NOT GREEN	LO_2 TANKING (PANEL 2) AIRBORNE FILL & DRAIN VALVE indicator NOT AMBER	21M-HGM16F-2-12 21M-HGM16F-2-9

Table 5-4. Malfunction Analysis Procedures, Facilities Remote Control Panel Indicators

1	2	3	4
ITEM NO.	MALFUNCTION INDICATION	MALFUNCTION ANALYSIS OR PROBABLE CAUSE	ACTION OR TROUBLESHOOTING REFERENCE
1	DIESEL GEN D-60 (D-61) OVERSPEED LOW LUBE OIL PRES HI-TEMP indicator RED **NOTE** Indication is normal for 8 seconds as diesel generator is being started (Tactical countdown may be initiated.)	DIESEL ENGINE CONTROL PANEL Diesel generator exceeded specified speed limit (OVER-SPEED indicator illuminated) Engine lube oil pressure below specified limits (LOW LUBE OIL PRESS indicator illuminated) Jacket water temperature exceeded the specified limit (HI-TEMP indicator illuminated)	Section IV SACCEM 21-SM65F-2-21* SACCEM 21-SM65F-2-21* SACCEM 21-SM65F-2-21* SACCEM 21-SM65F-2-20*
2	STORAGE AREA OXYGEN 19% indicator RED (Tactical countdown may be initiated.)	OXYGEN DETECTOR CABINET Detector unit malfunction (TROUBLE indicator illuminated) Leak in LN_2, GN_2, or Helium system	Section IV Check station indicator. Determine the area generating alarm and percent oxygen content. Locate and isolate leak referring to T.O. 21M-HGM16F-2-8, T.O. 21M-HGM16F-2-12, and T.O. 21M-HGM16F-2-29-1.
3	SILO CONTROL CABINET HI-TEMP indicator RED (Tactical countdown may be initiated.)	FTC NO. 2 Fan coil unit (FC-10) control circuit has been interrupted (FC-10 RUN indicator extinguished) SILO LEVEL 3 Chilled water input to cooling coil (CC-10) too warm (temperature indicator TI-10 above normal) Temperature control system coil CC-10 not operating properly (temperature indicator TI-11 above normal)	Section IV SACCEM 21-SM65F-2-20-* SACCEM 21-SM65F-2-19-* SACCEM 21-SM65F-2-20-*
	* Use the manual applicable to the base of operation		

Table 5-4. Malfunction Analysis Procedures, Facilities Remote Control Panel Indicators (CONT)

1	2	3	4
ITEM NO.	MALFUNCTION INDICATION	MALFUNCTION ANALYSIS OR PROBABLE CAUSE	ACTION OR TROUBLESHOOTING REFERENCE
4	SILO WATER CHILLER UNITS MALFUNCTION indicator RED **NOTE** Normal indication if nonessential bus is de-energized. (Tactical countdown may be initiated.)	SILO LEVEL 4 Compressor head pressure high (HIGH PRESSURE indicator illuminated) Chilled water output temperature too cold (CHILLED WATER TOO COLD indicator illuminated) Low oil pressure and(or) low oil level in compressor (OIL PRESSURE LOW indicator illuminated) SILO LEVEL 4 Water chiller unit not operating	SACCEM 21-SM65F-2-20-* SACCEM 21-SM65F-2-20-* SACCEM 21-SM65F-2-20-* Isolate cause of restricted water in condenser water system. SACCEM 21-SM65F-2-20-*
5	INSTRUCTION AIR RECEIVER LOW PRESSURE indicator RED (Tactical countdown may be initiated.)	INSTRUMENT AIR PREFAB Instrument air compressors not operating properly (gauge PI-710 less than 1100 PSI) Leak in instrument air system (gauge PI-710 less than 1100 PSI)	Section IV Isolate malfunction at instrument air compressors. Locate and isolate leak referring to SACCEM 21-SM65F-2-22-*
6	UTILITY WATER PRESSURE indicator RED **NOTE** Normal if the nonessential bus has been de-energized (Tactical countdown may be initiated.)	HYDROPNEUMATIC TANK TK-80 Loss of air pressure TK-80 gauge PI-82 less than normal) Water level low in TK-80 (sight gauge LG-80 less than ½-full	SACCEM 21-SM65F-2-22-* SACCEM 21-SM65F-2-19-*
	* Use the manual applicable to the base of operation		

Table 5-4. Malfunction Analysis Procedures, Facilities Remote Control Panel Indicators (CONT)

1	2	3	4
ITEM NO.	MALFUNCTION INDICATION	MALFUNCTION ANALYSIS OR PROBABLE CAUSE	ACTION OR TROUBLESHOOTING REFERENCE
7	WATER STORAGE TANK HIGH LEVEL indicator RED (Tactical countdown may be initiated.)	COMPLEX GRADE LEVEL Well control system float switch (where installed not operating properly)	SACCEM 21-SM65F-2-19-*
8	BATTERY CHARGER FAILURE indicator RED (Tactical countdown may be initiated.)	48 VOLT RECTIFIER Rectifier output voltage low (output DC Voltmeter not within specified limit) EMCC Loss of primary power source to 48 volt rectifier (48 volt battery charger circuit breaker tripped)	SACCEM 21-SM65F-2-21-* Reset circuit breaker and refer to SACCEM 21-SM65F-2-21-*
9	MAIN EXHAUST FAN NOT OPERATING indicator RED (Tactical countdown may be initiated.)	FTC NO. 2 Loss of control circuit (EF-30 RUN indicator extinguished) Temeprature in EF-30 plenum exceeded 125°F (firestat FST-30 or FST-31 tripped)	SACCEM 21-SM65F-2-20-* SACCEM 21-SM65F-2-21-* Reset firestats FST-30 or FST-31 referring to SACCEM 21-SM65F-2-22-*
10	STORAGE AREA OXYGEN 25% indicator RED	OXYGEN DETECTOR CABINET Detector unit malfunction (TROUBLE indicator illuminated) Leak in LO_2 system	Section IV Reset detector unit and refer to SACCEM 21-SM65F-2-22-* Check station indicator. Determine area generating alarm and percent oxygen content. Locate and isolate leak referring to T.O. 21M-HGM16F-2-8 and T.O. 21M-HGM16F-2-29-1
	* Use the manual applicable to the base of operation		

Table 5-4.   Malfunction Analysis Procedures, Facilities Remote Control Panel Indicators (CONT)

1	2	3	4
ITEM NO.	MALFUNCTION INDICATION	MALFUNCTION ANALYSIS OR PROBABLE CAUSE	ACTION OR TROUBLESHOOTING REFERENCE
11	SILO SUMP HIGH LEVEL indicator RED (Tactical countdown may be initiated.)	SUMP AREA SUMP pumps not operating properly	SACCEM 21-SM65F-2-19-*
12	AIR WASHER DUST COLLECTING UNITS NOT OPERATING indicator RED. (Tactical countdown may be initiated.)	FTC NO. 2 Loss of control circuit to SF-20 and SF-21 (SF-20 and SF-21 RUN indicators extinguished)	Section IV SACCEM 21-SM65F-2-20-*
13	LCC AIR RECEIVER indicator RED (Tactical countdown may be initiated.)	LCC AIR RECEIVER TANK Leak in LCC Air System (gauge PI-100 less than 450 PSI) SILO LEVEL 3 Loss of supply air to LCC air system (pressure controller malfunction or leak in instrument air system)	Locate and isolate leak referring to SACCEM 21-SM65F-2-22-* SACCEM 21-SM65F-2-22-*
14	DIESEL VAPOR HIGH LEVEL indicator RED	DIESEL ROOMS, SILO LEVEL 5 AND 6 Leak in diesel fuel system DIESEL FUEL VAPOR DETECOR CABINET Detector unit malfunction (TROUBLE indicator illuminated)	Section IV Locate and isolate leak referring to SACCEM 21-SM65F-2-21-* Reset detector cabinet and refer to SACCEM 21-SM65F-2-22-*
	* Use the manual applicable to the base of operation		

Table 5-4. Malfunction Analysis Procedures, Facilities Remote Control Panel Indicators (CONT)

1	2	3	4
ITEM NO.	MALFUNCTION INDICATION	MALFUNCTION ANALYSIS OR PROBABLE CAUSE	ACTION OR TROUBLESHOOTING REFERENCE
15	WATER STORAGE TANK LOW LEVEL indicator RED **NOTE** Normal if the non-essential bus has been de-energized (Tactical countdown may be initiated.)	COMPLEX GRADE LEVEL Well control system not operating properly (float switch or pressure switch malfunctioning) (where installed)	SACCEM 21-SM65F-2-19-*
16	LCC SEWAGE PUMP HIGH LEVEL indicator RED (Tactical countdown may be initiated.)	TUNNEL ENTRANCE AREA LCC sump pumps not operating properly	SACCEM 21-SM65F-2-19-*
17	DIESEL GENERATOR D-60 (D-61) ON indicator extinguished (Tactical countdown may be initiated.)	PRCP Loss of generator output voltage	SACCEM 21-SM65F-2-21-*
18	LAUNCH PLATFORM EXHAUST FAN ON indicator extinguished and(or) LAUNCH PLATFORM EXHAUST FAN OFF indicator RED **NOTE** This is a normal indication if STORAGE AREA OXYGEN 25% indicator is illuminated RED. (Tactical countdown may be initiated.)	FTC NO. 2 Loss of control circuit (LAUNCHER PLATFORM EXHAUST FAN EF-40 indicator extinguished.	SACCEM 21-SM65F-2-20-*
	* Use the manual applicable to the base of operation		

Table 5-4.   Malfunction Analysis Procedures, Facilities Remote Control Panel Indicators (CONT)

1	2	3	4
ITEM NO.	MALFUNCTION INDICATION	MALFUNCTION ANALYSIS OR PROBABLE CAUSE	ACTION OR TROUBLESHOOTING REFERENCE
19	LAUNCH PLATFORM DAMPERS OPEN indicator extinguished and(or) LAUNCH PLATFORM DAMPERS CLOSED indicator GREEN. **NOTE** This is a normal indication if STORAGE AREA OXYGEN 25% indicator is illuminated RED. (Tactical countdown may be initiated.)	FTC NO. 2 Loss of control circuit to (launcher platform exhaust fan EF-40 (EF-40 RUN indicator extinguished)	SACCEM 21-SM65F-2-20-*
20	GASEOUS OXYGEN VENT CLOSED indicator extinguished and(or) GASEOUS OXYGEN VENT OPEN indicator GREEN **NOTE** Normal indication during countdown and until 2 hour helium vent timer has expired. **NOTE** This is a normal indication if STORAGE AREA OXYGEN 25% indicator is illuminated RED. (Tactical countdown may be initiated.)	Malfunction has occurred in control circuit	SACCEM 21-SM65F-2-22-*
	* Use the manual applicable to the base of operation		

Table 5-4. Malfunction Analysis Procedures, Facilities Remote Control Panel Indicators (CONT)

1	2	3	4
ITEM NO.	MALFUNCTION INDICATION	MALFUNCTION ANALYSIS OR PROBABLE CAUSE	ACTION OR TROUBLESHOOTING REFERENCE
21	LCC SEWER VENT OPEN indicator extinguished and(or) LCC SEWER VENT CLOSED indicator RED **NOTE** Normal indication for 20 seconds after blast has ben detected. (Tactical countdown may be initiated.)	BLAST CLOSURES SYSTEM Loss of control circuit (blast detection unit malfunction) LCC LEVEL 1 Loss of air in LCC air receiver (PI-100 less than 300 PSI)	SACCEM 21-SM65F-2-22-* and SACCEM 35R-1-252 SACCEM 21-SM65F-2-22-*
22	LCC GRADE ENTRY DOOR CLOSED indicator extinguished and(or) GRADE ENTRY DOOR OPEN indicator RED (Tactical countdown may be initiated.)	GRADE ENTRY VESTIBULE Door limit switch not operating properly	SACCEM 21-SM65F-2-22-*
23	LAUNCH PLATFORM FAN COIL UNIT ON indicator extinguished and (or) LAUNCH PLATFORM FAN COIL UNIT OFF indicator RED (Tactical countdown may be initiated.)	FTC NO. 2 Control circuit has been interrupted (LAUNCHER PLATFORM FAN COIL UNIT RUN indicator extingushed)	SACCEM 21-SM65F-2-20-*
24	SILO AIR INTAKE CLOSURES OPEN indicator extinguished and (or) SILO AIR INTAKE CLOSURES CLOSED indicator RED	BLAST CLOSURES SYSTEM Loss of control circuit (blast closures system malfunction)	Section IV SACCEM 21-SM65F-2-22-* SACCEM 35R-1-252
	* Use the manual applicable to the base of operation		

Table 5-4. Malfunction Analysis Procedures, Facilities Remote Control Panel Indicators (CONT)

1	2	3	4
ITEM NO.	MALFUNCTION INDICATION	MALFUNCTION ANALYSIS OR PROBABLE CAUSE	ACTION OR TROUBLESHOOTING REFERENCE
24 (CONT)	**NOTE** Normal indication if blast has been detected and after missile lift up and locked indicator GREEN. (Tactical countdown may be initiated)	SILO LEVEL 2 Blast closures supply pressure gauge PI-3 less than 300 PSI	
25	LCC AIR INTAKE OPEN indicator extinguished and (or) LCC AIR INTAKE CLOSED indicator RED **NOTE** Normal indication if blast has been detected. (Tactical countdown may be initiated.)	BLAST CLOSURES SYSTEM Loss of control circuit (blast closures system malfunction) LCC LEVEL 1 Loss of air in LCC air receiver (PI-100 less than 300 PSI)	SACCEM 21-SM65F-2-22-* SACCEM 35R-1-252 SACCEM 21-SM65F-2-22-*
26	L C C VESTIBULE EX-TERIOR BLAST DOOR OPEN indicator RED and (or) LCC VESTIBULE EXTERIOR BLAST DOOR CLOSED indicator extinguished (NA OSTF-2) (Tactical countdown may be initiated.)	LCC VESTIBULE AREA Door limit switch not operating properly	SACCEM 21-SM65F-2-22-*
27	STARTING AIR RECEIVER NORMAL PRESS indicator extinguish	STARTING AIR RECEIVER Faulty pressure switch on starting air receiver. (PI-1 on TK-64 above 250 PSI) Leak in starting air system or instrument air supply	SACCEM 21-SM65F-2-22-* Locate and isolate leak in the air systems SACCEM 21-SM65F-2-22-*
	* Use the manual applicable to the base of operation		

Table 5-4.　Malfunction Analysis Procedures, Facilities Remote Control Panel Indicators (CONT)

1	2	3	4
ITEM NO.	MALFUNCTION INDICATION	MALFUNCTION ANALYSIS OR PROBABLE CAUSE	ACTION OR TROUBLESHOOTING REFERENCE
28	SILO AIR EXHAUST CLOSURES OPEN indicator extinguished and (or) SILO AIR EXHAUST CLOSURES CLOSED indicator RED **NOTE** Normal indication if blast has been detected or after MISSILE LIFT UP & LOCKED indicator GREEN. (Tactical countdown may be initiated.)	BLAST CLOSURES SYSTEM Loss of control circuit (blast detection unit malfunction)	SACCEM 21-SM65F-2-22-* SACCEM 35R-1-252
29	GASEOUS OXYGEN VENT FAN ON indicator illuminated. **NOTE** This is a normal indication if during countdown and until 2-hour helium vent timer has expired, or if STORAGE AREA OXYGEN 25% indicator is RED. (Tactical countdown may be initiated.)	Malfunction has occurred in the control circuit	SACCEM 21-SM65F-2-22-*
30	LCC AIR EXHAUST OPEN indicator extinguished and (or) LCC AIR EXHAUST CLOSED indicator RED	BLAST CLOSURES SYSTEM Loss of control circuit (blast detection unit malfunction)	SACCEM 21-SM65F-2-22-* SACCEM 35R-1-252
	* Use the manual applicable to the base of operation		

Table 5-4. Malfunction Analysis Procedures, Facilities Remote Control Panel Indicators (CONT)

1	2	3	4
ITEM NO.	MALFUNCTION INDICATION	MALFUNCTION ANALYSIS OR PROBABLE CAUSE	ACTION OR TROUBLESHOOTING REFERENCE
30 (CONT)	**NOTE** Normal indication if blast has been detected. (Tactical countdown may be initiated.)	LCC LEVEL 1 Loss of air in LCC air receiver (PI-100 less than 300 PSI.)	SACCEM 21-SM65F-2-22-*
31	LCC VESTIBULE INTERIOR BLAST DOOR OPEN indicator RED and (or) LCC VESTIBULE INTERIOR BLAST DOOR CLOSED indicator extinguished **NOTE** Normal if blast door is open. (Tactical countdown may be initiated.)	LCC VESTIBULE AREA Door limit switch not operating properly	SACCEM 21-SM65F-2-22-*
32	SILO BLAST DOOR OPEN indicator RED and (or) SILO BLAST DOOR CLOSED indicator indicator extinguished **NOTE** Normal if one or both blast doors are open. (Tactical countdown may be initiated.)	TUNNEL AREA Door limit switch not operating properly	SACCEM 21-SM65F-2-22-*
33	RP-1 AND FIRE FOG SYSTEM DAMPERS OPEN indicator RED and (or) RP-1 AND FIRE FOG SYSTEM DAMPERS CLOSED indicator extinguished	SUMP AREA Malfunction has occurred in control circuit. (Volume dampers VD-40 in open position)	SACCEM 21-SM65F-2-22-*
* Use the manual applicable to the base of operation			

T.O. 21M-HGM16F-1

Table 5-4. Malfunction Analysis Procedures, Facilities Remote Control Panel Indicators (CONT)

1	2	3	4
ITEM NO.	MALFUNCTION INDICATION	MALFUNCTION ANALYSIS OR PROBABLE CAUSE	ACTION OR TROUBLESHOOTING REFERENCE
33 (CONT)	**NOTE** RP-1 AND FIRE FOG SYSTEM DAMPERS CLOSED indicator is a normal indication during countdown. (Tactical countdown may be initiated.)		
34	TUNNEL BLAST DOOR CLOSED BLAST DOOR DOGGED VENT VALVE CLOSED indicator extinguished. **NOTE** Normally extinguished in standby. Vent valve V-600 is open. Illuminated GREEN in countdown with door closed and dogged and vent valve closed. (Tactical countdown may be initiated.)	TUNNEL AREA Faulty limit switch on door Faulty limit switch on door dog Faulty limit switch on vent valve V-600	SACCEM 21-SM65F-2-22-* SACCEM 21-SM65F-2-22-* SACCEM 21-SM65F-2-22-*
35	LCC STAIRWELL AIR EXHAUST OPEN indicator extinguished and (or) LCC STAIRWELL AIR EXHAUST CLOSED indicator RED (NA OSTF-2) **NOTE** Normal indication if blast has been detected. (Tactical countdown may be initiated.)	BLAST CLOSURES SYSTEM Loss of control circuit (blast detection unit malfunction) LCC LEVEL 1 Loss of air in LCC air receiver (PI-100 less than 300 PSI)	SACCEM 21-SM65F-2-22-* SACCEM 35R-1-252 SACCEM 21-SM65F-2-22-*
	* Use the manual applicable to the base of operation		

Table 5-4. Malfunction Analysis Procedures, Facilities Remote Control Panel Indicators (CONT)

1	2	3	4
ITEM NO.	MALFUNCTION INDICATION	MALFUNCTION ANALYSIS OR PROBABLE CAUSE	ACTION OR TROUBLESHOOTING REFERENCE
36	LCC ESCAPE HATCH DOOR OPEN indicator RED and (or) LCC ESCAPE HATCH DOOR CLOSED indicator extinguished (Tactical countdown may be initiated.)	LCC LEVEL 1 Faulty door limit switch or indicator circuit	SACCEM 21-SM65F-2-21-*
37	MISSILE ENCLOSURE FOG ON indicator RED and (or) MISSILE ENCLOSURE FOG OFF indicator extinguished **NOTE** Normal if fog system has been activated at FRCP or silo level 5.	MISSILE ENCLOSURE FOG SYSTEM CONTROL Malfunction in the fog system control circuit	FRCP Depress MISSILE ENCLOSURE FOG OFF pushbutton SILO LEVEL 5 Close water supply valve LCV-80 referring to SACCEM 21-SM65F-2-22-*
38	SILO GRADE ENTRY DOOR CLOSED indicator extinguished and (or) SILO GRADE ENTRY DOOR OPEN indicator RED (OSTF-2) (Tactical countdown may be initiated.)	SILO GRADE ENTRY VESTIBULE Door limit switch not operating properly	SACCEM 21-SM65F-2-22-*
39	SPRAY PUMP NORMAL indicator extinguished (OSTF-2) (Tactical countdown may be initiated.)	FTC NO. 2 Automatic circuit has transferred to alternate pump due to pressure drop.	SACCEM 21-SM65F-2-20-*
40	SETTLING TANK NORMAL indicator extinguished (OSTF-2) (Tactical countdown may be initiated.)	Sand settling tank plugged and in bypass position	SACCEM 21-SM65F-2-20-*
	* Use the manual applicable to the base of operation		

Table 5-4. Malfunction Analysis Procedures, Facilities Remote Control Panel Indicators
(CONT)

1	2	3	4
ITEM NO.	MALFUNCTION INDICATION	MALFUNCTION ANALYSIS OR PROBABLE CAUSE	ACTION OR TROUBLESHOOTING REFERENCE
41	POD AIR CONDITIONING MALFUNCTION indicator RED	FRCP	CAUTION Refer to T. O. 21M-HGM16F-2-4-1, Safety Precautions for proper procedures concerning IGS.
		High humidity (MISSILE POD HI HUMIDITY indicator RED)	SACCEM 21-SM65F-2-30-1
		Output airflow low (MISSILE POD AIR LO PRESSURE indicator RED)	SACCEM 21-SM65F-2-30-1
		Output temperature high (MISSILE POD AIR HI TEMPERATURE indicator RED)	SACCEM 21-SM65F-2-30-1
42	EMERGENCY WATER PUMP P-32 ON indicator RED NOTE Normal indication if water cooling tower has been destroyed. (Tactical countdown may be initiated but only on one generator.)	FTC NO. 2 Loss of water flow in condenser water system (EMERGENCY WATER PUMP P-32 RUN indicator illuminated)	Depress EMERGENCY WATER PUMP P-32 STOP pushbutton and verify water pump P-30 or P-31 RUN indicator illuminates and EMERGENCY WATER PUMP P-32 RUN indicator extinguishes. Refer to SACCEM 21-SM65F-2-20-*
	*Use the manual applicable to the base of operation		

SECTION VI

OPERATION LIMITATIONS

TABLE OF CONTENTS

LIST OF ILLUSTRATIONS

LIST OF TABLES

6-1. LIQUID OXYGEN AND LIQUID NITROGEN BOILOFF TIME NOMOGRAMS.

6-2. Figures 6-1 and 6-2 graphically illustrate the time required for liquid oxygen and liquid nitrogen to boiloff under different weather conditions when the missile is fully exposed (oxidizer tank not in silo). To calculate the boiloff time, the following information must be obtained:

    a. Percent relative humidity.

    b. Ambient temperature in degrees fahrenheit.

    c. Wind velocity in miles per hour.

6-3. To calculate the boiloff time for a fully exposed and full oxidizer (LO_2) tank the following procedures shall be utilized:

    a. Locate the percent relative humidity figure on percent relative humidity axis.

    b. Draw a vertical line from the percent relative humidity figure located in step a, to intersect with the proper ambient temperature curve.

    c. Draw a horizontal line from the percent relative humidity located in step b. to intersect with the proper wind velocity curve.

    d. Draw a vertical line, from the point of intersection created by step c, to intersect with the boiloff time axis.

    e. The point of intersection on the boiloff axis is the time required for complete boiloff.

6-4. Anytime the launcher platform, with a loaded missile aboard, is not down and locked and cannot be lowered, manual pressurization from the helium control charging unit must be accomplished when LO_2 (or LN_2) boiloff is complete. Missile fuel tank pressurization during and after boiloff is automatically maintained by the helium control charging unit. However, since oxidizer tank pressure will be maintained by LO_2 (or LN_2) boiloff, manual pressurization is required after completion of boiloff. The following boiloff times must be noted for rain conditions:

    a. During rain conditions with LN_2 loaded in the missile oxidizer tank, boiloff time is approximately 3.5 hours. Missile oxidizer tank pressure must be maintained from the helium control charging unit when boiloff is complete.

    b. During rain conditions with LO_2 loaded in the missile oxidizer tank, boiloff time is approximately 5.3 hours. Missile oxidizer tank pressure must be maintained from the helium control charging unit when boiloff is complete.

6-5. MINIMUM CREW REQUIREMENTS.

6-6. The standard missile crew is composed of five members: two Missile Officers, Air Force Specialty Code (AFSC) 3124B, (a missile combat crew commander and a deputy missile combat crew commander), and three airmen (Ballistic Missile Analyst Technician, AFSC 312X4D, Missile Facilities Technician, AFSC 541X0D, and Electrical Power Production Technician, AFSC 543X0). These five men comprise the minimum crew required to safely operate the weapon system. Additional crew members, as required, will be added at the discretion of the Commander.

6-7. SPGG HEAT AFTER AC-POWER INTERRUPTION.

6-8. Green and red indicators placarded B1 SPGG HEAT, B2 SPGG HEAT, and SUSTAINER SPGG HEAT are located on the AC power distribution panel (silo level 3). These indicators are provided to furnish personnel with positive indication of the heat environment of the SPGG'S. The indications provided by the lights are as follows:

    a. When a red indicator is illuminated, an over-temperature environment exists around the applicable SPGG. (Launch cannot be achieved.)

    b. When green indicators are illuminated, the correct heat environment exists around the SPGG'S and a launch can be accomplished.

    c. When one of the green indicators extinguishes, the SYSTEM IN STANDBY indicator extinguishes on missile ground power panel 1, and the ENGINES AND GROUND POWER indicator illuminates red in the standby patch.

6-9. Operating limitation after AC-power interruption is as follows:

a. If the green indicators illuminate when power is restored, countdown may be continued without a delayed launch.

b. If the green indicators fail to illuminate when power is restored, countdown (delayed launch) must be delayed for eight hours because this is the time required to stabilize the heat environment around the SPGG'S.

6-10. LIMITATION ON MISSILE BATTERY LIFE.

6-11. The missile battery may safely remain in the missile for a period of ten hours after activation. Vents for escapage of gas have been provided and the case will prevent liquid spillage for the ten-hour period.

6-12. Because of chemical deterioration of the elements within the battery it must be removed 10 hours after activation for the following reasons:

a. The design life of a battery after activation is 10 hours. Therefore, a successful launch cannot be expected if this limit is exceeded.

b. Escapage of liquid may occur and cause corrosion in the area around the battery.

NOTE

Two PLX's may be performed off of a battery, providing they occur within the 10-hour period.

6-13. WEATHER LIMITATIONS FOR NON-TACTICAL COUNTDOWNS.

6-14. To prevent air washer nozzles and pipes from freezing, do not initiate a nontactical countdown if outside air temperature is below 10°F (except at Squadron Complex 556). To prevent possible structual damage to the missile, do not initiate commit sequence during a non-tactical countdown if any of the following conditions exists:

a. Wind velocity or wind gust exceeds maximum allowable anemometer reading measured at a distance of 10 feet above ground. (Refer to classified supplement to this manual.)

b. Thunderstorms, hail, or lighting is present or forecasted within a 10-mile radius of the complex during the time the silo doors will be open.

6-15. COUNTDOWN HOLD LIMITATIONS.

6-16. The system has been designed for for a 1-hour hold capability during countdown when gas and fluid levels are serviced to requirements contained in column 6 of table 6-1. This capability includes a countdown to commit, countdown through launch, or abort with an accumulated hold in both countdowns, at ready for commit, of 1 hour.

6-17. ELECTRICAL POWER LIMITATIONS.

6-18. Limitations on launch complex electrical power require specific restrictions to be placed on the system. In addition, certain conditions inherent within the system must be noted. These restrictions and conditions are as follows:

a. Diesel generators shall be operated in parallel only to satisfy operational requirements, or when power loads exceed, or are reasonably expected to exceed, single generator operation limits.

b. With diesel generators operating in parallel the nonessential power bus shall be energized.

c. Diesel generators shall not be operated in parallel any time cooling water is being furnished from the utility water system. (Emergency water pump P-32 operating.)

d. Maximum period of parallel diesel generator operation shall not exceed 2-½ hours unless power load is in excess of, or reasonably expected to exceed, single generator operation limits.

e. Design load for a single diesel generator is 500 KW. Maximum load of 550 KW is permissible for a period not to exceed 2 hours in any 24-hour period. If the design load (500 KW) is exceeded for an extended period, damage to generator may occur.

f. After starting a diesel engine, it shall be allowed to operate for 30 minutes prior to shutdown.

g. No automatic shutdown feature is provided for a diesel generator high temperature condition. To prevent damage to diesel generator it must be manually shut down.

h. If countdown is performed with only one generator operating, nonessential bus must be de-energized prior to launcher platform up movement. Energize the nonessential bus after Missile lift up and locked.

i. If silo blast closures close and remain closed, a hazardous condition will be created in the silo due to lack of fresh air and normal exhaust. Diesel engine failure is probable.

6-19. LAUNCH COMPLEX GAS AND FLUID LEVEL REQUIREMENTS.

6-20. Table 6-1 presents the gas (helium and nitrogen) and fluid (LO_2 and LN_2) level requirements for the launch complex. Included in table 6-1 are the following:

a. The applicable storage tank (STORAGE TANK column).

b. The pressure or level gages on which the pressures or levels are indicated (READOUT GAGE column.)

c. The pressure or level to which the tanks are filled (FILL TANK TO column).

d. The pressure or level at which the tanks are resupplied (RESUPPLY TANK AT column).

e. The minimum pressure or level requirements for a single countdown with no holds, applicable to both a tactical and nontactical countdown (MINIMUM REUIREMENTS FOR 1 COUNTDOWN column.)

f. The minimum pressure or level requirements for two countdowns with a 1-hour hold (40-minute hold if LN_2 is in the LO_2 system) in the first countdown, applicable to both a tactical and nontactical countdown (MINIMUM REQUIREMENTS FOR 2 COUNTDOWNS column).

6-21. The LN_2 level requirements in the LO_2 storage and LO_2 topping tanks apply only when the launch complex is set up for a propellant loading exercise, using LN_2 in place of LO_2. Normally, gas pressures or fluid levels should never be allowed to fall to the minimum values listed in columns 5 and 6. In order to maintain a 10-day self-sufficiency period and full weapon system capability, the storage tanks should be replenished when the pressures or levels fall to the values listed in column 4.

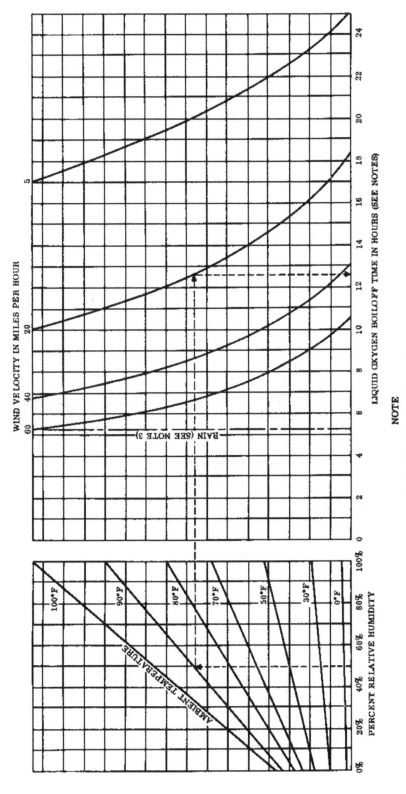

NOTE

1 ALL CONDITIONS FOR FULL OXIDIZER TANK EXPOSURE
  (MISSILE OXIDIZER TANK NOT IN SILO)

2 BOILOFF TIME BASED ON FULL OXIDIZER TANK

3 DURING RAIN CONDITIONS, BOILOFF TIME IS 5. 3 HOURS

4 DOTTED LINE SHOWS SAMPLE CALCULATION OF BOILOFF
  TIME (12. 6 HOURS BASED ON 50% RELATIVE HUMIDITY, 90°F
  AMBIENT TEMPERATURE, AND 20 MPH WIND VELOCITY)

5 (DELETED)

Figure 6-1.   Liquid Oxygen Boiloff Time Nomogram

40. 10—29(600)A

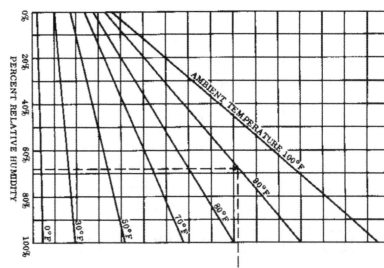

40.10-30(600)A

Figure 6-2. Liquid Nitrogen Boiloff Time Nomogram

PERCENT RELATIVE HUMIDITY

AMBIENT TEMPERATURE 100°F

90°F

80°F

70°F

50°F

30°F

0°F

LIQUID NITROGEN BOILOFF TIME IN HOURS (SEE NOTES)

RAIN (SEE NOTE 3)

WIND VELOCITY IN MILES PER HOUR

NOTE

1 ALL CONDITIONS FOR FULL OXIDIZER TANK EXPOSURE
   (MISSILE OXIDIZER TANK NOT IN SILO)
2 BOILOFF TIME BASED ON FULL OXIDIZER TANK
3 DURING RAIN CONDITION, BOILOFF TIME IS 3.5 HOURS
4 DOTTED LINE SHOWS SAMPLE CALCULATION OF BOILOFF
   TIME (8 HOURS BASED ON 68% RELATIVE HUMIDITY,
   90°F AMBIENT TEMPERATURE, AND 30 MPH WIND VELOCITY)
5 (DELETED)

1	2	3
STORAGE TANK	READOUT GAGE	FILL TANK TO
LO$_2$ storage tank LO$_2$ (VAFB only) LO$_2$ (except at VAFB) LN$_2$ (all sites)	LIQUID LEVEL LO$_2$ STORAGE gage LLI-1 on pressurization prefab instrument panel (figure 1-53)	21,850 ($\pm$250) GAL Full mark ($\pm$250 GAL) 15,250 ($\pm$250, GAL
LO$_2$ topping tank LO$_2$ (VAFB only) LO$_2$ (except at VAFB) LN$_2$ (all sites)	LIQUID LEVEL LO$_2$ TOPPING TANK gage LLI-2 on pressurization prefab instrument panel (figure 1-53)	3447 ($\pm$50) GAL Full mark ($\pm$50 GAL) 2400 ($\pm$50) GAL
Gaseous nitrogen tank (LO$_2$ transfer and gaseous nitrogen supply pressure)	VESSEL PRESSURE GN$_2$ STORAGE gage PI-1 on pressurization prefab instrument panel (figure 1-53)	4000 ($\pm$50) PSI***
LN$_2$ storage tank	LIQUID LEVEL LN$_2$ STORAGE VESSEL gage LLI-221 on LN$_2$ prefab panel	3550 ($\pm$50) GAL
LN$_2$ helium heat exchanger tank	LIQUID LEVEL LN$_2$ HE HEAT EXCHANGER gage LLI-220 on LN$_2$ prefab panel	1350 ($\pm$50) GAL

Notes: *Resupply level (or pressure) is the minimum level (or pressure) that provides a
              LN$_2$ resupply in the LO$_2$ tanks (storage and topping) provides only for a two cou
          **One-hour hold applies to a countdown with LO$_2$. Forty-minute hold if LN$_2$ is in
         ***An increase in pressure up to 4200 PSI is permissible after recharging due to w

Table 6-1. Launch Complex Gas and Fluid Level Requirements

4	5		6	
	MINIMUM REQUIREMENTS FOR 1 COUNTDOWN (LOAD, NO HOLD, AND COMMIT)		MINIMUM REQUIREMENTS FOR 2 COUNTDOWNS (LOAD, 1-HOUR HOLD, ABORT, RELOAD NO HOLD, AND COMMIT)**	
RESUPPLY TANK AT*	TACTICAL	NONTACTICAL	TACTICAL	NONTACTICAL
20,750 GAL	19,250 GAL	19,750 GAL	20,000 GAL	20,750 GAL
20,750 GAL	19,250 GAL	19,750 GAL	20,000 GAL	20,750 GAL
15,000 GAL	-	14,000 GAL	-	15,000 GAL
3100 GAL	800 GAL	1100 GAL	2650 GAL	3100 GAL
3100 GAL	800 GAL	1100 GAL	2650 GAL	3100 GAL
2200 GAL	-	800 GAL	-	2200 GAL
3200 PSI	2000 PSI	2200 PSI	2800 PSI	3200 PSI
2900 GAL	1000 GAL	1000 GAL	2600 GAL	2600 GAL
1200 GAL	900 GAL	900 GAL	1050 GAL	1050 GAL

.0-day self-sufficiency period and full weapon system capability.

ldown capability.

he LO$_2$ system. No hold during second countdown.

rming of the gas

Table 6-

1	2	3	4
STORAGE TANK	READOUT GAGE	FILL TANK TO	RESUPPLY TANK AT*
Inflight helium supply tank NO. 1	PRESSURE GAGE-1, AIRBORNE HELIUM SUPPLY NO. 1 on pneumatic system manifold regulator (figure 1-50)	6000 (±200) PSI	5400 PSI
Inflight helium supply tank NO. 2	PRESSURE GAGE-2, AIRBORNE HELIUM SUPPLY NO. 2 on pneumatic system manifold regulator (figure 1-50)	6000 (±200) PSI	5000 PSI
Gaseous nitrogen tanks	PRESSURE GAGE-20 PCU NITROGEN SUPPLY on pneumatic system manifold regulator (figure 1-50)	6000 (±200) PSI	4300 PSI
	PRESSURE GAGE PI-3 VESSEL PRESSURE LN_2 TRANS & NCU SUPPLY on pressurization prefab (figure 1-53)	4000 (±50) PSI***	2500 PSI
FUEL PRESSURIZATION TANK GN_2 FUEL LOADING PREFAB	PRESSURE GAGE PI-4 VESSEL PRESSURE GN_2 STORAGE on pressurization prefab (figure 1-53)	PI-4 PRESSURE EQUAL TO PI-3.	2000 PSI

Notes: *Resupply level (or pressure) is the minimum level (or pressure) that provides a 10-day self-suffi
LN_2 resupply in the LO_2 tanks (storage and topping) provides only for a two countdown capability.
**One-hour hold applies to a countdown with LO_2. Forty-minute hold if LN_2 is in the LO_2 system.
***An increase in pressure up to 4200 PSI is permissible after recharging due to warming of the gas

Launch Complex Gas and Fluid Level Requirements (Continued)

5		6	
MINIMUM REQUIREMENTS FOR 1 COUNTDOWN (LOAD, NO HOLD, AND COMMIT)		MINIMUM REQUIREMENTS FOR 2 COUNTDOWNS (LOAD, 1-HOUR HOLD, ABORT, RELOAD NO HOLD, AND COMMIT)**	
TACTICAL	NONTACTICAL	TACTICAL	NONTACTICAL
4000 PSI	4000 PSI	5300 PSI	5300 PSI
4700 PSI	4700 PSI	4850 PSI	4850 PSI
4000 PSI	4000 PSI	4100 PSI	4100 PSI
2200 PSI	2200 PSI	2300 PSI	2300 PSI
1500 PSI	1500 PSI	1500 PSI	1500 PSI

ıcy period and full weapon system capability.

› hold during second countdown.

SECTION VII

CREW DUTIES

TABLE OF CONTENTS

7-1. GENERAL. Each of the five crew members has certain specific duties to perform in maintaining and operating the launch complex. However, a crew member may be called upon to perform other than his normal tasks in the event an emergency situation should arise during standby, countdown or abort, or if a malfunction occurs during a tactical countdown.

7-2. MISSILE COMBAT CREW COMMANDER.

7-3. The missile combat crew commander (MCCC) is responsible for monitoring and controlling personnel and activities at the launch complex. His duties and responsibilities consist of but are not limited to the following:

    a. Monitor complex status.

    b. Control and monitor crew activities.

    c. Ensure compliance with current directives.

    d. Conduct required briefings.

    e. Ensure proper control of all personnel entering and within the launch complex.

    f. Ensure that communication requirements are met.

    g. Direct overall countdown and abort procedures.

    h. Perform procedures outlined in the MCCC checklists.

7-4. DEPUTY MISSILE COMBAT CREW COMMANDER.

7-5. The deputy missile combat crew commander (DMCCC) is responsible for assisting the MCCC in performance of his duties. In the absence of the MCCC from launch control center level 2, the DMCCC will assume the MCCC responsiblities. In addition, he will perform procedures outlined in the DMCCC checklists.

7-6. BALLISTIC MISSILE ANALYST TECHNICIAN.

7-7. The ballistic missile analyst technician (BMAT) is responsible for the following:

    a. Monitor complex status.

    b. Perform procedures outlined in the BMAT checklists.

    c. Troubleshoot and assist in maintenance tasks as directed.

    d. Prepare forms and reports as directed.

    e. Comply with current directives.

7-8. MISSILE FACILITIES TECHNICIAN.

7-9. The missile facilities technician (MFT) is responsible for the following:

    a. Monitor complex status.

    b. Perform procedures outlined in the MFT checklists.

    c. Troubleshoot and assist in maintenance tasks as directed.

    d. Prepare forms and reports as directed.

    e. Comply with current directives.

7-10. ELECTRICAL POWER PRODUCTION TECHNICIAN.

7-11. The electrical power production technician (EPPT) maintains and monitors electrical power generation and distribution. In addition he is responsible for the following:

    a. Monitor complex status.

    b. Perform procedures outlined in the EPPT checklists.

    c. Troubleshoot and assist in maintenance tasks as directed.

    d. Prepare forms and reports as directed.

    e. Comply with current directives.

GLOSSARY

ABORT: Stopping a missile countdown sequence and performing backout procedures to return the missile and associated aerospace ground equipment to a safe condition.

ACCELEROMETER: An electromechanical device used to measure missile accelerations.

ACTUATOR, HYDRAULIC: A hydraulically operated cylinder and piston assembly which transmits power to control movement or actuate a mechanical device.

AFB: Air Force Base.

AFSC: Air Force Specialty Code.

AGE: Aerospace ground equipment.

ALCO: Auxiliary launch control officer.

ANALOG SIGNALS: Signals using a varying voltage to steer the missile.

BOI: Break of inspection.

BOILOFF: Vaporization of liquid oxygen or liquid nitrogen as the temperature of the propellant mass rises during exposure to temperatures above the boiling point of liquid oxygen or liquid nitrogen.

CB: Circuit breaker.

CEF: Cross-range error function.

CO_2: Carbon dioxide.

COUNTDOWN: A progress of events where the missile and launch equipment are sequenced to a point of missile launch.

CSMOL: Control station manual operating level.

CTL: Crew training launch.

DEFCON: Defense condition.

DISCRETE SIGNALS: Signals using a fixed voltage to activate missile flight functions, such as staging and sustainer engine cutoff.

D/L: Dial line.

EMCC: Essential motor control center.

EWO: Emergency war order.

FAS: Facilities air supply.

FEEDBACK TRANSDUCER: A component that measures one type of energy action and feeds the measured quantity back to a device controlling the original for comparison.

FRCP: Facility remote control panel.

FTC: Facilities terminal cabinet.

GN_2: Gaseous nitrogen.

GO_2: Gaseous oxygen.

GYROSCOPE: An electromechanical device whose qualities to maintain rigidity in space and precession are used to furnish steering commands and to stabilize the guidance platform.

HCU: Helium control charging unit.

HEAD SUPPRESSION VALVE: The sustainer main oxidizer valve which varies the flow of oxidizer (liquid oxygen) to maintain current ration between fuel pump discharge pressure and oxidizer dome pressure to prevent excessive pressure at the turbopumps.

HEAT EXCHANGER: A device which regeneratively transfers heat from one fluid or substance to another.

HPU: Hydraulic pumping unit.

HYPERGOLIC: The ability of propellants to ignite spontaneously when mixed.

LAUNCH COMPLEX: Consists of one launch control center and one silo surrounded by a security fence.

LCC: Launch Control Center. Sometimes mistaken for Launch Control Console.

LCCMCC: Launch control center motor control center.

GLOSSARY (Continued)

LES: Launch enable system.

L/M: Line maintenance.

LP: Launcher platform.

LSR: Control-monitor group 3 of 4 and 4 of 4. (Launch Signal Responder.)

LN_2: Liquid nitrogen.

LO_2: Liquid oxygen.

MAMS: Missile Assembly and maintenance shops.

MEA: Missile enclosure area.

MEPU: Missile enclosure purge unit.

MFSS: Missile flight safety system.

MGS: Missile guidance system.

MLS: Missile lifting system.

MOCAM: Mobile checkout and maintenance.

MONOCOQUE: A type of airframe construction which uses no stiffening framework.

NEMCC: Nonessential motor control center.

OIC: Officer in charge.

ORI: Operational readiness inspection.

OSTF-2: Operational system test facility NO. 2.

PAS: Primary alerting system.

PETE: Pneumatic end-to-end test.

PLCP: Pneumatic local control panel.

PLM: Prelaunch monitor.

PLX: Propellant loading exercise.

PRCP: Power remote control panel .

PSC: Pressure system control.

PSMR: Pneumatic system manifold regulator.

PTS: Pneumatic test set.

READY STATE A: A known condition of readiness of the integrated weapon systems such that a countdown and tactical missile launching or training launch may be initiated immediately upon command.

READY STATE B: A known condition of readiness of the integrated weapon systems such that a propellant loading countdown can be initiated through commit, simulated launch, abort and return to a safe condition for the purpose of training or maintenance checkout.

REF: Range error function.

RPIE: Real property installed equipment.

R/V: Re-entry vehicle.

SAC: Strategic Air Command.

SERVOAMPLIFIER: An electronic device which amplifies and converts an alternating electrical input signal to direct current output to actuate electro-hydraulic servovalves.

SERVOMETER: An electric motor which acts in response to control signals.

SERVOVALVE: Electro-hydraulic valve which acts in response to control signals.

SPGG: Solid propellant gas generator.

STAGING: An operation whereby the booster engine section of the missile is unlocked from the adjacent missile tank section and jettisoned .

SUMMING NETWORK: An electrical circuit which contains two or more signals to form a control signal.

VAFB: Vandenberg Air Force Base.

NOW AVAILABLE!

TITAN I

MISSILE WEAPON SYSTEM

OPERATION AND ORGANIZATIONAL
MAINTENANCE MANUAL

United States Air Force

PROJECT MERCURY

FAMILIARIZATION MANUAL

Manned Satellite Capsule

Periscope Film LLC

MMS SUBCOURSE NUMBER 151

EDITION CODE 3

NIKE MISSILE
and Test Equipment

NIKE HERCULES

DECLASSIFIED

by U.S. Army Missile and Munitions Center and School
Periscope Film LLC

LMA 790-1

PROJECT APOLLO

lem
LUNAR EXCURSION MODULE

NOW AVAILABLE!

FIRST MANNED LUNAR LANDING
FAMILIARIZATION MANUAL

GRUMMAN AIRCRAFT ENGINEERING CORPORATION • BETHPAGE, L. I., N. Y.

NASA
PROJECT
GEMINI

FAMILIARIZATION
MANUAL
Manned Satellite Capsule

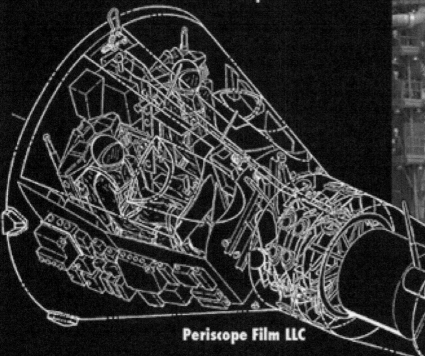

Periscope Film LLC